Lecture Notes in Statistics 102

Edited by P. Diggle, S. Fienberg, K. Krickeberg,
I. Olkin, N. Wermuth

T0235813

A.M. Mathai, Serge B. Provost,
and Takesi Hayakawa

Bilinear Forms and Zonal Polynomials

Springer-Verlag
New York Berlin Heidelberg London Paris
Tokyo Hong Kong Barcelona Budapest

A.M. Mathai
Department of Mathematics
McGill University
Montreal
Quebec, Canada H3A 2K6

Serge B. Provost
Department of Statistical and
Actuarial Sciences
University of Western Ontario
London
Ontario, Canada N6A 5B6

Takesi Hayakawa
Department of Economics
Hitotubashi University
2-1 Naka, Kunitachi
Tokyo 186
Japan

Library of Congress Cataloging-in-Publication Data Available
Printed on acid-free paper.

© 1995 Springer-Verlag New York, Inc.
All rights reserved. This work may not be translated or copied in whole or in part without the written permission of the publisher (Springer-Verlag New York, Inc., 175 Fifth Avenue, New York, NY 10010, USA), except for brief excerpts in connection with reviews or scholarly analysis. Use in connection with any form of information storage and retrieval, electronic adaptation, computer software, or by similar or dissimilar methodology now known or hereafter developed is forbidden. The use of general descriptive names, trade names, trademarks, etc., in this publication, even if the former are not especially identified, is not to be taken as a sign that such names, as understood by the Trade Marks and Merchandise Marks Act, may accordingly be used freely by anyone.

Camera ready copy provided by the author.
Printed and bound by Braun-Brumfield, Ann Arbor, MI.
Printed in the United States of America.

9 8 7 6 5 4 3 2 1

ISBN 0-387-94522-9 Springer-Verlag New York Berlin Heidelberg

To my grandparents and my godmother Gisèle Bédard
S.B.P.

To my wife Yuhko
T.H.

Preface

The book deals with bilinear forms in real random vectors and their generalizations as well as zonal polynomials and their applications in handling generalized quadratic and bilinear forms. The book is mostly self-contained. It starts from basic principles and brings the readers to the current research level in these areas. It is developed with detailed proofs and illustrative examples for easy readability and self-study. Several exercises are proposed at the end of the chapters. The complicated topic of zonal polynomials is explained in detail in this book. The book concentrates on the theoretical developments in all the topics covered. Some applications are pointed out but no detailed application to any particular field is attempted.

This book can be used as a textbook for a one-semester graduate course on quadratic and bilinear forms and/or on zonal polynomials. It is hoped that this book will be a valuable reference source for graduate students and research workers in the areas of mathematical statistics, quadratic and bilinear forms and their generalizations, zonal polynomials, invariant polynomials and related topics, and will benefit statisticians, mathematicians and other theoretical and applied scientists who use any of the above topics in their areas.

Chapter 1 gives the preliminaries needed in later chapters, including some Jacobians of matrix transformations. Chapter 2 is devoted to bilinear forms in Gaussian real random vectors, their properties, and techniques specially developed to deal with bilinear forms where the standard methods for handling quadratic forms become complicated. Distributional aspects and Laplacianness, a concept analogous to chi-squaredness for quadratic forms, are examined in detail.

Various distributional results on quadratic forms in elliptically contoured and spherically symmetric vectors are presented in Chapter 3. The central and noncentral distributions, the moments, Cochran's theorem as well as quadratic forms of random idempotent matrices are discussed, and the robust properties of certain test statistics are studied. Several results also apply to bilinear forms.

Chapter 4 proposes a systematic development of the theory of zonal polynomials, including worked examples and related topics. Many of these results are used in Chapter 5 which deals with the distribution of generalized bilinear and quadratic forms. The theoretical results and applications are brought up to the current research level. Invariant polynomials, which are an extension of zonal polynomials, are discussed in an appendix. Some fundamental results and useful coefficients relating to these polynomials are tabulated.

We are very grateful to Professor David R. Bellhouse, Chairman of the Department of Statistical and Actuarial Sciences at the University of Western Ontario, for making departmental personnel and equipment available to us. We would like to thank Dr Edmund Rudiuk and John Kawczak for their comments and suggestions on various parts of the book. We also wish to express our appreciation to Alicia Pleasence, Motoko Yuasa and Yoriko Fukushima who patiently and expertly typed most of Chapters 3, 4 and 5, including the Appendix. The financial support of the Natural Sciences and Engineering Research Council of Canada is gratefully acknowledged.

1 February 1995

Arak M. Mathai
Serge B. Provost
Takesi Hayakawa

Contents

Chapter 3
QUADRATIC AND BILINEAR FORMS
IN ELLIPTICALLY CONTOURED DISTRIBUTIONS

Chapter 4
ZONAL POLYNOMIALS

Chapter 5
GENERALIZED QUADRATIC FORMS

CHAPTER 1

Preliminaries

1.0 Introduction

For the study of generalized quadratic forms and bilinear forms we need the Jacobians associated with certain matrix transformations, properties of multivariate normal distribution, matrix-variate gamma and beta functions and hypergeometric functions. A brief outline of these will be given in this chapter.

1.1 Jacobians of Matrix Transformations

Let $X = (x_{ij})$ be a $p \times q$ matrix of functionally independent variables x_{ij}, $i = 1, \ldots, p$, $j = 1, \ldots, q$. The following notations will be used in our discussions

Notation 1.1.1 *Definiteness of matrices and integrals over matrices*

A' : transpose of the matrix A.

$A > 0$: the square matrix A is positive definite.

$A \geq 0$: the square matrix A is positive semidefinite.

$A < 0$: the square matrix A is negative definite.

$A \leq 0$: the square matrix A is negative semidefinite.

$|A|$: determinant of the square matrix A.

$\mathrm{tr}(A)$: trace of the square matrix A.

$\int_X$: integral over the elements of the matrix X.

$\int_{Y>0}$: integral over the positive definite matrix Y.

$\int_{A<Y<B} = \int_A^B$: integral over Y such that $A > 0,\ Y - A > 0,\ B - Y > 0$.

$\mathrm{Re}(\cdot)$: real part of $(\cdot)$.

$\mathrm{diag}(\lambda_1, \ldots, \lambda_p)$: diagonal matrix with the diagonal elements $\lambda_1, \ldots, \lambda_p$.

$A^{\frac{1}{2}}$: symmetric square root of the matrix A for $A = A'$ and $A > 0$.

$(\mathrm{d}\mathbf{X}) = (\mathrm{d}x_{ij})$: the matrix of differentials.

$\mathrm{d}X = \mathrm{d}x_{11} \ldots \mathrm{d}x_{1q} \mathrm{d}x_{21} \ldots \mathrm{d}x_{2q} \ldots \mathrm{d}x_{p1} \ldots \mathrm{d}x_{pq}$: the wedge product of all the pq differentials $\mathrm{d}x_{ij}$'s if $X = (x_{ij})$ is a $p \times q$ matrix of functionally independent pq real variables. In the wedge product note that $\mathrm{d}x_{ij}\mathrm{d}x_{mn} = -\mathrm{d}x_{mn}\mathrm{d}x_{ij}$ and thus $\mathrm{d}x_{ij}\mathrm{d}x_{ij} = 0$. When X is a $p \times p$ real symmetric matrix then $\mathrm{d}X$ is the product of the $p(p+1)/2$ differentials $\mathrm{d}x_{ij}$'s for $i \geq j$.

In order to distinguish between the wedge product $\mathrm{d}X$ and the matrix of differentials $(\mathrm{d}\mathbf{X})$ the X in the matrix of differentials will be put in bold. Constant vectors will be denoted by small bold letters and vector random variables by bold capital letters. There is room for a little confusion but the matrix of differentials and vector random variables will not be appearing together in our discussions later on.

Some properties of the trace that will be used frequently in the book are given below.

$$\mathrm{tr}(A + B) = \mathrm{tr}(A) + \mathrm{tr}(B) \tag{1.1.1}$$

and whenever AB and BA are defined

$$\mathrm{tr}(AB) = \mathrm{tr}(BA)\ . \tag{1.1.2}$$

For the $p \times p$ square matrix $A = (a_{ij})$

$$\mathrm{tr}(A) = a_{11} + \cdots + a_{pp} = \lambda_1 + \cdots + \lambda_p \tag{1.1.3}$$

where $\lambda_1, \ldots, \lambda_p$ are the eigenvalues of A. For the same A note also that

$$|A| = \lambda_1 \cdots \lambda_p\ .$$

1.1a Some Frequently Used Jacobians in the Real Case

If $\mathbf{x}$ is a $p \times 1$ vector of functionally independent real variables and if $A = (a_{ij})$ is a $p \times p$ real nonsingular matrix of constants then from basic calculus it is known that

$$\begin{aligned} \mathbf{y} = A\mathbf{x} \Rightarrow \mathrm{d}\mathbf{y} &= |A|\mathrm{d}\mathbf{x}\ , \\ &= a^p \mathrm{d}\mathbf{x} \end{aligned} \tag{1.1.4}$$

when $A = aI$ where I is the identity matrix and a is a scalar. This result can be extended to obtain Jacobians of linear transformations involving a $p \times q$ matrix X. These results will be stated here as theorems without proofs.

Theorem 1.1a.1 Let X be a $p\times q$ matrix of pq functionally independent real variables, A, B, C be matrices of constants and b a scalar where A is $p \times p$ and nonsingular, B is $q \times q$ and nonsingular and C is $p \times q$. Then

$$\begin{aligned} Y = bX + C &\Rightarrow \mathrm{d}Y = b^{pq}\mathrm{d}X \\ Y = AX + C &\Rightarrow \mathrm{d}Y = |A|^{q}\mathrm{d}X \\ Y = XB + C &\Rightarrow \mathrm{d}Y = |B|^{p}\mathrm{d}X \\ Y = AXB + C &\Rightarrow \mathrm{d}Y = |A|^{q}|B|^{p}\mathrm{d}X\ . \end{aligned} \tag{1.1.5}$$

Theorem 1.1a.2 Let X be a $p\times p$ upper triangular matrix of $p(p+1)/2$ functionally independent real variables, $B = (b_{ij})$ an upper triangular matrix of constants with $b_{jj} > 0, j = 1,\ldots,p$ and b a scalar quantity. Then

$$\begin{aligned} Y = X + X' &\Rightarrow \mathrm{d}Y = 2^{p}\mathrm{d}X \\ Y = XB &\Rightarrow \mathrm{d}Y = \Big\{\prod_{j=1}^{p} b_{jj}^{j}\Big\}\mathrm{d}X \\ Y = BX &\Rightarrow \mathrm{d}Y = \Big\{\prod_{j=1}^{p} b_{jj}^{p+1-j}\Big\}\mathrm{d}X \\ Y = bX &\Rightarrow \mathrm{d}Y = b^{\frac{p(p+1)}{2}}\mathrm{d}X\ . \end{aligned} \tag{1.1.6}$$

Take the transposes to get the corresponding results for lower triangular matrices.

Theorem 1.1a.3 Let X be a $p \times p$ real symmetric matrix of $p(p+1)/2$ functionally independent variables, A be a $p \times p$ nonsingular matrix of constants and b a scalar quantity. Then

$$\begin{aligned} Y = AXA' &\Rightarrow \mathrm{d}Y = |A|^{p+1}\mathrm{d}X \text{ for } X = X', \\ &\qquad = |A|^{p-1}\mathrm{d}X \text{ for } X = -X', \\ Y = bX &\Rightarrow \mathrm{d}Y = b^{\frac{p(p+1)}{2}}\mathrm{d}X \text{ for } X = X', \\ &\qquad = b^{\frac{p(p-1)}{2}}\mathrm{d}X \text{ for } X = -X'. \end{aligned} \tag{1.1.7}$$

Theorem 1.1a.4 Let X be a $p \times p$ nonsingular matrix of real variables. Then

$$Y = X^{-1} \Rightarrow$$

$$\mathrm{d}Y = |X|^{-2p}\mathrm{d}X \text{ for a general } X,$$

$$= |X|^{-(p+1)}\mathrm{d}X \text{ if } X = X',$$

$$= |X|^{-(p-1)}\mathrm{d}X \text{ if } X = -X',$$

$$= |X|^{-(p+1)}\mathrm{d}X \tag{1.1.8}$$

for X lower or upper triangular.

Theorem 1.1a.5 Let X be a real symmetric $p \times p$ matrix and $T = (t_{ij})$ be a real lower triangular matrix with $t_{jj} > 0, j = 1, \ldots, p$. Then

$$X = TT' \Rightarrow \mathrm{d}X = 2^p\Big\{\prod_{j=1}^{p} t_{jj}^{p+1-j}\Big\}\mathrm{d}T,$$

$$X = T'T \Rightarrow \mathrm{d}X = 2^p\Big\{\prod_{j=1}^{p} t_{jj}^{j}\Big\}\mathrm{d}T\ . \tag{1.1.9}$$

Theorem 1.1a.6 Let the real scalar variables $x_1, \ldots, x_p$ be transformed to the general polar coordinates $r, \theta_1, \ldots, \theta_{p-1}$ as follows, where $r > 0,\ 0 < \theta_j \le \pi, j = 1, \ldots, p-2,\ 0 < \theta_{p-1} \le 2\pi$.

$$\begin{aligned} x_1 &= r\sin\theta_1 \\ x_j &= r\cos\theta_1\cos\theta_2\cdots\cos\theta_{j-1}\sin\theta_j\ ,\quad j = 2, 3, \ldots, p-1 \\ x_p &= r\cos\theta_1\cos\theta_2\cdots\cos\theta_{p-1}\ . \end{aligned}$$

Then

$$\mathrm{d}x_1\ldots\mathrm{d}x_p = r^{p-1}\Big\{\prod_{j=1}^{p-1}|\cos\theta_j|^{p-j-1}\Big\}\mathrm{d}r\mathrm{d}\theta_1\ldots\mathrm{d}\theta_{p-1}\ . \tag{1.1.10}$$

Theorem 1.1a.7 Let X and Y be $p \times p$ symmetric positive definite matrices of functionally independent real variables such that $0 < X < I$. Then

$$Y = (I-X)^{-\frac{1}{2}}X(I-X)^{-\frac{1}{2}} \Rightarrow \mathrm{d}X = |I+Y|^{-(p+1)}\mathrm{d}Y$$

and

$$X = (I+Y)^{-\frac{1}{2}}Y(I+Y)^{-\frac{1}{2}} \Rightarrow \mathrm{d}Y = |I-X|^{-(p+1)}\mathrm{d}X,$$

where $(I-X)^{-\frac{1}{2}}$ and $(I+Y)^{-\frac{1}{2}}$ are the symmetric square roots of $(I-X)^{-1}$ and $(I+Y)^{-1}$ respectively.

Theorem 1.1a.8 Let X and Y be $p\times p$ symmetric matrices of functionally independent real variables and $u = \mathrm{tr}(X)$. Then

$$X = uY \Rightarrow \mathrm{d}X = u^{\frac{1}{2}p(p+1)-1}\mathrm{d}u\mathrm{d}Y,$$

where

$$dY = \left\{\prod_{i=1}^{p} dy_{ii}\right\} \prod_{2\le i<j\le p} dy_{ij}.$$

This can be proved by observing the following: $u = \mathrm{tr}(X) = u\mathrm{tr}(Y)$, $\mathrm{tr}(Y) = 1$. This means that

$$x_{11} = u\left(1 - \sum_{i=2}^{p} y_{ii}\right), \quad x_{ii} = uy_{ii}, \; i = 2,\dots,p,$$

$$x_{ij} = uy_{ij}, \; 2 \le i < j \le p.$$

Theorem 1.1a.9 Let X be a $p\times p$ real symmetric matrix of $\frac{1}{2}p(p+1)$ functionally independent variables. Let X be decomposed as

$$X = H\begin{bmatrix} \lambda_1 & \mathbf{u}' \\ \mathbf{u} & W \end{bmatrix} H',$$

where λ_1 is the largest latent root of X, H is an orthogonal matrix with $(p-1)$ functionally independent first column vectors and W is a $(p-1)\times(p-1)$ real symmetric matrix such that $\lambda_1 I_{p-1} > W > 0$. Then

$$dX = \frac{|\lambda_1 I_{p-1} - W|}{(1-\sum_{i=2}^{p} h_{ij}^2)^{\frac{1}{2}}} d\lambda_1 \left\{\prod_{i=2}^{p} dh_{ij}\right\} dW.$$

Theorem 1.1a.10 Let X be a $p\times p$ real positive definite symmetric matrix. Let $X = H\Lambda H'$, where $H = [\mathbf{h}_1,\dots,\mathbf{h}_p]$ is an orthogonal matrix with $\mathbf{h}_j, j = 1,\dots,p$ being the various columns such that $\mathbf{h}_1$ has all positive elements and $\Lambda = \mathrm{diag}(\lambda_1,\dots,\lambda_p)$, $\lambda_1 > \dots > \lambda_p > 0$. Then

$$dX = \left\{\prod_{i<j}(\lambda_i - \lambda_j)\right\} d\Lambda d(\tilde{H}),$$

where

$$d(\tilde{H}) = \prod_{i<j} \mathbf{h}_i' d\mathbf{h}_j.$$

Theorem 1.1a.11 Let X be a $p\times n$ matrix , T be a $p\times p$ lower triangular matrix with positive diagonal elements and L be an $n\times p$ matrix such that $L'L = I_p$. All the matrices are of functionally independent real variables. Then

$$X = TL' \Rightarrow$$

$$dX = \left\{\prod_{i=1}^{p} t_{ii}^{p-i}\right\} dT d(\tilde{L}),$$

where

$$\mathrm{d}(\tilde{L}) = \left\{ \prod_{1\leq i<j\leq p} \mathbf{l}_i'\mathrm{d}\mathbf{l}_j \right\} \prod_{i=1}^{p}\prod_{j=1}^{n-p} \mathbf{l}_i'\mathrm{d}\mathbf{l}_j,$$

and

$$[L, B] = [\mathbf{l}_1, \ldots, \mathbf{l}_p, \mathbf{b}_1, \ldots, \mathbf{b}_{n-p}] \in O(n).$$

Theorem 1.1a.12 Let X be a $p \times n$ matrix, S be a $p \times p$ positive definite symmetric matrix and L be an $n \times p$ matrix with $L'L = I_p$, all are of functionally independent real variables. Then

$$X = S^{\frac{1}{2}} L' \Rightarrow$$
$$\mathrm{d}X = 2^{-p} |S|^{\frac{1}{2}(n-p-1)} \mathrm{d}S \mathrm{d}(\tilde{L}).$$

More results on Jacobians of matrix transformations are available from Deemer and Olkin (1951).

1.2 Singular and Nonsingular Normal Distributions

1.2a Normal Distribution in the Real Case

Let $\mathbf{X}$ be a $p\times 1$ real random vector with the mean value vector $\boldsymbol{\mu}$ and the covariance matrix Σ. That is,

$$\boldsymbol{\mu} = E(\mathbf{X}); \ \Sigma = \mathrm{Cov}(\mathbf{X}) = E(\mathbf{X} - E(\mathbf{X}))(\mathbf{X} - E(\mathbf{X}))' = \Sigma'$$

where E denotes the expected value. If $\mathbf{X}$ has a multivariate normal distribution then it is denoted as

$$\mathbf{X} \sim N_p(\boldsymbol{\mu}, \Sigma)$$

where $\sim$ indicates *distributed as*. If Σ is nonsingular then $\Sigma = \Sigma' > 0$ and $\mathbf{X}$ in this case is said to have a *nonsingular normal distribution* and if Σ is singular then $\mathbf{X}$ is *singular normal*. In the nonsingular case $\mathbf{X}$ has a density and it is given by

$$f(\mathbf{x}) = \frac{e^{-\frac{1}{2}(\mathbf{x}-\boldsymbol{\mu})'\Sigma^{-1}(\mathbf{x}-\boldsymbol{\mu})}}{(2\pi)^{\frac{p}{2}}|\Sigma|^{\frac{1}{2}}}, \tag{1.2.1}$$

for $-\infty < \mathbf{x} < \infty$, $-\infty < \boldsymbol{\mu} < \infty$, $\Sigma > 0$ where , for example, the notation $-\infty < \mathbf{x} < \infty$ means that each component x_i of the $p \times 1$ vector $\mathbf{x} = (x_1, \ldots, x_p)'$ is such that $-\infty < x_i < \infty$, $i = 1, \ldots, p$.

Definition 1.2.1 *Real nonsingular multinormal density* A real $p\times 1$ vector $\mathbf{X}$ is said to have a nonsingular multinormal density if its density is of the form given in (1.2.1).

Let Σ be singular of rank $r < p$. Then there exists a $p \times r$ matrix B such that $\Sigma = BB'$ and we may write

$$\mathbf{X} = \boldsymbol{\mu} + B\mathbf{Y} \Rightarrow$$
$$E(\mathbf{X}) = \boldsymbol{\mu} + BE(\mathbf{Y})$$

and

$$\Sigma = \mathrm{Cov}(\mathbf{X}) = B\mathrm{Cov}(\mathbf{Y})B' \tag{1.2.2}$$

which means that there exists an $r \times 1$ vector $\mathbf{Y}$ which has a nonsingular normal distribution with the parameters $E(\mathbf{Y}) = \mathbf{0}$ and $\mathrm{Cov}(\mathbf{Y}) = I$ where I denotes the $r \times r$ identity matrix. That is,

$$\mathbf{Y} \sim N_r(\mathbf{0}, I) . \tag{1.2.3}$$

Definition 1.2.2 *Real singular normal distribution* A real $p \times 1$ vector $\mathbf{X}$ is said to have a singular normal distribution if $\mathbf{X}$ is as given in (1.2.2) with $\mathbf{Y}$ having a nonsingular real normal distribution given in (1.2.3).

With the help of the nonsingular $\mathbf{Y}$ one can handle the case when $\mathbf{X}$ is singular normal.

1.2b The Moment Generating Function for the Real Normal Distribution

Let $\mathbf{t}' = (t_1, \ldots, t_p)$ be a vector of arbitrary parameters $t_1, \ldots, t_p$. Then the moment generating function (m.g.f.) of the real $p \times 1$ random vector $\mathbf{X}$, denoted by $M_{\mathbf{X}}(\mathbf{t})$ is defined as follows.

Definition 1.2.3 *Moment generating function in the multivariate case*

$$M_{\mathbf{X}}(\mathbf{t}) = E(\mathrm{e}^{\mathbf{t}'\mathbf{X}}) \tag{1.2.4}$$

where E denotes the expected value. Thus for the nonsingular normal

$$\begin{aligned} M_{\mathbf{X}}(t) &= \int_{\Re^p} \frac{\mathrm{e}^{\mathbf{t}'\boldsymbol{x}} \mathrm{e}^{-\frac{1}{2}(\boldsymbol{x}-\boldsymbol{\mu})'\Sigma^{-1}(\boldsymbol{x}-\boldsymbol{\mu})}}{(2\pi)^{\frac{p}{2}} |\Sigma|^{\frac{1}{2}}} \mathrm{d}\boldsymbol{x} \\ &= \mathrm{e}^{\mathbf{t}'\boldsymbol{\mu} + \frac{1}{2}\mathbf{t}'\Sigma\mathbf{t}} \end{aligned} \tag{1.2.5}$$

which is worked out by making the transformation

$$\Sigma^{-\frac{1}{2}}(\mathbf{X} - \boldsymbol{\mu}) = \mathbf{U} \Rightarrow \mathrm{d}\mathbf{U} = |\Sigma|^{-\frac{1}{2}} \mathrm{d}\mathbf{X} ,$$

simplifying the exponent by using the fact

$$\mathbf{U}'\mathbf{U} - 2\mathbf{t}'\Sigma^{\frac{1}{2}}\mathbf{U} = (\mathbf{U} - \Sigma^{\frac{1}{2}}\mathbf{t})'(\mathbf{U} - \Sigma^{\frac{1}{2}}\mathbf{t}) - \mathbf{t}'\Sigma\mathbf{t}$$

and then noting that for a real multinormal density $N_p(\Sigma^{\frac{1}{2}}\mathbf{t}, I)$, the total integral is unity.

When $\mathbf{X}$ is singular normal use the definition in (1.2.4) and simplify with the help of the density of $\mathbf{Y}$ in (1.2.3) to see that the m.g.f. reduces to the same form as in (1.2.5) irrespective of whether the real $\mathbf{X}$ is singular or nonsingular normal.

1.3 Quadratic Forms in Normal Variables

Let the $p \times 1$ real random vector $\mathbf{X}$ have a normal distribution, singular or nonsingular, that is, $\mathbf{X} \sim N_p(\boldsymbol{\mu}, \Sigma)$, $\Sigma \geq 0$. Let $Q_0(\mathbf{X}) = \mathbf{X}'A\mathbf{X}$, $A = A'$ be a quadratic form where $A = (a_{ij})$ is a matrix of constants. That is,

$$Q_0(\mathbf{X}) = \mathbf{X}'A\mathbf{X},\ A = A',\ \mathbf{X} \sim N_p(\boldsymbol{\mu}, \Sigma),\ \Sigma \geq 0. \tag{1.3.1}$$

Theoretical results on Q_0 as well as applications and generalizations are available from Mathai and Provost (1992). A quadratic expression in $\mathbf{X}$ or a second degree polynomial in the components of $\mathbf{X}$ is given by

$$Q(\mathbf{X}) = \mathbf{X}'A\mathbf{X} + \mathbf{a}'\mathbf{X} + d,\ A = A',\ \mathbf{X} \sim N_p(\boldsymbol{\mu}, \Sigma),\ \Sigma \geq 0 \tag{1.3.2}$$

where d is a scalar constant and $\mathbf{a}$ is a constant vector. Some of the representations and the m.g.f. of (1.3.2) are needed for our discussion in this book. Hence some of these results will be reproduced here.

1.3a Representations of a Quadratic Form

Let the $p \times 1$ real random vector $\mathbf{X}$ be such that $\boldsymbol{\mu} = E(\mathbf{X})$ and $\Sigma = \mathrm{Cov}(\mathbf{X}) > 0$. Then one can obtain the following representation.

$$\begin{aligned} Q_0(\mathbf{X}) &= \mathbf{X}'A\mathbf{X},\ A = A',\ \Sigma > 0 \\ &= \sum_{j=1}^{p} \lambda_j (U_j + b_j)^2 \text{ for } \boldsymbol{\mu} \neq \mathbf{0} \\ &= \sum_{j=1}^{p} \lambda_j U_j^2 \text{ for } \boldsymbol{\mu} = \mathbf{0} \end{aligned} \tag{1.3.3}$$

where $\lambda_1, \ldots, \lambda_p$ are the eigenvalues of $\Sigma^{\frac{1}{2}} A \Sigma^{\frac{1}{2}}$,

$$\begin{aligned} \mathbf{b}' &= (b_1, \ldots, b_p) = (P'\Sigma^{-\frac{1}{2}}\boldsymbol{\mu})', \\ PP' &= I,\ P'\Sigma^{\frac{1}{2}} A \Sigma^{\frac{1}{2}} P = \mathrm{diag}(\lambda_1, \ldots, \lambda_p), \\ E(\mathbf{U}) &= \mathbf{0},\ \mathrm{Cov}(\mathbf{U}) = I \end{aligned}$$

and

$$\mathbf{U} = (U_1, \ldots, U_p)'.$$

When Σ is singular of rank $r < p$, write $\Sigma = BB'$, where B is a $p \times r$ matrix of rank r. Then one can derive the following representation by using (1.2.2) and (1.2.3).

$$\begin{aligned} Q_0(\mathbf{X}) &= \mathbf{X}'A\mathbf{X},\ A = A',\ \Sigma \geq 0 \\ &= \sum_{j=1}^{r} \lambda_j \Big(Z_j + \frac{b_j}{\lambda_j}\Big)^2 + \Big(a - \sum_{j=1}^{r} \frac{b_j^2}{\lambda_j}\Big), \\ &\lambda_j \neq 0, j = 1, \ldots, r \\ &= \sum_{j=1}^{r} \lambda_j Z_j^2 \text{ for } \boldsymbol{\mu} = \mathbf{0} \end{aligned} \tag{1.3.4}$$

where

$$\begin{aligned} \mathbf{b}' &= (b_1, \ldots, b_r) = \boldsymbol{\mu}'ABP, \\ \mathbf{Z}' &= (Z_1, \ldots, Z_r),\ E(\mathbf{Z}) = \mathbf{0}, \\ \mathrm{Cov}(\mathbf{Z}) &= I_r,\ P'B'ABP = \mathrm{diag}(\lambda_1, \ldots, \lambda_r), \\ PP' &= I,\ a = \boldsymbol{\mu}'A\boldsymbol{\mu}\ . \end{aligned}$$

Note that when $\mathbf{X} \sim N_p(\boldsymbol{\mu}, \Sigma)$ then the U_j's and Z_j's are mutually independent real standard normal variables and thus (1.3.3) and (1.3.4) in that case give linear functions of independent noncentral chi-square variables when $\boldsymbol{\mu} \neq \mathbf{0}$ and that of central chi-square variables when $\boldsymbol{\mu} = \mathbf{0}$.

1.3b Representations of the m.g.f. of a Quadratic Expression

Consider a general quadratic expression $Q(\mathbf{X})$ in the real $p \times 1$ normal vector $\mathbf{X} \sim N_p(\boldsymbol{\mu}, \Sigma)$.

$$Q(\mathbf{X}) = \mathbf{X}'A\mathbf{X} + \mathbf{a}'\mathbf{X} + d,\ A = A' \tag{1.3.5}$$

where d is a scalar constant and $\mathbf{a}$ a constant vector. A number of different representations for the m.g.f. of $Q(\mathbf{X})$ can be obtained for the nonsingular as well as singular cases, that is when Σ is nonsingular or singular. All the representations in the nonsingular case can be obtained from those of the singular case but some of the representations in the singular case cannot be obtained from those in the nonsingular case. Hence for convenience these will be listed here separately. Let $Q = Q(\mathbf{X})$ be as in (1.3.5) and let $M_Q(t)$ denote its m.g.f. when $\mathbf{X} \sim N_p(\boldsymbol{\mu}, \Sigma)$. Then for $\Sigma > 0$

$$\begin{aligned} M_Q(t) &= |I - 2tA\Sigma|^{-\frac{1}{2}} \exp\Big\{ -\frac{1}{2}\Big(\boldsymbol{\mu}'\Sigma^{-1}\boldsymbol{\mu} - 2td\Big) \\ &\quad + \frac{1}{2}(\boldsymbol{\mu} + t\Sigma\mathbf{a})'(I - 2tA\Sigma)^{-1}\Sigma^{-1}(\boldsymbol{\mu} + t\Sigma\mathbf{a})\Big\} \\ &= \Big\{\prod_{j=1}^{p}(1 - 2t\lambda_j)^{-\frac{1}{2}}\Big\} \exp\Big\{t(d + \boldsymbol{\mu}'A\boldsymbol{\mu} + \mathbf{a}'\boldsymbol{\mu}) \\ &\quad + \frac{t^2}{2}\sum_{j=1}^{p} b_j^{*2}(1 - 2t\lambda_j)^{-1}\Big\} \end{aligned} \tag{1.3.6}$$

where

$$\mathbf{b}^* = (b_1^*, \ldots, b_p^*)' = P'(\Sigma^{\frac{1}{2}}\mathbf{a} + 2\Sigma^{\frac{1}{2}}A\boldsymbol{\mu}) ,$$
$$PP' = I, \; P'\Sigma^{\frac{1}{2}}A\Sigma^{\frac{1}{2}}P = \text{diag}(\lambda_1, \ldots, \lambda_p) .$$

For $\mathbf{X} \sim N_p(\boldsymbol{\mu}, \Sigma)$, $\Sigma \geq 0$, $\Sigma = BB'$, B is $p \times r$ of rank $r =$ rank of Σ and $B'AB \neq O$,

$$\begin{aligned}
M_Q(t) &= |I - 2tB'AB|^{-\frac{1}{2}} \exp\Big\{ t(\boldsymbol{\mu}'A\boldsymbol{\mu} + \mathbf{a}'\boldsymbol{\mu} + d) \\
&\quad + \frac{t^2}{2}(B'\mathbf{a} + 2B'A\boldsymbol{\mu})'(I - 2tB'AB)^{-1}(B'\mathbf{a} + 2B'A\boldsymbol{\mu}) \Big\} \\
&= \Big\{ \prod_{j=1}^{r} (1 - 2t\lambda_j)^{-\frac{1}{2}} \Big\} \exp\Big\{ -\frac{1}{8}\sum_{j=1}^{m} \frac{b_j^{*2}}{\lambda_j^2} \\
&\quad + t\Big(\boldsymbol{\mu}'A\boldsymbol{\mu} + \mathbf{a}'\boldsymbol{\mu} + d - \frac{1}{4}\sum_{j=1}^{m}\frac{b_j^{*2}}{\lambda_j}\Big) + \frac{t^2}{2}\Big(\sum_{j=m+1}^{r} b_j^{*2}\Big) \\
&\quad + \frac{1}{8}\sum_{j=1}^{m}\frac{b_j^{*2}}{\lambda_j^2}(1 - 2t\lambda_j)^{-1} \Big\}, \; \lambda_j \neq 0, \; j = 1, \ldots, m , \\
&\qquad \lambda_j = 0 , \; j = m+1, \ldots, r
\end{aligned} \tag{1.3.7}$$

where

$$\mathbf{b}^* = P'(2B'A\boldsymbol{\mu} + B'\mathbf{a}) = (b_1^*, \ldots, b_r^*)' ,$$
$$PP' = I, \; P'B'ABP = \text{diag}(\lambda_1, \ldots, \lambda_r) .$$

The cumulant generating function of $Q(\mathbf{X})$, when $\mathbf{X} \sim N_p(\boldsymbol{\mu}, \Sigma)$, is available by taking the logarithms of $M_Q(t)$ of (1.3.6) and (1.3.7).

1.4 Matrix-variate Gamma and Beta Functions

The reader may be familiar with the gamma and beta functions in the scalar case. Here we will extend the ideas to real symmetric positive definite matrices.

1.4a Matrix-variate Gamma, Real Case

The gamma function $\Gamma(\alpha)$ in the real scalar case can be defined in many ways and it has the following integral representation for $\text{Re}(\alpha) > 0$.

$$\Gamma(\alpha) = \int_0^\infty x^{\alpha-1} e^{-x} dx, \; \text{Re}(\alpha) > 0 . \tag{1.4.1}$$

What will be the value of the corresponding integral if $x^{\alpha-1}$ is replaced by $|X|^{\alpha - \frac{p+1}{2}}$ and e^{-x} by $e^{-\text{tr}(X)}$ where $X = (x_{ij})$ is a $p \times p$ real symmetric positive definite matrix?

This will be examined here, denoting it by $\Gamma_p(\alpha)$. That is,

$$\Gamma_p(\alpha) = \int_{X>0} |X|^{\alpha-\frac{p+1}{2}} e^{-\mathrm{tr}(X)} dX \tag{1.4.2}$$

where dX denotes the wedge product of the $p(p+1)/2$ differentials dx_{ij}'s. The integral in (1.4.2) can be evaluated by using Theorem 1.1a.5. Let $T = (t_{ij})$ be a lower triangular matrix with $t_{jj} > 0, j = 1, \ldots, p$, $t_{ij} = 0, i < j$ such that $X = TT'$. Then from Theorem 1.1a.5

$$X = TT' \Rightarrow dX = 2^p \left\{ \prod_{j=1}^{p} t_{jj}^{p+1-j} \right\} dT .$$

Under this transformation

$$|X| = |TT'| = \prod_{j=1}^{p} t_{jj}^2$$

and

$$\mathrm{tr}(X) = \mathrm{tr}(TT') = t_{11}^2 + \left(t_{21}^2 + t_{22}^2\right) + \cdots + \left(t_{p1}^2 + \cdots + t_{pp}^2\right).$$

The integral in (1.4.2) will split into p integrals over t_{jj}'s, $j = 1, \ldots, p$ and $p(p-1)/2$ integrals over t_{ij}'s, $i > j$. Note that when $TT' > 0$ and $t_{jj} > 0,\ j = 1, \ldots, p$, one has $-\infty < t_{ij} < \infty, i > j$. Thus

$$\begin{aligned}\Gamma_p(\alpha) &= \int_{X>0} |X|^{\alpha-\frac{p+1}{2}} e^{-\mathrm{tr}(X)} dX \\ &= \left\{ \prod_{j=1}^{p} 2 \int_0^\infty \left(t_{jj}^2\right)^{\alpha-\frac{p+1}{2}} e^{-t_{jj}^2} \left(t_{jj}\right)^{p+1-j} dt_{jj} \right\} \\ &\quad \times \left\{ \prod_{i>j} \int_{-\infty}^{\infty} e^{-t_{ij}^2} dt_{ij} \right\}.\end{aligned}$$

Note that

$$2 \int_0^\infty \left(t_{jj}^2\right)^{\alpha-\frac{p+1}{2}} e^{-t_{jj}^2} dt_{jj} = \Gamma(\alpha - \frac{j-1}{2}),\ \mathrm{Re}(\alpha) > \frac{j-1}{2}$$

for $j = 1, \ldots, p$ and

$$\int_{-\infty}^{\infty} e^{-t_{ij}^2} dt_{ij} = \sqrt{\pi}$$

for each of the $p(p-1)/2$ integrals. Hence we have the following.

Notation 1.4.1 *Matrix-variate gamma, real case*

$$\Gamma_p(\alpha) = \pi^{\frac{p(p-1)}{4}} \Gamma(\alpha)\Gamma(\alpha - \frac{1}{2}) \cdots \Gamma(\alpha - \frac{p-1}{2}),\ \mathrm{Re}(\alpha) > (p-1)/2 . \tag{1.4.3}$$

We will call $\Gamma_p(\alpha)$ the matrix-variate gamma function in the real case, real in the sense that in the integral representation of (1.4.2) the matrix X is real symmetric positive definite. If X is a hermitian positive definite matrix then one has the corresponding matrix-variate gamma in the complex case.

1.4b Matrix-variate Gamma Density, Real Case

From (1.4.2) we can create a density function. Let

$$f(X) = \frac{|X|^{\alpha-\frac{p+1}{2}} e^{-\mathrm{tr}(X)}}{\Gamma_p(\alpha)}, \quad X = X' > 0\ , \ \mathrm{Re}(\alpha) > \frac{p-1}{2} \tag{1.4.4}$$

and $f(X) = 0$ elsewhere. Then evidently $f(X)$ is a density function. Note that if $B = B' > 0$ is a constant matrix then

$$\begin{aligned} f_X(X) &= \frac{|B|^{\alpha}|X|^{\alpha-\frac{p+1}{2}} e^{-\mathrm{tr}(BX)}}{\Gamma_p(\alpha)}, \\ & X = X' > 0,\ B = B' > 0,\ \mathrm{Re}(\alpha) > \frac{p-1}{2} \end{aligned} \tag{1.4.5}$$

and $f_X(X) = 0$ elsewhere is a density function with parameters (α, B). This can be seen by using the fact that

$$\mathrm{tr}(BX) = \mathrm{tr}(B^{\frac{1}{2}} X B^{\frac{1}{2}})$$

and then evaluating the integral by using (1.4.2) after making the transformation

$$U = B^{\frac{1}{2}} X B^{\frac{1}{2}} \Rightarrow \mathrm{d}U = |B|^{\frac{p+1}{2}} \mathrm{d}X,\ |X| = |B|^{-1}|U|\ .$$

From (1.4.5) we may also get the following identity.

$$|B|^{-\alpha} \equiv \frac{1}{\Gamma(\alpha)} \int_{X>0} |X|^{\alpha-\frac{p+1}{2}} e^{-\mathrm{tr}(BX)} \mathrm{d}X \tag{1.4.6}$$

for $\mathrm{Re}(\alpha) > \frac{p-1}{2}$, $X = X' > 0$, $B = B' > 0$ where X and B are $p \times p$ and B is free of X.

Definition 1.4b.1 *Matrix-variate real gamma density* The density in (1.4.5) is called a matrix-variate real gamma density with parameters (α, B).

Definition 1.4b.2 *Wishart density, real nonsingular case* The density in (1.4.5) with $\alpha = \frac{n}{2}$ where $n \geq p$ and an integer, $B = (2\Sigma)^{-1}$ where Σ is the covariance matrix associated with a real nonsingular normal vector, is called a nonsingular real Wishart density in the central case with the parameter matrix Σ, degrees of freedom n and dimensionality p and it is denoted by $X \sim W_p(n, \Sigma)$.

1.4c The m.g.f. of a Matrix-variate Real Gamma Variable

When $X = (X_{ij})$ is a real symmetric matrix of functionally independent variables there are only $p(p+1)/2$ variables in X. They are X_{ij}'s for $i \geq j$. Consider a real symmetric parameter matrix $T = (\delta_{ij}t_{ij})$, $t_{ij} = t_{ji}$ where $\delta_{ij} = 1$ if $i = j$ and it is $\frac{1}{2}$ for $i \neq j$. That is, T is of the form

$$T = \begin{bmatrix} t_{11} & \frac{1}{2}t_{12} & \dots & \frac{1}{2}t_{1p} \\ \vdots & \vdots & \ddots & \vdots \\ \frac{1}{2}t_{p1} & \frac{1}{2}t_{p2} & \dots & t_{pp} \end{bmatrix}, \; T = T' \; . \tag{1.4.7}$$

Then

$$\mathrm{tr}(TX) = t_{11}X_{11} + (t_{21}X_{21} + t_{22}X_{22}) + \cdots + (t_{p1}X_{p1} + \cdots + t_{pp}X_{pp}) \; .$$

Thus the m.g.f. of X is given by

$$M_X(T) = E\left(\mathrm{e}^{\mathrm{tr}(TX)}\right)$$

where T is a parameter matrix of the form in (1.4.7). If X has the real gamma density as in (1.4.5) then

$$\begin{aligned} M_X(T) &= \frac{|B|^\alpha}{\Gamma_p(\alpha)} \int_{X>0} \mathrm{e}^{\mathrm{tr}(TX)} |X|^{\alpha - \frac{p+1}{2}} \mathrm{e}^{-\mathrm{tr}(BX)} \mathrm{d}X \\ &= \frac{|B|^\alpha}{\Gamma_p(\alpha)} \int_{X>0} |X|^{\alpha - \frac{p+1}{2}} \mathrm{e}^{-\mathrm{tr}((B-T)X)} \mathrm{d}X \text{ for } B - T > 0 \; . \end{aligned}$$

Make the transformation

$$V = (B-T)^{\frac{1}{2}} X (B-T)^{\frac{1}{2}} \Rightarrow \mathrm{d}V = |B-T|^{\frac{p+1}{2}} \mathrm{d}X$$

and integrate out V by using (1.4.2) to get

$$\begin{aligned} M_X(T) &= |B|^\alpha |B-T|^{-\alpha} \\ &= |I - B^{-1}T|^{-\alpha} \text{ for } I - B^{-1}T > 0 \; . \end{aligned} \tag{1.4.8}$$

1.4d Matrix-variate Beta, Real Case

A type-1 real scalar beta random variable has the density

$$g_1(x) = \frac{x^{\alpha-1}(1-x)^{\beta-1}}{B(\alpha,\beta)} \; , \quad 0 < x < 1, \; \mathrm{Re}(\alpha) > 0, \; \mathrm{Re}(\beta) > 0 \tag{1.4.9}$$

and $g_1(x) = 0$ elsewhere where $B(\alpha, \beta)$ is the beta function

$$\begin{aligned} B(\alpha,\beta) &= \int_0^1 x^{\alpha-1}(1-x)^{\beta-1} \mathrm{d}x \; , \quad \mathrm{Re}(\alpha) > 0, \; \mathrm{Re}(\beta) > 0 \\ &= \frac{\Gamma(\alpha)\Gamma(\beta)}{\Gamma(\alpha+\beta)} \; . \end{aligned} \tag{1.4.10}$$

One can obtain a matrix analogue of the beta function and its integral representation.

Definition 1.4d.1 *Matrix-variate beta function, real case* It is denoted by $B_p(\alpha,\beta)$ and it is defined as

$$B_p(\alpha,\beta)=\frac{\Gamma_p(\alpha)\Gamma_p(\beta)}{\Gamma_p(\alpha+\beta)}\,,\quad \mathrm{Re}(\alpha)>\frac{p-1}{2},\ \mathrm{Re}(\beta)>\frac{p-1}{2}\,. \tag{1.4.11}$$

One can establish an integral representation of $B_p(\alpha,\beta)$, analogous to the one in (1.4.10).

Theorem 1.4d.1 For $\mathrm{Re}(\alpha)>\frac{p-1}{2}$, $\mathrm{Re}(\beta)>\frac{p-1}{2}$,

$$\int_{0<X<I}|X|^{\alpha-\frac{p+1}{2}}|I-X|^{\beta-\frac{p+1}{2}}\mathrm{d}X=B_p(\alpha,\beta) \tag{1.4.12}$$

where X is real $p\times p$ symmetric positive definite and $0<X<I$ means that $X>0$ and $I-X>0$.

PROOF Take the integral representation of a matrix-variate gamma from (1.4.4) and write

$$\begin{aligned}\Gamma_p(\alpha)\Gamma_p(\beta)&=\left\{\int_{X>0}|X|^{\alpha-\frac{p+1}{2}}\mathrm{e}^{-\mathrm{tr}(X)}\mathrm{d}X\right\}\\&\quad\times\left\{\int_{Y>0}|Y|^{\beta-\frac{p+1}{2}}\mathrm{e}^{-\mathrm{tr}(Y)}\mathrm{d}Y\right\}\\&=\int_{X>0}\int_{Y>0}|X|^{\alpha-\frac{p+1}{2}}|Y|^{\beta-\frac{p+1}{2}}\mathrm{e}^{-\mathrm{tr}(X+Y)}\mathrm{d}X\mathrm{d}Y\,.\end{aligned}$$

Put $U=X+Y$ for fixed Y which means $\mathrm{d}U=\mathrm{d}X$ and $Y=U-X$. Then

$$\Gamma_p(\alpha)\Gamma_p(\beta)=\int_{X>0}\int_{U-X>0}|X|^{\alpha-\frac{p+1}{2}}|U-X|^{\beta-\frac{p+1}{2}}\mathrm{e}^{-\mathrm{tr}(U)}\mathrm{d}U\mathrm{d}X\,.$$

Put $X=U^{\frac{1}{2}}VU^{\frac{1}{2}}$ for fixed $U\Rightarrow \mathrm{d}X=|U|^{\frac{p+1}{2}}\mathrm{d}V$ and then

$$\begin{aligned}\Gamma_p(\alpha)\Gamma_p(\beta)&=\left\{\int_{U>0}|U|^{\alpha+\beta-\frac{p+1}{2}}\mathrm{e}^{-\mathrm{tr}(U)}\mathrm{d}U\right\}\\&\quad\times\left\{\int_{0<V<I}|V|^{\alpha-\frac{p+1}{2}}|I-V|^{\beta-\frac{p+1}{2}}\mathrm{d}V\right\}.\end{aligned}$$

Noting that the integral over U yields $\Gamma_p(\alpha+\beta)$ the result is established.

Definition 1.4d.2 *A real matrix-variate type-1 beta density* A real $p\times p$ symmetric positive definite matrix X is said to have a type-1 beta density if the density is of the form

$$g(X) = \frac{|X|^{\alpha-\frac{p+1}{2}}|I-X|^{\beta-\frac{p+1}{2}}}{B_p(\alpha,\beta)}, \quad 0<X<I,$$
$$\mathrm{Re}(\alpha) > \frac{p-1}{2}, \ \mathrm{Re}(\beta) > \frac{p-1}{2} \tag{1.4.13}$$

and $g(X) = 0$ elsewhere where $B_p(\alpha, \beta)$ is given in (1.4.11).

Note that for $U = I - X$ in (1.4.13) the parameters α and β are interchanged. A real type-2 scalar beta random variable has the density

$$g_2(x) = \frac{x^{\alpha-1}(1+x)^{-(\alpha+\beta)}}{B(\alpha,\beta)}, \quad x>0, \ \mathrm{Re}(\alpha)>0, \ \mathrm{Re}(\beta)>0 \tag{1.4.14}$$

and $g_2(x) = 0$ elsewhere. The corresponding matrix analogue is given by the following.

Definition 1.4d.3 *A real matrix-variate type-2 beta density* This is defined as

$$g_X(X) = \frac{|X|^{\alpha-\frac{p+1}{2}}|I+X|^{-(\alpha+\beta)}}{B_p(\alpha,\beta)}, \quad X = X' > 0,$$
$$\mathrm{Re}(\alpha) > \frac{p-1}{2}, \ \mathrm{Re}(\beta) > \frac{p-1}{2} \tag{1.4.15}$$

and $g_X(X) = 0$ elsewhere where X is real $p \times p$ and $B_p(\alpha, \beta)$ is the matrix-variate beta function given in (1.4.11).

Note that for $U = X^{-1}$ in (1.4.15) the parameters α and β are interchanged. It can be shown that if the $p \times p$ real symmetric positive definite matrix X is type-1 beta distributed then $Y = (I-X)^{-\frac{1}{2}}X(I-X)^{-\frac{1}{2}}$ is type-2 beta distributed and if the $p \times p$ real symmetric positive definite matrix Y is type-2 beta distributed then $X = (I+Y)^{-1}$ as well as $(I+Y)^{-\frac{1}{2}}Y(I+Y)^{-\frac{1}{2}}$ are type-1 beta distributed. The Jacobians in these cases are evaluated as follows.

$$\begin{aligned} Y &= (I-X)^{-\frac{1}{2}}X(I-X)^{-\frac{1}{2}} = (I-X)^{-\frac{1}{2}}X^{\frac{1}{2}}X^{\frac{1}{2}}(I-X)^{-\frac{1}{2}} \\ &= (X^{-1}-I)^{-\frac{1}{2}}(X^{-1}-I)^{-\frac{1}{2}} = (X^{-1}-I)^{-1} \Rightarrow \\ Y^{-1} &= X^{-1} - I \Rightarrow |Y|^{-(p+1)}\mathrm{d}Y = |X|^{-(p+1)}\mathrm{d}X \end{aligned}$$

(see also Theorem 1.1a.4). That is,

$$\begin{aligned} \mathrm{d}X &= |X|^{-(p+1)}|Y|^{-(p+1)}\mathrm{d}Y = |I+Y^{-1}|^{-(p+1)}|Y|^{-(p+1)}\mathrm{d}Y \\ &= |I+Y|^{-(p+1)}\mathrm{d}Y. \\ X &= (I+Y)^{-1} \Rightarrow Y = X^{-1} - I \Rightarrow \mathrm{d}Y = |X|^{-(p+1)}\mathrm{d}X. \\ X &= (I+Y)^{-\frac{1}{2}}Y(I+Y)^{-\frac{1}{2}} = (I+Y)^{-\frac{1}{2}}Y^{\frac{1}{2}}Y^{\frac{1}{2}}(I+Y)^{-\frac{1}{2}} \\ &= (Y^{-1}+I)^{-1} \Rightarrow X^{-1} = Y^{-1} + I \Rightarrow \\ \mathrm{d}Y &= |Y|^{(p+1)}|X|^{-(p+1)}\mathrm{d}X \\ &= |X^{-1}-I|^{-(p+1)}|X|^{-(p+1)}\mathrm{d}X = |I-X|^{-(p+1)}\mathrm{d}X\,. \end{aligned}$$

A constant multiple of X in (1.4.15) is the matrix-variate F random variable in the real case.

1.5 Hypergeometric Series, Real Case

A hypergeometric series in the real scalar case with r upper parameters $a_1, \ldots, a_r$ and s lower parameters $b_1, \ldots, b_s$ is denoted by the following.

Notation 1.5.1 *Hypergeometric series*

$$ {}_rF_s = {}_rF_s((a);(b);x) = {}_rF_s(a_1, \ldots, a_r; b_1, \ldots, b_s; x) $$

Definition 1.5.1 *Hypergeometric series* For x a real scalar variable

$$ \begin{aligned} {}_rF_s &= {}_rF_s(a_1, \ldots, a_r; b_1, \ldots, b_s; x) \\ &= \sum_{k=0}^{\infty} \frac{(a_1)_k \cdots (a_r)_k}{(b_1)_k \cdots (b_s)_k} \frac{x^k}{k!} \end{aligned} \tag{1.5.1} $$

where, for example, $(a)_k = a(a+1)\cdots(a+k-1)$ with $(a)_0 = 1,\ a \neq 0$. The lower parameters $b_1, \ldots, b_s$ are such that none of them is a negative integer or zero and if a b_j is a negative integer or zero it is assumed that there is an a_m such that $(a_m)_k$ vanishes before $(b_j)_k$ vanishes. The series in (1.5.1) is convergent for all x when $s \geq r$. When $r = s+1$ the series is convergent for $|x| < 1$. Convergence conditions can be worked out for $x = 1$ and $x = -1$.

Generalizations of hypergeometric series to the matrix-variate case and other details will be considered in Chapter 4 and hence further discussion is postponed.

CHAPTER 2

Quadratic and Bilinear Forms In Normal Vectors

2.0 Introduction

Mathai and Provost (1992) dealt with quadratic forms in random variables, their distributions, moments and various properties including chisquaredness. In this chapter we will concentrate on bilinear forms in random variables and their properties. Generalizations to matrix variables will be dealt with in detail in Chapters 4 and 5 but some aspects of matrix variables will also be considered in this chapter. Even though a bilinear form can be considered to be a particular case of a quadratic form, the general methods of tackling quadratic forms can sometimes become too complicated to handle bilinear forms. In these cases specific techniques are to be developed for dealing with bilinear forms. This will be seen from the discussions later on in this chapter. The material in this as well as in the remaining chapters will complement that in Mathai and Provost (1992).

Consider scalar variables $x_1, \ldots, x_p$ and $y_1, \ldots, y_q$. In vector notation we may write $\mathbf{x}' = (x_1, \ldots, x_p)$ and $\mathbf{y}' = (y_1, \ldots, y_q)$, where a prime denotes a transpose. The variables may be mathematical variables or random variables. A function of $x_1, \ldots, x_p$ and $y_1, \ldots, y_q$ which is homogeneous, and linear in x_j's as well as in y_j's is generally known as a bilinear form. A formal definition can be given as follows:

Definition 2.0.1 *Bilinear form* Let $\mathbf{x}, \mathbf{y}, \mathbf{z}$ be elements of a vector space and α a scalar quantity. A function $f(\mathbf{x}, \mathbf{y})$ is called a bilinear form if it satisfies the following conditions.

$$f(\mathbf{x}+\mathbf{z}, \mathbf{y}) = f(\mathbf{x}, \mathbf{y}) + f(\mathbf{z}, \mathbf{y}) \tag{2.0.1}$$

$$f(\mathbf{x}, \mathbf{z}+\mathbf{y}) = f(\mathbf{x}, \mathbf{y}) + f(\mathbf{x}, \mathbf{z}) \tag{2.0.2}$$

$$f(\alpha\mathbf{x}, \mathbf{y}) = \alpha f(\mathbf{x}, \mathbf{y}) = f(\mathbf{x}, \alpha\mathbf{y}) \tag{2.0.3}$$

for all $\mathbf{x}, \mathbf{y}, \mathbf{z}$ and α. In addition, if $f(\mathbf{x}, \mathbf{y}) = f(\mathbf{y}, \mathbf{x})$ then the bilinear form is called symmetric.

Similarly one can define a quadratic form by the condition

$$f(\alpha\mathbf{x}) = \alpha^2 f(\mathbf{x})$$

for all scalars α and all vectors $\mathbf{x}$ in a given vector space.

Example 2.0.1 Check whether the following are bilinear forms in $\mathbf{x}' = (x_1, x_2, x_3)$ and $\mathbf{y}' = (y_1, y_2)$.

$$Q_1 = x_1y_1 + x_1y_2 + x_2y_1 + x_2y_2 + x_3y_1 + x_3y_2 \tag{2.0.4}$$

$$Q_2 = 2x_1y_1 - 3x_2y_2 + 4x_1 + 5 \tag{2.0.5}$$

$$Q_3 = x_1y_1 - 4x_3y_2 + y_1 - 6. \tag{2.0.6}$$

SOLUTION Q_1 is linear in $\mathbf{x}$ as well as in $\mathbf{y}$ and it is a homogeneous function. Note that Q_1 satisfies conditions (2.0.1), (2.0.2) and (2.0.3) of Definition 2.0.1 and hence Q_1 is a bilinear form in $\mathbf{x}$ and $\mathbf{y}$. But Q_2 and Q_3 do not satisfy (2.0.3) of Definition 2.0.1 and hence these are not bilinear forms.

Note that one can write Q_1, Q_2, and Q_3 in the following forms by using vector and matrix notations.

$$\begin{aligned} Q_1 &= (x_1, x_2, x_3)\begin{bmatrix} 1 & 1 \\ 1 & 1 \\ 1 & 1 \end{bmatrix}\begin{bmatrix} y_1 \\ y_2 \end{bmatrix} = \mathbf{x}'A_1\mathbf{y} \\ &= (y_1, y_2)\begin{bmatrix} 1 & 1 & 1 \\ 1 & 1 & 1 \end{bmatrix}\begin{bmatrix} x_1 \\ x_2 \\ x_3 \end{bmatrix} = \mathbf{y}'A_1'\mathbf{x}, \end{aligned}$$

$$A_1' = \begin{bmatrix} 1 & 1 & 1 \\ 1 & 1 & 1 \end{bmatrix}.$$

$$\begin{aligned} Q_2 &= (x_1, x_2, x_3)\begin{bmatrix} 2 & 0 \\ 0 & -3 \\ 0 & 0 \end{bmatrix}\begin{bmatrix} y_1 \\ y_2 \end{bmatrix} + (x_1, x_2, x_3)\begin{bmatrix} 4 \\ 0 \\ 0 \end{bmatrix} + 5 \\ &= \mathbf{x}'A_2\mathbf{y} + \mathbf{x}'\mathbf{b_2} + d_2 = \mathbf{y}'A_2'\mathbf{x} + \mathbf{b_2'}\mathbf{x} + d_2, \end{aligned}$$

$$A_2' = \begin{bmatrix} 2 & 0 & 0 \\ 0 & -3 & 0 \end{bmatrix}, \ \mathbf{b_2'} = (4, 0, 0), \ d_2 = 5.$$

$$Q_3 = (x_1, x_2, x_3)\begin{bmatrix}1 & 0\\ 0 & 0\\ 0 & -4\end{bmatrix}\begin{bmatrix}y_1\\ y_2\end{bmatrix} + (y_1, y_2)\begin{bmatrix}1\\ 0\end{bmatrix} - 6$$
$$= \mathbf{x}'A_3\mathbf{y} + \mathbf{y}'\mathbf{c}_3 + d_3 = \mathbf{y}'A_3'\mathbf{x} + \mathbf{c}_3'\mathbf{y} + d_3,$$
$$A_3' = \begin{bmatrix}1 & 0 & 0\\ 0 & 0 & -4\end{bmatrix}, \quad \mathbf{c}_3{}' = (1,0), \; d_3 = -6.$$

2.1 Various Representations

From Example 2.0.1 note that in general bilinear forms in $\mathbf{x}' = (x_1, \ldots, x_p)$ and $\mathbf{y}' = (y_1, \ldots, y_q)$ can be written as follows:

$$Q = \mathbf{x}'A\mathbf{y} = \mathbf{y}'A'\mathbf{x}, \; A = (a_{ij}) \tag{2.1.1}$$
$$= \sum_{i=1}^{p}\sum_{j=1}^{q} a_{ij}x_iy_j$$

where A is a $p \times q$ rectangular matrix of the elements a_{ij}'s which are free of x_j, $j = 1, \ldots, p$ and y_k, $k = 1, \ldots, q$. These a_{ij}'s could be numbers or functions of other variables and hence are constants as far as $\mathbf{x}$ and $\mathbf{y}$ are concerned. If $p = q$ and $A = A'$ then Q is a symmetric bilinear form. Note that Q can also be written as a quadratic form in the vector $\mathbf{z}' = (\mathbf{x}', \mathbf{y}') = (x_1, \ldots, x_p, y_1, \ldots, y_q)$. In this case

$$Q = \mathbf{x}'A\mathbf{y} = \mathbf{y}'A'\mathbf{x}$$
$$= \mathbf{z}'A^*\mathbf{z} = (\mathbf{x}', \mathbf{y}')\begin{bmatrix}O & A\\ O & O\end{bmatrix}\begin{bmatrix}\mathbf{x}\\ \mathbf{y}\end{bmatrix} \tag{2.1.2}$$
$$= (\mathbf{x}', \mathbf{y}')\begin{bmatrix}O & O\\ A' & O\end{bmatrix}\begin{bmatrix}\mathbf{x}\\ \mathbf{y}\end{bmatrix} \tag{2.1.3}$$
$$= (\mathbf{x}', \mathbf{y}')\begin{bmatrix}O & \frac{1}{2}A\\ \frac{1}{2}A' & O\end{bmatrix}\begin{bmatrix}\mathbf{x}\\ \mathbf{y}\end{bmatrix} \tag{2.1.4}$$

where the matrix A^* is different in the different representations, that is,

$$A^* = \begin{bmatrix}O & A\\ O & O\end{bmatrix}, \; \begin{bmatrix}O & O\\ A' & O\end{bmatrix}, \; \begin{bmatrix}O & \frac{1}{2}A\\ \frac{1}{2}A' & O\end{bmatrix}$$

in (2.1.2), (2.1.3) and (2.1.4) respectively. If the properties of quadratic form are used to study bilinear forms then the representation in (2.1.4) can be seen to be the most convenient one because the matrix A^* is symmetric in this case.

Note that in Q_2 and Q_3 of Example 2.0.1 we have a linear expression in $\mathbf{x}$ alone and a linear expression in $\mathbf{y}$ alone respectively. We can give a general bilinear expression in $\mathbf{x}$ and $\mathbf{y}$ as follows:

Definition 2.1.1 *A general bilinear expression* Let $\mathbf{x}' = (x_1, \ldots, x_p)$, $\mathbf{y}' = (y_1, \ldots, y_q)$, $A = (a_{ij})$, $i = 1, \ldots, p$, $j = 1, \ldots, q$, $\mathbf{b}' = (b_1, \ldots, b_p)$, $\mathbf{c}' = (c_1, \ldots, c_q)$

and d a scalar constant where the elements in $A, \boldsymbol{b}$ and $\mathbf{c}$ are constants. Then a general bilinear expression in $\mathbf{x}$ and $\mathbf{y}$ is defined as

$$\begin{aligned} Q &= \mathbf{x}'A\mathbf{y} + \boldsymbol{b}'\mathbf{x} + \mathbf{c}'\mathbf{y} + d \\ &= \mathbf{y}'A'\mathbf{x} + \mathbf{x}'\boldsymbol{b} + \mathbf{y}'\mathbf{c} + d. \end{aligned} \tag{2.1.5}$$

In terms of $\mathbf{z}' = (\mathbf{x}', \mathbf{y}')$ we can represent (2.1.5) as a quadratic expression by using the following procedure.

$$\begin{aligned} Q &= \mathbf{z}'A^*\mathbf{z} + \mathbf{z}'\boldsymbol{b}^* + d \\ &= \mathbf{z}'A^{*'}\mathbf{z} + \boldsymbol{b}^{*'}\mathbf{z} + d \end{aligned} \tag{2.1.6}$$

where, for convenience, we take

$$A^* = \begin{bmatrix} O & \frac{1}{2}A \\ \frac{1}{2}A' & O \end{bmatrix}, \quad \mathbf{b}^* = \begin{bmatrix} \mathbf{b} \\ \mathbf{c} \end{bmatrix}.$$

Thus Q in (2.1.5) can be looked upon as a quadratic expression or a second degree polynomial in $\mathbf{z}$ as given in (2.1.6). Hence the theory of quadratic form given in Mathai and Provost (1992) is applicable to the bilinear forms as well, by taking the representation in (2.1.6). But as will be explained later on, this procedure becomes too complicated when dealing with bilinear forms. Alternate precedures are needed to simplify matters.

We will be mainly dealing with bilinear forms in random variables and particularly in singular and nonsingular Gaussian or normal random vectors. The distribution of a quadratic form in normal vectors reduces to that of a linear combination of independent central or noncentral chi-square random variables when the quadratic form is positive definite. What will be the corresponding result when dealing with bilinear forms? It will be shown later that they fall in the categories of gamma difference, Laplacian and generalized Laplacian. As particular cases, chi-squares will also come in. We will develop the necessary distribution theory for gamma difference in the next section.

2.2 Density of a Gamma Difference

Let X_1 and X_2 be two independently distributed real gamma variables with the parameters (α_1, β_1) and (α_2, β_2) respectively, $\alpha_i > 0,\ \beta_i > 0,\ i = 1, 2$. A real random variable X is said to have a gamma density with the parameters (α, β) if the density is of the following form.

$$f_X(x) = \frac{1}{\beta^\alpha \Gamma(\alpha)} x^{\alpha-1} e^{-(x/\beta)},\ \alpha > 0,\ \beta > 0,\ x > 0$$

and $f_X(x) = 0$ otherwise. Then the moment generating function (m.g.f.), denoted by $M_X(t)$ where t is a parameter, is seen to be the following:

$$M_X(t) = E\left(e^{(tX)}\right) = \int_0^\infty e^{(tx)} f_X(x) dx = (1 - \beta t)^{-\alpha}.$$

Let $U = X_1 - X_2$. Since X_1 and X_2 are independently distributed the m.g.f. of U is given by

$$M_U(t) = (1 - \beta_1 t)^{-\alpha_1}(1 + \beta_2 t)^{-\alpha_2} \tag{2.2.1}$$

for $|t\beta_1| < 1$, $|t\beta_2| < 1$. Thus for $\alpha_1 = \alpha_2 = \alpha$, $\beta_1 = \beta_2 = \beta$, (2.2.1) becomes

$$M_U(t) = (1 - \beta^2 t^2)^{-\alpha}. \tag{2.2.2}$$

We will call the density corresponding to the $M_U(t)$ of (2.2.1) the density of a gamma difference and the one corresponding to (2.2.2) will be called a generalized Laplace density. Note that U in (2.2.2) for $\alpha = 1$ has the Laplace density given by

$$f_U(u) = \frac{1}{2\beta} e^{-(|x|/\beta)}, \quad -\infty < x < \infty, \ \beta > 0. \tag{2.2.3}$$

We can evaluate the density of U by the technique of transformation of variables. Then the density of U, denoted by $f_U(u)$, is given by

$$f_U(u) = \begin{cases} c e^{(u/\beta_2)} \int_u^\infty v^{\alpha_1 - 1}(v - u)^{\alpha_2 - 1} \\ \quad \times e^{-(\beta_0 v)} dv & \text{for } u > 0, \\ c e^{(u/\beta_2)} \int_0^\infty v^{\alpha_1 - 1}(v - u)^{\alpha_2 - 1} \\ \quad \times e^{-(\beta_0 v)} dv & \text{for } u \le 0 \end{cases} \tag{2.2.4}$$

where

$$c = \frac{1}{\Gamma(\alpha_1)\Gamma(\alpha_2)\beta_1^{\alpha_1}\beta_2^{\alpha_2}}, \quad \beta_0 = \frac{1}{\beta_1} + \frac{1}{\beta_2}. \tag{2.2.5}$$

Thus

$$f_U(u) = \begin{cases} c_1 u^{\left(\frac{\alpha_1+\alpha_2}{2} - 1\right)} e^{-\left(\frac{u}{2}\right)\left(\frac{1}{\beta_1} - \frac{1}{\beta_2}\right)} \\ \quad \times W_{\frac{\alpha_1-\alpha_2}{2}, \frac{1-\alpha_1-\alpha_2}{2}}(\beta_0 u) \ \text{ for } u > 0, \\ c_2(-u)^{\left(\frac{\alpha_1+\alpha_2}{2} - 1\right)} e^{\left(\frac{u}{2}\right)\left(\frac{1}{\beta_2} - \frac{1}{\beta_1}\right)} \\ \quad \times W_{\frac{\alpha_2-\alpha_1}{2}, \frac{1-\alpha_1-\alpha_2}{2}}(-\beta_0 u) \ \text{ for } u \le 0 \end{cases} \tag{2.2.6}$$

where β_0 is defined in (2.2.5), $W_{\cdot,\cdot}(\cdot)$ is a Whittaker function,

$$c_1^{-1} = \Gamma(\alpha_1)\beta_1^{\left(\frac{\alpha_1-\alpha_2}{2}\right)}\beta_2^{\left(\frac{\alpha_2-\alpha_1}{2}\right)}(\beta_1 + \beta_2)^{\left(\frac{\alpha_1+\alpha_2}{2}\right)}$$

and

$$c_2^{-1} = \Gamma(\alpha_2)\beta_1^{\left(\frac{\alpha_1-\alpha_2}{2}\right)}\beta_2^{\left(\frac{\alpha_2-\alpha_1}{2}\right)}(\beta_1 + \beta_2)^{\left(\frac{\alpha_1+\alpha_2}{2}\right)}.$$

Some particular cases can be reduced to simpler forms.

2.2a Some Particular Cases

(i) When $\alpha_1 = \alpha_2 = \alpha$

$$f_U(u) = \frac{|u|^{\alpha-1} e^{\left(\frac{|u|}{2}\right)\left(\frac{1}{\beta_1}-\frac{1}{\beta_2}\right)}}{\Gamma(\alpha)(\beta_1+\beta_2)^{\alpha}} W_{0,\frac{1}{2}-\alpha}(\beta_0 |u|), \quad -\infty < u < \infty \tag{2.2.7}$$

where

$$\begin{aligned} W_{0,\frac{1}{2}-\alpha}(x) = e^{-(x/2)}\Big\{ & x^{1-\alpha}\frac{\Gamma(2\alpha-1)}{\Gamma(\alpha)}{}_1F_1(1-\alpha; 2(1-\alpha); x) \\ & + x^{\alpha}\frac{\Gamma(1-2\alpha)}{\Gamma(1-\alpha)}{}_1F_1(\alpha; 2\alpha; x)\Big\} \\ & \text{for } |\alpha| \neq \frac{m}{2},\ m = 1, 2, \ldots \end{aligned} \tag{2.2.8}$$

and ${}_1F_1$ is a hypergeometric series, see also Gradshteyn and Ryshik (1980).

(ii) When $\alpha_1 = \alpha_2 = \alpha = 1$, $\beta_1 = \beta_2 = \beta$, $f_U(u)$ is the Laplace density given in (2.2.3).

(iii) When α_1 and α_2 are positive integers

$$f_U(u) = \begin{cases} c_3 u^{\alpha_1-1} e^{-(u/\beta_1)} \\ \times \sum_{r=0}^{\alpha_1-1}\left[\frac{(\alpha_2)_r}{r!(\alpha_1-1-r)!}(\beta_0 u)^{-r}\right] & \text{for } u > 0 \\ c_4(-u)^{\alpha_2-1} e^{(u/\beta_2)} \\ \times \sum_{r=0}^{\alpha_2-1}\left[\frac{(\alpha_1)_r}{r!(\alpha_2-1-r)!}(-\beta_0 u)^{-r}\right] & \text{for } u \le 0 \end{cases} \tag{2.2.9}$$

where

$$c_3^{-1} = \beta_0^{\alpha_2}\beta_1^{\alpha_1}\beta_2^{\alpha_2}, \quad c_4^{-1} = \beta_0^{\alpha_1}\beta_1^{\alpha_1}\beta_2^{\alpha_2}$$

and for example, $(a)_r = a(a+1)\cdots(a+r-1)$, $(a)_0 = 1$, $a \neq 0$.

(iv) When $\alpha_1 = \alpha_2 = \alpha$, $\beta_1 = \beta_2 = \beta$ and α is a positive integer

$$f_U(u) = \frac{1}{(2\beta)^{\alpha}}|u|^{\alpha-1}e^{-(|u|/\beta)}\sum_{r=0}^{\alpha-1}\left[\frac{(\alpha)_r}{r!(\alpha-1-r)!}\left(\frac{2|u|}{\beta}\right)^{-r}\right]$$
$$\text{for } -\infty < u < \infty. \tag{2.2.10}$$

Let us examine some bilinear functions of scalar normal random variables to see that they fall in the category of gamma differences and generalized Laplacian.

Example 2.2.1 Consider a bivariate normal with the mean vector null and the covariance matrix

$$V = \begin{pmatrix} 4 & 2 \\ 2 & 4 \end{pmatrix}.$$

Let $\binom{X_i}{Y_i}, i = 1, 2, 3$ be a simple random sample of size 3 from this bivariate normal. Then show that the following are distributed as gamma differences.

$$\begin{aligned}U_1 = \frac{1}{3}[2(X_1Y_1 + X_2Y_2 + X_3Y_3)\\ - (X_1Y_2 + X_2Y_1 + X_1Y_3 + X_3Y_1 + X_2Y_3 + X_3Y_2)]\end{aligned}$$

and

$$\begin{aligned}U_2 = 4X_1Y_1 + X_2Y_2 + 4X_3Y_3 + 2X_1Y_2 + 2X_2Y_1\\ + 4X_1Y_3 + 4X_3Y_1 + 2X_2Y_3 + 2X_3Y_2.\end{aligned}$$

SOLUTION The result on U_1 can be directly established by writing U_1 as $Q_2 = \mathbf{Z}'A\mathbf{Z}$, $\mathbf{Z}' = (X_1, X_2, X_3, Y_1, Y_2, Y_3)$. Then one has

$$A = \begin{pmatrix} O & \frac{1}{2}B_2 \\ \frac{1}{2}B_2' & O \end{pmatrix}, \quad B_2 = \frac{1}{3}\begin{pmatrix} 2 & -1 & -1 \\ -1 & 2 & -1 \\ -1 & -1 & 2 \end{pmatrix}.$$

Then the covariance matrix of $\mathbf{Z}$, denoted by Σ, is given by

$$\Sigma = \begin{pmatrix} 4I & 2I \\ 2I & 4I \end{pmatrix}, \text{and thus } 2A\Sigma = \begin{pmatrix} 2B_2 & 4B_2 \\ 4B_2' & 2B_2' \end{pmatrix}.$$

The m.g.f. of $\mathbf{Z}$ is available by direct integration or from Mathai and Provost (1992). Denoting it by $M_{Q_2}(t)$ one has the following, see also a general result in this direction in Section 2.8:

$$\begin{aligned}M_{Q_2}(t) &= |I - 2tA\Sigma|^{-\frac{1}{2}}\\ &= \left\{(1-6t)^2(1+2t)^2\right\}^{-\frac{1}{2}}\\ &= (1-\beta_1 t)^{-\alpha_1}(1+\beta_2 t)^{-\alpha_2}.\end{aligned}$$

Now compare it with (2.2.1) to see that U_1 is distributed as a gamma difference with the parameters $\alpha_1 = 1, \alpha_2 = 1, \ \beta_1 = 6, \beta_2 = 2$.

Writing U_2 in the form $\mathbf{X}'B_2\mathbf{Y}$, $\mathbf{X}' = (X_1, X_2, X_3)$, $\mathbf{Y}' = (Y_1, Y_2, Y_3)$ one has

$$B_2 = \begin{pmatrix} 4 & 2 & 4 \\ 2 & 1 & 2 \\ 4 & 2 & 4 \end{pmatrix}$$

and then the eigenvalues of B_2 are 9, 0, 0. Thus proceeding as in the case of U_1 above we see that U_2 is a gamma difference with the parameters $\alpha_1 = \frac{1}{2}, \ \alpha_2 = \frac{1}{2}, \ \beta_1 = 54$, $\beta_2 = 18$.

Example 2.2.2 Consider a simple random sample of size n from a bivariate normal with zero correlation. That is, $\binom{X_i}{Y_i}, \ i = 1, \dots, n$ are independently and identically distributed with $X_i \sim N(\mu_x, \ \sigma_x^2)$, $Y_i \sim N(\mu_y, \ \sigma_y^2)$, $\mathrm{Cov}(X_i, Y_i) = 0$ for all i. Let

$U = \sum_{i=1}^{n}(X_i - \bar{X})(Y_i - \bar{Y})$, $\bar{X} = \sum_{i=1}^{n} X_i/n$, $\bar{Y} = \sum_{i=1}^{n} Y_i/n$. Then show that U is a generalized Laplacian.

SOLUTION Write $U = \mathbf{Z}'A\mathbf{Z}$ where $\mathbf{Z}$ is a column vector of X_i, $i = 1, \ldots, n$, Y_i, $i = 1, \ldots, n$ and A is a matrix of the form

$$A = \begin{pmatrix} O & \frac{1}{2}B \\ \frac{1}{2}B' & O \end{pmatrix}$$

and the covariance matrix of Z is of the form

$$\Sigma = \begin{pmatrix} \sigma_x^2 I_n & O \\ O & \sigma_y^2 I_n \end{pmatrix}.$$

Note that $U = \mathbf{Z}'A\mathbf{Z} = \mathbf{X}'B\mathbf{Y}$, $\mathbf{Y}' = (X_1, \ldots, X_n)$, $\mathbf{Y}' = (Y_1, \ldots, Y_n)$, $B = I_n - \frac{\mathbf{1}\mathbf{1}'}{n}$, $B = B^2$ with rank $n-1$, $\mathbf{1}' = (1, 1, \ldots, 1)$, and I_n is an identity matrix of order n. The eigenvalues of B are 1 repeated $n-1$ times and a 0. Thus the m.g.f. reduces to the following:

$$|I - 2tA\Sigma|^{-(1/2)} = (1 - \sigma_x^2\sigma_y^2 t^2)^{-(n-1)/2}.$$

Thus U is a generalized Laplacian of (2.2.2) with the parameters $\alpha = \frac{n-1}{2}$ and $\beta = \sigma_x\sigma_y$.

Note that the above example is on a covariance structure of sample coming from a bivariate normal with zero correlation.

2.3 Noncentral Gamma Difference

A noncentral chi-square with m degrees of freedom and noncentrality parameter λ has the m.g.f.

$$M(t) = (1 - 2t)^{-m/2} \exp\left\{-\lambda + \lambda(1 - 2t)^{-1}\right\}$$

for $| 2t | < 1$. Thus if Y_i is a noncentral chi-square with degrees of freedom m_i and noncentrality parameter λ_i, $i = 1, 2$ then when Y_1 and Y_2 are independently distributed, $V = Y_1 - Y_2$ has the m.g.f.

$$M_V(t) = (1 - 2t)^{-m_1/2}(1 + 2t)^{-m_2/2} \times \exp\left\{-(\lambda_1 + \lambda_2) + \lambda_1(1 - 2t)^{-1} + \lambda_2(1 + 2t)^{-1}\right\}.$$

Now consider a random variable U having the m.g.f.

$$M_U(t) = (1 - \beta_1 t)^{-\alpha_1}(1 + \beta_2 t)^{-\alpha_2} \times \exp\left\{-(\lambda_1 + \lambda_2) + \lambda_1(1 - \beta_1 t)^{-1} + \lambda_2(1 + \beta_2 t)^{-1}\right\} ; \tag{2.3.1}$$

then U can be called a *noncentral gamma difference* where $\beta_i > 0$, $\alpha_i > 0$, $| \beta_i t | < 1$, $\lambda_i > 0$, $i = 1, 2$. It is easy to verify that (2.3.1) is in fact a m.g.f.. When $\lambda_1 = \lambda_2 = \lambda$, $\beta_1 = \beta_2 = \beta$, $\alpha_1 = \alpha_2 = \alpha$ in (2.3.1) one has

$$M_U(t) = (1 - \beta^2 t^2)^{-\alpha} \exp\left\{-2\lambda + 2\lambda(1 - \beta^2 t^2)^{-1}\right\}. \tag{2.3.2}$$

Then U in (2.3.2) can be called a *noncentral generalized Laplace variable* (NGL). Thus we define the following:

Definition 2.3.1 *Central and noncentral gamma differences* The random variable corresponding to the m.g.f. in (2.3.1) is called a noncentral gamma difference with the parameters $(\alpha_1, \alpha_2, \beta_1, \beta_2, \lambda_1, \lambda_2)$ and when $\lambda_1 = 0 = \lambda_2$ it is called a central gamma difference.

Definition 2.3.2 *Central and noncentral generalized Laplacian* The random variable corresponding to the m.g.f. in (2.3.2) will be called a noncentral generalized Laplacian (NGL) with the parameters (α, β, λ). When $\lambda = 0$ it is called a generalized Laplacian or a central generalized Laplacian with the parameters (α, β).

Some properties of bilinear forms can be studied by using (2.3.1) and (2.3.2) and their particular cases. Some examples will be given after looking at the distribution of U coming from (2.3.1). From (2.3.1) one has

$$M_U(t) = e^{-(\lambda_1+\lambda_2)} \sum_{r_1=0}^{\infty} \sum_{r_2=0}^{\infty} \left[\frac{(\lambda_1)^{r_1}(\lambda_2)^{r_2}}{r_1! r_2!} \right.$$

$$\left. \times (1-\beta_1 t)^{-(\alpha_1+r_1)}(1+\beta_2 t)^{-(\alpha_2+r_2)} \right]; \tag{2.3.3}$$

and then on taking the inverse Laplace transform of (2.3.3) one has

$$g(u) = e^{-(\lambda_1+\lambda_2)} \sum_{r_1=0}^{\infty} \sum_{r_2=0}^{\infty} \left[\frac{(\lambda_1)^{r_1}(\lambda_2)^{r_2}}{r_1! r_2!} g_{r_1,r_2}(u) \right] \tag{2.3.4}$$

where $g(u)$ is the density of U of (2.3.1) and $g_{r_1,r_2}(u)$ is the $f_U(u)$ of (2.2.6) with α_i replaced by $\alpha_i + r_i$, $i = 1, 2$. The particular cases are then available from (2.3.4). The density for a chi-square difference is available from (2.3.4) by replacing β_i by 2 and α_i by $m_i/2$, $i = 1, 2$ with $m_i = 1, 2, \dots$ where the m_i's denote the degrees of freedom of the chi-squares.

The same procedure is applicable even if one has a linear combination of independent gamma or chi-square variables. One can proceed as follows. Let

$$W = a_1 X_1 + \cdots + a_k X_k$$

where the X_i's are independently distributed real gamma variables with the parameters (α_i, β_i), and the a_i's are constants, $i = 1, \dots, k$. Then the m.g.f. of W is the following:

$$M_W(t) = M_{W_1}(t) M_{W_2}(-t) \tag{2.3.5}$$

so that $W = W_1 - W_2$ with W_1 and W_2 independently distributed with

$$M_{W_1}(t) = \prod_{i=1}^{m} (1-\gamma_i t)^{-\alpha_i} \tag{2.3.6}$$

and

$$M_{W_2}(t) = \prod_{i=m+1}^{k} (1-\gamma_i)^{-\alpha_i} \tag{2.3.7}$$

for some γ_i such as $|t\gamma_i| < 1$, $i = 1,\ldots,k$. If $a_j > 0$, $j = 1,\ldots,m$ and $a_j < 0$, $j = m+1,\ldots,k$ then $\gamma_i = a_i\beta_i$, $i = 1,\ldots,k$. It is possible that one of the products (2.3.6) or (2.3.7) could be empty depending upon the a_i's. Note that W_1 as well as W_2 could be looked upon as linear combinations of independently distributed real gamma variables. The structure of these variables will then be the following:

$$W_1 = a_1X_1 + \cdots + a_mX_m \tag{2.3.8}$$

and

$$W_2 = a_{m+1}X_{m+1} + \cdots + a_kX_k \tag{2.3.9}$$

where the X_i's are all mutually independently distributed real gamma variables with the parameters (α_i, β_i). This is for the case $a_i > 0$, $i = 1,\ldots,m$ and $a_i < 0$, $i = m+1,\ldots,k$. The density for a linear function of independent gamma variables is available from Mathai and Provost (1992). Various representations of the density are given there. Thus the densities of W_1 and W_2 are available. Then by the transformation of variable technique one has the density of $W = W_1 - W_2$. Note that since W_1 and W_2 contain non-overlapping and mutually independent variables, W_1 and W_2 are independently distributed. For the sake of illustration the density of W, denoted by $f(w)$, coming from (2.3.8) and (2.3.9) is reproduced here from Mathai and Provost (1992).

$$f(w) = \begin{cases} \sum_{n=0}^{\infty}\sum_{\nu=0}^{\infty}\{\theta_n\theta'_\nu/\Gamma(\alpha+n)\} \\ \quad \times b^{-(\alpha+n+\alpha'+\nu)/2} w^{(\alpha+n+\alpha'+\nu-2)/2} \\ \quad \times e^{w(\gamma'^{-1}-\gamma^{-1})/2} \\ \quad \times W_{\frac{\alpha+n-\alpha'-\nu}{2},\frac{1-\alpha'-\nu-\alpha-n}{2}}(bw) \\ \quad \text{for } w > 0, \\ \sum_{n=0}^{\infty}\sum_{\nu=0}^{\infty}\{\theta_n\theta'_\nu/\Gamma(\alpha+\nu)\} \\ \quad \times b^{-(\alpha+\nu+\alpha'+n)/2}(-w)^{(\alpha'+\nu+\alpha+n-2)/2} \\ \quad \times e^{w(\gamma'^{-1}-\gamma^{-1})/2} \\ \quad \times W_{\frac{\alpha+\nu-\alpha'-n}{2},\frac{1-\alpha-n-\alpha'-\nu}{2}}(-bw) \\ \quad \text{for } w \le 0, \end{cases}$$

where

$$\begin{aligned} \theta_n &= \sum_{k_1+\ldots+k_m=n}\prod_{i=1}^{m}\left[\gamma_i^{-\alpha_i}(\alpha_i)_{k_i}(\gamma^{-1}-\gamma_i^{-1})^{k_i}/k_i!\right], \\ \theta'_\nu &= \sum_{\nu_{m+1}+\ldots+\nu_k=\nu}\prod_{i=m+1}^{k}\left[\gamma_i^{-\alpha_i}(\alpha_i)_{\nu_i}\left((\gamma')^{-1}-\gamma_i^{-1}\right)^{\nu_i}/\nu_i!\right], \end{aligned} \tag{2.3.10}$$

$$\alpha = \alpha_1 + \cdots + \alpha_m,\ \alpha' = \alpha_{m+1} + \cdots + \alpha_k,\ \gamma_i = a_i\beta_i,$$

γ is some average of the γ_i's, $i = 1,\ldots,m$, γ' is some average of the γ_i's, $i = m+1,\ldots,k$, $b = \gamma^{-1} + \gamma'^{-1}$, $|1-\gamma/\gamma_i| < 1$, $i = 1,\ldots,m$, $|1-\gamma'/\gamma_i| < 1$, $i = m+1,\ldots,k$

and $W_{\cdot,\cdot}(\cdot)$ is a Whittaker function, see for example, Mathai and Saxena (1973). Some techniques of dealing with linear combinations, convolutions and differences are available from Mathai and Saxena (1973) or from Mathai (1993e).

Example 2.3.1 Let $\mathbf{X} \sim N_3(\boldsymbol{\mu}, \Sigma)$ where $\boldsymbol{\mu}' = (1, -\sqrt{3}, 1)$ and

$$\Sigma = \begin{pmatrix} \frac{2}{3} & \frac{1}{\sqrt{6}} & -\frac{1}{\sqrt{18}} \\ \frac{1}{\sqrt{6}} & \frac{1}{2} & \frac{1}{\sqrt{12}} \\ -\frac{1}{\sqrt{18}} & \frac{1}{\sqrt{12}} & \frac{5}{6} \end{pmatrix}.$$

Let

$$U = \mathbf{X}'A\mathbf{X} = \frac{X_2^2}{2} - \frac{X_3^2}{2} + \frac{2}{\sqrt{6}}X_1X_2 + \frac{2}{\sqrt{2}}X_1X_3 + \frac{2}{\sqrt{12}}X_2X_3$$

where

$$A = \begin{pmatrix} 0 & \frac{1}{\sqrt{6}} & \frac{1}{\sqrt{2}} \\ \frac{1}{\sqrt{6}} & \frac{1}{2} & \frac{1}{\sqrt{12}} \\ \frac{1}{\sqrt{2}} & \frac{1}{\sqrt{12}} & -\frac{1}{2} \end{pmatrix}.$$

Then show that U is a NGL with parameters $(\alpha, \beta, \lambda) = (\frac{1}{2}, 2, \frac{1}{2} - \frac{\sqrt{2}}{3})$.

SOLUTION Note that $A\Sigma = \Sigma A = A$ and the eigenvalues of A are $1, -1, 0$. Thus the NGL parameters are $\alpha = \frac{1}{2}, \beta = 2$. Note that the m.g.f. of U reduces to the following form.

$$M_U(t) = \{(1-2t)(1+2t)\}^{-\frac{1}{2}} \exp\left\{-2\lambda + 2\lambda(1-4t^2)^{-1}\right\}$$

with $\lambda = \frac{1}{2} - \frac{\sqrt{2}}{3}$. Hence the result. (Some more details are available in Example 2.5.4 later on.)

2.4 Moments and Cumulants of Bilinear Forms

Here we consider bilinear forms of the type $\mathbf{X}'A_i\mathbf{Y}$, A_i is $p \times q$, $i = 1, 2$ and quadratic forms of the type $\mathbf{Y}'B_i\mathbf{Y}$, $B_i = B_i'$ where $\mathbf{X}$ and $\mathbf{Y}$ have a joint $(p+q)$-variate nonsingular normal distribution with $\text{Cov}(\mathbf{X}, \mathbf{Y}) = \Sigma_{12}$ not necessarily null. Explicit forms for the cumulants of $\mathbf{X}'A_i\mathbf{Y}$, joint cumulants of $\mathbf{X}'A_1\mathbf{Y}$, $\mathbf{X}'A_2\mathbf{Y}$ and joint cumulants of $\mathbf{X}'A_1\mathbf{Y}$, $\mathbf{Y}'B_2\mathbf{Y}$ will be derived.

Moments, cumulants and joint cumulants of bilinear forms can be obtained from those of the corresponding quadratic form. But this procedure becomes quite complicated which will be seen later. Also one can derive the results on bilinear expressions in the singular normal case and then derive the nonsingular normal case as particular cases. But most of the users of these results may have a situation of a nonsingular normal case. Hence this case will be considered first and then others will be derived as generalizations.

2.4a Joint Moments and Cumulants of Quadratic and Bilinear Forms

First we will consider the joint moments and cumulants of a quadratic form and a bilinear form. From these we can derive the cumulants of a bilinear form as well as those of a quadratic form. Let the $p \times 1$ vector $\mathbf{X}$ and $q \times 1$ vector $\mathbf{Y}$ have a joint real nonsingular normal distribution $N_{p+q}(\mathbf{0}, \Sigma)$, $\Sigma > 0$. Let A_1 be a $p \times q$ real matrix and A_2 be a $q \times q$ real symmetric matrix of constants. Let $Q_1 = \mathbf{X}' A_1 \mathbf{Y}$ and $Q_2 = \mathbf{Y}' B_2 \mathbf{Y}$. The joint m.g.f. of Q_1 and Q_2, denoted by $M_{Q_1,Q_2}(t_1, t_2)$, is the following expected value.

$$\begin{aligned} M_{Q_1,Q_2}(t_1,t_2) &= E\left[\exp\{t_1 Q_1 + t_2 Q_2\}\right] \\ &= \left\{|\Sigma|^{\frac{1}{2}} (2\pi)^{\left(\frac{p+q}{2}\right)}\right\}^{-1} \\ &\quad \times \int_{\mathbf{x}}\int_{\mathbf{y}} \exp\left\{-\frac{1}{2}(\mathbf{x}', \mathbf{y}')\tilde{\Sigma}\begin{pmatrix}\mathbf{x}\\ \mathbf{y}\end{pmatrix}\right\} d\mathbf{x}d\mathbf{y} \\ &= \frac{|\tilde{\Sigma}|^{-\frac{1}{2}}}{|\Sigma|^{\frac{1}{2}}} \end{aligned}$$

where

$$\tilde{\Sigma} = \begin{pmatrix} \Sigma^{11} & \Sigma^{12} - t_1 A_1 \\ \Sigma^{21} - t_1 A_1' & \Sigma^{22} - 2t_2 B_2 \end{pmatrix},$$

$$\Sigma^{-1} = \begin{pmatrix} \Sigma^{11} & \Sigma^{12} \\ \Sigma^{21} & \Sigma^{22} \end{pmatrix}, \quad \Sigma = \begin{pmatrix} \Sigma_{11} & \Sigma_{12} \\ \Sigma_{21} & \Sigma_{22} \end{pmatrix}.$$

Here Σ_{11} is the covariance matrix of $\mathbf{X}$, that is, $\Sigma_{11} = \mathrm{Cov}(\mathbf{X})$, $\Sigma_{22} = \mathrm{Cov}(\mathbf{Y})$, $\Sigma_{12} = \mathrm{Cov}(\mathbf{X}, \mathbf{Y})$. Multiply Σ and Σ^{-1} to get

$$\Sigma^{21}(\Sigma^{11})^{-1} = -\Sigma_{22}^{-1}\Sigma_{21}, \quad (\Sigma^{11})^{-1}\Sigma^{12} = -\Sigma_{12}\Sigma_{22}^{-1},$$

$$\Sigma_{22}^{-1} = \Sigma^{22} - \Sigma^{21}(\Sigma^{11})^{-1}\Sigma^{12}, \quad |\Sigma| = |\Sigma_{22}|\,|\Sigma^{11}|^{-1}. \tag{2.4.1}$$

By using (2.4.1) one can simplify $\tilde{\Sigma}$ as follows:

$$\begin{aligned} |\tilde{\Sigma}| &= |\Sigma^{11}| \left|(\Sigma^{22} - 2t_2 B_2) - (\Sigma^{21} - t_1 A_1')(\Sigma^{11})^{-1}(\Sigma^{12} - t_1 A_1)\right| \\ &= |\Sigma^{11}|\,|\Sigma_{22}|^{-1} \Big| I - 2t_2 \Sigma_{22}^{1/2} B_2 \Sigma_{22}^{1/2} \\ &\quad - t_1\left(\Sigma_{22}^{1/2} A_1' \Sigma_{12}\Sigma_{22}^{-1/2} + \Sigma_{22}^{-1/2}\Sigma_{21} A_1 \Sigma_{22}^{1/2}\right) \\ &\quad - t_1^2 \Sigma_{22}^{1/2} A_1' \left(\Sigma_{11} - \Sigma_{12}\Sigma_{22}^{-1}\Sigma_{21}\right) A_1 \Sigma_{22}^{1/2} \Big| \end{aligned}$$

where $\Sigma_{22}^{1/2}$ denotes the symmetric square root of the positive definite symmetric matrix

Σ_{22}. One can also write $\Sigma_{22} = BB'$ and replace one of $\Sigma_{22}^{1/2}$ by B and the other by B'. For notational convenience we will use the symmetric square root. For convenience, we will write the joint m.g.f. in the following form.

$$M_{Q_1,Q_2}(t_1,t_2) = \mid I - t_1E_1 - t_2E_2 - t_1^2E_3 \mid^{-\frac{1}{2}} \tag{2.4.2}$$

where

$$\begin{aligned} E_1 = E_1' &= \Sigma_{22}^{-\frac{1}{2}}\Sigma_{21}A_1\Sigma_{22}^{\frac{1}{2}} + \Sigma_{22}^{\frac{1}{2}}A_1'\Sigma_{12}\Sigma_{22}^{-\frac{1}{2}} \\ E_2 = E_2' &= 2\Sigma_{22}^{\frac{1}{2}}B_2\Sigma_{22}^{\frac{1}{2}} \\ E_3 = E_3' &= \Sigma_{22}^{\frac{1}{2}}A_1'(\Sigma_{11} - \Sigma_{12}\Sigma_{22}^{-1}\Sigma_{21})A_1\Sigma_{22}^{\frac{1}{2}}. \end{aligned}$$

The joint cumulants will be evaluated first. The cumulant generating function is available by taking the logarithm of the joint m.g.f. and expanding it. That is,

$$\ln M_{Q_1,Q_2}(t_1,t_2) = \frac{1}{2}\left\{\sum_{k=1}^{\infty}\frac{1}{k}\text{tr}\,(F^k)\right\} \tag{2.4.3}$$

where

$$F = t_1E_1 + t_2E_2 + t_1^2E_3. \tag{2.4.4}$$

Without loss of generality it can be assumed that $\|F\| < 1$ where $\|(\ )\|$ denotes the norm of ( ). For example, the covariance between Q_1 and Q_2 is available from (2.4.3) by taking the coefficient of t_1t_2 in the expansion on the right-side of (2.4.3). This can come only from F^2 where

$$\begin{aligned} F^2 &= t_1^2E_1^2 + t_2^2E_2^2 + t_1^4E_3^2 + t_1t_2(E_1E_2 + E_2E_1) \\ &\quad + t_1^3(E_1E_3 + E_3E_1) + t_2t_1^2(E_2E_3 + E_3E_2). \end{aligned} \tag{2.4.5}$$

A convenient notation to use for the joint cumulants is K_{r_1,r_2}.

Notation 2.4.1 K_{r_1,r_2} : *r_1-th and r_2-th joint cumulant*

From (2.4.5) we get $K_{1,1}$ as the following: Observe that for any two matrices A and B,

$$\text{tr}(AB) = \text{tr}(BA), \quad \text{tr}(A) = \text{tr}(A')$$

whenever the products are defined. These properties will be frequently made use of in the discussions to follow.

$$\begin{aligned} K_{1,1} &= \text{Cov}(Q_1,Q_2) = \frac{1}{4}\text{tr}(E_1E_2 + E_2E_1) \\ &= \frac{1}{2}\text{tr}(E_1E_2) \\ &= \frac{1}{2}\text{tr}\left(\Sigma_{22}^{-\frac{1}{2}}\Sigma_{21}A_1\Sigma_{22}^{\frac{1}{2}} + \Sigma_{22}^{\frac{1}{2}}A_1'\Sigma_{12}\Sigma_{22}^{-\frac{1}{2}}\right)\left(2\Sigma_{22}^{\frac{1}{2}}B_2\Sigma_{22}^{\frac{1}{2}}\right) \\ &= 2\text{tr}(\Sigma_{21}A_1\Sigma_{22}B_2). \end{aligned} \tag{2.4.6}$$

Note that $K_{1,1} = 0$ when $\Sigma_{21} = O$ or when $A_1\Sigma_{22}B_2' = O$.

Now we will find a convenient way of computing the coefficient of $t_1^{s_1}t_2^{s_2}$, for arbitrary s_1 and s_2, from the expansion (2.4.3). Rewrite F as

$$F = t_1E_1 + t_2E_2 + t_3E_3, \quad t_3 = t_1^2. \tag{2.4.7}$$

The terms containing $t_1^{r_1}t_2^{r_2}t_3^{r_3}$ where $r_1 + r_2 + r_3 = r$ can come only from F^r. From (2.4.4), (2.4.5) and from the higher powers of F one can obtain the coefficient of $t_1^{r_1}t_2^{r_2}t_3^{r_3}$ as

$$\frac{1}{2r}\operatorname{tr}\left(\sum_{(r_1,r_2,r_3)}(E_1E_2E_3)\right)$$

where the following notation is used.

Notation 2.4.2 $\sum_{(r_1,r_2,r_3)}(E_1E_2E_3)$:

stands for the sum of products of permutations of E_1, E_2, E_3 taking any number of each of them at a time so that the sum of the exponents of E_i in each term is r_i, $i = 1, 2, 3$. Thus we have the following result.

Theorem 2.4.1 Let $Q_1 = \mathbf{X}'A_1\mathbf{Y}$, $Q_2 = \mathbf{Y}'B_2\mathbf{Y}$, $\binom{\mathbf{X}}{\mathbf{Y}} \sim N_{p+q}(\mathbf{0}, \Sigma)$, $\Sigma > 0$, where $B_2 = B_2'$ and A_1 be real matrices of constants, A_1 be $p \times q$ and B_2 be $q \times q$. Then the $(r_1 + 2r_3, r_2)$-th joint cumulant of Q_1 and Q_2, denoted $K_{r_1+2r_3,r_2}$ is given by

$$K_{r_1+2r_3,r_2} = \frac{(r_1 + 2r_3)!r_2!}{2r}\operatorname{tr}\Big(\sum_{(r_1,r_2,r_3)}(E_1E_2E_3)\Big) \tag{2.4.8}$$

where E_1, E_2, E_3 are defined in (2.4.2), $r = r_1+r_2+r_3$ and $\sum_{(r_1,r_2,r_3)}$ is given in Notation 2.4.2 above.

Example 2.4.1 If the bilinear form Q_1 and the quadratic form Q_2 are as given in Theorem 2.4.1 then compute the joint cumulant $K_{2,2}$.

SOLUTION The possible partitions of (r_1, r_2, r_3) are $(0, 2, 1)$ and $(2, 2, 0)$.

$$\sum_{(0,2,1)}(E_1E_2E_3) = E_2^2E_3 + E_3E_2^2 + E_2E_3E_2, \quad r = 3, r_1 + 2r_3 = 2, r_2 = 2.$$

$$\sum_{(2,2,0)}(E_1E_2E_3) = E_1^2E_2^2 + E_2^2E_1^2 + E_1E_2^2E_1 + E_2E_1^2E_2$$

$$+E_1E_2E_1E_2 + E_2E_1E_2E_1, \quad r = 4, \ r_1 + 2r_3 = 2, \ r_2 = 2.$$

Thus

$$\begin{aligned}
K_{2,2} &= \left(\frac{2}{3}\right) \operatorname{tr}\left(E_2^2E_3 + E_3E_2^2 + E_2E_3E_2\right) \\
&+ \left(\frac{1}{2}\right) \operatorname{tr}(E_1^2E_2^2 + E_2^2E_1^2 + E_1E_2^2E_1 \\
&+ E_2E_1^2E_2 + E_1E_2E_1E_2 + E_2E_1E_2E_1) \\
&= \operatorname{tr}\left(2E_2E_3E_2 + 2E_1^2E_2^2 + (E_1E_2)^2\right). \qquad (2.4.9)
\end{aligned}$$

Definition 2.4.1 *Equicorrelated vectors* Two $p \times 1$ vector random variables $\mathbf{x}$ and $\mathbf{y}$ are said to be equicorrelated with unit variance if $\operatorname{Cov}(\mathbf{x}) = \Sigma_{11} = I = \operatorname{Cov}(\mathbf{y}) = \Sigma_{22}$ and $\operatorname{Cov}(\mathbf{x}, \mathbf{y}) = \Sigma_{12} = \rho I$ where ρ is a scalar quantity and I is the identity matrix.

Example 2.4.2 If $\mathbf{X}$ and $\mathbf{Y}$ are as defined in Theorem 2.4.1 but (a) independently distributed; (b) equicorrelated as in Definition 2.4.1, compute $K_{2,2}$.

SOLUTION When $\mathbf{X}$ and $\mathbf{Y}$ are independently distributed the $\operatorname{Cov}(\mathbf{X}, \mathbf{Y}) = \Sigma_{12} = O$ where O denotes a null matrix. Then the joint cumulants of Q_1 and Q_2 are available from (2.4.8) by replacing E_1 by a null matrix, E_3 by $\Sigma_{22}^{\frac{1}{2}} A_1' \Sigma_{11} A_1 \Sigma_{22}^{\frac{1}{2}}$ and E_2 remaining the same. In the sum $\sum_{(r_1, r_2, r_3)}$, replace E_1 by an identity matrix and put $r_1 = 0$. Thus when X and Y are independently distributed

$$K_{2,2} = 8\operatorname{tr}\left(\Sigma_{22}B_2\Sigma_{22}A_1'\Sigma_{11}A_1\Sigma_{22}B_2\right).$$

When $\mathbf{X}$ and $\mathbf{Y}$ are equicorrelated we have $E_1 = \rho(A_1 + A_1')$, $E_2 = 2B_2$ and $E_3 = (1-\rho^2)(A_1'A_1)$. Thus $K_{2,2}$ simplifies to the following:

$$\begin{aligned}
K_{2,2} = 8\operatorname{tr}\Big[&(1-\rho^2) A_1'A_1B_2^2 + \left(\frac{1}{2}\right) \rho^2 \left\{(A_1 + A_1')B_2\right\}^2 \\
&+ \rho^2 (A_1 + A_1')^2 B_2^2 \Big].
\end{aligned}$$

2.4a.1 The First Few Joint Cumulants of a Bilinear Form and a Quadratic Form in the Nonsingular Normal Case

$$\begin{aligned}
K_{1,1} &= 2 \operatorname{tr}\left(\Sigma_{21}A_1\Sigma_{22}B_2\right) \\
&= 0 \quad \text{(under independence)} \\
&= 2\rho \operatorname{tr}(A_1B_2) \quad \text{(under equicorrelation)}.
\end{aligned}$$

$$\begin{aligned}
K_{1,2} &= 8\ \mathrm{tr}\,(A_1'\Sigma_{12}B_2\Sigma_{22}B_2\Sigma_{22}) \\
&= 0 \ \ \text{(under independence)} \\
&= 8\rho\ \mathrm{tr}\,(A_1'B_2^2) \ \ \text{(under equicorrelation).}
\end{aligned}$$

$$\begin{aligned}
K_{2,1} &= 2\big(\mathrm{tr}(B_2\Sigma_{22}A_1'\Sigma_{11}A_1\Sigma_{22}) + \mathrm{tr}(\Sigma_{21}A_1\Sigma_{22}A_1'\Sigma_{12}B_2) \\
&\quad + 2\ \mathrm{tr}(\Sigma_{21}A_1\Sigma_{21}A_1\Sigma_{22}B_2)\big) \\
&= 2\ \mathrm{tr}(B_2\Sigma_{22}A_1'\Sigma_{11}A_1\Sigma_{22}) \ \ \text{(under independence)} \\
&= 2\big(\mathrm{tr}(B_2A_1'A_1) + \rho^2\mathrm{tr}(A_1A_1'B_2) \\
&\quad + 2\rho^2\mathrm{tr}(A_1^2B_2)\big) \ \ \text{(under equicorrelation).}
\end{aligned}$$

$$\begin{aligned}
K_{2,2} &= \left[2\ \mathrm{tr}\,(E_2^2E_3) + 2\ \mathrm{tr}\,(E_1^2E_2^2) + \mathrm{tr}\,((E_1E_2)^2)\right] \\
&= 8\ \mathrm{tr}(\Sigma_{22}B_2\Sigma_{22}A_1'\Sigma_{11}A_1\Sigma_{22}B_2) \\
&\quad \text{(under independence)} \\
&= 8\Big[\,(1-\rho^2)\,\mathrm{tr}\,(A_1'A_1B_2^2) + \left(\frac{1}{2}\right)\rho^2\mathrm{tr}\,((A_1+A_1')B_2)^2 \\
&\quad + \rho^2\mathrm{tr}\,\big((A_1+A_1')^2B_2^2\big)\,\Big] \ \ \text{(under equicorrelation).}
\end{aligned}$$

Example 2.4.3 Let $\mathbf{X}' = (X_1, X_2)$, $\mathbf{Y}' = (Y_1, Y_2)$. Let $\binom{\mathbf{X}}{\mathbf{Y}}$ be normally distributed with the mean value vector null and the covariance matrix

$$\Sigma = \begin{bmatrix} 1 & 1 & 1 & 1 \\ 1 & 2 & 1 & 1 \\ 1 & 1 & 2 & 1 \\ 1 & 1 & 1 & 2 \end{bmatrix}.$$

Let $Q_1 = X_1Y_1 + X_1Y_2 + 2X_2Y_1 + X_2Y_2$ and $Q_2 = 2Y_1^2 + 2Y_1Y_2 + Y_2^2$. Compute the joint cumulant $K_{1,2}$.

SOLUTION Writing in terms of our notation we get

$$Q_1 = [X_1, X_2]\begin{bmatrix} 1 & 1 \\ 2 & 1 \end{bmatrix}\begin{bmatrix} Y_1 \\ Y_2 \end{bmatrix}$$

$$Q_2 = [Y_1, Y_2]\begin{bmatrix} 2 & 1 \\ 1 & 1 \end{bmatrix}\begin{bmatrix} Y_1 \\ Y_2 \end{bmatrix}$$

$$\Sigma_{11} = \begin{bmatrix} 1 & 1 \\ 1 & 1 \end{bmatrix}, \quad \Sigma_{22} = \begin{bmatrix} 2 & 1 \\ 1 & 2 \end{bmatrix}$$

$$A_1 = \begin{bmatrix} 1 & 1 \\ 2 & 1 \end{bmatrix}, \quad B_2 = \begin{bmatrix} 2 & 1 \\ 1 & 1 \end{bmatrix}$$

$$A_1'\Sigma_{12} = \begin{bmatrix} 1 & 2 \\ 1 & 1 \end{bmatrix}\begin{bmatrix} 1 & 1 \\ 1 & 1 \end{bmatrix} = \begin{bmatrix} 3 & 3 \\ 2 & 2 \end{bmatrix}$$

$$(B_2\Sigma_{22})^2 = \left(\begin{bmatrix} 2 & 1 \\ 1 & 1 \end{bmatrix}\begin{bmatrix} 2 & 1 \\ 1 & 2 \end{bmatrix}\right)^2 = \begin{bmatrix} 37 & 32 \\ 24 & 21 \end{bmatrix}.$$

But

$$\begin{aligned} K_{1,2} &= 8\ \mathrm{tr}(A_1'\Sigma_{12}B_2\Sigma_{22}B_2\Sigma_{22}) \\ &= 8\ \mathrm{tr}\left(\begin{bmatrix} 3 & 3 \\ 2 & 2 \end{bmatrix}\begin{bmatrix} 37 & 32 \\ 24 & 21 \end{bmatrix}\right) \\ &= 2312. \end{aligned}$$

If the cumulants of the quadratic form Q_2 are to be obtained from Theorem 2.4.1 then in (2.4.8) put $r_1 = 0$, $r_3 = 0$ and replace E_1 and E_2 by identity matrices in the sum $\sum_{(r_1,r_2,r_3)}$. Thus we have the following

Corollary 2.4.1 The r-th cumulant of the quadratic form Q_2, denoted by $K_r^{(Q_2)}$, is given by

$$K_r^{(Q_2)} = \left[\frac{r!}{2r}\right] \mathrm{tr}\Big(\sum_{(r)}(E_2)\Big) = 2^{r-1}(r-1)!\mathrm{tr}\left(\Sigma_{22}^{1/2}B_2\Sigma_{22}^{1/2}\right)^r. \tag{2.4.10}$$

We will list a few of them explicitly here.

2.4a.2 The First Few Cumulants of a Quadratic Form in the Nonsingular Normal Case

$$K_1 = \mathrm{tr}\,(B_2\Sigma_{22}).$$
$$K_2 = 2\ \mathrm{tr}\,(B_2\Sigma_{22})^2.$$
$$K_3 = 8\ \mathrm{tr}\,(B_2\Sigma_{22})^3.$$
$$K_4 = 48\ \mathrm{tr}\,(B_2\Sigma_{22})^4.$$

If the cumulants of the bilinear form Q_1 are to be obtained from Theorem 2.4.1 then in (2.4.8) put $r_2 = 0$ and replace E_2 by I in the sum.

Corollary 2.4.2 The (r_1+2r_3)-th cumulant of Q_1, denoted by $K^{(Q_1)}_{r_1+2r_3}$, is given by

$$K^{(Q_1)}_{r_1+2r_3} = \left[\frac{(r_1+2r_3)!}{2(r_1+r_3)}\right] \operatorname{tr}\Big(\sum_{(r_1,r_3)} (E_1E_3)\Big). \tag{2.4.11}$$

The first four cumulants of Q_1 will be listed here explicitly. These will be useful when approximating the density of Q_1 with the help of the first few moments.

2.4a.3 The First Few Cumulants of a Bilinear Form in the Nonsingular Normal Case

$$\begin{aligned} K_1^{(Q_1)} &= \left(\frac{1}{2}\right)\operatorname{tr}\Big(\sum_{(1,0)}(E_1E_3)\Big) = \left(\frac{1}{2}\right)\operatorname{tr}(E_1) \\ &= \operatorname{tr}(\Sigma_{21}A_1) = \text{expected value of } Q_1 \\ &= 0 \ \text{(under independence)} \\ &= \rho\operatorname{tr}(A_1) \ \text{(under equicorrelation).} \end{aligned}$$

$$\begin{aligned} K_2^{(Q_1)} &= \left(\frac{2!}{2(2)}\right)\operatorname{tr}\Big(\sum_{(2,0)}(E_1E_3)\Big) + \left(\frac{2!}{2(1)}\right)\operatorname{tr}\Big(\sum_{(0,1)}(E_1E_3)\Big) \\ &= \left(\frac{1}{2}\right)\operatorname{tr}(E_1^2) + \operatorname{tr}(E_3) \\ &= \operatorname{tr}(A_1'\Sigma_{11}A_1\Sigma_{22}) \ \text{(under independence)} \\ &= \rho^2\operatorname{tr}(A_1^2) + \operatorname{tr}(A_1'A_1) \ \text{(under equicorrelation).} \end{aligned}$$

$$\begin{aligned} K_3^{(Q_1)} &= \left(\frac{3!}{2(3)}\right)\operatorname{tr}\Big(\sum_{(3,0)}(E_1E_3)\Big) + \left(\frac{3!}{2(2)}\right)\operatorname{tr}\Big(\sum_{(1,1)}(E_1E_3)\Big) \\ &= \operatorname{tr}(E_1^3) + 3\operatorname{tr}(E_1E_3) \\ &= 0 \ \text{(under independence)} \\ &= \rho^3\operatorname{tr}((A_1+A_1')^3) + 3\rho(1-\rho^2)\operatorname{tr}(A_1^2A_1') \\ &\quad + 3\rho(1-\rho^2)\operatorname{tr}(A_1'^2A_1) \\ &\quad \text{(under equicorrelation).} \end{aligned}$$

$$\begin{aligned} K_4^{(Q_1)} &= \left(\frac{4!}{2(4)}\right)\operatorname{tr}\Big(\sum_{(4,0)}(E_1E_3)\Big) + \left(\frac{4!}{2(3)}\right)\operatorname{tr}\Big(\sum_{(2,1)}(E_1E_3)\Big) \\ &\quad + \left(\frac{4!}{2(2)}\right)\operatorname{tr}\Big(\sum_{(0,2)}(E_1E_3)\Big) \\ &= 3\operatorname{tr}(E_1^4) + 12\operatorname{tr}(E_3E_1^2) + 6\operatorname{tr}(E_3^2) \end{aligned}$$

$$= 6 \operatorname{tr}(A_1'\Sigma_{11}A_1\Sigma_{22}A_1'\Sigma_{11}A_1\Sigma_{22})$$
$$\text{(under independence)}$$
$$= 3\left[\rho^4 \operatorname{tr}(A_1 + A_1')^4 + 4\rho^2(1-\rho^2)\operatorname{tr}\left(A_1'A_1(A_1+A_1')^2\right) + 2\left(1-\rho^2\right)^2 \operatorname{tr}(A_1'A_1)^2\right] \text{ (under equicorrelation).}$$

Some explicit forms will be given in Section 2.4c.1.

Example 2.4.4 In Example 2.4.3 compute the variance of the bilinear form Q_1.

SOLUTION In our notation

$$\operatorname{Var}(Q_1) = \left(\frac{1}{2}\right)\operatorname{tr}\left(E_1^2\right) + \operatorname{tr}(E_3)$$
$$= \operatorname{tr}(A_1\Sigma_{21}A_1\Sigma_{21}) + \operatorname{tr}(A_1\Sigma_{22}A_1'\Sigma_{11})$$

where

$$A_1 = \begin{bmatrix} 1 & 1 \\ 2 & 1 \end{bmatrix}, \quad \Sigma_{21} = \begin{bmatrix} 1 & 1 \\ 1 & 1 \end{bmatrix},$$
$$\Sigma_{22} = \begin{bmatrix} 2 & 1 \\ 1 & 2 \end{bmatrix}, \quad \Sigma_{11} = \begin{bmatrix} 1 & 1 \\ 1 & 2 \end{bmatrix}.$$
$$(A_1\Sigma_{21})^2 = \begin{bmatrix} 10 & 10 \\ 15 & 15 \end{bmatrix}, \quad \operatorname{tr}(A_1\Sigma_{21})^2 = 25.$$
$$A_1\Sigma_{22} = \begin{bmatrix} 3 & 3 \\ 5 & 4 \end{bmatrix}, \quad A_1'\Sigma_{11} = \begin{bmatrix} 3 & 5 \\ 2 & 3 \end{bmatrix}.$$

Thus

$$\operatorname{tr}(A_1\Sigma_{22}A_1'\Sigma_{11}) = 52$$

and

$$\operatorname{Var}(Q_1) = 77.$$

2.4b Joint Cumulants of Bilinear Forms

Let $\binom{\mathbf{X}}{\mathbf{Y}} \sim N_{p+q}(\mathbf{0}, \Sigma)$, $\Sigma > 0$ with $\operatorname{Cov}(\mathbf{X}) = \Sigma_{11}$, $\operatorname{Cov}(\mathbf{Y}) = \Sigma_{22}$, $\Sigma_{12} = \operatorname{Cov}(\mathbf{X}, \mathbf{Y}) = \Sigma_{21}'$. Let $Q_1 = \mathbf{X}'A_1\mathbf{Y}$ and $Q_2 = \mathbf{X}'A_2\mathbf{Y}$ where A_1 and A_2 are $p \times q$ real matrices of constants. We will use the same notation Q_2 to denote a bilinear form here. This will not create any confusion with the notation in Section 2.4a since we will not be considering results where the Q_2's of Sections 2.4a and 2.4b both will be involved. The joint m.g.f. of Q_1 and Q_2, denoted again by $M_{Q_1,Q_2}(t_1, t_2)$, is available by following through similar steps as in (2.4.1) to (2.4.2). Then

$$M_{Q_1,Q_2}(t_1, t_2) = \frac{|\,\Sigma^*\,|^{-\frac{1}{2}}}{|\,\Sigma\,|^{\frac{1}{2}}} \tag{2.4.12}$$

where

$$| \Sigma^* | = \begin{vmatrix} \Sigma^{11} & \Sigma^{12} - t_1 A_1 - t_2 A_2 \\ \Sigma^{21} - t_1 A_1' - t_2 A_2' & \Sigma^{22} \end{vmatrix}$$

$$= \left|\Sigma^{11}\right| \left|\Sigma_{22}^{-1}\right| \Big| I_q - \Sigma_{21}(t_1 A_1 + t_2 A_2)$$

$$- \Sigma_{22}(t_1 A_1' + t_2 A_2')\Sigma_{12}\Sigma_{22}^{-1}$$

$$- \Sigma_{22}(t_1 A_1' + t_2 A_2')(\Sigma_{11} - \Sigma_{12}\Sigma_{22}^{-1}\Sigma_{21})(t_1 A_1 + t_2 A_2)\Big|.$$

Thus

$$M_{Q_1,Q_2}(t_1, t_2) = \Big| I_q - \Sigma_{22}^{-\frac{1}{2}}\Sigma_{21}(t_1 A_1 + t_2 A_2)\Sigma_{22}^{\frac{1}{2}}$$

$$- \Sigma_{22}^{\frac{1}{2}}(t_1 A_1' + t_2 A_2')\Sigma_{12}\Sigma_{22}^{-\frac{1}{2}} - \Sigma_{22}^{-\frac{1}{2}}(t_1 A_1 + t_2 A_2)'$$

$$\times (\Sigma_{11} - \Sigma_{12}\Sigma_{22}^{-1}\Sigma_{21})(t_1 A_1 + t_2 A_2)\Sigma_{22}^{\frac{1}{2}}\Big|^{-\frac{1}{2}}$$

$$= \Big| I_p - \Sigma_{11}^{-\frac{1}{2}}\Sigma_{12}(t_1 A_1' + t_2 A_2')\Sigma_{11}^{\frac{1}{2}}$$

$$- \Sigma_{11}^{\frac{1}{2}}(t_1 A_1 + t_2 A_2)\Sigma_{21}\Sigma_{11}^{-\frac{1}{2}} - \Sigma_{11}^{\frac{1}{2}}(t_1 A_1 + t_2 A_2)$$

$$\times (\Sigma_{22} - \Sigma_{21}\Sigma_{11}^{-1}\Sigma_{12})(t_1 A_1 + t_2 A_2)'\Sigma_{11}^{\frac{1}{2}}\Big|^{-\frac{1}{2}}. \tag{2.4.13}$$

From (2.4.13) one can represent the joint m.g.f. as follows:

$$M_{Q_1,Q_2}(t_1, t_2) = | I_q - t_1 F_1 - t_2 F_2 - t_3 F_3 - t_4 F_4 - t_5 F_5 |^{-\frac{1}{2}} \tag{2.4.14}$$

where

$$t_3 = t_1^2, \quad t_4 = t_2^2, \quad t_5 = t_1 t_2$$

$$F_1 = F_1' = \Sigma_{22}^{-\frac{1}{2}}\Sigma_{21}A_1\Sigma_{22}^{\frac{1}{2}} + \Sigma_{22}^{\frac{1}{2}}A_1'\Sigma_{12}\Sigma_{22}^{-\frac{1}{2}}$$

$$F_2 = F_2' = \Sigma_{22}^{-\frac{1}{2}}\Sigma_{21}A_2\Sigma_{22}^{\frac{1}{2}} + \Sigma_{22}^{\frac{1}{2}}A_2'\Sigma_{12}\Sigma_{22}^{-\frac{1}{2}}$$

$$F_3 = F_3' = \Sigma_{22}^{\frac{1}{2}}A_1'DA_1\Sigma_{22}^{\frac{1}{2}}, \quad D = \Sigma_{11} - \Sigma_{12}\Sigma_{22}^{-1}\Sigma_{21}$$

$$F_4 = F_4' \;=\; \Sigma_{22}^{\frac{1}{2}} A_2' D A_2 \Sigma_{22}^{\frac{1}{2}}$$

$$F_5 = F_5' \;=\; \Sigma_{22}^{\frac{1}{2}} A_1' D A_2 \Sigma_{22}^{\frac{1}{2}} + \Sigma_{22}^{\frac{1}{2}} A_2' D A_1 \Sigma_{22}^{\frac{1}{2}}. \tag{2.4.15}$$

Consider the expansion of $\ln M_{Q_1,Q_2}(t_1,t_2)$ as in (2.4.3) with

$$F = t_1 F_1 + \cdots + t_5 F_5.$$

The coefficient of $t_1^{r_1}\cdots t_5^{r_5}$, $r = r_1 + \cdots + r_5$ is coming from F^r only. This is given by

$$\left(\frac{1}{2r}\right)\operatorname{tr}\left(\sum_{(r_1,\ldots,r_5)}(F_1\ldots F_5)\right)$$

where the notation is explained in Notation 2.4.2. Note that $t_3 = t_1^2$, $t_4 = t_2^2$ and $t_5 = t_1 t_2$ and hence the joint cumulants of the type $(r_1 + 2r_3 + r_5, r_2 + 2r_4 + r_5)$ can be obtained from this coefficient. Again denoting the (s_1, s_2)-th joint cumulant of Q_1 and Q_2 by K_{s_1,s_2} we have

Theorem 2.4.2. If the bilinear forms are as defined in Section 2.4b and the F_j, $j = 1,\ldots,5$ are as given in (2.4.15) then

$$K_{r_1+2r_3+r_5,\,r_2+2r_4+r_5} = \left[\frac{(r_1+2r_3+r_5)!(r_2+2r_4+r_5)!}{2r}\right] \times \operatorname{tr}\Big(\sum_{(r_1,\ldots,r_5)}(F_1\ldots F_5)\Big) \tag{2.4.16}$$

where $r = r_1 + \cdots + r_5$.

The first few cumulants will be needed if someone is trying to approximate the distributions of Q_1 and Q_2 or the joint distribution of Q_1 and Q_2. Hence a few of these will be listed here explicitly. Take $K_{1,0}$, $K_{0,1}$, $K_{2,0}$, $K_{0,2}$ from Corollary 2.4.2 or from the explicit forms given there.

2.4b.1 The First Few Joint Cumulants of Two Bilinear Forms

$$\begin{aligned}
K_{1,1} &= \left(\frac{1}{2(2)}\right)\operatorname{tr}(F_1F_2 + F_2F_1) + \left(\frac{1}{2(1)}\right)\operatorname{tr}(F_5)\\
&= \left(\frac{1}{2}\right)\operatorname{tr}(F_1F_2) + \left(\frac{1}{2}\right)\operatorname{tr}(F_5)\\
&= \operatorname{tr}(\Sigma_{21}A_1\Sigma_{21}A_2) + \operatorname{tr}(\Sigma_{22}A_1'\Sigma_{11}A_2) = \operatorname{Cov}(Q_1, Q_2)\\
&= \operatorname{tr}(\Sigma_{22}A_1'\Sigma_{11}A_2)\ \ \text{(under independence)}\\
&= \rho^2\operatorname{tr}(A_1A_2) + \operatorname{tr}(A_1'A_2)\ \ \text{(under equicorrelation).}
\end{aligned}$$

$$
\begin{aligned}
K_{2,1} &= \left(\frac{2!}{2(3)}\right) \operatorname{tr}\left(F_1^2F_2 + F_2F_1^2 + F_1F_2F_1\right) \\
&\quad + \left(\frac{2!}{2(2)}\right) \operatorname{tr}(F_1F_5 + F_5F_1) + \left(\frac{2!}{2(2)}\right) \operatorname{tr}(F_2F_3 + F_3F_2) \\
&= \operatorname{tr}(F_1^2F_2) + \operatorname{tr}(F_1F_5) + \operatorname{tr}(F_2F_3) \\
&= 0 \quad \text{(under independence)} \\
&= \rho^3 \operatorname{tr}\left[(A_1 + A_1')^2(A_2 + A_2')\right] \\
&\quad + \rho\left(1-\rho^2\right)\left[\operatorname{tr}((A_1 + A_1')(A_1'A_2 + A_2'A_1)) \right. \\
&\quad \left. + \operatorname{tr}((A_2 + A_2')(A_1'A_2 + A_2'A_1))\right] \quad \text{(under equicorrelation).}
\end{aligned}
$$

$$
\begin{aligned}
K_{1,2} &= \operatorname{tr}\left(F_2^2F_1\right) + \operatorname{tr}(F_2F_5) + \operatorname{tr}(F_1F_4) \\
&= 0 \quad \text{(under independence)} \\
&= \rho^3 \operatorname{tr}\left[(A_2 + A_2')^2(A_1 + A_1')\right] \\
&\quad + \rho\left(1-\rho^2\right)\left[\operatorname{tr}((A_2 + A_2')(A_1'A_2 + A_2'A_1)) \right. \\
&\quad \left. + \operatorname{tr}((A_1 + A_1')A_2'A_2)\right] \quad \text{(under equicorrelation).}
\end{aligned}
$$

$$
\begin{aligned}
K_{2,2} &= 2\operatorname{tr}\left(F_1^2F_2^2\right) + 2\operatorname{tr}\left(F_1^2F_4\right) + \operatorname{tr}\left(F_5^2\right) + 2\operatorname{tr}(F_3F_4) \\
&\quad + 2\operatorname{tr}\left(F_2^2F_3\right) + 4\operatorname{tr}(F_1F_2F_5) + \operatorname{tr}\left(F_1F_2\right)^2 .
\end{aligned} \tag{2.4.17}
$$

Since the expressions are lengthy $K_{2,2}$ will not be written explicitly for the cases of independence and equicorrelation.

Corollary 2.4.3 Various cumulants of Q_1 are available from Theorem 2.4.2 by putting $r_2 = 0$, $r_4 = 0$, $r_5 = 0$ and replacing F_2, F_4 and F_5 by identity matrices in the sum $\sum(F_1 \dots F_5)$.

Corollary 2.4.4 When **X** and **Y** are independently distributed various joint cumulants of Q_1 and Q_2 are available from Theorem 2.4.2 by putting $F_1 = O$, $F_2 = O$ and replacing the D of (2.4.15) by Σ_{11}. In the sum $\sum(F_1 \dots F_5)$ put $r_1 = 0$, $r_2 = 0$ and replace F_1 and F_2 by identity matrices. Under independence the substitutions are the following:

$$
\begin{aligned}
F_1 &= O\,; \\
F_2 &= O\,; \\
F_3 &= \Sigma_{22}^{\frac{1}{2}} A_1' \Sigma_{11} A_1 \Sigma_{22}^{\frac{1}{2}}\,; \\
F_4 &= \Sigma_{22}^{\frac{1}{2}} A_2' \Sigma_{11} A_2 \Sigma_{22}^{\frac{1}{2}}\,; \\
F_5 &= \Sigma_{22}^{\frac{1}{2}} A_1' \Sigma_{11} A_2 \Sigma_{22}^{\frac{1}{2}} + \Sigma_{22}^{\frac{1}{2}} A_2' \Sigma_{11} A_1 \Sigma_{22}^{\frac{1}{2}}\,.
\end{aligned}
$$

Corollary 2.4.5 When **X** and **Y** are equicorrelated in the sense of Definition 2.4.1

the joint cumulants of Q_1 and Q_2 are available from Theorem 2.4.2 by putting $q = p$ and

$$\begin{aligned}
F_1 &= \rho(A_1 + A_1') ; \\
F_2 &= \rho(A_2 + A_2') ; \\
F_3 &= \left(1-\rho^2\right) A_1'A_1 ; \\
F_4 &= \left(1-\rho^2\right) A_2'A_2 ; \\
F_5 &= \left(1-\rho^2\right) \left(A_1'A_2 + A_2'A_1\right) .
\end{aligned}$$

Example 2.4.5 In Example 2.4.3 let Q_1 be as it is there but $Q_2 = 2X_1Y_1 + X_2Y_1 + X_2Y_2$. Compute the joint cumulant $K_{1,1}$ of the bilinear forms Q_1 and Q_2 if the covariance matrix is of the form

$$\Sigma = \begin{bmatrix} 1 & 1 & 0 & 0 \\ 1 & 2 & 0 & 0 \\ 0 & 0 & 2 & 1 \\ 0 & 0 & 1 & 2 \end{bmatrix} .$$

SOLUTION

$$Q_1 = [X_1, X_2] \begin{bmatrix} 1 & 1 \\ 2 & 1 \end{bmatrix} \begin{bmatrix} Y_1 \\ Y_2 \end{bmatrix} \Rightarrow A_1 = \begin{bmatrix} 1 & 1 \\ 2 & 1 \end{bmatrix} .$$

$$Q_2 = [X_1, X_2] \begin{bmatrix} 2 & 0 \\ 1 & 1 \end{bmatrix} \begin{bmatrix} Y_1 \\ Y_2 \end{bmatrix} \Rightarrow A_2 = \begin{bmatrix} 2 & 0 \\ 1 & 1 \end{bmatrix} .$$

When $\Sigma_{12} = O$ we have

$$K_{1,1} = \mathrm{tr}\left(\Sigma_{22}A_1'\Sigma_{11}A_2\right).$$

But

$$\Sigma_{22}A_1' = \begin{bmatrix} 2 & 1 \\ 1 & 2 \end{bmatrix} \begin{bmatrix} 1 & 2 \\ 1 & 1 \end{bmatrix} = \begin{bmatrix} 3 & 5 \\ 3 & 4 \end{bmatrix}$$

and

$$\Sigma_{11}A_2 = \begin{bmatrix} 1 & 1 \\ 1 & 2 \end{bmatrix} \begin{bmatrix} 2 & 0 \\ 1 & 1 \end{bmatrix} = \begin{bmatrix} 3 & 1 \\ 4 & 2 \end{bmatrix} .$$

Hence

$$K_{1,1} = \mathrm{tr}\left(\begin{bmatrix} 3 & 5 \\ 3 & 4 \end{bmatrix} \begin{bmatrix} 3 & 1 \\ 4 & 2 \end{bmatrix} \right) = 40.$$

2.4c Moments and Cumulants in the Singular Normal Case

Let $\mathbf{Z} = \binom{\mathbf{X}}{\mathbf{Y}} \sim N_{p+q}(\boldsymbol{\mu}, \Sigma)$. Let $\mathbf{X}$ be $p \times 1$ and $\mathbf{Y}$ be $q \times 1$. If Σ is strictly positive definite then $\mathbf{Z}$ has a nonsingular normal distribution and if Σ is positive semidefinite then $\mathbf{Z}$ is singular normal. Let Σ be singular with rank $r \leq (p+q)$. Then we can find a $(p+q) \times r$ matrix B such that $\Sigma = BB'$. Let Σ be partitioned as follows:

$$\Sigma = \begin{bmatrix} \Sigma_{11} & \Sigma_{12} \\ \Sigma_{21} & \Sigma_{22} \end{bmatrix}$$

where let Σ_{11} be $p \times p$. The singularity in Σ may be in Σ_{11} alone or in Σ_{22} alone or in both. For convenience, we can assume that $r \geq p$ and that the $p \times p$ matrix Σ_{11} is nonsingular. Then the singularity is in Σ_{22}. We can show that there exists an $r \times 1$ vector $\mathbf{Z}^*$ such that

$$\mathbf{Z} = \boldsymbol{\mu} + B\mathbf{Z}^* \Rightarrow \mathbf{Z}^* \sim N_r(\mathbf{0}, I) \tag{2.4.18}$$

and $\mathbf{Z}^*$ is nonsingular normal.

If Q is the bilinear form $Q = \mathbf{X}'A\mathbf{Y}$ then one can write it as a quadratic form, see also Section 2.1, as follows:

$$Q = \mathbf{X}'A\mathbf{Y} = \mathbf{Z}'A^*\mathbf{Z}, \quad \mathbf{Z}' = (\mathbf{X}', \mathbf{Y}'),$$
$$A^* = \begin{bmatrix} O & \frac{1}{2}A \\ \frac{1}{2}A' & O \end{bmatrix}.$$

By direct integration, using the nonsingular normal density of $\mathbf{Z}^*$, we can evaluate the m.g.f. of Q as the following, see also Mathai and Provost (1992).

$$M_Q = |I - 2tB'A^*B|^{-\frac{1}{2}}.$$

Now expand as in (2.4.3) to see that the r-th cumulant of Q, denoted by K_r^Q and treated as a quadratic form in $\mathbf{Z}$, will be the following.

$$\begin{aligned} K_r^Q &= 2^{r-1}(r-1)!\text{tr}\left((B'A^*B)^r\right) \\ &= 2^{r-1}(r-1)!\text{tr}\left((A^*\Sigma)^r\right) \end{aligned} \tag{2.4.19}$$

which is the same as the one for the nonsingular case also. Hence the various cases discussed under the nonsingular case can also be obtained from the discussion of the singular case. But

$$\begin{aligned} A^*\Sigma &= \begin{bmatrix} O & \frac{1}{2}A \\ \frac{1}{2}A' & O \end{bmatrix} \begin{bmatrix} \Sigma_{11} & \Sigma_{12} \\ \Sigma_{21} & \Sigma_{22} \end{bmatrix} \\ &= \left(\frac{1}{2}\right) \begin{bmatrix} A\Sigma_{21} & A\Sigma_{22} \\ A'\Sigma_{11} & A'\Sigma_{12} \end{bmatrix}. \end{aligned}$$

Hence one way of evaluating the various cumulants of Q will be to take the various powers and then the traces. This process becomes complicated. The first few from this procedure are the following.

2.4c.1 The First Few Cumulants of a Bilinear Form in the Singular and Nonsingular Normal Cases

$$\begin{aligned} K_1^Q &= E(Q) \\ &= \left(\frac{1}{2}\right)(\text{tr}(A\Sigma_{21}) + \text{tr}(A'\Sigma_{12})) = \text{tr}(A\Sigma_{21}) \\ &= 0 \ \text{(under independence)} \\ &= \rho \ \text{tr}(A) \ \text{(under equicorrelation)}. \end{aligned}$$

$$
\begin{aligned}
K_2^Q &= \mathrm{Var}(Q) = \left(\frac{1}{2}\right) \mathrm{tr} \begin{bmatrix} A\Sigma_{21} & A\Sigma_{22} \\ A'\Sigma_{11} & A'\Sigma_{12} \end{bmatrix}^2 \\
&= \left(\frac{1}{2}\right) (\mathrm{tr}(A\Sigma_{21})^2 + \mathrm{tr}(A\Sigma_{22}A'\Sigma_{11}) \\
&\quad + \mathrm{tr}(A'\Sigma_{11}A\Sigma_{22} + \mathrm{tr}(A'\Sigma_{12}A'\Sigma_{12})) \\
&= \mathrm{tr}(A\Sigma_{21})^2 + \mathrm{tr}(A\Sigma_{22}A'\Sigma_{11}) \\
&= \mathrm{tr}(A\Sigma_{22}A'\Sigma_{11}) \quad \text{(under independence)} \\
&= \rho^2 \ \mathrm{tr}\left(A^2\right) + \mathrm{tr}(AA') \quad \text{(under equicorrelation).}
\end{aligned}
$$

$$
\begin{aligned}
K_3^Q &= \mathrm{tr}\left((A\Sigma_{21})^3\right) + 3 \ \mathrm{tr}(A\Sigma_{21}A\Sigma_{22}A'\Sigma_{11}) \\
&\quad + 3 \ \mathrm{tr}(A\Sigma_{22}A'\Sigma_{12}A'\Sigma_{11}) + \mathrm{tr}\left((A'\Sigma_{12})^3\right) \\
&= 0 \quad \text{(under independence)} \\
&= \rho^3 \ \mathrm{tr}\left(A^3\right) + \rho^3 \ \mathrm{tr}\left(A'^3\right) + 3\rho \ \mathrm{tr}\left(A^2A'\right) \\
&\quad + 3\rho \ \mathrm{tr}\left(AA'^2\right) \quad \text{(under equicorrelation).}
\end{aligned}
$$

$$
\begin{aligned}
K_4^Q &= 3\Big\{ \mathrm{tr}\left((A\Sigma_{21})^4\right) + 4 \ \mathrm{tr}\left((A\Sigma_{21})^2 A\Sigma_{22}A'\Sigma_{11}\right) \\
&\quad + 4 \ \mathrm{tr}\left(A\Sigma_{21}A\Sigma_{22}A'\Sigma_{12}A'\Sigma_{11}\right) \\
&\quad + 2 \ \mathrm{tr}\left((A\Sigma_{22}A'\Sigma_{11})^2\right) + \mathrm{tr}\left((A'\Sigma_{12})^4\right) \\
&\quad + 4 \ \mathrm{tr}\left((A'\Sigma_{12})^2 A'\Sigma_{11}A\Sigma_{22}\right) \Big\} \\
&= 6 \ \mathrm{tr}\left((A\Sigma_{22}A'\Sigma_{11})^2\right) \quad \text{(under independence)} \\
&= 3\Big\{ \rho^4 \ \mathrm{tr}\left(A^4\right) + \rho^4 \ \mathrm{tr}\left(A'^4\right) \\
&\quad + 4\rho^2 \ \mathrm{tr}\left(A^3A'\right) + 4\rho^2 \ \mathrm{tr}\left(A'^3A\right) \\
&\quad + 4\rho^2 \ \mathrm{tr}\left(A^2A'^2\right) + 2 \ \mathrm{tr}\left((AA')^2\right) \Big\} \\
&\quad \text{(under equicorrelation).}
\end{aligned}
$$

We may note that these agree with the list given in Section 2.4a.3.

Remark 2.4.1 The joint cumulants of a quadratic form and a bilinear form or two bilinear forms in the singular case can be extracted by using the joint cumulant generating function of two quadratic forms given in Mathai and Provost (1992) and then going through the procedure discussed above. This process becomes complicated. The first few of them in the singular case also can be shown to be the same as the ones given in Sections 2.4a.1 and 2.4b.1.

2.4d Cumulants of Bilinear Expressions

A general bilinear expression is given in (2.1.5) and (2.1.6), that is,

$$Q = \mathbf{Z}'A^*\mathbf{Z} + \mathbf{Z}'\mathbf{b}^* + d.$$

If $\mathbf{Z}' = (\mathbf{X}', \mathbf{Y}')$ where $\mathbf{Z} \sim N_{p+q}(\boldsymbol{\mu}, \Sigma)$, $\Sigma \geq 0$ and $\mathbf{X}$ is $p \times 1$ then the m.g.f. of the quadratic expression given in (2.1.6) can be shown to be the following, see also Mathai and Provost (1992).

$$M_Q(t) = |I - 2tA^*\Sigma|^{-\frac{1}{2}}\exp\Big\{ -\left(\frac{1}{2}\right)\boldsymbol{\mu}'\Sigma^{-1}\boldsymbol{\mu} + td$$
$$+\left(\frac{1}{2}\right)\left(\Sigma^{-\frac{1}{2}}\boldsymbol{\mu} + t\Sigma^{\frac{1}{2}}\mathbf{b}^*\right)'(I - 2t\Sigma^{\frac{1}{2}}A^*\Sigma^{\frac{1}{2}})^{-1}\left(\Sigma^{-\frac{1}{2}}\boldsymbol{\mu} + t\Sigma^{\frac{1}{2}}\mathbf{b}^*\right)\Big\}$$

for the nonsingular normal case and

$$M_Q(t) = |I - 2tB'A^*B|^{-\frac{1}{2}}\exp\Big\{t(\boldsymbol{\mu}'A^*\boldsymbol{\mu} + \mathbf{b}'^*\boldsymbol{\mu} + d)$$
$$+\left(\frac{t^2}{2}\right)(B'\mathbf{b}^* + 2B'A^*\boldsymbol{\mu})'(I - 2tB'A^*B)^{-1}(B'\mathbf{b}^* + 2B'A^*\boldsymbol{\mu})\Big\}$$

for the singular normal case where $\Sigma = BB'$. In both these cases one can show that the r-th cumulant of Q, denoted by K_r, reduces to the following:

$$K_r = 2^{r-1}r!\Big\{\left(\frac{1}{r}\right)\operatorname{tr}(A^*\Sigma)^r + \left(\frac{1}{4}\right)\mathbf{b}^{*\prime}(\Sigma A^*)^{r-2}\Sigma\mathbf{b}^*$$
$$+\boldsymbol{\mu}'(A^*\Sigma)^{r-1}A^*\boldsymbol{\mu} + \mathbf{b}^{*\prime}(\Sigma A^*)^{r-1}\boldsymbol{\mu}\Big\},\ r \geq 2$$

and

$$K_1 = \operatorname{tr}(A^*\Sigma) + \boldsymbol{\mu}'A^*\boldsymbol{\mu} + \mathbf{b}^{*\prime}\boldsymbol{\mu} + d,\ r = 1.$$

Thus

$$K_2 = 2\operatorname{tr}(A^*\Sigma)^2 + \mathbf{b}^{*\prime}\Sigma\mathbf{b}^*$$
$$+ 4\boldsymbol{\mu}'A^*\Sigma A^*\boldsymbol{\mu} + 4\mathbf{b}^{*\prime}\Sigma A^*\boldsymbol{\mu}$$

and

$$K_3 = 8\operatorname{tr}(A^*\Sigma)^3 + 6\mathbf{b}^{*\prime}\Sigma A^*\Sigma\mathbf{b}^*$$
$$+ 24\boldsymbol{\mu}'A^*\Sigma A^*\Sigma A^*\boldsymbol{\mu} + 24\mathbf{b}^{*\prime}\Sigma A^*\Sigma A^*\boldsymbol{\mu}.$$

2.4d.1 Some Special Cases of Bilinear Expressions

Some special cases of interest are when the bilinear expression contains only a linear term in $\mathbf{X}$ alone or a linear term in $\mathbf{Y}$ alone. Let

$$Q_1^* = \mathbf{X}'A\mathbf{Y} + \mathbf{b}'\mathbf{X} + \mathbf{c}'\mathbf{Y} + d;$$

$$Q_2^* = \mathbf{X}'A\mathbf{Y} + \mathbf{b}'\mathbf{X} + d;$$

$$Q_3^* = \mathbf{X}'A\mathbf{Y} + \mathbf{c}'\mathbf{Y} + d;$$

$$Q_4^* = \mathbf{b}'\mathbf{X} + \mathbf{c}'\mathbf{Y} + d.$$

The first few cumulants of these will be listed here explicitly for convenience. Let the r-th cumulant of Q_i^* be denoted by $K_r^{Q_i}$. Let

$$E(\mathbf{X}) = \boldsymbol{\mu}_1,\ E(\mathbf{Y}) = \boldsymbol{\mu}_2,\ \boldsymbol{\mu}' = (\boldsymbol{\mu}_1', \boldsymbol{\mu}_2')$$

and

$$\mathbf{b}^{*\prime} = (\mathbf{b}', \mathbf{c}'),\quad A^* = \begin{bmatrix} O & \frac{1}{2}A \\ \frac{1}{2}A' & O \end{bmatrix}.$$

Then

$$\boldsymbol{\mu}'A^*\boldsymbol{\mu} = \boldsymbol{\mu}_1'A\boldsymbol{\mu}_2,\ \mathbf{b}^{*\prime}\boldsymbol{\mu} = \mathbf{b}'\boldsymbol{\mu}_1 + \mathbf{c}'\boldsymbol{\mu}_2.$$

Take K_r^Q from Section 2.4c.1 for the bilinear form $Q = \mathbf{X}'A\mathbf{Y}$. Then

$$K_1^{Q_1} = K_1^Q + \boldsymbol{\mu}_1'A\boldsymbol{\mu}_2 + \mathbf{b}'\boldsymbol{\mu}_1 + \mathbf{c}'\boldsymbol{\mu}_2 + d.$$

$$K_1^{Q_2} = K_1^Q + \boldsymbol{\mu}_1'A\boldsymbol{\mu}_2 + \mathbf{b}'\boldsymbol{\mu}_1 + d.$$

$$K_1^{Q_3} = K_1^Q + \boldsymbol{\mu}_1'A\boldsymbol{\mu}_2 + \mathbf{c}'\boldsymbol{\mu}_2 + d.$$

$$K_1^{Q_4} = \mathbf{b}'\boldsymbol{\mu}_1 + \mathbf{c}'\boldsymbol{\mu}_2 + d.$$

Note the following:

$$\mathbf{b}^{*\prime}\Sigma\mathbf{b}^* = \mathbf{b}'\Sigma_{11}\mathbf{b} + \mathbf{c}'\Sigma_{22}\mathbf{c} + 2\mathbf{b}'\Sigma_{12}\mathbf{c}.$$

$$\begin{aligned}\boldsymbol{\mu}'A^*\Sigma A^*\boldsymbol{\mu} &= \left(\frac{1}{4}\right)\{\boldsymbol{\mu}_1'A\Sigma_{22}A'\boldsymbol{\mu}_1 \\ &\quad + \boldsymbol{\mu}_2'A'\Sigma_{11}A\boldsymbol{\mu}_2 + 2\boldsymbol{\mu}_1'A\Sigma_{21}A\boldsymbol{\mu}_2\}.\\ \mathbf{b}^{*\prime}\Sigma A^*\boldsymbol{\mu} &= \left(\frac{1}{2}\right)\{\mathbf{b}'\Sigma_{12}A'\boldsymbol{\mu}_1 + \mathbf{b}'\Sigma_{11}A\boldsymbol{\mu}_2 \\ &\quad + \mathbf{c}'\Sigma_{22}A'\boldsymbol{\mu}_1 + \mathbf{c}'\Sigma_{21}A\boldsymbol{\mu}_2\}.\end{aligned}$$

From these we can write the second cumulant as follows:

$$\begin{aligned}K_2^{Q_1} &= K_2^Q + \mathbf{b}'\Sigma_{11}\mathbf{b} + \mathbf{c}'\Sigma_{22}\mathbf{c} + 2\mathbf{b}'\Sigma_{12}\mathbf{c} \\ &\quad + \boldsymbol{\mu}_1'A\Sigma_{22}A'\boldsymbol{\mu}_1 + \boldsymbol{\mu}_2'A'\Sigma_{11}A\boldsymbol{\mu}_2 + 2\boldsymbol{\mu}_1'A\Sigma_{21}A\boldsymbol{\mu}_2 \\ &\quad + 2\{\mathbf{b}'\Sigma_{12}A'\boldsymbol{\mu}_1 + \mathbf{b}'\Sigma_{11}A\boldsymbol{\mu}_2 \\ &\quad + \mathbf{c}'\Sigma_{22}A'\boldsymbol{\mu}_1 + \mathbf{c}'\Sigma_{21}A\boldsymbol{\mu}_2.\}\end{aligned}$$

$$K_2^{Q_2} = K_2^Q + \mathbf{b}'\Sigma_{11}\mathbf{b} + \boldsymbol{\mu}_1' A\Sigma_{22}A'\boldsymbol{\mu}_1 + \boldsymbol{\mu}_2' A'\Sigma_{11}A\boldsymbol{\mu}_2 + 2\boldsymbol{\mu}_1' A\Sigma_{21}A\boldsymbol{\mu}_2 + 2\{\mathbf{b}'\Sigma_{12}A'\boldsymbol{\mu}_1 + \mathbf{b}'\Sigma_{11}A\boldsymbol{\mu}_2\}.$$

$$K_2^{Q_3} = K_2^Q + \mathbf{c}'\Sigma_{22}\mathbf{c} + \boldsymbol{\mu}_1' A\Sigma_{22}A'\boldsymbol{\mu}_1 + \boldsymbol{\mu}_2' A'\Sigma_{11}A\boldsymbol{\mu}_2 + 2\boldsymbol{\mu}_1' A\Sigma_{21}A\boldsymbol{\mu}_2 + 2\{\mathbf{c}'\Sigma_{22}A'\boldsymbol{\mu}_1 + \mathbf{c}'\Sigma_{21}A\boldsymbol{\mu}_2\}.$$

$$K_2^{Q_4} = \mathbf{b}'\Sigma_{11}\mathbf{b} + \mathbf{c}'\Sigma_{22}\mathbf{c} + 2\mathbf{b}'\Sigma_{12}\mathbf{c}.$$

Writing the third cumulant of a bilinear expression explicitly will take up too much space and hence it will not be given here.

Example 2.4.6 Evaluate the variance of the bilinear expression

$$Q_1 = 2X_1Y_1 + X_1Y_2 + X_2Y_1 + 3X_2Y_2 + X_2 - Y_1 + 7$$

where $E(\mathbf{X}') = E(X_1, X_2) = (1, -1)$, $E(\mathbf{Y}') = E(Y_1, Y_2) = (0, 1)$ and $\binom{\mathbf{X}}{\mathbf{Y}}$ is jointly normally distributed with the covariance matrix

$$\Sigma = \begin{bmatrix} 1 & 1 & 1 & 1 \\ 1 & 1 & 1 & 1 \\ 1 & 1 & 1 & 1 \\ 1 & 1 & 1 & 2 \end{bmatrix}.$$

SOLUTION According to the notation of Section 2.4d we have the following:

$$\boldsymbol{\mu}_1 = \begin{bmatrix} 1 \\ -1 \end{bmatrix}, \quad \boldsymbol{\mu}_2 = \begin{bmatrix} 0 \\ 1 \end{bmatrix}, \quad A = \begin{bmatrix} 2 & 1 \\ 1 & 3 \end{bmatrix},$$

$$\mathbf{b} = \begin{bmatrix} 0 \\ 1 \end{bmatrix}, \quad \mathbf{c} = \begin{bmatrix} -1 \\ 0 \end{bmatrix}, \quad d = 7,$$

$$\Sigma_{11} = \begin{bmatrix} 1 & 1 \\ 1 & 1 \end{bmatrix}, \quad \Sigma_{12} = \begin{bmatrix} 1 & 1 \\ 1 & 1 \end{bmatrix}, \quad \Sigma_{22} = \begin{bmatrix} 1 & 1 \\ 1 & 2 \end{bmatrix}.$$

Note also that Σ is singular. Let us compute the various quantities.

$$\text{tr}\left((A\Sigma_{21})^2\right) = 49, \quad \text{tr}(A\Sigma_{22}A'\Sigma_{11}) = 65.$$

$$K_2^Q = \text{tr}\left((A\Sigma_{21})^2\right) + \text{tr}(A\Sigma_{22}A'\Sigma_{11}) = 114.$$

$$\mathbf{b}'\Sigma_{11}\mathbf{b} = 1, \quad \mathbf{c}'\Sigma_{22}\mathbf{c} = 1, \quad 2\mathbf{b}'\Sigma_{12}\mathbf{c} = -2.$$

$$\boldsymbol{\mu}_1' A\Sigma_{22}A'\boldsymbol{\mu}_1 = 5, \quad \boldsymbol{\mu}_2' A'\Sigma_{11}A\boldsymbol{\mu}_2 = 16.$$

$$2\boldsymbol{\mu}_1' A\Sigma_{21}A\boldsymbol{\mu}_2 = -8, \quad 2\mathbf{b}'\Sigma_{12}A'\boldsymbol{\mu}_1 = -2.$$

$$2\mathbf{b}'\Sigma_{11}A\boldsymbol{\mu}_2 = 8, \quad 2\mathbf{c}'\Sigma_{22}A'\boldsymbol{\mu}_1 = 2.$$
$$2\mathbf{c}'\Sigma_{21}A\boldsymbol{\mu}_2 = -8.$$

The sum of all the above quantities starting from K_2^Q gives the variance of Q_1. That is,

$$\mathrm{Var}(Q_1) = 127.$$

2.5 Laplacianness of Bilinear Forms

In Mathai and Provost (1992) it was seen that a quadratic form in nonsingular or singular Gaussian random variables is distributed as a linear combination of central or noncentral chi-square random variables. What is the corresponding result if we are considering a bilinear form in Gaussian random variables? We can show that under some mild conditions a bilinear form in normal variables is distributed as a Laplacian or generalized Laplacian. This will be the topic of discussion in this section.

Let

$$\mathbf{Z} = \begin{pmatrix} \mathbf{X} \\ \mathbf{Y} \end{pmatrix} \sim N_{p+q}(\boldsymbol{\mu}, \Sigma), \ \mathbf{X} \sim N_p(\boldsymbol{\mu}_1, \Sigma_{11}),$$

$$\boldsymbol{\mu} = \begin{pmatrix} \boldsymbol{\mu}_1 \\ \boldsymbol{\mu}_2 \end{pmatrix}, \quad \Sigma = \begin{pmatrix} \Sigma_{11} & \Sigma_{12} \\ \Sigma_{21} & \Sigma_{22} \end{pmatrix}.$$

When $\mathbf{X}$ and $\mathbf{Y}$ are noncorrelated we have $\Sigma_{12} = O$. In general, $\Sigma = \Sigma'$ and $\Sigma > 0$ (positive definite) if $\mathbf{Z}$ is nonsingular normal, and $\Sigma \geq 0$ if $\mathbf{Z}$ is singular normal. Consider a general quadratic expression $Q(\mathbf{Z})$ in $\mathbf{Z}$. Then

$$Q(\mathbf{Z}) = \mathbf{Z}'A\mathbf{Z} + \mathbf{a}'\mathbf{Z} + d, \quad A = A' \tag{2.5.1}$$

where A , $\mathbf{a}$ and d do not involve $\mathbf{X}$ or $\mathbf{Y}$.

The necessary and sufficient conditions (hereafter called the NS conditions) for $Q(\mathbf{Z})$ to be distributed as a noncentral generalized Laplace variable (hereafter called NGL variable) as the U in (2.3.2) will be investigated. This will be called noncentral generalized Laplacianness (NGL) of $Q(\mathbf{Z})$. The parameters of the NGL will be denoted by $(\alpha,\ \beta,\ \lambda)$. Nonsingular normal and singular normal cases will be considered. As particular cases of these general results we will derive a number of results for bilinear and quadratic expressions in $\mathbf{X}$ and $\mathbf{Y}$. Let

$$Q_1(\mathbf{Z}) = \mathbf{X}'B_1\mathbf{X} + \mathbf{b_1}'\mathbf{X} + \mathbf{c_1}'\mathbf{Y} + d_1, \quad B_1 = B_1' \tag{2.5.2}$$

$$Q_2(\mathbf{Z}) = \mathbf{X}'B_2\mathbf{Y} + \mathbf{b_2}'\mathbf{X} + \mathbf{c_2}'\mathbf{Y} + d_2, \tag{2.5.3}$$

$$Q_3(\mathbf{Z}) = \mathbf{Y}'B_3\mathbf{Y} + \mathbf{b_3}'\mathbf{X} + \mathbf{c_3}'\mathbf{Y} + d_3, \quad B_3 = B_3'. \tag{2.5.4}$$

Then Q_1 represents a general quadratic expression in $\mathbf{X}$, Q_3 represents a general quadratic expression in $\mathbf{Y}$ and $Q_2(\mathbf{Z})$ represents a general bilinear expression.

$$Q_1(\mathbf{Z}) = Q(\mathbf{Z}) \quad \text{for} \quad \mathbf{a}' = (\mathbf{b_1}', \mathbf{c_1}'),$$
$$d = d_1, \quad B_1 = B_1', \quad A = \begin{pmatrix} B_1 & O \\ O & O \end{pmatrix}$$

and

$$Q_2(\mathbf{Z}) = Q(\mathbf{Z}), \quad \text{for} \quad \mathbf{a}' = (\mathbf{b_2}', \mathbf{c_2}'),$$
$$d = d_2, \quad A = \begin{pmatrix} O & \frac{1}{2}B_2 \\ \frac{1}{2}B_2' & O \end{pmatrix}.$$

Note that Q_1 and Q_3 have similar structures. Hence we will consider only Q_1 and Q_2. Some results are needed for establishing the results on generalized Laplacianness. These will be stated as lemmas.

Lemma 2.5.1 Let $P_i(x)$, $i = 1, 2, 3, 4$ be polynomials in x with rational coefficients such that

$$P_1(x)e^{P_2(x)/P_4(x)} = P_3(x) \tag{2.5.5}$$

for all x where $P_2(0)/P_4(0) = 0$, $P_1(0) = P_3(0) = 1$. Then $P_1(x) = P_3(x)$ and $P_2(x) = P_4(x)$. This lemma may be seen from Laha (1956) and Driscoll and Gundberg (1986).

Lemma 2.5.2 Let

$$(1 - \beta^2 t^2)^{-\alpha} = \prod_{j=1}^{p} (1 - 2t\lambda_j)^{-\frac{1}{2}} \tag{2.5.6}$$

for $|\beta^2 t^2| < 1$, $|2t\lambda_j| < 1$, λ_j's real, $j = 1, 2, \ldots, p$, $\alpha > 0$, $\beta > 0$, then

(i) $2\alpha = r$ for some positive integer r;

(ii) exactly r of the λ_j's are equal to $\beta/2$, exactly r of them are equal to $-\beta/2$ and the remaining $p - 2r$ of the λ_j's are equal to zero.

PROOF Take logarithms, expand and equate the coefficients of t, $t^2, \ldots$ on both sides of (2.5.6) to get

$$0 = \sum_{j=1}^{p} \lambda_j^k, \quad k = 1, 3, 5, \ldots \tag{2.5.7}$$

$$\alpha = \sum_j \left(\frac{\lambda_j}{\beta}\right)^2 \tag{2.5.8}$$

$$\alpha = 2^2 \sum_j \left(\frac{\lambda_j}{\beta}\right)^4 \tag{2.5.9}$$

$$\alpha = 2^4 \sum_j \left(\frac{\lambda_j}{\beta}\right)^6 \tag{2.5.10}$$

and so on. From (2.5.8) and (2.5.9) one has, for $\delta_j = \lambda_j/\beta$,

$$\sum_j \delta_j^2 \left[(2\delta_j)^2 - 1\right] = 0. \tag{2.5.11}$$

From (2.5.9) and (2.5.10) one has

$$\sum_j \delta_j^2 \left[(2\delta_j)^4 - 1\right] = 0. \tag{2.5.12}$$

From (2.5.11) and (2.5.12) one has

$$\sum_j \delta_j^2 \left[(2\delta_j)^2 - 1\right]^2 = 0$$

$$\Rightarrow \delta_j = 0 \text{ or } \delta_j^2 = \frac{1}{4} \Rightarrow \delta_j = \pm\frac{1}{2} \Rightarrow \lambda_j = \pm\beta/2,\ \beta > 0. \tag{2.5.13}$$

From (2.5.13), and (2.5.7) for $k = 1$ one has the number of positive λ_j's equal to the number of negative λ_j's. Let each be r in number.

$$\begin{aligned}\sum_{j=1}^{p} \lambda_j^k &= 0 + \sum_{j=1}^{2r} \lambda_j^k \\ &= \left(\frac{\beta}{2}\right)^k \left[(1)^k + \cdots + (1)^k + \cdots + (-1)^k + \cdots + (-1)^k\right].\end{aligned}$$

Substituting in (2.5.8) one has

$$\alpha = \sum_{j=1}^{2r} \left(\frac{\beta}{2\beta}\right)^2 = \frac{2r}{4} \Rightarrow 2\alpha = r.$$

This completes the proof.

We will establish a set of NS conditions for noncentral generalized Laplacianness of quadratic expressions for the nonsingular and singular normal cases with the help of Lemmas 2.5.1 and 2.5.2.

2.5a Quadratic and Bilinear Forms in the Nonsingular Normal Case

Let $Q(\mathbf{Z}) = \mathbf{Z}'A\mathbf{Z} \sim N_{p+q}(\boldsymbol{\mu}, \Sigma)$, $\Sigma > 0$. Then from Section 2.4d or by direct integration one can obtain the m.g.f. of Q as

$$\begin{aligned} M_Q(t) = \left|I - 2t\Sigma^{\frac{1}{2}} A \Sigma^{\frac{1}{2}}\right|^{-\frac{1}{2}} \exp\Big\{ &-\left(\frac{1}{2}\right) \boldsymbol{\mu}'\Sigma^{-1}\boldsymbol{\mu} \\ &+ \left(\frac{1}{2}\right) \boldsymbol{\mu}'\Sigma^{-\frac{1}{2}} \left(I - 2t\Sigma^{\frac{1}{2}} A \Sigma^{\frac{1}{2}}\right)^{-1} \Sigma^{-\frac{1}{2}}\boldsymbol{\mu}\Big\} \end{aligned} \tag{2.5.14}$$

for $\| 2t\Sigma^{\frac{1}{2}} A \Sigma^{\frac{1}{2}} \| < 1$, where $\| (\ ) \|$ denotes a norm of ( ) and $\Sigma^{\frac{1}{2}}$ denotes the symmetric square root of Σ. Any square root of Σ can be used but for convenience we will use the symmetric square root. If a nonsymmetric square root is used by writing $\Sigma = BB'$ then $\Sigma^{\frac{1}{2}}A\Sigma^{\frac{1}{2}}$ should be written as $B'AB$. From (2.3.2) the m.g.f. of a real NGL is

$$M_U(t) = (1-\beta^2t^2)^{-\alpha}\exp\left\{-2\lambda + 2\lambda(1-\beta^2t^2)^{-1}\right\} \tag{2.5.15}$$

for $\lambda > 0,\ \beta > 0,\ \alpha > 0$. What are the NS conditions for (2.5.14) and (2.5.15) to be equal ?

Theorem 2.5.1 The NS conditions for $Q(\mathbf{Z})$ of (2.5.14) to be a NGL of (2.5.15) are

$(i)_a$ $2\alpha = r$ for some positive integer r;

$(ii)_a$ The eigenvalues of $\Sigma^{\frac{1}{2}}A\Sigma^{\frac{1}{2}}$ are such that exactly r of them are $\beta/2$, r of them are $-\beta/2$ and the remaining are equal to zero or $A\Sigma A = \left(\frac{4}{\beta^2}\right)(A\Sigma)^3A$ and $\mathrm{tr}(A\Sigma) = 0$;

$(iii)_a$ $\boldsymbol{\mu}'A\boldsymbol{\mu} = 0$;

$(iv)_a$ $\lambda = \left(\frac{1}{\beta^2}\right)\boldsymbol{\mu}'A\Sigma A\boldsymbol{\mu}$.

PROOF *Necessity* From (2.5.14), (2.5.15) and Lemmas 2.1 and 2.2, conditions $(i)_a$ and $(ii)_a$ are necessary. Now equate the exponential parts in (2.5.14) and (2.5.15) to get

$$2\lambda\sum_{k=1}^{\infty}(\beta^2t^2)^k = \left(\frac{1}{2}\right)\boldsymbol{\mu}'\Sigma^{-\frac{1}{2}}\left[\sum_{k=1}^{\infty}\left(2t\Sigma^{\frac{1}{2}}A\Sigma^{\frac{1}{2}}\right)^k\right]\Sigma^{-\frac{1}{2}}\boldsymbol{\mu}. \tag{2.5.16}$$

The coefficient of t on both sides of (2.5.16) gives condition $(iii)_a$. The coefficient of t^2 on both sides gives $(iv)_a$. Compare the coefficients of t^{2m+1} on both sides for $m = 1, 2, \ldots$

$$\begin{aligned}
0 &= \boldsymbol{\mu}'\Sigma^{-\frac{1}{2}}\left(\Sigma^{\frac{1}{2}}A\Sigma^{\frac{1}{2}}\right)^{2m+1}\Sigma^{-\frac{1}{2}}\boldsymbol{\mu}\\
&= \boldsymbol{\mu}'\Sigma^{-\frac{1}{2}}P\big[\mathrm{diag}(\frac{\beta}{2},\ldots,\frac{\beta}{2},-\frac{\beta}{2},-\frac{\beta}{2},\\
&\qquad \ldots,-\frac{\beta}{2},0,\ldots,0)\big]^{2m+1}P'\Sigma^{-\frac{1}{2}}\boldsymbol{\mu},\quad PP' = I\\
&= \left(\frac{\beta}{2}\right)^{2m+1}\boldsymbol{\mu}'\Sigma^{-\frac{1}{2}}P\big[\mathrm{diag}(1,\ldots,1,-1,\\
&\qquad \ldots,-1,0,\ldots,0)\big]P'\Sigma^{-\frac{1}{2}}\boldsymbol{\mu}\\
&= \left(\frac{\beta}{2}\right)^{2m}\boldsymbol{\mu}'\Sigma^{-\frac{1}{2}}\left(\Sigma^{\frac{1}{2}}A\Sigma^{\frac{1}{2}}\right)\Sigma^{-\frac{1}{2}}\boldsymbol{\mu}\\
&= \left(\frac{\beta}{2}\right)^{2m}\boldsymbol{\mu}'A\boldsymbol{\mu}.
\end{aligned}$$

Compare the coefficient of t^{2m} to get

$$2\lambda\beta^{2m} = \left(\frac{1}{2}\right)(2)^{2m}\left(\mu'\Sigma^{-\frac{1}{2}}\right)\left(\Sigma^{\frac{1}{2}}A\Sigma^{\frac{1}{2}}\right)^{2m}\left(\Sigma^{-\frac{1}{2}}\mu\right)$$

$$= \left(\frac{1}{2}\right)\beta^{2m}\left(\mu'\Sigma^{-\frac{1}{2}}\right)P\left[\text{diag}(1,\ldots,1,0,\ldots,0)\right]P'\left(\Sigma^{-\frac{1}{2}}\mu\right)$$

$$= 2\beta^{2m-2}\left(\mu'\Sigma^{-\frac{1}{2}}\right)\left(\Sigma^{\frac{1}{2}}A\Sigma^{\frac{1}{2}}\right)^{2}\left(\Sigma^{-\frac{1}{2}}\mu\right)$$

$$= 2\beta^{2m-2}\mu'A\Sigma A\mu.$$

This is condition $(iv)_a$. Thus $(i)_a$ to $(iv)_a$ are necessary. By retracing the steps we see that the conditions are also sufficient.

Remark 2.5.1 When $A \geq 0$ then $\mu'A\mu = 0 \Rightarrow A\mu = \mathbf{0}$. In this case $\lambda = 0$. Thus the conditions in this case reduce to $(i)_a$, $(ii)_a$ and $A\mu = \mathbf{0}$, $\lambda = 0$. But $\text{tr}(A\Sigma)$ in this case is zero only if A is null. Hence when $A \geq 0$, $\mathbf{Z}'A\mathbf{Z}$ cannot be a NGL.

When $\beta = 2$, $\alpha = \nu/2$, $\nu = 1, 2, \ldots$, $\lambda = \left(\frac{1}{2}\right)\mu'\Sigma^{-1}\mu$, (2.5.15) gives the m.g.f. of a noncentral chi-square difference with ν degrees of freedom. Hence we can give the following corollary.

Corollary 2.5.1 The NS conditions for $Q(\mathbf{Z})$ of (2.5.14) to be distributed as a noncentral chi-square difference are $(i)_a$ to $(iv)_a$ of Theorem 2.5.1 with $\beta = 2$, $\alpha = \nu/2$.

Corollary 2.5.2 The NS conditions for the bilinear form $\mathbf{X}'B_2\mathbf{Y}$ where $\binom{\mathbf{X}}{\mathbf{Y}} \sim N_{p+q}(\mu, \Sigma)$, $\mu = \binom{\mu_1}{\mu_2}$, $\Sigma = \begin{pmatrix}\Sigma_{11} & \Sigma_{12}\\ \Sigma_{21} & \Sigma_{22}\end{pmatrix} > 0$, $\mathbf{X} \sim N_p(\mu_1, \Sigma_{11})$, to be a NGL are the following:

$(i)_b \equiv (i)_a$;

$(ii)_b$

$$\left(\frac{4}{\beta^2}\right)A\Sigma A\Sigma A\Sigma A = A\Sigma A$$

and

$$\text{tr}\left(\Sigma^{\frac{1}{2}}A\Sigma^{\frac{1}{2}}\right) = \text{tr}(A\Sigma) = 0, \quad A = \begin{pmatrix} O & \frac{1}{2}B_2 \\ \frac{1}{2}B_2' & O\end{pmatrix};$$

or

$$B_2\Sigma_{22}B_2' = \left(\frac{1}{\beta^2}\right)\Big\{\left[(B_2\Sigma_{21})^2 + B_2\Sigma_{22}B_2'\Sigma_{11}\right](B_2\Sigma_{22}B_2')$$
$$+ \left[B_2\Sigma_{21}B_2\Sigma_{22} + B_2\Sigma_{22}B_2'\Sigma_{12}\right](B_2'\Sigma_{12}B_2')\Big\};$$

$$B_2\Sigma_{21}B_2 = \left(\frac{1}{\beta^2}\right)\Big\{ \left[(B_2\Sigma_{21})^2 + B_2\Sigma_{22}B_2'\Sigma_{11}\right](B_2\Sigma_{21}B_2)$$
$$+ [B_2\Sigma_{21}B_2\Sigma_{22} + B_2\Sigma_{22}B_2'\Sigma_{12}](B_2'\Sigma_{11}B_2)\Big\};$$
$$B_2'\Sigma_{12}B_2' = \left(\frac{1}{\beta^2}\right)\Big\{ [B_2'\Sigma_{11}B_2\Sigma_{21} + B_2'\Sigma_{12}B_2'\Sigma_{11}](B_2\Sigma_{22}B_2')$$
$$+ \left[B_2'\Sigma_{11}B_2\Sigma_{22} + (B_2'\Sigma_{12})^2\right](B_2'\Sigma_{12}B_2')\Big\};$$
$$B_2'\Sigma_{11}B_2 = \left(\frac{1}{\beta^2}\right)\Big\{[B_2'\Sigma_{11}B_2\Sigma_{21} + B_2'\Sigma_{12}B_2'\Sigma_{11}](B_2\Sigma_{21}B_2)$$
$$+ \left[B_2'\Sigma_{11}B_2\Sigma_{22} + (B_2'\Sigma_{12})^2\right](B_2'\Sigma_{11}B_2)\Big\}$$

and

$$\text{tr}(B_2\Sigma_{21}) = 0;$$

$(iii)_b$ $$\boldsymbol{\mu}_1{}'B_2\boldsymbol{\mu}_2 = 0;$$

$(iv)_b$

$$\lambda = \left(\frac{1}{4\beta^2}\right)\{\boldsymbol{\mu}'B_2\Sigma_{22}B_2'\boldsymbol{\mu}_1 + \boldsymbol{\mu}_2{}'B_2'\Sigma_{11}B_2\boldsymbol{\mu}_2$$
$$+ 2\boldsymbol{\mu}_1{}'B_2\Sigma_{21}B_2\boldsymbol{\mu}_2\}.$$

PROOF Take $A = \begin{pmatrix} O & \frac{1}{2}B_2 \\ \frac{1}{2}B_2' & O \end{pmatrix}$. Then $Q(\mathbf{Z})$ of Theorem 2.5.1 reduces to the bilinear form under consideration. Conditions $(iii)_b$ and $(iv)_b$ follow by direct substitution in $(iii)_a$ and $(iv)_a$ respectively. Now consider condition $(ii)_a$. Since $\left(\Sigma^{\frac{1}{2}}A\Sigma^{\frac{1}{2}}\right)$ is symmetric with tr $\left(\Sigma^{\frac{1}{2}}A\Sigma^{\frac{1}{2}}\right) = 0$, $(ii)_a$ will imply that $\left(\frac{4}{\beta^2}\right)\left(\Sigma^{\frac{1}{2}}A\Sigma A\Sigma^{\frac{1}{2}}\right)^2$ is idempotent and of rank $2r$ which will then imply $(ii)_b$ for A as defined above.

Some particular cases will be considered here. Hereafter noncorrelated case will mean that $\mathbf{X}$ and $\mathbf{Y}$ are noncorrelated, that is $\Sigma_{12} = O$.

2.5b NS Conditions for the Noncorrelated Normal Case

Now $Q(\mathbf{Z}) = \mathbf{X}'B_2\mathbf{Y}$ with $\Sigma_{12} = O$. The NS conditions can be obtained from Corollary 2.5.2 as a particular case. These reduce to the following:

$(ii)_b$

$$\left(\frac{1}{\beta^2}\right)B_2\Sigma_{22}B_2'\Sigma_{11}B_2\Sigma_{22}B_2' = B_2\Sigma_{22}B_2'$$

and

$$\left(\frac{1}{\beta^2}\right) B_2'\Sigma_{11}B_2\Sigma_{22}B_2'\Sigma_{11}B_2 = B_2'\Sigma_{11}B_2;$$

$(iv)_b$

$$\lambda = \left(\frac{1}{4\beta^2}\right)\{\mu_1{}'B_2\Sigma_{22}B_2'\mu_1 + \mu_2{}'B_2'\Sigma_{11}B_2\mu_2\}.$$

Example 2.5.1 Consider a simple random sample of size n from a bivariate normal with zero correlation. That is, $\binom{X_i}{Y_i}$, $i = 1, \ldots, n$ are independently and identically distributed with $X_i \sim N(\mu_x, \sigma_x^2)$, $Y_i \sim N(\mu_y, \sigma_y^2)$, $\text{Cov}(X_i, Y_i) = 0$ for all i. Let $U = \sum_{i=1}^n (X_i - \bar{X})(Y_i - \bar{Y})$, $\bar{X} = \sum_{i=1}^n X_i/n$, $\bar{Y} = \sum_{i=1}^n Y_i/n$. Then U is a NGL with parameters $(\alpha, \beta, \lambda) = (\frac{n-1}{2}, \sigma_x\sigma_y, 0)$.

SOLUTION

Note that this is Example 2.2.2. We are redoing it with the help of Theorem 2.5.1. Now $U = \mathbf{X}'B_2\mathbf{Y}$ where $\mathbf{X}' = (X_1, \ldots, X_n)$, $\mathbf{Y}' = (Y_1, \ldots, Y_n)$, $B_2 = I_n - \frac{\mathbf{11}'}{n}$, $B_2 = B_2^2$ with rank $n-1$, and $\mathbf{1}' = (1, 1, \ldots, 1)$, I_n is the identity matrix of order n. Then

$$\Sigma_{11} = \sigma_x^2 I_n,\ \Sigma_{22} = \sigma_y^2 I_n,\ \Sigma_{12} = O,\ \mu_1 = \mu_x\mathbf{1},\ \mu_2 = \mu_y\mathbf{1}$$

in the notations of Corollary 2.5.2. Condition $(ii)_b$ then implies that $\beta = \sigma_x\sigma_y$ and condition $(i)_a$ gives $2\alpha = n - 1$. Note that

$$\mu_1{}'B_2\Sigma_{22}B_2'\mu_1 = \sigma_y^2\mu_x^2\mathbf{1}'\left[I - \frac{\mathbf{11}'}{n}\right] I \left[I - \frac{\mathbf{11}'}{n}\right]\mathbf{1} = 0$$

since $\mathbf{1}'\mathbf{1} = n$. Similarly $\mu_2{}'B_2'\Sigma_{11}B_2\mu_2 = 0$. Thus from $(iv)_b$ one has $\lambda = 0$.

Corollary 2.5.3 The NS conditions for the quadratic form $Q(\mathbf{Z}) = \mathbf{X}'B_1\mathbf{X}$, $B_1 = B_1'$, $\mathbf{X} \sim N_p(\mu_1, \Sigma_{11})$, $\Sigma_{11} > 0$, to be distributed as a NGL are the following:

$(i)_c \equiv (i)_a$;

$(ii)_c$

$$\left(\frac{4}{\beta^2}\right)(B_1\Sigma_{11})^3 B_1 = B_1\Sigma_{11}B_1$$

and

$$\text{tr}(B_1\Sigma_{11}) = 0;$$

$(iii)_c$ $\quad \mu_1{}'B_1\mu_1 = 0;$

$(iv)_c$ $\quad \lambda = \left(\frac{1}{\beta^2}\right)\mu_1{}'B_1\Sigma_{11}B_1\mu_1.$

This result follows from Theorem 2.5.1 by taking $A = \begin{pmatrix} B_1 & O \\ O & O \end{pmatrix}$.

Remark 2.5.2 In Collorary 2.5.3 if the Laplacian is central, that is $\lambda = 0$, then $\lambda = 0 \Rightarrow \boldsymbol{\mu}_1{}'B_1\Sigma_{11}^{\frac{1}{2}}\Sigma_{11}^{\frac{1}{2}}B_1\boldsymbol{\mu}_1 = 0 \Rightarrow \mathbf{b}'\mathbf{b} = 0 \Rightarrow \mathbf{b} = \mathbf{0},\ \mathbf{b} = \Sigma_{11}^{\frac{1}{2}}B_1\boldsymbol{\mu}_1$. Then $B_1\boldsymbol{\mu}_1 = \mathbf{0}$ since $|\Sigma_{11}| \neq 0$. This is the condition available in Beckett et al (1980). Thus in this case $(iii)_c$ reduces to $B_1\boldsymbol{\mu}_1 = \mathbf{0}$. If $B_1 \geq 0$ then $\boldsymbol{\mu}_1{}'B_1\boldsymbol{\mu}_1 = 0 \Rightarrow B_1\boldsymbol{\mu}_1 = \mathbf{0}$. In this case λ must be zero. But in this case $\text{tr}(B_1\Sigma_{11})$ cannot be zero unless B_1 is null. Hence $\mathbf{X}'B_1\mathbf{X}$ cannot be a generalized Laplacian if $B_1 \geq 0$.

Example 2.5.2 Let $X_1 \sim N(\sqrt{2}, \sigma^2)$, $X_2 \sim N(-\sqrt{3}, \sigma^2)$, $X_3 \sim N(1, \sigma^2)$ and be mutually independently distributed. Let $U = X_2^2 - X_3^2 + \frac{4}{\sqrt{6}}X_1X_2 + \frac{4}{\sqrt{2}}X_1X_3 + \frac{2}{\sqrt{3}}X_2X_3$. Then show that U is a NGL with parameters $(\alpha, \beta, \lambda) = (\frac{1}{2}, 4\sigma^2, 0)$.

SOLUTION

In the notation of Corollary 2.5.3 we have $\boldsymbol{\mu}_1 = (\sqrt{2}, -\sqrt{3}, 1)'$, $\Sigma_{11} = \sigma^2 I$,

$$B_1 = \begin{pmatrix} 0 & \frac{2}{\sqrt{6}} & \frac{2}{\sqrt{2}} \\ \frac{2}{\sqrt{6}} & 1 & \frac{1}{\sqrt{3}} \\ \frac{2}{\sqrt{2}} & \frac{1}{\sqrt{3}} & -1 \end{pmatrix}, \quad B_1^2 = \begin{pmatrix} \frac{8}{3} & \frac{4}{\sqrt{6}} & -\frac{4}{3\sqrt{2}} \\ \frac{4}{\sqrt{6}} & 2 & \frac{2}{\sqrt{3}} \\ -\frac{4}{3\sqrt{2}} & \frac{2}{\sqrt{3}} & \frac{10}{3} \end{pmatrix}.$$

Note that B_1 has the eigenvalues $2, -2, 0$. Hence the eigenvalues of $\Sigma_{11}^{\frac{1}{2}}B_1\Sigma_{11}^{\frac{1}{2}}$ are $2\sigma^2, -2\sigma^2, 0$. Thus condition $(i)_c$ is satisfied with $\alpha = \frac{1}{2}$. Note that $\boldsymbol{\mu}_1{}'B\boldsymbol{\mu}_1 = 0$ and hence $(iii)_c$ is satisfied. Diagonalize the B_1's on both sides of condition $(ii)_c$, take out 2 from each diagonal matrix and substitute for β to see that $(ii)_c$ is satisfied. Since $B_1\boldsymbol{\mu}_1 = \mathbf{0}$ we have $\lambda = 0$. Hence the result.

2.5c Quadratic and Bilinear Forms in the Singular Normal Case

Let $Q(\mathbf{Z}) = \mathbf{Z}'A\mathbf{Z}$, $A = A'$, $\mathbf{Z} = \binom{\mathbf{X}}{\mathbf{Y}} \sim N_{p+q}(\boldsymbol{\mu}, \Sigma)$, $\Sigma \geq 0$, $\Sigma = BB'$, B is $(p+q) \times r$ of rank r, $\boldsymbol{\mu}' = (\boldsymbol{\mu}_1{}', \boldsymbol{\mu}_2{}')$, $\mathbf{X} \sim N_p(\boldsymbol{\mu}_1, \Sigma_{11})$, $\Sigma_{11} \geq 0$, $\text{Cov}(\mathbf{X}, \mathbf{Y}) = \Sigma_{12} = \Sigma_{21}'$. We will investigate the NS conditions for $Q(\mathbf{Z})$ to be NGL. By direct integration, see also Section 2.4d, one can see that the m.g.f. of $Q(\mathbf{Z})$ is available as

$$M_Q(t) = |I_r - 2tB'AB|^{-\frac{1}{2}}\exp\Big\{t\boldsymbol{\mu}'A\boldsymbol{\mu} + 2t^2\boldsymbol{\mu}'AB(I - 2tB'AB)^{-1}B'A\boldsymbol{\mu}\Big\} \tag{2.5.17}$$

and the m.g.f. of the NGL is available from (2.5.15).

Theorem 2.5.2 The NS conditions for $Q(\mathbf{Z})$ to be NGL are the following:

$(i)_d$ $\quad 2\alpha = s$ where s is a positive integer;

$(ii)_d$ The eigenvalues of $B'AB$ are such that exactly s of them are equal to $\beta/2$, exactly s of them are equal to $-\beta/2$ and the remaining $r-2s$ of them are equal to zero or $(\Sigma A)^2\Sigma = \left(\frac{4}{\beta^2}\right)(\Sigma A)^4\Sigma$ and $\text{tr}(A\Sigma) = 0$;

$(iii)_d$ $\mu'A\mu = 0 = \mu'A\Sigma A\Sigma A\mu$;

$(iv)_d$ $\lambda = \left(\frac{1}{\beta^2}\right)\mu'A\Sigma A\mu$

and if $B'AB$ is singular then

$(v)_d$ $\mu'A\Sigma A\mu = \left(\frac{4}{\beta^2}\right)\mu'A\Sigma A\Sigma A\Sigma A\mu$ or $\Sigma A\mu = \left(\frac{4}{\beta^2}\right)(\Sigma A)^3\mu$.

PROOF *Necessity* From Lemmas 2.5.1 and 2.5.2 we see that conditions $(i)_d$ and $(ii)_d$ are necessary. Equate and expand the exponents of (2.5.17) and (2.5.15) and then compare the coefficients of t^{2m} on both sides as well as t^{2m+1} on both sides to get the following:

$$2\lambda\beta^{2m} = 2(2^{2m-2})\mu'AB(B'AB)^{2m-2}B'A\mu, \quad m = 1, 2, \ldots \tag{2.5.18}$$

$$= 2(2^{2m-2})\mathbf{b}'\left[\text{diag}\left(\frac{\beta}{2},\ldots,\frac{\beta}{2},-\frac{\beta}{2},\ldots,-\frac{\beta}{2},0,\ldots,0\right)\right]^{2m-2}\mathbf{b},$$

$$\mathbf{b} = P'B'A\mu, \quad PP' = I,$$

$$P'B'ABP = \text{diag}\left(\frac{\beta}{2},\ldots,\frac{\beta}{2},-\frac{\beta}{2},\ldots,-\frac{\beta}{2},0,\ldots,0\right).$$

From (2.5.18) for $m = 1$ we get $(iv)_d$. But (2.5.18) simplifies to $(iv)_d$ if there is no zero eigenvalue for $B'AB$, that is, if $|B'AB| \neq 0$. If $B'AB$ is singular then (2.5.18) reduces to the following:

$$2\lambda\beta^4 = 2(2)^2\left(\frac{\beta}{2}\right)^2\mathbf{b}'[\text{diag}(1,\ldots,1,0,\ldots,0)]\mathbf{b} \tag{2.5.19}$$

$$\Rightarrow \lambda = \left(\frac{4}{\beta^4}\right)\mu'A\Sigma A\Sigma A\Sigma A\mu. \tag{2.5.20}$$

Equating the values of λ from (2.5.18) and (2.5.20) one has

$$\mu'A\Sigma A\mu = \left(\frac{4}{\beta^2}\right)\mu'A\Sigma A\Sigma A\Sigma A\mu \tag{2.5.21}$$

$$\Rightarrow 0 = \hat{\mathbf{b}}'\left[I - \left(\frac{4}{\beta^2}\right)(B'AB)^2\right]\hat{\mathbf{b}}, \quad \hat{\mathbf{b}} = B'A\mu. \tag{2.5.22}$$

But since

$$\left[\left(\frac{4}{\beta^2}\right)(B'AB)^2\right]^2 = \left[\left(\frac{4}{\beta^2}\right)(B'AB)^2\right]$$

one has from (2.5.22), $\mathbf{c}'\mathbf{c} = 0 \Rightarrow \mathbf{c} = \mathbf{0}$ where

$$\mathbf{c} = \left(I - \left(\frac{4}{\beta^2}\right)(B'AB)^2\right)\hat{\mathbf{b}}.$$

Now writing in terms of $BB' = \Sigma$ one has

$$\Sigma A\mu = \left(\frac{4}{\beta^2}\right)(\Sigma A)^3\mu. \tag{2.5.23}$$

Now equating the λ's from (2.5.20) and (2.5.18) for $m = 1$ as well as from (2.5.23) one has condition $(v)_d$. From odd powers of t one has

$$0 = \boldsymbol{\mu}'A\boldsymbol{\mu}; \tag{2.5.24}$$

$$\begin{aligned} 0 &= \boldsymbol{\mu}'AB(B'AB)^{2m+1}B'A\boldsymbol{\mu},\ m = 1, 2, \ldots \qquad (2.5.25)\\ &= \left(\frac{\beta}{2}\right)^{2m+1} \mathbf{b}'[\mathrm{diag}(1, \ldots, 1, -1, \ldots, -1, 0, \ldots, 0)]^{2m+1}\mathbf{b} \end{aligned}$$

where $\mathbf{b}$ is defined in (2.5.18). That is,

$$\begin{aligned} 0 &= \left(\frac{\beta}{2}\right) \mathbf{b}'[\mathrm{diag}(1, \ldots, 1, -1, \ldots, -1, 0, \ldots, 0)]\mathbf{b}\\ &= \boldsymbol{\mu}'AB(B'AB)B'A\boldsymbol{\mu} = \boldsymbol{\mu}'A\Sigma A\Sigma A\boldsymbol{\mu}. \end{aligned}$$

This completes the proof of necessity. Sufficiency can be seen by retracing the steps.

Corollary 2.5.4 The NS conditions for $Q(\mathbf{Z})$ of Theorem 2.5.2 to be distributed as a noncentral chi-square difference with degrees of freedom ν and noncentrality parameter λ are $(i)_d$ to $(iv)_d$ if $B'AB$ is nonsingular and $(i)_d$ to $(v)_d$ if $B'AB$ is singular, with $s = \nu$, $\beta = 2$.

Corollary 2.5.5 The NS conditions for $Q(\mathbf{Z}) = \mathbf{X}'B_2\mathbf{Y}$ to be NGL are $(i)_d$ to $(iv)_d$ if $B'AB$ is nonsingular and $(i)_d$ to $(v)_d$ if $B'AB$ is singular with

$$A = \begin{pmatrix} 0 & \frac{1}{2}B_2 \\ \frac{1}{2}B_2' & 0 \end{pmatrix}.$$

Writing the conditions $(i)_d$ to $(v)_d$ in terms of B_2, Σ_{11}, Σ_{22}, $\boldsymbol{\mu}_1$ and $\boldsymbol{\mu}_2$ would take up too much space. Hence we will list some particular cases of Corollary 2.5.5 here.

2.5d Noncorrelated Singular Normal Case

Corollary 2.5.6 The NS conditions for $Q(\mathbf{Z}) = \mathbf{X}'B_2\mathbf{Y}$ to be distributed as a NGL, when $\mathbf{X}$ and $\mathbf{Y}$ are noncorrelated, that is , when $\Sigma_{12} = O$, are the following:

$(i)_e \equiv (i)_d$;

$(ii)_e = (ii)_d$ with A as above or $(ii)_b$;

$(iii)_e \quad \mu_1{}'B_2\mu_2 = 0 = \mu_1{}'B_2\Sigma_{22}B_2'\Sigma_{11}B_2\mu_2$;

$(iv)_e \quad \lambda = \left(\frac{1}{4\beta^2}\right)\{\mu_1{}'B_2\Sigma_{22}B_2'\mu_1 + \mu_2{}'B_2'\Sigma_{11}B_2\mu_2\}$

and if $B'AB$ is singular then

$(v)_e$

$$\Sigma_{11}B_2\mu_2 = \left(\frac{1}{\beta^2}\right)(\Sigma_{11}B_2\Sigma_{22}B_2'\Sigma_{11}B_2\mu_2)$$

and

$$\Sigma_{22}B_2'\mu_1 = \left(\frac{1}{\beta^2}\right)(\Sigma_{22}B_2'\Sigma_{11}B_2\Sigma_{22}B_2'\mu_1).$$

Example 2.5.3 Let $\binom{X_1}{X_2}$ and $\binom{Y_1}{Y_2}$ be a simple random sample of size two from a bivariate normal $N(\mu_1, \Sigma_{11})$ with

$$\mu_1 = \begin{pmatrix} 0 \\ 0 \end{pmatrix}, \quad \Sigma_{11} = \begin{pmatrix} 1 & 1 \\ 1 & 1 \end{pmatrix}.$$

Let $U = X_1Y_1 + X_2Y_2 + 2X_1Y_2 + 2X_2Y_1$. Then show that U is a NGL with the parameters $(\alpha, \beta, \lambda) = (\frac{1}{2}, 3, 0)$.

SOLUTION

In the notations of Corollary 2.5.5 we have

$$B_2 = \begin{pmatrix} 1 & 2 \\ 2 & 1 \end{pmatrix}, \quad \Sigma_{11} = \Sigma_{22} = \begin{pmatrix} 1 & 1 \\ 1 & 1 \end{pmatrix}, \quad \Sigma_{12} = O,$$

$$A = \begin{pmatrix} O & \frac{1}{2}B_2 \\ \frac{1}{2}B_2' & O \end{pmatrix}, \quad \Sigma = \begin{pmatrix} \Sigma_{11} & O \\ O & \Sigma_{22} \end{pmatrix}.$$

The eigenvalues of $A\Sigma$ are $\frac{3}{2}, -\frac{3}{2}, 0, 0$. Hence $\alpha = \frac{1}{2}$, $\beta = 3$. Note that all the conditions of Corollary 2.5.5 are satisfied.

2.5e The NS Conditions for a Quadratic Form to be NGL

From Theorem 2.5.2 we can also get the NS conditions for a quadratic form to be a NGL. This will be stated here as a corollary.

Corollary 2.5.7 The NS conditions for $Q(\mathbf{Z}) = \mathbf{X}'B_1\mathbf{X}$, $B_1 = B_1'$, $\mathbf{X} \sim N_p(\mu_1, \Sigma_{11})$, $\Sigma_{11} \geq 0$ with $\Sigma_{12} = O$ to be a NGL are the following:

$(i)_f \equiv (i)_d$;

$(ii)_f = (ii)_d$ with

$$A = \begin{pmatrix} B_1 & O \\ O & O \end{pmatrix}$$

or

$$\left(\frac{4}{\beta^2}\right)(\Sigma_{11}B_1)^4\Sigma_{11} = (\Sigma_{11}B_1)^2\Sigma_{11}, \text{ and } \mathrm{tr}(B_1\Sigma_{11}) = 0;$$

$(iii)_f$ $\boldsymbol{\mu}_1{}'B_1\boldsymbol{\mu}_1 = 0 = \boldsymbol{\mu}_1{}'B_1\Sigma_{11}B_1\Sigma_{11}B_1\mu_1;$

$(iv)_f$ $\lambda = \left(\frac{1}{\beta^2}\right)(\boldsymbol{\mu}_1{}'B_1\Sigma_{11}B_1\boldsymbol{\mu}_1);$

$(v)_f$ $\Sigma_{11}B_1\boldsymbol{\mu}_1 = \left(\frac{4}{\beta^2}\right)(\Sigma_{11}B_1)^3\boldsymbol{\mu}_1.$

Our Example 2.3.1 was on a quadratic form in normal variables. Let us redo this example by using Corollary 2.5.7.

Example 2.5.4 Let $\mathbf{X} \sim N_3(\mu_1, \Sigma_{11})$ where $\boldsymbol{\mu}' = (1, -\sqrt{3}, 1)$ and

$$\Sigma_{11} = \begin{pmatrix} \frac{2}{3} & \frac{1}{\sqrt{6}} & -\frac{1}{\sqrt{18}} \\ \frac{1}{\sqrt{6}} & \frac{1}{2} & \frac{1}{\sqrt{12}} \\ -\frac{1}{\sqrt{18}} & \frac{1}{\sqrt{12}} & \frac{5}{6} \end{pmatrix}.$$

Let

$$U = \mathbf{X}'B_1\mathbf{X} = \frac{X_2^2}{2} - \frac{X_3^2}{2} + \frac{2}{\sqrt{6}}X_1X_2 + \frac{2}{\sqrt{2}}X_1X_3 + \frac{2}{\sqrt{12}}X_2X_3$$

where

$$B_1 = \begin{pmatrix} 0 & \frac{1}{\sqrt{6}} & \frac{1}{\sqrt{2}} \\ \frac{1}{\sqrt{6}} & \frac{1}{2} & \frac{1}{\sqrt{12}} \\ \frac{1}{\sqrt{2}} & \frac{1}{\sqrt{12}} & -\frac{1}{2} \end{pmatrix}.$$

Show that U is a NGL with the parameters $(\alpha, \beta, \lambda) = \left(\frac{1}{2}, 2, \frac{1}{2} - \frac{\sqrt{2}}{3}\right)$.

SOLUTION

Note that $B_1\Sigma_{11} = \Sigma_{11}B_1 = B_1$ and the eigenvalues of B_1 are $1, -1, 0$. Thus $\alpha = \frac{1}{2}$, $\beta = 2$. Also $B = PD_1P'$ where $D_1 = \mathrm{diag}(1, -1, 0)$, $PP' = I$,

$$P' = \begin{pmatrix} \frac{1}{\sqrt{3}} & \frac{1}{\sqrt{2}} & \frac{1}{\sqrt{6}} \\ \frac{1}{\sqrt{3}} & 0 & -\frac{2}{\sqrt{6}} \\ \frac{1}{\sqrt{3}} & -\frac{1}{\sqrt{2}} & \frac{1}{\sqrt{6}} \end{pmatrix}.$$

Note that

$$B_1^3 = B_1,$$

$$B_1\Sigma_{11}B_1\Sigma_{11}B_1 = (B_1\Sigma_{11})(B_1\Sigma_{11})(B_1)$$
$$= B_1^3 = B_1.$$

Also

$$\mu_1{}'B_1\mu_1 = \mu_1{}'PD_1P'\mu_1.$$

But $P'\mu_1$ has the first two elements equal to $\frac{\sqrt{2}-2}{\sqrt{6}}$ and hence

$$\mu_1{}'PD_1P'\mu_1 = 0.$$

Thus condition $(iii)_f$ is satisfied. Also

$$\left(\frac{1}{\beta^2}\right)(\mu_1{}'B_1\Sigma_{11}B_1\mu_1) = \left(\frac{1}{4}\right)(\mu_1{}'B_1^2\mu_1) = \left(\frac{1}{4}\right)\left\{2\left(\frac{\sqrt{2}-2}{\sqrt{6}}\right)^2\right\}$$
$$= \frac{1}{2} - \frac{\sqrt{2}}{3} = \lambda.$$

Note that

$$\Sigma_{11} = \Sigma_{11}' \geq 0, \ \ \Sigma_{11} = PD_2P',$$
$$D_2 = \text{diag}(1,1,0).$$

Condition $(v)_f$ is also satisfied, hence the result.

2.6 Generalizations to Bilinear and Quadratic Expressions

2.6a Bilinear and Quadratic Expressions in the Nonsingular Normal Case

Consider a general quadratic expression $Q(\mathbf{Z})$ as given in (2.5.1). We will investigate the NS conditions for $Q(\mathbf{Z})$ to be distributed as a NGL. This will generalize the results in Section 2.5. But the interesting part will be to see the types of results one can obtain when $Q(\mathbf{Z})$ is a general bilinear expression. By direct integration or from Section 2.4d we have

$$M_Q(t) = \left|I - 2t\Sigma^{\frac{1}{2}}A\Sigma^{\frac{1}{2}}\right|^{-\frac{1}{2}} \exp\left\{t(\mu' A\mu + \mathbf{a}'\mu + d)\right. \tag{2.6.1}$$
$$\left. + \left(\frac{t^2}{2}\right)(\mathbf{a}+2A\mu)'\Sigma^{\frac{1}{2}}\left(I - 2t\Sigma^{\frac{1}{2}}A\Sigma^{\frac{1}{2}}\right)^{-1}\Sigma^{\frac{1}{2}}(\mathbf{a}+2A\mu)\right\}.$$

Theorem 2.6.1 The NS conditions for $Q(\mathbf{Z}) = \mathbf{Z}'A\mathbf{Z} + \mathbf{a}'\mathbf{Z} + d$, $A = A'$, $\mathbf{Z} \sim N_{p+q}(\mu, \Sigma)$, $\Sigma > 0$ to be distributed as a NGL are the following:

$(i)_g \equiv (i)_d$;

$(ii)_g \equiv (ii)_d$;

$(iii)_g \quad \lambda = \left(\frac{1}{\beta^2}\right)\left\{\mathbf{a}'\Sigma A\boldsymbol{\mu} + \boldsymbol{\mu}'A\Sigma A\boldsymbol{\mu} + \frac{1}{4}\mathbf{a}'\Sigma\mathbf{a}\right\}$;

$(iv)_g \quad d + \boldsymbol{\mu}'A\boldsymbol{\mu} + \boldsymbol{\mu}'\mathbf{a} = 0$;

$(v)_g \quad d = \left(\frac{1}{\beta^2}\right)(\mathbf{a}'\Sigma A\Sigma\mathbf{a})$

and if the rank of A is less than $p+q$ then,

$(vi)_g \quad \mathbf{a}'\Sigma\mathbf{a} = \left(\frac{4}{\beta^2}\right)(\mathbf{a}'\Sigma A\Sigma A\Sigma\mathbf{a})$ or $\mathbf{a} = \left(\frac{4}{\beta^2}\right)(A\Sigma)^2\mathbf{a}$.

PROOF By comparing the coefficients of t on both sides of the exponents of (2.6.1) and (2.5.15) we get $(iv)_g$. Comparing the coefficients of t^{2m+1}, $m \geq 2$, and under $(i)_g$ and $(ii)_g$ we get

$$0 = \boldsymbol{\mu}'A\boldsymbol{\mu} + \left(\frac{2}{\beta}\right)^2(\boldsymbol{\mu}'A\Sigma A\Sigma\mathbf{a}) + \left(\frac{1}{\beta^2}\right)(\mathbf{a}'\Sigma A\Sigma\mathbf{a}). \tag{2.6.2}$$

Comparing the coefficients of t^2 we get $(iii)_g$. Comparing the coefficients of t^{2m}, $m \geq 2$ and under $(i)_g$ and $(ii)_g$ we get

$$\begin{aligned}\lambda\beta^2 &= \boldsymbol{\mu}'A\Sigma A\boldsymbol{\mu} + \boldsymbol{\mu}'A\Sigma\mathbf{a} \\ &\quad + \left(\frac{1}{4}\right)\mathbf{a}'\Sigma^{\frac{1}{2}}P[\text{diag}(1,\ldots,1,0,\ldots,0)]P'\Sigma^{\frac{1}{2}}\mathbf{a}\end{aligned} \tag{2.6.3}$$

where

$$\begin{aligned}PP' &= I, \\ P'(\Sigma^{\frac{1}{2}}A\Sigma^{\frac{1}{2}})P &= \text{diag}\left(\frac{\beta}{2},\ldots,\frac{\beta}{2},-\frac{\beta}{2},\ldots,-\frac{\beta}{2},0,\ldots,0\right).\end{aligned}$$

Thus if A is of full rank then (2.6.3) goes back to $(iii)_g$. If not $(iii)_g$ and (2.6.3) give condition $(vi)_g$. Proceeding as in the proof of Theorem 2.5.2 one has condition $(vi)_g$ in the form

$$\mathbf{a} = \left(\frac{4}{\beta^2}\right)(A\Sigma)^2\mathbf{a}. \tag{2.6.4}$$

Now substitute this in (2.6.2) to see that under $(iv)_g$ this goes back to condition $(v)_g$. Sufficiency is established by retracing the steps.

Remark 2.6.1 When A is of full rank $p+q$

$$\boldsymbol{\mu}'A\Sigma A\boldsymbol{\mu} = \left(\frac{\beta^2}{4}\right)(\boldsymbol{\mu}'\Sigma^{-1}\boldsymbol{\mu}).$$

This equality is obtained by letting $\mathbf{a} = \Sigma^{-1}\mu$ in $(vi)_g$.

A very useful and interesting particular case of this theorem is when $Q(\mathbf{Z})$ is a general bilinear expression of the form

$$Q(\mathbf{Z}) = \mathbf{X}'B_2\mathbf{Y} + \mathbf{b_2}'\mathbf{X} + \mathbf{c_2}'\mathbf{Y} + d_2. \tag{2.6.5}$$

Comparing with (2.6.1) we have

$$A = \begin{pmatrix} O & \frac{1}{2}B_2 \\ \frac{1}{2}B_2' & O \end{pmatrix}, \quad \mathbf{a}' = (b_2', c_2'), \quad d = d_2. \tag{2.6.6}$$

Corollary 2.6.1 The NS conditions for $Q(\mathbf{Z})$ of (2.6.2) to be a NGL are as given in Theorem 2.6.1 with A, $\mathbf{a}$ and d given in (2.6.6).

An explicit listing of the conditions $(i)_g$ to $(vi)_g$, in terms of B_2, $\mathbf{b_2}$, $\mathbf{c_2}$, d_2 and the submatrices of Σ, would take up too much space. Hence we will only consider some particular cases here.

Corollary 2.6.2 When $\mathbf{X}$ and $\mathbf{Y}$ are noncorrelated the NS conditions of $Q(\mathbf{Z})$ of (2.6.5) to be a NGL are the following:

$(i)_h \equiv (i)_g$;

$(ii)_h \equiv (ii)_d$;

$(iii)_h$

$$\lambda = \left(\frac{1}{4\beta^2}\right)\{2(\mathbf{b_2}'\Sigma_{11}B_2\boldsymbol{\mu}_2 + \mathbf{c_2}'\Sigma_{22}B_2'\boldsymbol{\mu}_1) + \boldsymbol{\mu}_1'B_2\Sigma_{22}B_2'\boldsymbol{\mu}_1$$
$$+ \boldsymbol{\mu}_2'B_2'\Sigma_{11}B_2\boldsymbol{\mu}_2 + \mathbf{b_2}'\Sigma_{11}\mathbf{b_2} + \mathbf{c_2}'\Sigma_{22}\mathbf{c_2}\};$$

$(iv)_h$ $0 = d_2 + \boldsymbol{\mu}_1'B_2\boldsymbol{\mu}_2 + \boldsymbol{\mu}_1'\mathbf{b_2} + \boldsymbol{\mu}_2'\mathbf{c_2}$;

$(v)_h$ $d_2 = \left(\frac{1}{\beta^2}\right)(\mathbf{b_2}'\Sigma_{11}B_2\Sigma_{22}\mathbf{c_2})$

and when the rank of $\begin{pmatrix} O & \frac{1}{2}B_2 \\ \frac{1}{2}B_2' & O \end{pmatrix}$ is less than $p+q$ then

$(vi)_h$

$$\mathbf{b_2} = \left(\frac{1}{\beta^2}\right)(B_2\Sigma_{22}B_2'\Sigma_{11}\mathbf{b_2})$$

and

$$\mathbf{c_2} = \left(\frac{1}{\beta^2}\right)(B_2'\Sigma_{11}B_2\Sigma_{22}\mathbf{c_2}).$$

Remark 2.6.2 Note that the special cases $\mathbf{b_2} = \mathbf{0}$; $\mathbf{c_2} = \mathbf{0}$; $\mathbf{b_2} = \mathbf{0}$ and $\mathbf{c_2} = \mathbf{0}$ are of interest on their own but these will not be listed here in order to save space. The case $B_2 = O$ will not hold condition $(ii)_h$ and hence purely linear forms in $\mathbf{X}$ or $\mathbf{Y}$ or $\mathbf{X}$ and $\mathbf{Y}$ cannot be Laplacian which is also evident because they should lead to normal variables.

Corollary 2.6.3 Let

$$\begin{pmatrix}\mathbf{X}\\ \mathbf{Y}\end{pmatrix} \sim N_{p+q}(\mu, \Sigma), \quad \mathbf{X} \sim N_p(\boldsymbol{\mu}_1, \Sigma_{11}),$$

$$\Sigma = \begin{pmatrix}\Sigma_{11} & \Sigma_{12}\\ \Sigma_{21} & \Sigma_{22}\end{pmatrix}, \quad \Sigma_{11} > 0$$

then the NS conditions for

$$Q(\mathbf{Z}) = \mathbf{X}'B_1\mathbf{X} + \mathbf{b_1}'\mathbf{X} + \mathbf{c_1}'\mathbf{Y} + d_1, \quad B_1 = B_1'$$

to be NGL are the following:

$(i)_j \equiv (i)_g$;

$(ii)_j \equiv (ii)_d$, that is,

$$B_1\Sigma_{11}B_1 = \left(\frac{4}{\beta^2}\right)(B_1\Sigma_{11}B_1\Sigma_{11}B_1\Sigma_{11}B_1)$$

and

$$\mathrm{tr}(B_1\Sigma_{11}) = 0;$$

$(iii)_j$

$$\lambda = \left(\frac{1}{\beta^2}\right)\Big[\mathbf{b_1}'\Sigma_{11}B_1\boldsymbol{\mu}_1 + \mathbf{c_1}'\Sigma_{21}B_1\boldsymbol{\mu}_1 + \boldsymbol{\mu}_1'B_1\Sigma_{11}B_1\boldsymbol{\mu}_1 + \left(\frac{1}{4}\right)(\mathbf{b_1}'\Sigma_{11}\mathbf{b_1} + 2\mathbf{b_1}'\Sigma_{12}\mathbf{c_1} + \mathbf{c_1}'\Sigma_{22}\mathbf{c_1})\Big];$$

$(iv)_j$ $d_1 + \boldsymbol{\mu}_1'B_1\boldsymbol{\mu}_1 + \boldsymbol{\mu}_1'\mathbf{b_1} + \boldsymbol{\mu}_2'\mathbf{c_1} = 0$;

$(v)_j$ $d_1 = \left(\frac{1}{\beta^2}\right)[\mathbf{b_1}'\Sigma_{11}B_1\Sigma_{11}\mathbf{b_1} + \mathbf{c_1}'\Sigma_{21}B_1\Sigma_{12}\mathbf{c_1} + 2\mathbf{b_1}'\Sigma_{11}B_1\Sigma_{12}\mathbf{c_1})]$;

$(vi)_j$ $\mathbf{b_1} = \left(\frac{4}{\beta^2}\right)[(B_1\Sigma_{11})^2\mathbf{b_1} + B_1\Sigma_{11}B_1\Sigma_{12}\mathbf{c_1}]$ and $\mathbf{c_1} = \mathbf{0}$.

Since $\mathbf{c_1} = \mathbf{0}$ the terms containing $\mathbf{c_1}$ can be deleted from all the above conditions.

Remark 2.6.3 From $(vi)_j$ we see that $\mathbf{c_1} = \mathbf{0}$. This will also wipe out the effect of Σ_{12} from all the conditions. Thus the conditions $(i)_j$ to $(vi)_j$ with $\mathbf{c_1} = \mathbf{0}$ remain the same for the following cases also (a) **X** and **Y** noncorrelated, that is, $\Sigma_{12} = O$; (b) **X** and **Y** equicorrelated , that is, $\Sigma_{11} = I$, $\Sigma_{22} = I$, $\Sigma_{12} = \Sigma_{21} = \rho I$, $p = q$, $-1 < \rho < 1$, where I is the identity matrix.

2.6b Bilinear and Quadratic Expressions in the Singular Normal Case

By following through the same steps as in Theorem 2.6.1 one can get the corresponding conditions for the Laplacianness of a quadratic expression in the singular normal case. This will be stated as a theorem.

Theorem 2.6.2 Let $Q(\mathbf{Z}) = \mathbf{Z}'A\mathbf{Z} + \mathbf{a}'\mathbf{Z} + d$, $A = A'$ with $\mathbf{Z} \sim N_{p+q}(\boldsymbol{\mu}, \Sigma)$, $\Sigma \geq 0$ where d is a scalar constant, $\mathbf{a}$ is a $(p+q) \times 1$ vector of constants and A is a matrix of constants. Then the NS conditions for $Q(\mathbf{Z})$ to be a NGL are the following:

$(i)_k \equiv (i)_d$;

$(ii)_k \equiv (ii)_d$ or $\Sigma(A\Sigma)^2 = \left(\frac{4}{\beta^2}\right)\Sigma(A\Sigma)^4$ and $\mathrm{tr}(A\Sigma) = 0$;

$(iii)_k$

$$\boldsymbol{\mu}'A\boldsymbol{\mu} + \mathbf{a}'\boldsymbol{\mu} + d = 0$$

and

$$(\Sigma\mathbf{a} + 2\Sigma A\boldsymbol{\mu})'A(\Sigma\mathbf{a} + 2\Sigma A\boldsymbol{\mu}) = 0;$$

$(iv)_k$ $\lambda = \left(\frac{1}{\beta^2}\right)\left[\boldsymbol{\mu}'A\Sigma A\boldsymbol{\mu} + \mathbf{a}'\Sigma A\mu + \left(\frac{1}{4}\right)(\mathbf{a}'\Sigma\mathbf{a})\right]$;

and if the rank of $B'AB$ is less than r, where $\Sigma = BB'$ of rank r and B is $(p+q) \times r$ of rank r, then

$(v)_k$ $\Sigma(\mathbf{a} + 2A\boldsymbol{\mu}) = \left(\frac{4}{\beta^2}\right)(\Sigma A)^2\Sigma(\mathbf{a} + 2A\boldsymbol{\mu})$.

As particular cases of the above theorem we can look at the NS conditions for a general bilinear expression in singular normal variables to be a NGL. We can also look at the noncorrelated case under the singular normal. These will be listed here as corollaries.

Corollary 2.6.4 Let

$$\mathbf{Z} = \begin{pmatrix} \mathbf{X} \\ \mathbf{Y} \end{pmatrix} \sim N_{p+q}(\boldsymbol{\mu}, \Sigma), \quad \Sigma \geq 0$$

then the NS conditions for the bilinear expression

$$Q(\mathbf{Z}) = \mathbf{X}'B_2\mathbf{Y} + \mathbf{b_2}'\mathbf{X} + \mathbf{c_2}'\mathbf{Y} + d_2$$

to be a NGL are $(i)_k$ to $(v)_k$ for

$$A = \begin{pmatrix} O & \frac{1}{2}B_2 \\ \frac{1}{2}B_2' & O \end{pmatrix}, \quad \mathbf{a}' = (\mathbf{b_2}', \mathbf{c_2}'), \quad d = d_2.$$

The explicit forms of the conditions in terms of B_2, $\mathbf{b_2}$, $\mathbf{c_2}$, d_2 and the submatrices of

Σ would take up too much space and hence they will not be listed here. Some particular cases are the following.

Corollary 2.6.5 Let $\mathbf{X}$ and $\mathbf{Y}$ be noncorrelated. Then the NS conditions for the bilinear expression $Q(\mathbf{Z})$ of Corollary 2.6.4 to be a NGL are the following:

$(i)_\ell \equiv (i)_k$;

$(ii)_\ell$

$$\Sigma_{11}B_2\Sigma_{22}B_2'\Sigma_{11} = \left(\frac{1}{\beta^2}\right)(\Sigma_{11}B_2\Sigma_{22}B_2'\Sigma_{11}B_2\Sigma_{22}B_2'\Sigma_{11}),$$
$$\Sigma_{22}B_2'\Sigma_{11}B_2\Sigma_{22} = \left(\frac{1}{\beta^2}\right)(\Sigma_{22}B_2'\Sigma_{11}B_2\Sigma_{22}B_2'\Sigma_{11}B_2\Sigma_{22})$$

and

$$\mathrm{tr}(B_2\Sigma_{21}) = 0;$$

$(iii)_\ell$

$$\boldsymbol{\mu}_1{}'B_2\boldsymbol{\mu}_2 + \mathbf{b_2}'\boldsymbol{\mu}_1 + \mathbf{c_2}'\boldsymbol{\mu}_2 + d_2 = 0$$
$$(\mathbf{b_2} + B_2\boldsymbol{\mu}_2)'\Sigma_{11}B_2\Sigma_{22}(\mathbf{c_2} + B_2'\boldsymbol{\mu}_1) = 0;$$

$(iv)_\ell$

$$\lambda = \left(\frac{1}{\beta^2}\right)\left[\left(\frac{1}{4}\right)(\boldsymbol{\mu}_1{}'B_2\Sigma_{22}B_2'\boldsymbol{\mu}_1 + \boldsymbol{\mu}_2{}'B_2'\Sigma_{11}B_2\boldsymbol{\mu}_2)\right.$$
$$+\left(\frac{1}{2}\right)(\mathbf{b_2}'\Sigma_{11}B_2\boldsymbol{\mu}_2 + \mathbf{c_2}'\Sigma_{22}B_2'\boldsymbol{\mu}_1)$$
$$\left.+\left(\frac{1}{4}\right)(\mathbf{b_2}'\Sigma_{11}\mathbf{b_2} + \mathbf{c_2}'\Sigma_{22}\mathbf{c_2})\right]$$

and if the rank of $\begin{pmatrix} O & \Sigma_{11}B_2 \\ \Sigma_{22}B_2' & O \end{pmatrix}$ is less than that of $\Sigma = \begin{pmatrix} \Sigma_{11} & O \\ O & \Sigma_{22} \end{pmatrix}$ then

$(v)_\ell$

$$\Sigma_{11}(\mathbf{b_2} + B_2\boldsymbol{\mu}_2) = \left(\frac{1}{\beta^2}\right)\Sigma_{11}B_2\Sigma_{22}B_2'\Sigma_{11}(\mathbf{b_2} + B_2\boldsymbol{\mu}_2)$$

and

$$\Sigma_{22}(\mathbf{c_2} + B_2'\boldsymbol{\mu}_1) = \left(\frac{1}{\beta^2}\right)\Sigma_{22}B_2'\Sigma_{11}B_2\Sigma_{22}(\mathbf{c_2} + B_2'\boldsymbol{\mu}_1).$$

Corollary 2.6.6 When $\mathbf{X}$ and $\mathbf{Y}$ are noncorrelated the NS conditions for a quadratic expression

$$Q(\mathbf{Z}) = \mathbf{X}'B_1\mathbf{X} + \mathbf{b_1}'\mathbf{X} + \mathbf{c_1}'\mathbf{Y} + d_1, \;\; B_1 = B_1'$$

to be a NGL are the following:

$(i)_m \equiv (i)_k$;

$(ii)_m$ $\Sigma_{11}(B_1\Sigma_{11})^2 = \left(\frac{4}{\beta^2}\right)\Sigma_{11}(B_1\Sigma_{11})^4$ and $\mathrm{tr}(B_1\Sigma_{11}) = 0$;

$(iii)_m$

$$\boldsymbol{\mu}_1{}'B_1\boldsymbol{\mu}_1 + \mathbf{b}_1{}'\boldsymbol{\mu}_1 + \mathbf{c}_1{}'\boldsymbol{\mu}_2 + d_1 = 0$$
$$(\mathbf{b}_1 + 2B_1\boldsymbol{\mu}_1)'\Sigma_{11}B_1\Sigma_{11}(\mathbf{b}_1 + 2B_1\boldsymbol{\mu}_1) = 0;$$

$(iv)_m$

$$\lambda = \left(\frac{1}{\beta^2}\right)\left[\boldsymbol{\mu}_1{}'B_1\Sigma_{11}B_1\boldsymbol{\mu}_1 + \mathbf{b}_1{}'\Sigma_{11}B_1\boldsymbol{\mu}_1 + \left(\frac{1}{4}\right)(\mathbf{b}_1{}'\Sigma_{11}\mathbf{b}_1 + \mathbf{c}_1{}'\Sigma_{22}\mathbf{c}_1)\right];$$

$(v)_m$

$$\Sigma_{11}(\mathbf{b}_1 + 2B_1\boldsymbol{\mu}_1) = \left(\frac{4}{\beta^2}\right)(\Sigma_{11}B_1)^2\Sigma_{11}(\mathbf{b}_1 + 2B_1\boldsymbol{\mu}_1)$$

and

$$\Sigma_{22}\mathbf{c}_1 = \mathbf{0}.$$

Remark 2.6.4 If $|\Sigma_{22}| \neq 0$, that is if all the singularity is in Σ_{11}, then condition $(v)_m$ will give $\mathbf{c}_1 = \mathbf{0}$. Then in $(iii)_m$ to $(v)_m$ the terms containing $\mathbf{c}_1$ are to be deleted and the condition $\mathbf{c}_1 = \mathbf{0}$ is to be added.

2.7 Independence of Bilinear and Quadratic Expressions

We will start with bilinear forms and quadratic forms, then two bilinear forms and then generalize to bilinear and quadratic expressions. For simplicity we will start with the nonsingular normal case first.

2.7a Independence of a Bilinear and a Quadratic Form

First we will consider a bilinear form and a quadratic form and then two bilinear forms in normal variables and check for the NS conditions for these to be independently distributed. A result which we will frequently use in the proofs of some of the results later on is on a property of the trace of a product of two matrices. This will be stated here as a lemma.

Lemma 2.7.1 For two arbitrary $p \times p$ matrices A and B

$$\mathrm{tr}((AB)^2) + 2\mathrm{tr}((AB)(B'A')) = 0 \Rightarrow AB = O. \tag{2.7.1}$$

This result was established by Kawada (1950). The proof follows by observing that the left side of (2.7.1) can be written as a sum of positive definite quadratic forms of the type

$$\sum_{ij}(c_{ij}, c_{ji})\begin{pmatrix} 1 & 1/2 \\ 1/2 & 1 \end{pmatrix}\begin{pmatrix} c_{ij} \\ c_{ji} \end{pmatrix}$$

which can be zero only if $c_{ij} = 0$ for all i and j, where c_{ij} is the (i,j)-th element of AB. We will frequently make use of this result as well as the fact that for any two matrices A and B

$$\mathrm{tr}(AB) = \mathrm{tr}(BA)$$

when AB and BA are defined, and when A is a square matrix $\mathrm{tr}(A) = \mathrm{tr}(A')$, where A' is the transpose of A.

Theorem 2.7.1 Let $\binom{\mathbf{X}}{\mathbf{Y}} \sim N_{p+q}(\mathbf{0}, \Sigma)$, $\Sigma > 0$. Let $Q_1 = \mathbf{X}'A_1\mathbf{Y}$ and $Q_2 = \mathbf{Y}'A_2\mathbf{Y}$, $A_2 = A_2'$ and A_1 be $p \times q$. Let $\mathrm{Cov}(\mathbf{X}) = \Sigma_{11}$, $\mathrm{Cov}(\mathbf{Y}) = \Sigma_{22}$, $\mathrm{Cov}(\mathbf{X}, \mathbf{Y}) = \Sigma_{12} = \Sigma_{21}'$. Then the NS conditions for Q_1 and Q_2 to be independently distributed are $A_1\Sigma_{22}A_2 = O$ and $A_1'\Sigma_{12}A_2 = O$.

PROOF Let Q_1 and Q_2 be independently distributed. Then

$$M_{Q_1,0}(t_1, 0)M_{0,Q_2}(0, t_2) = M_{Q_1,Q_2}(t_1, t_2) \tag{2.7.2}$$

where $M_{Q_1,Q_2}(t_1, t_2)$ denotes the joint m.g.f. of Q_1 and Q_2. Take logarithms on both sides of (2.7.2), expand and then equate the coefficient of $t_1^2 t_2^2$ on both sides to get the following.

$$2\mathrm{tr}(E_2E_3E_2) + 2\mathrm{tr}(E_1^2E_2^2) + \mathrm{tr}((E_1E_2)^2) = 0 \tag{2.7.3}$$

where E_1, E_2, E_3 are given in (2.4.2). Observe that E_1, E_2, E_3 are symmetric matrices and further E_3 can be written as $B'B$ where $B = (\Sigma_{11} - \Sigma_{12}\Sigma_{22}^{-1}\Sigma_{21})^{\frac{1}{2}}A_1\Sigma_{22}^{\frac{1}{2}}$. Hence (2.7.3) reduces to

$$\left[2\mathrm{tr}\left((E_2'B')(BE_2)\right)\right] + \left[\mathrm{tr}\left((E_1E_2)^2\right) + 2\mathrm{tr}\left((E_1E_2)(E_1E_2)'\right)\right] = 0. \tag{2.7.4}$$

But the quantities in each bracket is nonnegative and hence each is zero. That is, $E_1E_2 = O$ and $BE_2 = O$. But $BE_2 = O \Rightarrow A_1\Sigma_{22}A_2 = O$ since $\Sigma_{11} - \Sigma_{12}\Sigma_{22}^{-1}\Sigma_{21} > 0$, $\Sigma_{22} > 0$. Then

$$\begin{aligned} E_1E_2 &= 2\left(\Sigma_{22}^{-\frac{1}{2}}\Sigma_{21}A_1\Sigma_{22}A_2\Sigma_{22}^{\frac{1}{2}} + \Sigma_{22}^{\frac{1}{2}}A_1'\Sigma_{12}A_2\Sigma_{22}^{\frac{1}{2}}\right) \\ &= 2\left(\Sigma_{22}^{\frac{1}{2}}A_1'\Sigma_{12}A_2\Sigma_{22}^{\frac{1}{2}}\right). \end{aligned}$$

Thus $E_1E_2 = O \Rightarrow A_1'\Sigma_{12}A_2 = O$. This establishes the necessity. Now check

$M_{Q_1,0}(t_1,0)\, M_{0,Q_2}(0,t_2)$ and $M_{Q_1,Q_2}(t_1,t_2)$ seperately under the conditions $A_1\Sigma_{22}A_2 = O$ and $A_1'\Sigma_{12}A_2 = O$ to see that they are equal. This completes the proof.

Corollary 2.7.1 The NS condition for the independence of Q_1 and Q_2 is (a): $A_1\Sigma_{22}A_2 = O$ when $\mathbf{X}$ and $\mathbf{Y}$ are independently distributed; and (b): $A_1A_2 = O$ when $\mathbf{X}$ and $\mathbf{Y}$ are equicorrelated in the sense described in Definition 2.4.1.

Remark 2.7.1 In the proof of Theorem 2.7.1 we have made use of the explicit expressions for the cumulants. By writing bilinear forms as quadratic forms one can also make use of the results on quadratic forms for establishing this theorem. This will be done later. In this case write

$$Q_1 = \mathbf{X}'A_1\mathbf{Y} = \left(\frac{1}{2}\right)(\mathbf{X}',\mathbf{Y}')\begin{pmatrix} O & A_1 \\ A_1' & O \end{pmatrix}\begin{pmatrix}\mathbf{X}\\ \mathbf{Y}\end{pmatrix}$$

and

$$Q_2 = \mathbf{Y}'A_2\mathbf{Y} = (\mathbf{X}',\mathbf{Y}')\begin{pmatrix} O & O \\ O & A_2 \end{pmatrix}\begin{pmatrix}\mathbf{X}\\ \mathbf{Y}\end{pmatrix}.$$

2.7b Independence of Two Bilinear Forms

Theorem 2.7.2 Let $\binom{\mathbf{X}}{\mathbf{Y}} \sim N_{p+q}(\mathbf{0},\ \Sigma)$, $\Sigma > 0$, $\mathrm{Cov}(\mathbf{X}) = \Sigma_{11}$, $\mathrm{Cov}(\mathbf{Y}) = \Sigma_{22}$, $\mathrm{Cov}(\mathbf{X},\mathbf{Y}) = \Sigma_{12} = \Sigma_{21}'$. Consider $Q_i = \mathbf{X}'A_i\mathbf{Y}$, where A_i is a $p\times q$ real matrix of constants, $i = 1, 2$. Then the necessary and sufficient conditions for the independence of Q_1 and Q_2 are

$$A_1\Sigma_{22}A_2' = O,\quad A_1'\Sigma_{11}A_2 = O,\quad A_2\Sigma_{21}A_1 = O,\quad A_1\Sigma_{21}A_2 = O.$$

PROOF *Necessity* Let $M_{Q_1,Q_2}(t_1,t_2)$ denote the joint m.g.f. of Q_1 and Q_2. This is available from Section 2.4b. If Q_1 and Q_2 are independently distributed then

$$M_{Q_1,0}(t_1,0)M_{0,Q_2}(0,t_2) = M_{Q_1,Q_2}(t_1,t_2). \tag{2.7.5}$$

Take logarithms on both sides of (2.7.5), expand and compare the coefficient of $t_1^2t_2^2$ on both sides to get $K_{2,2} = 0$ where $K_{2,2}$ is given in Section 2.4b.1. Since $K_{2,2} = 0$ for all Σ_{12} it should hold for $\Sigma_{12} = 0$ also. Then from Section 2.4b.1 we have

$$\mathrm{tr}\,(F_5^2) + 2\mathrm{tr}(F_3F_4) = 0. \tag{2.7.6}$$

But note that

$$\mathrm{tr}\,(F_5^2) = \mathrm{tr}\left[(F_5)(F_5)'\right] \geq 0.$$
$$\mathrm{tr}(F_3F_4) = \mathrm{tr}\left[\left(D^{\frac{1}{2}}A_1\Sigma_{22}A_2'D^{\frac{1}{2}}\right)\left(D^{\frac{1}{2}}A_1\Sigma_{22}A_2'D^{\frac{1}{2}}\right)'\right] \geq 0.$$

Hence

$$\left(D^{\frac{1}{2}}A_1\Sigma_{22}A_2'D^{\frac{1}{2}}\right) = O \Rightarrow A_1\Sigma_{22}A_2' = O.$$

Then from symmetry one has $A_1'\Sigma_{11}A_2 = O$. Now impose these conditions on $K_{2,2}$ of Section 2.4b.1 and simplify. After some algebra $K_{2,2}$ can be written as follows:

$$\begin{aligned} K_{2,2} &= 2\mathrm{tr}\left(U_1'\left(\Sigma_{22}^{\frac{1}{2}}A_2'\Sigma_{11}A_2\Sigma_{22}^{\frac{1}{2}}\right)U_1\right) \\ &+ 2\mathrm{tr}\left(U_2'\left(\Sigma_{22}^{\frac{1}{2}}A_1'\Sigma_{11}A_1\Sigma_{22}^{\frac{1}{2}}\right)U_2\right) \\ &+ 2\left[\mathrm{tr}\left((U_1U_2)^2\right) + 2\mathrm{tr}\left((U_1U_2)(U_2U_1)\right)\right] \end{aligned} \tag{2.7.7}$$

where

$$U_1 = \Sigma_{22}^{-\frac{1}{2}}\Sigma_{21}A_1\Sigma_{22}^{\frac{1}{2}}, \quad U_2 = \Sigma_{22}^{-\frac{1}{2}}\Sigma_{21}A_2\Sigma_{22}^{\frac{1}{2}}.$$

Note that $U_1U_2' = O$. Also independence of Q_1 and Q_2 implies that $K_{2,2} = 0$ for all $A_1, A_2, \Sigma_{11}, \Sigma_{22}$. Put $U_2 = O$, that is, select an A_2 such that $\Sigma_{21}A_2 = O$. Then

$$\begin{aligned} K_{2,2} = 0 &\Rightarrow \Sigma_{11}^{\frac{1}{2}}A_2\Sigma_{22}^{\frac{1}{2}}U_1 = O \\ &\Rightarrow A_2\Sigma_{21}A_1 = O. \end{aligned}$$

Similarly $A_1\Sigma_{21}A_2 = O$. Under the conditions $A_2\Sigma_{21}A_1 = O$ and $A_1\Sigma_{21}A_2 = O$, note that $U_1U_2 = O$, $U_2U_1 = O$ and $K_{2,2} = 0$. Hence, the conditions $A_2\Sigma_{21}A_1 = O$ and $A_1\Sigma_{21}A_2 = O$ are necessary.

To see the sufficiency take $M_{Q_1,0}(t_1,0)M_{0,Q_2}(0,t_2)$ and $M_{Q_1,Q_2}(t_1,t_2)$ separately and put the conditions $A_1\Sigma_{22}A_2' = O$, $A_1'\Sigma_{11}A_2 = O$, $A_2\Sigma_{21}A_1 = O$, and $A_1\Sigma_{21}A_2 = O$. Under these conditions the F_5 of $M_{Q_1,Q_2}(t_1,t_2)$ reduces to

$$\begin{aligned} F_5 &= -\Sigma_{22}^{\frac{1}{2}}A_1'\Sigma_{12}\Sigma_{22}^{-1}\Sigma_{21}A_2\Sigma_{22}^{\frac{1}{2}} \\ &- \Sigma_{22}^{\frac{1}{2}}A_2'\Sigma_{12}\Sigma_{22}^{-1}\Sigma_{21}A_1\Sigma_{22}^{\frac{1}{2}}. \end{aligned} \tag{2.7.8}$$

$$\begin{aligned} M_{Q_1,0}(t_1,0)M_{0,Q_2}(0,t_2) = \Big| &I - t_1F_1 - t_2F_2 - t_1^2F_3 - t_2^2F_4 \\ &+ t_1^2t_2F_3F_2 + t_1t_2F_1F_2 \\ &+ t_1t_2^2F_1F_4 + t_1^2t_2^2F_3F_4\Big|^{-\frac{1}{2}}. \end{aligned}$$

But under the necessary conditions $F_1F_4 = O$, $F_3F_2 = O$, $F_3F_4 = O$ and $F_1F_2 = -F_5$ of (2.7.8) which establishes sufficiency.

Corollary 2.7.2 When **X** and **Y** are independently distributed then Q_1 and Q_2 are independently distributed iff $A_1\Sigma_{22}A_2' = O$ and $A_1'\Sigma_{11}A_2 = O$.

Corollary 2.7.3 When **X** and **Y** are equicorrelated in the sense $q = p$, $\Sigma_{11} = \sigma_1^2I$, $\Sigma_{22} = \sigma_2^2I$, $\Sigma_{21} = \rho I = \Sigma_{12}$ then Q_1 and Q_2 are independently distributed iff $A_1A_2' = O$, $A_1'A_2 = O$, $A_1A_2 = O$, $A_2A_1 = O$.

The above procedure of working out the NS conditions directly from the joint m.g.f. of two bilinear forms is quite involved. We shall derive the above result as well as many other results by using the independence of two quadratic expressions in normal variables, singular or nonsingular. The set of NS conditions for two quadratic expressions in singular or nonsingular normal variables to be independently distributed are given in Mathai and Provost (1992). These will be restated here and then as corollaries we will derive the NS conditions for the independence of two bilinear expressions, one bilinear and one quadratic expressions and so on.

2.7c Independence of Quadratic Expressions: Nonsingular Normal Case

Theorem 2.7.3 Let $Z \sim N_{p+q}(\boldsymbol{\mu}, \Sigma), \Sigma > 0$ and

$$Q_i(\mathbf{Z}) = \mathbf{Z}'A_i\mathbf{Z} + \mathbf{a}_i{}'\mathbf{Z} + d_i,\ A_i = A_i',\ i = 1,2.$$

The NS conditions for the independence of Q_1 and Q_2 are the following:

$(i)_n$ $A_1\Sigma A_2 = O = A_2\Sigma A_1$;

$(ii)_n$ $A_2\Sigma\mathbf{a}_1 = A_1\Sigma\mathbf{a}_2 = \mathbf{0}$;

$(iii)_n$ $(\mathbf{a}_1 + 2A_1\boldsymbol{\mu})'\Sigma(\mathbf{a}_2 + 2A_2\boldsymbol{\mu}) = 0.$

If Q_1 and Q_2 are both bilinear expressions then we can write

$$Q_1(\mathbf{Z}) = \mathbf{X}'B_1\mathbf{Y} + \mathbf{b}_1{}'\mathbf{X} + \mathbf{c}_1{}'\mathbf{Y} + d_1$$

and

$$Q_2(\mathbf{Z}) = \mathbf{X}'B_2\mathbf{Y} + \mathbf{b}_2{}'\mathbf{X} + \mathbf{c}_2{}'\mathbf{Y} + d_2,\quad \mathbf{Z}' = (\mathbf{X}', \mathbf{Y}'). \tag{2.7.9}$$

Corollary 2.7.4 The NS conditions for the independence of the bilinear expressions Q_1 and Q_2 of (2.7.9) when $\mathbf{Z}$ has a nonsingular normal distribution are the following:

$(i)_p$

$$B_1\Sigma_{22}B_2' = O,\quad B_1\Sigma_{21}B_2 = O,$$
$$B_1'\Sigma_{11}B_2 = O,\quad B_2\Sigma_{21}B_1 = O;$$

$(ii)_p$

$$B_2\Sigma_{21}\mathbf{b}_1 + B_2\Sigma_{22}\mathbf{c}_1 = \mathbf{0},\ B_2'\Sigma_{11}\mathbf{b}_1 + B_2'\Sigma_{12}\mathbf{c}_1 = \mathbf{0},$$
$$B_1\Sigma_{21}\mathbf{b}_2 + B_1\Sigma_{22}\mathbf{c}_2 = \mathbf{0},\ B_1'\Sigma_{11}\mathbf{b}_2 + B_1'\Sigma_{12}\mathbf{c}_2 = \mathbf{0};$$

$(iii)_p$

$$\begin{aligned}(\mathbf{b}_1 + B_1\boldsymbol{\mu}_2)'\Sigma_{11}(\mathbf{b}_2 + B_2\boldsymbol{\mu}_2) + (\mathbf{b}_1 + B_1\boldsymbol{\mu}_2)'\Sigma_{12}(\mathbf{c}_2 + B_2'\boldsymbol{\mu}_1)&\\ + (\mathbf{c}_1 + B_1'\boldsymbol{\mu}_1)'\Sigma_{22}(\mathbf{c}_2 + B_2'\boldsymbol{\mu}_1)&\\ + (\mathbf{c}_1 + B_1'\boldsymbol{\mu}_1)'\Sigma_{21}(\mathbf{b}_2 + B_2\boldsymbol{\mu}_2) &= 0.\end{aligned}$$

These are obtained by making the substitutions

$$A_i = \begin{pmatrix} O & \frac{1}{2}B_i \\ \frac{1}{2}B_i' & O \end{pmatrix}, \quad \mathbf{a}_i = \begin{pmatrix} \mathbf{b}_i \\ \mathbf{c}_i \end{pmatrix}, \quad i = 1, 2.$$

Special cases of interest are the conditions for the independence of bilinear and linear forms. Note that the conditions for the independence of linear and linear forms in $\mathbf{X}$ or $\mathbf{Y}$ are available in the literature. Some special cases will be listed here.

Corollary 2.7.5 The NS conditions for the independence of the expressions in the following cases, under the nonsingular normal case, are listed below.

Case(1) : $Q_1 = \mathbf{X}'B_1\mathbf{X} + \mathbf{b_1}'\mathbf{X} + \mathbf{c_1}'\mathbf{Y} + d_1, \quad Q_2 = \mathbf{b_2}'\mathbf{X} + \mathbf{c_2}'\mathbf{Y} + d_2;$
Case(2) : Same Q_1, $\quad Q_2 = \mathbf{b_2}'\mathbf{X} + d_2;$
Case(3) : Same Q_1, $\quad Q_2 = \mathbf{c_2}'\mathbf{Y} + d_2;$
Case(4) : $Q_1 = \mathbf{b_1}'\mathbf{X} + \mathbf{c_1}'\mathbf{Y} + d_1, \quad Q_2 = \mathbf{b_2}'\mathbf{X} + \mathbf{c_2}'\mathbf{Y} + d_2;$
Case(5) : $Q_1 = \mathbf{b_1}'\mathbf{X} + d_1, \quad Q_2 = \mathbf{c_2}'\mathbf{Y} + d_2.$

Case(1) The NS conditions are the following:

$(ii)_q$

$$B_1\Sigma_{21}\mathbf{b_2} + B_1\Sigma_{22}\mathbf{c_2} = \mathbf{0}, \quad B_1'\Sigma_{11}\mathbf{b_2} + B_1'\Sigma_{12}\mathbf{c_2} = \mathbf{0};$$

$(iii)_q$

$$(\mathbf{b_1} + B_1\boldsymbol{\mu_2})'\Sigma_{11}\mathbf{b_2} + (\mathbf{b_1} + B_1\boldsymbol{\mu_2})'\Sigma_{12}\mathbf{c_2}$$
$$+(\mathbf{c_1} + B_1'\boldsymbol{\mu_1})'\Sigma_{22}\mathbf{c_2} + (\mathbf{c_1} + B_1'\boldsymbol{\mu_1})'\Sigma_{21}\mathbf{b_2} = 0.$$

Case(2) The NS conditions are the following:

$(ii)_q$ $\quad B_1\Sigma_{21}\mathbf{b_2} = \mathbf{0}, \quad B_1'\Sigma_{11}\mathbf{b_2} = \mathbf{0};$

$(iii)_q$ $\quad (\mathbf{b_1} + B_1\boldsymbol{\mu_2})'\Sigma_{11}\mathbf{b_2} + (\mathbf{c_1} + B_1'\boldsymbol{\mu_1})'\Sigma_{21}\mathbf{b_2} = 0.$

Case(3) The NS conditions are the following:

$(ii)_q$ $\quad B_1\Sigma_{22}\mathbf{c_2} = \mathbf{0}, \quad B_1'\Sigma_{12}\mathbf{c_2} = \mathbf{0};$

$(iii)_q$ $\quad (\mathbf{b_1} + B_1\boldsymbol{\mu_2})'\Sigma_{12}\mathbf{c_2} + (\mathbf{c_1} + B_1'\boldsymbol{\mu_1})'\Sigma_{22}\mathbf{c_2} = 0.$

Case(4) The NS conditions are the following:

$(ii)_q$ Nil

$(iii)_q$ $\quad \mathbf{b_1}'\Sigma_{11}\mathbf{b_2} + \mathbf{b_1}'\Sigma_{12}\mathbf{c_2} + \mathbf{c_1}'\Sigma_{22}\mathbf{c_2} + \mathbf{c_1}'\Sigma_{21}\mathbf{b_2} = 0.$

Case(5) The NS conditions are the following:

$(iii)_q$ $\quad \mathbf{b_1}'\Sigma_{12}\mathbf{c_2} = 0.$

Remark 2.7.2 Put $\Sigma_{12} = O$ to get the corresponding results for the noncorrelated case and put $\Sigma_{11} = I = \Sigma_{22}$, $\Sigma_{12} = \Sigma_{21} = \rho I$, $p = q$, $B_i = B_i'$, $i = 1, 2$ to get the results for the equicorrelated case. Note also that other special cases, which are not listed above, are available by the corresponding substitutions in Corollary 2.7.5.

Corollary 2.7.6 Let Q_1 be a bilinear expression as in Theorem 2.7.3 and $Q_2 = \mathbf{X}'B_2\mathbf{X} + \mathbf{b_2}'\mathbf{X} + \mathbf{c_2}'\mathbf{Y} + d_2$, $B_2 = B_2'$. Let $\mathbf{Z}' = (\mathbf{X}', \mathbf{Y}')$. Then the NS conditions for the independence of Q_1 and Q_2 are the following:

$(i)_r$

$$B_1\Sigma_{21}B_2 = O, \quad B_1'\Sigma_{11}B_2 = O,$$
$$B_2\Sigma_{11}B_1 = O, \quad B_2\Sigma_{12}B_1' = O;$$

$(ii)_r$

$$B_2(\Sigma_{11}\mathbf{b_1} + \Sigma_{12}\mathbf{c_1}) = \mathbf{0},$$
$$B_1(\Sigma_{21}\mathbf{b_2} + \Sigma_{22}\mathbf{c_2}) = \mathbf{0},$$
$$B_1'(\Sigma_{11}\mathbf{b_2} + \Sigma_{12}\mathbf{c_2}) = \mathbf{0};$$

$(iii)_r$

$$(\mathbf{b_1} + B_1\boldsymbol{\mu}_2)'\Sigma_{11}(\mathbf{b_2} + 2B_2\boldsymbol{\mu}_1) + (\mathbf{c_1} + B_1'\boldsymbol{\mu}_1)'\Sigma_{22}\mathbf{c_2}$$
$$+(\mathbf{b_1} + B_1\boldsymbol{\mu}_2)'\Sigma_{12}\mathbf{c_2} + (\mathbf{c_1} + B_1'\boldsymbol{\mu}_1)'\Sigma_{21}(\mathbf{b_2} + 2B_2\boldsymbol{\mu}_1) = 0.$$

These conditions are available from the following substitutions in Theorem 2.7.3.

$$A_1 = \begin{pmatrix} O & \frac{1}{2}B_1 \\ \frac{1}{2}B_1' & O \end{pmatrix}, \quad A_2 = \begin{pmatrix} B_2 & O \\ O & O \end{pmatrix},$$
$$\mathbf{a_i} = \begin{pmatrix} \mathbf{b_i} \\ \mathbf{c_i} \end{pmatrix}, \quad i = 1, 2, \quad B_2 = B_2'.$$

Note that similar conditions are available if Q_2 is quadratic in $\mathbf{Y}$. Some special cases are given below. Other special cases are available by similar substitutions in Corollary 2.7.6.

Case(1) : $Q_1 = \mathbf{b_1}'\mathbf{X} + \mathbf{c_1}'\mathbf{Y} + d_1$, $Q_2 = \mathbf{X}'B_2\mathbf{X} + \mathbf{b_2}'\mathbf{X} + \mathbf{c_2}'\mathbf{Y} + d_2$;

Case(2) : $Q_1 = \mathbf{b_1}'\mathbf{X} + d_1$, Q_2=same;

Case(3) : $Q_1 = \mathbf{c_1}'\mathbf{Y} + d$, Q_2=same.

Case(1) The NS conditions are the following:

$(ii)_r$

$$B_2(\Sigma_{11}\mathbf{b_1} + \Sigma_{12}\mathbf{c_1}) = \mathbf{0};$$

$(iii)_r$

$$\mathbf{b_1}'\Sigma_{11}(\mathbf{b_2} + 2B_2\boldsymbol{\mu}_1) + \mathbf{c_1}'\Sigma_{22}\mathbf{c_2} + \mathbf{b_1}'\Sigma_{12}\mathbf{c_2}$$
$$+ \mathbf{c_1}'\Sigma_{21}(\mathbf{b_2} + 2B_2\boldsymbol{\mu}_1) = 0.$$

Case(2) The NS conditions are the following:

$(ii)_r$ $B_2\Sigma_{11}\mathbf{b_1} = \mathbf{0}$;

$(iii)_r$ $\mathbf{b_1}'\Sigma_{11}(\mathbf{b_2} + 2B_2\boldsymbol{\mu}_1) + \mathbf{b_1}'\Sigma_{12}\mathbf{c_2} = 0$.

Case(3) The NS conditions are the following:

$(ii)_r \quad B_2\Sigma_{12}\mathbf{c}_1 = \mathbf{0};$

$(iii)_r \quad \mathbf{c}_1'\Sigma_{22}\mathbf{c}_2 + \mathbf{c}_1'\Sigma_{21}(\mathbf{b}_2 + 2B_2\boldsymbol{\mu}_1) = 0.$

Remark 2.7.3 Other results of interest are when Q_1 is quadratic in $\mathbf{X}$ and Q_2 is quadratic in $\mathbf{Y}$, each with mixed linear terms and special cases of these. Special cases of all these results for the noncorrealted and equicorrelated situations also give rise to some interesting results.

In the singular normal case also one can proceed as in the nonsingular normal case and obtain a number of particular cases. Some of these will be stated next. Note that the NS conditions for the nonsingular normal case can be obtained from those in the singular normal case but not vice versa.

2.7d Independence in the Singular Normal Case

Consider again two quadratic forms Q_1 and Q_2 in $\mathbf{Z}$ as in Theorem 2.7.3 but with $\Sigma \geq 0$. Then we have two quadratic expressions under the singular normal situation. The NS conditions for the independence of Q_1 and Q_2 are given in Mathai and Provost (1992). This will be stated as the next theorem.

Theorem 2.7.4 The NS conditions for Q_1 and Q_2 of Theorem 2.7.3, under the singular normal case, to be independently distributed are the following:
$(i)_s \quad \Sigma A_1 \Sigma A_2 \Sigma = O;$

$(ii)_s \quad \Sigma A_2 \Sigma(2A_1\boldsymbol{\mu} + \mathbf{a}_1) = \Sigma A_1 \Sigma(2A_2\boldsymbol{\mu} + \mathbf{a}_2) = \mathbf{0};$

$(iii)_s \quad (\mathbf{a}_1 + 2A_1\boldsymbol{\mu})'\Sigma(\mathbf{a}_2 + 2A_2\boldsymbol{\mu}) = 0.$
We will consider the cases when Q_1 and Q_2 are bilinear expressions in $\mathbf{X}$ and $\mathbf{Y}$ or bilinear and quadratic expressions and their special cases. The main results will be listed as corollaries. Various conditions will be listed explicitly so that they will be readily used.

Corollary 2.7.7 Let $Q_i(\mathbf{Z}) = \mathbf{X}'B_i\mathbf{Y} + \mathbf{b}_i'\mathbf{X} + \mathbf{c}_i'\mathbf{Y} + d_i,\ i = 1, 2$ and $\mathbf{Z}$ be distributed as singular normal. Then the NS conditions for the independence of Q_1 and Q_2 are the following:

$(i)_t \quad \begin{pmatrix} G_1 & G_2 \\ G_3 & G_4 \end{pmatrix}\begin{pmatrix} \Sigma_{11} & \Sigma_{12} \\ \Sigma_{21} & \Sigma_{22} \end{pmatrix} = O$ where

$$\begin{aligned} G_1 &= \Sigma_{12}B_1'\Sigma_{12}B_2' + \Sigma_{11}B_1\Sigma_{22}B_2', \\ G_2 &= \Sigma_{12}B_1'\Sigma_{11}B_2 + \Sigma_{11}B_1\Sigma_{21}B_2, \\ G_3 &= \Sigma_{22}B_1'\Sigma_{12}B_2' + \Sigma_{21}B_1\Sigma_{22}B_2', \\ G_4 &= \Sigma_{22}B_1'\Sigma_{11}B_2 + \Sigma_{21}B_1\Sigma_{21}B_2; \end{aligned}$$

$(ii)_t$

$$\begin{aligned}&(\Sigma_{12}B_2'\Sigma_{11}+\Sigma_{11}B_2\Sigma_{21})(\mathbf{b}_1+B_1\boldsymbol{\mu}_2)\\&+(\Sigma_{12}B_2'\Sigma_{12}+\Sigma_{11}B_2\Sigma_{22})(\mathbf{c}_1+B_1'\boldsymbol{\mu}_1)=\mathbf{0},\\&(\Sigma_{22}B_2'\Sigma_{11}+\Sigma_{21}B_2\Sigma_{21})(\mathbf{b}_1+B_1\boldsymbol{\mu}_2)\\&+(\Sigma_{22}B_2'\Sigma_{12}+\Sigma_{21}B_2\Sigma_{22})(\mathbf{c}_1+B_1'\boldsymbol{\mu}_1)=\mathbf{0},\end{aligned}$$

and two more conditions with B_1 and B_2, $(\mathbf{b}_1, \mathbf{c}_1)$ and $(\mathbf{b}_2, \mathbf{c}_2)$ interchanged.

$(iii)_t$

$$\begin{aligned}&(\mathbf{b}_1+B_1\boldsymbol{\mu}_2)'\Sigma_{11}(\mathbf{b}_2+B_2\boldsymbol{\mu}_2)+(\mathbf{b}_1+B_1\boldsymbol{\mu}_2)'\Sigma_{12}(\mathbf{c}_2+B_2'\boldsymbol{\mu}_1)\\&+(\mathbf{c}_1+B_1'\boldsymbol{\mu}_1)'\Sigma_{21}(\mathbf{b}_2+B_2\boldsymbol{\mu}_2)+(\mathbf{c}_1+B_1'\boldsymbol{\mu}_1)'\Sigma_{22}(\mathbf{c}_2+B_2'\boldsymbol{\mu}_1)=0.\end{aligned}$$

Some special cases of interest will be explicitly listed here.

Case(1) : $Q_1 = \mathbf{X}'B_1\mathbf{Y} + \mathbf{b}_1{}'\mathbf{X} + \mathbf{c}_1{}'\mathbf{Y} + d_1$, $Q_2 = \mathbf{b}_2{}'\mathbf{X} + \mathbf{c}_2{}'\mathbf{Y} + d_2$.

Case(2) : Q_1=same, $Q_2 = \mathbf{b}_2{}'\mathbf{X} + d_2$.

Case(3) : Q_1=same, $Q_2 = \mathbf{c}_2{}'\mathbf{Y} + d_2$.

Case(1) The NS conditions for independence are the following:

$(ii)_t$

$$\begin{aligned}&(\Sigma_{12}B_1'\Sigma_{11}+\Sigma_{11}B_1\Sigma_{21})\mathbf{b}_2+(\Sigma_{12}B_1'\Sigma_{12}+\Sigma_{11}B_1\Sigma_{22})\mathbf{c}_2=\mathbf{0},\\&(\Sigma_{22}B_1'\Sigma_{11}+\Sigma_{21}B_1\Sigma_{21})\mathbf{b}_2+(\Sigma_{22}B_1'\Sigma_{12}+\Sigma_{21}B_1\Sigma_{22})\mathbf{c}_2=\mathbf{0};\end{aligned}$$

$(iii)_t$

$$\begin{aligned}&(\mathbf{b}_1+B_1\boldsymbol{\mu}_2)'\Sigma_{11}\mathbf{b}_2+(\mathbf{b}_1+B_1\boldsymbol{\mu}_2)'\Sigma_{12}\mathbf{c}_2\\&+(\mathbf{c}_1+B_1'\boldsymbol{\mu}_1)'\Sigma_{21}\mathbf{b}_2+(\mathbf{c}_1+B_1'\boldsymbol{\mu}_1)'\Sigma_{22}\mathbf{c}_2=0.\end{aligned}$$

Case(2) The NS conditions are the following:

$(ii)_t$

$$(\Sigma_{12}B_1'\Sigma_{11}+\Sigma_{11}B_1\Sigma_{21})\mathbf{b}_2=\mathbf{0}$$

and

$$(\Sigma_{22}B_1'\Sigma_{11}+\Sigma_{21}B_1\Sigma_{21})\mathbf{b}_2=\mathbf{0};$$

$(iii)_t$

$$(\mathbf{b}_1+B_1\boldsymbol{\mu}_2)'\Sigma_{11}\mathbf{b}_2+(\mathbf{c}_1+B_1'\boldsymbol{\mu}_1)'\Sigma_{21}\mathbf{b}_2=0.$$

Case(3) The NS conditions are the following:

$(ii)_t$

$$(\Sigma_{12}B_1'\Sigma_{12}+\Sigma_{11}B_1\Sigma_{22})\mathbf{c}_2=\mathbf{0}$$

and

$$(\Sigma_{22}B_1'\Sigma_{12} + \Sigma_{21}B_1\Sigma_{22})\mathbf{c_2} = \mathbf{0};$$

$(iii)_t$

$$(\mathbf{b_1} + B_1\boldsymbol{\mu}_2)'\Sigma_{12}\mathbf{c_2} + (\mathbf{c_1} + B_1'\boldsymbol{\mu}_1)'\Sigma_{22}\mathbf{c_2} = 0.$$

Corollary 2.7.8 Let $Q_1(\mathbf{Z}) = \mathbf{X}'B_1\mathbf{X} + \mathbf{b_1}'\mathbf{X} + \mathbf{c_1}'\mathbf{Y} + d_1$, $B_1 = B_1'$, $Q_2(\mathbf{Z}) = \mathbf{X}'B_2\mathbf{Y} + \mathbf{b_2}'\mathbf{X} + \mathbf{c_2}'\mathbf{Y} + d_2$ be a quadratic and a bilinear expression respectively, where $\mathbf{Z} = \binom{\mathbf{X}}{\mathbf{Y}}$ has a singular normal distribution as in Theorem 2.7.4. Then the NS conditions for the independence of Q_1 and Q_2 are the following:

$(i)_u$

$$\begin{pmatrix} H_1 & H_2 \\ H_3 & H_4 \end{pmatrix} \Sigma = O,$$

$$\begin{aligned} H_1 &= \Sigma_{11}B_1\Sigma_{12}B_2', \\ H_2 &= \Sigma_{11}B_1\Sigma_{11}B_2 \\ H_3 &= \Sigma_{21}B_1\Sigma_{12}B_2', \\ H_4 &= \Sigma_{21}B_1\Sigma_{11}B_2; \end{aligned}$$

$(ii)_u$

$$\begin{aligned} &(\Sigma_{12}B_2'\Sigma_{11} + \Sigma_{11}B_2\Sigma_{21})(\mathbf{b_1} + 2B_1\boldsymbol{\mu}_1) \\ &+(\Sigma_{12}B_2'\Sigma_{12} + \Sigma_{11}B_2\Sigma_{22})\mathbf{c_1} = \mathbf{0} \\ &= \Sigma_{11}B_1\Sigma_{11}(\mathbf{b_2} + B_2\boldsymbol{\mu}_2) \\ &+\Sigma_{11}B_1\Sigma_{12}(\mathbf{c_2} + B_2'\boldsymbol{\mu}_1) \end{aligned}$$

and

$$\begin{aligned} &(\Sigma_{22}B_2'\Sigma_{11} + \Sigma_{21}B_2\Sigma_{21})(\mathbf{b_1} + 2B_1\boldsymbol{\mu}_1) \\ &+ (\Sigma_{22}B_2'\Sigma_{12} + \Sigma_{21}B_2\Sigma_{22})\mathbf{c_1} = \mathbf{0} \\ &= \Sigma_{21}B_1\Sigma_{11}(\mathbf{b_2} + B_2\boldsymbol{\mu}_2) \\ &+ \Sigma_{21}B_1\Sigma_{12}(\mathbf{c_2} + B_2'\boldsymbol{\mu}_1); \end{aligned}$$

$(iii)_u$

$$\begin{aligned} &(\mathbf{b_1} + 2B_1\boldsymbol{\mu}_1)'\Sigma_{11}(\mathbf{b_2} + B_2\boldsymbol{\mu}_2) \\ &+(\mathbf{b_1} + 2B_1\boldsymbol{\mu}_1)'\Sigma_{12}(\mathbf{c_2} + B_2'\boldsymbol{\mu}_1) \\ &+\mathbf{c_1}'\Sigma_{21}(\mathbf{b_2} + B_2\boldsymbol{\mu}_2) + \mathbf{c_1}'\Sigma_{22}(\mathbf{c_2} + B_2'\boldsymbol{\mu}_1) = 0. \end{aligned}$$

Some special cases of interest are the following.

Case(1) : $Q_2 = \mathbf{b_2}'\mathbf{X} + \mathbf{c_2}'\mathbf{Y} + d_2$, $Q_1 = \mathbf{X}'B_1\mathbf{X} + \mathbf{b_1}'\mathbf{X} + \mathbf{c_1}'\mathbf{Y} + d_1$.

Case(2) : $Q_2 = \mathbf{b_2}'\mathbf{X} + d_2$, Q_1=same.

Case(3) : $Q_2 = \mathbf{c_2}'\mathbf{Y} + d_2$, Q_1=same.

Case(4) : $Q_2 = \mathbf{b_2}'\mathbf{X} + \mathbf{c_2}'\mathbf{Y} + d_2$, $Q_1 = \mathbf{X}'B_1\mathbf{X} + \mathbf{c_1}'\mathbf{Y} + d_1$.

Case(1) The NS conditions are the following:

$(ii)_u$

$$\Sigma_{11}B_1\Sigma_{11}\mathbf{b_2} + \Sigma_{11}B_1\Sigma_{12}\mathbf{c_2} = \mathbf{0}$$

and

$$\Sigma_{21}B_1\Sigma_{11}\mathbf{b_2} + \Sigma_{21}B_1\Sigma_{12}\mathbf{c_2} = \mathbf{0};$$

$(iii)_u$

$$(\mathbf{b_1} + 2B_1\boldsymbol{\mu}_1)'\Sigma_{11}\mathbf{b_2} + (\mathbf{b_1} + 2B_1\boldsymbol{\mu}_1)'\Sigma_{12}\mathbf{c_2} + \mathbf{c_1}'\Sigma_{21}\mathbf{b_2} + \mathbf{c_1}'\Sigma_{22}\mathbf{c_2} = 0.$$

Case(2) The NS conditions are the following:

$(ii)_u$

$$\Sigma_{11}B_1\Sigma_{11}\mathbf{b_2} = \mathbf{0}$$

and

$$\Sigma_{21}B_1\Sigma_{11}\mathbf{b_2} = \mathbf{0};$$

$(iii)_u$

$$(\mathbf{b_1} + 2B_1\boldsymbol{\mu}_1)'\Sigma_{11}\mathbf{b_2} + \mathbf{c_1}'\Sigma_{21}\mathbf{b_2} = 0.$$

Case(3) The NS conditions are the following:

$(ii)_u$

$$\Sigma_{11}B_1\Sigma_{12}\mathbf{c_2} = \mathbf{0}$$

and

$$\Sigma_{21}B_1\Sigma_{12}\mathbf{c_2} = \mathbf{0};$$

$(iii)_u$

$$(\mathbf{b_1} + 2B_1\boldsymbol{\mu}_1)'\Sigma_{12}\mathbf{c_2} + \mathbf{c_1}'\Sigma_{22}\mathbf{c_2} = 0.$$

Case(4) The NS conditions are the following:

$(ii)_u$

$$\Sigma_{11}B_1\Sigma_{11}\mathbf{b_2} + \Sigma_{11}B_1\Sigma_{12}\mathbf{c_2} = \mathbf{0}$$

and

$$\Sigma_{21}B_1\Sigma_{11}\mathbf{b_2} + \Sigma_{21}B_1\Sigma_{12}\mathbf{c_2} = \mathbf{0};$$

$(iii)_u$

$$2\boldsymbol{\mu}_1'B_1\Sigma_{11}\mathbf{b_2} + 2\boldsymbol{\mu}_1'B_1\Sigma_{12}\mathbf{c_2} + \mathbf{c_1}'\Sigma_{21}\mathbf{b_2} + \mathbf{c_1}'\Sigma_{22}\mathbf{c_2} = 0.$$

2.8 Bilinear Forms and Noncentral Gamma Difference

In the previous sections we considered the Laplacianness and independence of bilinear forms as well as quadratic forms. In this section we will examine the NS conditions for a bilinear form to be distributed as a gamma difference. Generalization to matrix-variate cases will also be dealt with.

Consider the bilinear form $Q = \mathbf{X}'B_2\mathbf{Y}$ with

$$\begin{pmatrix}\mathbf{X}\\ \mathbf{Y}\end{pmatrix} \sim N_{p+q}(\boldsymbol{\mu}, \Sigma),\ \Sigma = \begin{pmatrix}\Sigma_{11} & \Sigma_{12}\\ \Sigma_{21} & \Sigma_{22}\end{pmatrix},\ \Sigma > 0$$

where $\mathbf{X}$ is $p \times 1$ and $\mathbf{Y}$ is $q \times 1$. Writing Q as a quadratic form in $\mathbf{Z} = \binom{\mathbf{X}}{\mathbf{Y}}$ one has

$$Q = \mathbf{X}'B_2\mathbf{Y} = \mathbf{Z}'A\mathbf{Z},\ A = \begin{bmatrix} O & \frac{1}{2}B_2\\ \frac{1}{2}B_2' & O\end{bmatrix},\ A = A'.$$

Then from the standard result on the m.g.f. of Q one has

$$M_Q(t) = |I - 2tA\Sigma|^{-\frac{1}{2}}. \tag{2.8.1}$$

Thus when $p = q, B_2 = B_2', \Sigma_{11} = I = \Sigma_{22}, \Sigma_{12} = \rho I = \Sigma_{21}$ one has

$$\begin{aligned}|I - 2tA\Sigma| &= \begin{vmatrix} I - t\rho B_2 & -tB_2\\ -tB_2 & I - t\rho B_2\end{vmatrix}\\ &= |(I - t\rho B_2)|\left|(I - t\rho B_2) - t^2B_2(I - t\rho B_2)^{-1}B_2\right|\\ &= \left|(I - t\rho B_2)^2 - t^2(I - t\rho B_2)B_2(I - t\rho B_2)^{-1}B_2\right|.\end{aligned}$$

Let P be an orthogonal matrix, $PP' = I$, such that $P'B_2P = \mathrm{diag}(\lambda_1, \ldots, \lambda_p)$. Then

$$\begin{aligned}|I - 2tA\Sigma| &= \prod_{j=1}^{p}\left[(1 - t\rho\lambda_j)^2 - t^2\lambda_j^2\right] \qquad (2.8.2)\\ &= \prod_{j=1}^{p}\left[(1 - a_1\lambda_jt)(1 + a_2\lambda_jt)\right],\ a_1 = 1 + \rho,\ a_2 = 1 - \rho.\end{aligned}$$

If $\Sigma_{11} = \sigma_1^2I,\ \Sigma_{22} = \sigma_2^2I,\ \Sigma_{12} = \sigma_1\sigma_2\rho I$ then a_1 and a_2 will be changed to $a_1 = \sigma_1\sigma_2(1 + \rho),\ a_2 = \sigma_1\sigma_2(1 - \rho)$. This result will be stated as a lemma.

Lemma 2.8.1 Let

$$Q = \mathbf{X}'B_2\mathbf{Y},\ \begin{pmatrix}\mathbf{X}\\ \mathbf{Y}\end{pmatrix} \sim N_{p+q}(\boldsymbol{\mu}, \Sigma)$$

where

$$B_2 = B_2',\ \boldsymbol{\mu} = \mathbf{0},\ \Sigma = \begin{pmatrix}\sigma_1^2I & \rho\sigma_1\sigma_2I\\ \rho\sigma_1\sigma_2I & \sigma_2^2I\end{pmatrix}.$$

Then

$$M_Q(t) = \prod_{j=1}^{p} \left[(1 - a_1\lambda_j t)^{-\frac{1}{2}} (1 + a_2\lambda_j t)^{-\frac{1}{2}} \right]$$

where

$$a_1 = \sigma_1\sigma_2(1+\rho), \quad a_2 = \sigma_1\sigma_2(1-\rho),$$

and

$$M_Q(t) = \left[(1 - a_1 t)^{-\frac{r}{2}} (1 + a_2 t)^{-\frac{r}{2}} \right] \tag{2.8.3}$$

if $B_2 = B_2^2$ and r =rank(B_2).

2.8a Bilinear Forms in the Equicorrelated Case

Let

$$Q = \mathbf{Z}'A\mathbf{Z} = \mathbf{X}'B_2\mathbf{Y},$$

$$A = \begin{pmatrix} O & \frac{1}{2}B_2 \\ \frac{1}{2}B_2' & O \end{pmatrix}, \quad \mathbf{Z} = \begin{pmatrix} \mathbf{X} \\ \mathbf{Y} \end{pmatrix} \sim N_{p+q}(\mathbf{0}, \Sigma),$$

$$p = q, \; B_2 = B_2', \; \Sigma = \begin{pmatrix} I & \rho I \\ \rho I & I \end{pmatrix}.$$

Let $B_2 \neq O$. Then one can prove the following result.

Theorem 2.8.1 The necessary and sufficient conditions for the bilinear form $Q = \mathbf{Z}'A\mathbf{Z} = \mathbf{X}'B_2\mathbf{Y}$, as defined above, to be distributed as a central gamma difference of Definition 2.3.1 with the parameters $(\alpha_1, \alpha_2, \beta_1, \beta_2)$ are the following: (i) $\alpha_1 = \alpha_2 = r/2$ where r is the number of nonzero eigenvalues of B_2; (ii) All the nonzero eigenvalues of B_2 are positive and equal such that $\frac{\beta_1}{\beta_2} = \frac{1+\rho}{1-\rho}$ or all the nonzero eigenvalues of B_2 are negative and equal such that $\frac{\beta_1}{\beta_2} = \frac{1-\rho}{1+\rho}$. If $\rho = 0$ then this gamma difference will be a generalized Laplacian under (i) and (ii).

PROOF *Necessity* From Lemma 2.8.1 and (2.3.1) one has

$$\prod_{j=1}^{r} [1 - (1+\rho)\lambda_j t]^{-\frac{1}{2}} [1 + (1-\rho)\lambda_j t]^{-\frac{1}{2}} = (1 - \beta_1 t)^{-\alpha_1} (1 + \beta_2 t)^{-\alpha_2} \tag{2.8.4}$$

where r is the rank of B_2, $\alpha_j > 0$, $\beta_j > 0$, $j = 1, 2$, $-1 < \rho < 1$. That is,

$$\prod_{j=1}^{r} [1 - (1+\rho)\lambda_j t][1 + (1-\rho)\lambda_j t] = (1 - \beta_1 t)^{2\alpha_1} (1 + \beta_2 t)^{2\alpha_2}. \tag{2.8.5}$$

For $\lambda_j > 0$ put $t = [(1+\rho)\lambda_j]^{-1}$ and then $t = -[(1-\rho)\lambda_j]^{-1}$. Then

$$\beta_1 = (1+\rho)\lambda_j \text{ and } \beta_2 = (1-\rho)\lambda_j \text{ for all } \lambda_j > 0. \tag{2.8.6}$$

Similarly

$$\beta_1 = -(1-\rho)\lambda_m \text{ and } \beta_2 = -(1+\rho)\lambda_m \text{ for all } \lambda_m < 0. \tag{2.8.7}$$

If possible let $\lambda_j > 0$ and $\lambda_m < 0$ for some j and m, $j \neq m$. Then from (2.8.6) and (2.8.7) one has

$$\frac{\lambda_j}{\lambda_m} = \frac{1+\rho}{\rho-1} = \frac{\rho-1}{1+\rho} \Rightarrow (1+\rho)^2 = (\rho-1)^2 \Rightarrow \rho = 0. \tag{2.8.8}$$

Thus for $\rho \neq 0$, (2.8.4) cannot hold unless all eigenvalues are of the same sign and equal. Thus from (2.8.6) and (2.8.7) one has $2\alpha_1 = r = 2\alpha_2$. If $\rho = 0$ then (2.8.5) reduces to

$$\prod_{j=1}^{r}(1-\lambda_j^2 t^2) = (1-\beta_1 t)^{2\alpha_1}(1+\beta_2 t)^{2\alpha_2},\ b_j > 0,\ j = 1,2. \tag{2.8.9}$$

Put $t = \lambda_j^{-1}$ for $\lambda_j > 0 \Rightarrow \beta_1 = \lambda_j$ for all j. Put $t = \lambda_m^{-1}$ for $\lambda_m < 0 \Rightarrow \beta_2 = -\lambda_m$ for all m. Put $t = [|\lambda_k|]^{-1} \Rightarrow \beta_1 = |\lambda_k|$ for all $k, \Rightarrow \beta_1 = \beta_2$ and the λ_j's are all equal in absolute value. Then from (2.8.9), $\alpha_1 = \alpha_2 = r/2$. Sufficiency is established by retracing the steps.

Corollary 2.8.1 For the bilinear form $U = \mathbf{X}'B_2\mathbf{Y}$ defined in Lemma 2.8.1 the NS conditions for U to be distributed as a gamma difference with the parameters $\sigma_1\sigma_2(1+\rho)\lambda$, $\sigma_1\sigma_2(1-\rho)\lambda$, $r/2$, $r/2$ is that $\lambda B_2 = B_2^2$ and of rank r, where λ is the common repeated eigenvalue of B_2.

Example 2.8.1 Let $\binom{X}{Y}$ be bivariate normal with parameters $\mu_x = 0 = \mu_y$, $\sigma_x^2 = 1 = \sigma_y^2$ and ρ. What is the distribution of $U = XY$? Writing XY as a quadratic form

$$Q = (X, Y)\begin{bmatrix} 0 & \frac{1}{2} \\ \frac{1}{2} & 0 \end{bmatrix}\binom{X}{Y}$$

one has, in the notations of Theorem 2.8.1, $B_2 = 1$, $\lambda_1 = 1$, $\beta_1 = 1+\rho$, $\beta_2 = 1-\rho$. Hence $U = XY$ is distributed as a gamma difference with the parameters $\beta_1 = 1+\rho$, $\beta_2 = 1-\rho$, $\alpha_1 = 1 = \alpha_2$ for $\rho \neq 0$. If $\rho = 0$ then it is a Laplacian. If $\sigma_1 \neq 1, \sigma_2 \neq 1$ then for $\rho \neq 0$, U is a gamma difference with the parameters $\beta_1 = \sigma_1\sigma_2(1+\rho)$, $\beta_2 = \sigma_1\sigma_2(1-\rho)$, $\alpha_1 = 1 = \alpha_2$.

Example 2.8.2 Let $\binom{X_i}{Y_i}, i = 1, \ldots, n$ be independently and identically distributed as the $\binom{X}{Y}$ of Example 2.8.1. Let $U_1 = \sum_{i=1}^n X_iY_i$, $U_2 = \sum_{i=1}^n (X_i - \bar{X})(Y_i - \bar{Y})$, $\bar{X} = \sum_{i=1}^n X_i/n$, $\bar{Y} = \sum_{i=1}^n Y_i/n$. Then U_1 is a gamma difference with the parameters $\beta_1 = \sigma_1\sigma_2(1+\rho)$, $\beta_2 = \sigma_1\sigma_2(1-\rho)$, $\alpha_1 = n = \alpha_2$ for $\rho \neq 0$ and it is a generalized Laplacian with $\alpha = n$, $\beta = \sigma_1\sigma_2$ for $\rho = 0$. In the case of U_2 the distribution will still be gamma difference for $\rho \neq 0$ and generalized Laplacian for $\rho = 0$ with the same parameters β_1, β_2 but $\alpha_1 = n-1 = \alpha_2$. These results are seen by observing the following.

Let $\mathbf{X}' = (X_1, \ldots, X_n)$, $\mathbf{Y}' = (Y_1, \ldots, Y_n)$,

$$\mathbf{Z} = \begin{pmatrix} \mathbf{X} \\ \mathbf{Y} \end{pmatrix} \sim N_{2n}(\mathbf{0}, \Sigma),\ \Sigma = \begin{pmatrix} \Sigma_{11} & \Sigma_{12} \\ \Sigma_{21} & \Sigma_{22} \end{pmatrix}$$

where $\Sigma_{11} = \sigma_1^2 I$, $\Sigma_{12} = \sigma_1\sigma_2\rho I$, $\Sigma_{22} = \sigma_2^2 I$, I is the identity matrix of order n. $U_1 = \mathbf{X}'B_2\mathbf{Y}$ with $B_2 = I$ and $U_2 = \mathbf{X}'B_2\mathbf{Y}$ with $B_2 = I - \frac{\mathbf{1}\mathbf{1}'}{n}$ of rank $n-1$, $\mathbf{1}' = (1, \ldots, 1)$. The eigenvalues of B_2 in U_1 are all unities and those in U_2 are such that $n-1$ of them are unities and one is zero. Thus the results follow. In the case of U_2 the distribution remains the same even if the normal population has a non-null mean vector.

Example 2.8.3 Consider a bivariate normal with mean vector null and covariance matrix

$$V = \begin{pmatrix} 4 & 2 \\ 2 & 4 \end{pmatrix}.$$

Let $\binom{X_i}{Y_i}$, $i = 1, 2, 3$ be a sample of size 3 from this bivariate normal. Then the following are gamma differences with the parameters $(\alpha_1, \alpha_2, \beta_1, \beta_2)$ given in the brackets.

$$\begin{aligned}
U_1 &= X_1Y_1 + X_2Y_2;\quad (1, 1, 6, 2), \\
U_2 &= \left(\frac{1}{2}\right)(X_1Y_1 + X_2Y_2 - X_1Y_2 - X_2Y_1);\quad \left(\frac{1}{2}, \frac{1}{2}, 6, 2\right), \\
U_3 &= \left(\frac{1}{3}\right)[2(X_1Y_1 + X_2Y_2 + X_3Y_3) \\
&\quad - (X_1Y_2 + X_2Y_1 + X_1Y_3 + X_3Y_1 + X_2Y_3 + X_3Y_2)];\quad (1, 1, 6, 2), \\
U_4 &= 4X_1Y_1 + X_2Y_2 + 2X_1Y_2 + 2X_2Y_1;\quad \left(\frac{1}{2}, \frac{1}{2}, 30, 10\right),
\end{aligned}$$

and

$$\begin{aligned}
U_5 &= 4X_1Y_1 + X_2Y_2 + 4X_3Y_3 + 2X_1Y_2 + 2X_2Y_1 \\
&\quad + 4X_1Y_3 + 4X_3Y_1 + 2X_2Y_3 + 2X_3Y_2;\quad \left(\frac{1}{2}, \frac{1}{2}, 54, 18\right).
\end{aligned}$$

SOLUTION As an example let us consider U_5. Writing the covariance matrix in the form

$$\begin{bmatrix} \sigma_1^2 & \rho\sigma_1\sigma_2 \\ \rho\sigma_1\sigma_2 & \sigma_2^2 \end{bmatrix} = \begin{bmatrix} 4 & 2 \\ 2 & 4 \end{bmatrix}$$

one has $\sigma_1 = 2 = \sigma_2$, $\rho = \frac{1}{2}$. Let $\mathbf{X}' = (X_1, X_2, X_3)$, $\mathbf{Y}' = (Y_1, Y_2, Y_3)$. Then writing

$$U_5 = \mathbf{X}'B_2\mathbf{Y} \Rightarrow B_2 = \begin{bmatrix} 4 & 2 & 4 \\ 2 & 1 & 2 \\ 4 & 2 & 4 \end{bmatrix}.$$

Note that

$$B_2^2 = 9B_2 \Rightarrow \lambda = 9,$$
$$\sigma_1\sigma_2(1+\rho)\lambda = 4(\frac{3}{2})9 = 54,$$
$$\sigma_1\sigma_2(1-\rho)\lambda = 4(\frac{1}{2})9 = 18.$$

The eigenvalues of B_2 are 0, 0, 9 since the rank of B_2 is evidently one and $\text{tr}(B_2) = 9$. Hence $\alpha_1 = \frac{1}{2} = \alpha_2$, $\beta_1 = 54$, $\beta_2 = 18$. In a similar fashion all others can be established and these are left as exercises for the reader.

Theorem 2.8.2 The NS conditions for the quadratic form $\mathbf{X}'B_1\mathbf{X}$, $B_1 = B_1'$, $\mathbf{X} \sim N_p(\mathbf{0}, \Sigma_{11})$, $\Sigma_{11} > 0$ to be distributed as a gamma difference with parameters $(\alpha_1, \alpha_2, \beta_1, \beta_2)$ are the following: (i) $\alpha_1 = \nu_1/2$, $\alpha_2 = \nu_2/2$ for some positive integers ν_1 and ν_2; (ii) All the positive eigenvalues of $\Sigma_{11}B_1$ are equal to $\beta_1/2$ and all the negative eigenvalues are equal to $-\beta_2/2$, at least one of them is there in each set.

PROOF *Necessity* Equating the m.g.f.'s of a gamma difference and the quadratic form one has

$$\prod_{j=1}^{r}(1-2t\lambda_j)^{-\frac{1}{2}} = (1-\beta_1 t)^{-\alpha_1}(1+\beta_2 t)^{-\alpha_2},\ \beta_j > 0,\ j = 1, 2$$

where r is the number of nonzero eigenvalues of $B_1\Sigma_{11}$. By proceeding as in the proof of Theorem 2.8.1 the result is established.

2.8b Noncentral Case

Let

$$\begin{pmatrix} X \\ Y \end{pmatrix} \sim N_2(\boldsymbol{\mu}_{(1)}, V_2),$$
$$V_2 = \begin{pmatrix} \sigma_1^2 & \rho\sigma_1\sigma_2 \\ \rho\sigma_2\sigma_1 & \sigma_2^2 \end{pmatrix} > 0,\ \boldsymbol{\mu}_{(1)} = \begin{pmatrix} \mu_1 \\ \mu_2 \end{pmatrix}. \tag{2.8.10}$$

The m.g.f. of a quadratic form

$$Q = \mathbf{Z}'A\mathbf{Z},\ A = A',\ \mathbf{Z} \sim N_{p+q}(\boldsymbol{\mu}, V),\ V > 0$$

is given by

$$M_Q(t) = |I - 2tAV|^{-\frac{1}{2}}$$
$$\times \exp\left[-\left(\frac{1}{2}\right)(\boldsymbol{\mu}'V^{-1}\boldsymbol{\mu}) + \left(\frac{1}{2}\right)\left(\boldsymbol{\mu}'V^{-\frac{1}{2}}(I-2tAV)^{-1}V^{-\frac{1}{2}}\boldsymbol{\mu}\right)\right].$$

If $Q = XY$ then

$$A = \begin{pmatrix} 0 & \frac{1}{2} \\ \frac{1}{2} & 0 \end{pmatrix},$$

$$2AV = \begin{pmatrix} \rho\sigma_1\sigma_2 & \sigma_2^2 \\ \sigma_1^2 & \rho\sigma_1\sigma_2 \end{pmatrix}$$

and

$$|I - 2tAV|^{-\frac{1}{2}} = (1 - a_1 t)^{-\frac{1}{2}}(1 + a_2 t)^{-\frac{1}{2}},$$
$$a_1 = (1+\rho)\sigma_1\sigma_2, \quad a_2 = (1-\rho)\sigma_1\sigma_2.$$

The exponent simplifies to

$$-\left(\frac{1}{2}\right)(b_1^2 + b_2^2) + \left(\frac{1}{2}\right) b_1^2(1 - a_1 t)^{-1} + \left(\frac{1}{2}\right) b_2^2(1 + a_2 t)^{-1}$$

where

$$\begin{pmatrix} b_1 \\ b_2 \end{pmatrix} = P'V^{-\frac{1}{2}}\mu,$$
$$P'V^{\frac{1}{2}}AV^{\frac{1}{2}}P = \text{diagonal}, \quad PP' = I.$$

Note that if a random variable U has the m.g.f. given by

$$M_U(t) = (1 - \beta_1 t)^{-\alpha_1}(1 + \beta_2 t)^{-\alpha_2}$$
$$\times \exp[-(\lambda_1 + \lambda_2) + \lambda_1(1 - \beta_1 t)^{-1} + \lambda_2(1 + \beta_2 t)^{-1}]$$

for $\alpha_i > 0$, $\beta_i > 0$, $\lambda_i > 0$, $i = 1, 2$ we call this U a noncentral gamma difference with the parameters $(\alpha_1,\ \alpha_2,\ \beta_1,\ \beta_2,\ \lambda_1,\ \lambda_2)$.

Example 2.8.4 Let $\binom{X_i}{Y_i}$, $i = 1, \dots, n$ be a simple random sample from the bivariate normal defined in (2.8.10). Then show that $Q_n = \sum_{i=1}^{n} X_i Y_i$ is a noncentral gamma difference with the parameters $(\alpha_1,\ \alpha_2,\ \beta_1,\ \beta_2,\ \lambda_1,\ \lambda_2) = (\frac{n}{2},\ \frac{n}{2},\ (1+\rho)\sigma_1\sigma_2,\ (1-\rho)\sigma_1\sigma_2,\ \lambda_1,\ \lambda_2)$ where

$$\begin{pmatrix} \lambda_1 \\ \lambda_2 \end{pmatrix} = P'V^{-\frac{1}{2}}\boldsymbol{\mu}, \ \boldsymbol{\mu}' = (\mu_1, \dots, \mu_1, \mu_2, \dots, \mu_2), \ V = \begin{pmatrix} V_{11} & V_{12} \\ V_{21} & V_{22} \end{pmatrix},$$

$V_{11} = \sigma_1^2 I$, $V_{22} = \sigma_2^2 I$, $V_{12} = \rho\sigma_1\sigma_2 I$, I is the identity matrix, $PP' = I$, $P'V^{\frac{1}{2}}AV^{\frac{1}{2}}P$ = diagonal,

$$A = \begin{pmatrix} O & \frac{1}{2}I \\ \frac{1}{2}I & O \end{pmatrix}.$$

SOLUTION Let

$$\mathbf{X}' = (X_1, \dots, X_n), \quad \mathbf{Y}' = (Y_1, \dots, Y_n)$$

$$Q_n = \mathbf{X}'\mathbf{Y} = (\mathbf{X}', \mathbf{Y}')A\binom{\mathbf{X}}{\mathbf{Y}}$$

where A is as given above and I is the identity matrix of order n. The covariance matrix of $\mathbf{Z} = \binom{\mathbf{X}}{\mathbf{Y}}$ is V as given above.

$$\begin{aligned} 2AV &= \begin{bmatrix} O & I \\ I & O \end{bmatrix} \begin{bmatrix} \sigma_1^2 I & \rho\sigma_1\sigma_2 I \\ \rho\sigma_1\sigma_2 I & \sigma_2^2 I \end{bmatrix} \\ &= \begin{bmatrix} \rho\sigma_1\sigma_2 I & \sigma_2^2 I \\ \sigma_1^2 I & \rho\sigma_1\sigma_2 I \end{bmatrix}. \end{aligned}$$

$$\begin{aligned} |I - 2tAV| &= \begin{vmatrix} (1 - t\rho\sigma_1\sigma_2)I & -t\sigma_2^2 I \\ -t\sigma_1^2 I & (1 - t\rho\sigma_1\sigma_2)I \end{vmatrix} \\ &= [1 - (1+\rho)\sigma_1\sigma_2 t]^n [1 + (1-\rho)\sigma_1\sigma_2 t]^n. \end{aligned}$$

Hence $\alpha_1 = \frac{n}{2}$, $\alpha_2 = \frac{n}{2}$, $\beta_1 = (1+\rho)\sigma_1\sigma_2$ and $\beta_2 = (1-\rho)\sigma_1\sigma_2$. From the exponent we have λ_1 and λ_2 and thus the result is established.

2.9 Rectangular Matrices

In the previous sections we considered certain covariance structures and bilinear and quadratic forms. Here we will consider rectangular submatrices of a symmetric positive definite matrix. Let the $p \times p$ matrix S have the matrix-variate gamma density

$$f(S) = [\Gamma_p(\alpha)]^{-1} |S|^{\alpha - \frac{p+1}{2}} e^{-\mathrm{tr}(S)}, \; S = S' > 0, \; R(\alpha) > \frac{p-1}{2} \tag{2.9.1}$$

where $R(\cdot)$ denotes the real part of $(\cdot)$,

$$\Gamma_p(\alpha) = \pi^{\frac{p(p-1)}{4}} \Gamma_p(\alpha)\Gamma_p\left(\alpha - \frac{1}{2}\right) \cdots \Gamma_p\left(\alpha - \frac{p-1}{2}\right), \; R(\alpha) > \frac{p-1}{2},$$

$$S = \begin{pmatrix} S_{11} & S_{12} \\ S_{21} & S_{22} \end{pmatrix}, \; \underset{r \times r}{S_{11}}, \; \underset{q \times q}{S_{22}}, \; r + q = p, \; r \geq q.$$

Let us consider the m.g.f. $M_{S_{21}}(T)$ of S_{21} and the density of S_{21}. That is,

$$M_{S_{21}}(T) = E\left[e^{\mathrm{tr}(TS_{21})}\right] \tag{2.9.2}$$

where T is $r \times q$ of full rank q. We can integrate out over S or over S_{11}, S_{22}, S_{21}. Then

$$M_{S_{21}}(T) = \int_{S_{11}, S_{22}, S_{21}} \frac{|S|^{\alpha - \frac{p+1}{2}} e^{-\mathrm{tr}(S) + \mathrm{tr}(TS_{21})}}{\Gamma_p(\alpha)} \mathrm{d}S_{11}\, \mathrm{d}S_{22}\, \mathrm{d}S_{21}. \tag{2.9.3}$$

But

$$|S| = |S_{22}||S_{11} - S_{12}S_{22}^{-1}S_{21}| \text{ and } S_{11} > S_{12}S_{22}^{-1}S_{21}.$$

Integrate out S_{11} by using a matrix-variate gamma integral, see also Chapter 1, to get

$$\int_{S_{11}>S_{12}S_{22}^{-1}S_{21}} |S_{11}-S_{12}S_{22}^{-1}S_{21}|^{(\alpha-\frac{q}{2})-\frac{r+1}{2}} e^{-\mathrm{tr}(S_{11})} \mathrm{d}S_{11}$$
$$= \Gamma_r(\alpha - \frac{q}{2}) e^{-\mathrm{tr}(S_{12}S_{22}^{-1}S_{21})}. \tag{2.9.4}$$

Write

$$\mathrm{tr}(TS_{21}) = \left(\frac{1}{2}\right)[\mathrm{tr}(TS_{21}) + \mathrm{tr}(S_{12}T')]$$

and integrate out S_{21}. Here make the substitution

$$S_{22}^{-\frac{1}{2}}S_{21} = U \Rightarrow |S_{22}|^{-\frac{r}{2}}\mathrm{d}S_{21} = \mathrm{d}U,$$

see also Chapter 1. Then

$$\int_{S_{21}} e^{-\mathrm{tr}(S_{12}S_{22}^{-1}S_{21})+\frac{1}{2}[\mathrm{tr}(TS_{21})+\mathrm{tr}(S_{12}T')]}\mathrm{d}S_{21}$$
$$= |S_{22}|^{\frac{r}{2}} \int_U e^{-\mathrm{tr}(U'U)+\frac{1}{2}\left[\mathrm{tr}(TS_{22}^{\frac{1}{2}}U+U'S_{22}^{\frac{1}{2}}T')\right]}\mathrm{d}U$$
$$= |S_{22}|^{\frac{r}{2}} \int_U e^{-\mathrm{tr}[(U-C)'(U-C)]+\mathrm{tr}(C'C)}\mathrm{d}U, \; C = \left(\frac{1}{2}\right) S_{22}^{\frac{1}{2}}T'$$
$$= \pi^{\frac{qr}{2}} |S_{22}|^{\frac{r}{2}} e^{\frac{1}{4}\mathrm{tr}(TS_{22}T')}.$$

Note that $\mathrm{tr}[(U-C)'(U-C)]$ gives the sum of squares of qr elements and each of these gives the integral of the type

$$\int_{-\infty}^{\infty} e^{-z^2}\mathrm{d}z = \sqrt{\pi}.$$

Now integrating over S_{22} by using a matrix-variate gamma integral one gets

$$\int_{S_{22}>0} |S_{22}|^{\alpha+\frac{r}{2}-\frac{p+1}{2}} e^{-\mathrm{tr}(S_{22})+\frac{1}{4}\mathrm{tr}(TS_{22}T')}\mathrm{d}S_{22}$$
$$= \Gamma_q(\alpha)\left|I - \frac{1}{4}T'T\right|^{-\alpha}, \quad \left\|\frac{1}{4}T'T\right\| < 1.$$

Hence

$$M_{S_{21}}(T) = \left|I - \frac{1}{4}T'T\right|^{-\alpha}, \quad \left\|\frac{1}{4}T'T\right\| < 1, \tag{2.9.5}$$

observing that

$$\Gamma_r\left(\alpha - \frac{q}{2}\right)\Gamma_q(\alpha)\pi^{\frac{qr}{2}} = \Gamma_p(\alpha).$$

If S is Wishart distributed $W_p(2\alpha, \Sigma)$ with $\Sigma = \begin{pmatrix} \Sigma_{11} & O \\ O & \Sigma_{22} \end{pmatrix}$, $\Sigma_{11} > 0$, $\Sigma_{22} > 0$ and with the density

$$f(S) = \frac{|S|^{\alpha-\frac{p+1}{2}} e^{-\frac{1}{2}\mathrm{tr}(\Sigma^{-1}S)}}{2^{\alpha p}\Gamma_p(\alpha)|\Sigma|^{\alpha}}, \quad S = \begin{bmatrix} S_{11} & S_{12} \\ S_{21} & S_{22} \end{bmatrix}, \quad R(\alpha) > \frac{p-1}{2}$$

then proceeding as before one has the m.g.f. of S_{21} given by

$$M_{S_{21}}(T) = \left| I - \Sigma_{22}^{\frac{1}{2}} T' \Sigma_{11} T \Sigma_{22}^{\frac{1}{2}} \right|^{-\alpha}, \quad \|\Sigma_{22} T' \Sigma_{11} T\| < 1. \tag{2.9.6}$$

Since the structure in (2.9.6) corresponds to the m.g.f. of a generalized Laplacian in the scalar case we will call the corresponding variable the matrix-variate generalized Laplacian.

Definition 2.9.1 *Rectangular matrix-variate generalized Laplacian* A $q \times r$ matrix with $r \geq q$ having the m.g.f. as in the right side of (2.9.6) with $\Sigma_{22} = \Sigma_{22}' > 0$, $\Sigma_{11} = \Sigma_{11}' > 0$, and where the parameter matrix T is $r \times q$ of full rank q will be called a *rectangular matrix-variate generalized Laplacian.*

2.9a Matrix-variate Laplacian

If $S \sim W_p(2\alpha, \Sigma)$ with $\Sigma = \begin{pmatrix} \Sigma_{11} & \Sigma_{12} \\ \Sigma_{21} & \Sigma_{22} \end{pmatrix}$, $\Sigma > 0$ then denoting

$$\Sigma^{-1} = \begin{bmatrix} \Sigma^{11} & \Sigma^{12} \\ \Sigma^{21} & \Sigma^{22} \end{bmatrix}, \quad S = \begin{bmatrix} S_{11} & S_{12} \\ S_{21} & S_{22} \end{bmatrix}$$

one has

$$\mathrm{tr}\,(\Sigma^{-1}S) = \mathrm{tr}\,(\Sigma^{11}S_{11}) + \mathrm{tr}\,(\Sigma^{22}S_{22}) + \mathrm{tr}\,(\Sigma^{12}S_{21}) + \mathrm{tr}\,(\Sigma^{21}S_{12}).$$

Then proceeding as before one has

$$\begin{aligned} M_{S_{21}}(T) &= \frac{|\Sigma^{11}|^{-\alpha}|\Sigma^{22}|^{-\alpha}}{|\Sigma|^{\alpha}} \\ &\quad \times \left| I - (\Sigma^{22})^{-\frac{1}{2}}(T' - \Sigma^{21})(\Sigma^{11})^{-1}(T - \Sigma^{12})(\Sigma^{22})^{-\frac{1}{2}} \right|^{-\alpha} \\ &= \Big| I - \Sigma_{22}^{\frac{1}{2}} T' [\Sigma_{11} - \Sigma_{12}\Sigma_{22}^{-1}\Sigma_{21}] T \Sigma_{22}^{\frac{1}{2}} \\ &\quad - \Sigma_{22}^{-\frac{1}{2}} \Sigma_{21} T \Sigma_{22}^{\frac{1}{2}} - \Sigma_{22}^{\frac{1}{2}} T' \Sigma_{12} \Sigma_{22}^{-\frac{1}{2}} \Big|^{-\alpha}, \text{ for} \\ &\left\| (\Sigma^{22})^{-\frac{1}{2}}(T' - \Sigma^{21})(\Sigma_{11})^{-1}(T - \Sigma^{12})(\Sigma^{22})^{-\frac{1}{2}} \right\| < 1, \end{aligned} \tag{2.9.7}$$

by observing that $\Sigma_{12}\Sigma_{22}^{-1} = -(\Sigma^{11})^{-1}\Sigma^{12}$ and $(\Sigma^{11})^{-1} = \Sigma_{11} - \Sigma_{12}\Sigma_{22}^{-1}\Sigma_{21}$.

Theorem 2.9.1 Let

$$S = \begin{bmatrix} S_{11} & S_{12} \\ S_{21} & S_{22} \end{bmatrix} \sim W_p(2\alpha, \Sigma),\ \Sigma = \begin{bmatrix} \Sigma_{11} & \Sigma_{12} \\ \Sigma_{21} & \Sigma_{22} \end{bmatrix} = \Sigma' > 0$$

where S_{11} and Σ_{11} are $r \times r$, S_{21} is $q \times r$, $r \geq q$, $r + q = p$. Then S_{21} is a rectangular matrix-variate generalized Laplacian as in the Definition 2.9.1 iff $\Sigma_{12} = O$.

PROOF The m.g.f. of S_{21} is given in (2.9.7). If $\Sigma_{12} = O$ then this m.g.f. reduces to that in (2.9.6). Hence the condition is sufficient. In order to see the necessity consider the equation

$$\Big| I - \Sigma_{22}^{\frac{1}{2}} T' [\Sigma_{11} - \Sigma_{12}\Sigma_{22}^{-1}\Sigma_{21}] T \Sigma_{22}^{\frac{1}{2}}$$

$$- \Sigma_{22}^{-\frac{1}{2}} \Sigma_{21} T \Sigma_{22}^{\frac{1}{2}} - \Sigma_{22}^{\frac{1}{2}} T' \Sigma_{12} \Sigma_{22}^{-\frac{1}{2}} \Big|^{-\alpha}$$

$$= |I - AT'BT|^{-\alpha},\ A = A' > 0,\ B = B' > 0.$$

Take logarithms on both sides, expand and write in power series involving traces by using the format, for $||G|| < 1$,

$$\ln |I - G|^{-\alpha} = \alpha \left[\mathrm{tr}(G) + \frac{1}{2}\mathrm{tr}(G^2) + \cdots \right].$$

Then equate the linear terms in T on both sides to get

$$\mathrm{tr}\left[\Sigma_{22}^{-\frac{1}{2}} \Sigma_{21} T \Sigma_{22}^{\frac{1}{2}}\right] + \mathrm{tr}\left[\Sigma_{22}^{\frac{1}{2}} T' \Sigma_{12} \Sigma_{22}^{-\frac{1}{2}}\right] = 0$$

$\Rightarrow \mathrm{tr}[\Sigma_{21} T] = 0 \Rightarrow \Sigma_{21} = O$ since T is arbitrary. This establishes the necessity. Imposing the condition $\Sigma_{12} = O$ one can note that $A = \Sigma_{22}$ and $B = \Sigma_{11}$.

2.9b The Density of S_{21}

The density of S_{21} can be evaluated either as the inverse Laplace transform of $M_{S_{21}}(-T)$ in (2.9.7) or by direct integration from the Wishart density. Let $g(S_{21})$ be the density of S_{21}. Then

$$g(S_{21}) = \int_{S_{11}>0} \int_{S_{22}>0} \frac{|S|^{\alpha - \frac{p+1}{2}} e^{-\frac{1}{2}\mathrm{tr}(\Sigma^{-1}S)}}{2^{\alpha p} |\Sigma|^{\alpha} \Gamma_p(\alpha)} \mathrm{d}S_{11}\, \mathrm{d}S_{22}.$$

Write $|S| = |S_{22}||S_{11} - S_{12}S_{22}^{-1}S_{21}|$ then $S_{11} > S_{12}S_{22}^{-1}S_{21}$, $S_{22} > 0$. Integrate out S_{11} to get

$$g(S_{21}) = \frac{\Gamma_r\left(\alpha - \frac{q}{2}\right) |\Sigma^{11}|^{-\alpha + \frac{q}{2}}}{|\Sigma|^{\alpha} \Gamma_p(\alpha)} \int_{S_{22}>0} |S_{22}|^{\alpha - \frac{p+1}{2}} e^{-\mathrm{tr}(\Sigma^{22} S_{22})}$$

$$\times e^{-\mathrm{tr}(\Sigma^{12}S_{21}) - \mathrm{tr}(\Sigma^{21}S_{12}) - \mathrm{tr}\left[(\Sigma^{11})^{\frac{1}{2}} S_{12} S_{22}^{-1} S_{21} (\Sigma^{11})^{\frac{1}{2}}\right]} \mathrm{d}S_{22}.$$

Write $U_2 = (\Sigma^{22})^{\frac{1}{2}} S_{22} (\Sigma^{22})^{\frac{1}{2}}$. Then

$$g(S_{21}) = \frac{\Gamma_r\left(\alpha - \frac{q}{2}\right)}{\Gamma_p(\alpha)} \frac{|\Sigma^{11}|^{-\alpha+\frac{q}{2}} |\Sigma^{22}|^{-\alpha+\frac{r}{2}}}{|\Sigma|^{\alpha}}$$

$$\times \int_{U_2>0} |U_2|^{\alpha-\frac{p+1}{2}} e^{-\mathrm{tr}(U_2)-\mathrm{tr}\left(\Sigma^{12} S_{21}\right)} \tag{2.9.8}$$

$$\times e^{-\mathrm{tr}\left(\Sigma^{21} S_{12}\right)-\mathrm{tr}\left[(\Sigma^{11})^{\frac{1}{2}} S_{12} (\Sigma^{22})^{\frac{1}{2}} U_2^{-1} (\Sigma^{22})^{\frac{1}{2}} S_{21} (\Sigma^{11})^{\frac{1}{2}}\right]} \mathrm{d}U_2.$$

The gammas can be simplified to the following:

$$\frac{\Gamma_r\left(\alpha - \frac{q}{2}\right)}{\Gamma_p(\alpha)} = \frac{1}{\pi^{\frac{qr}{2}} \Gamma_q(\alpha)}.$$

The integral in (2.9.8) can be evaluated as a G-function of matrix argument or in terms of a Bessel function of matrix argument. But for studying the properties of $g(S_{21})$ the integral representation in (2.9.8) will be more convenient. In order to verify that the m.g.f. of S_{21} agrees with (2.9.7) write

$$M_{S_{21}}(T) = E\left[e^{\mathrm{tr}(TS_{21})}\right] = E\left[e^{\frac{1}{2}\left\{\mathrm{tr}(TS_{21})+\mathrm{tr}(S_{12}T')\right\}}\right].$$

Express this expected value with the help of (2.9.8). Integrate out S_{21} first and then U_2 to see that the result agrees with that in (2.9.7).

2.9c A Particular Case

What is the distribution of S_{21} when $r = q = 1,\ p = 2$? In this case $g(S_{21})$ reduces to a very simple form. This can be seen by using the result that for $R(\alpha) > 0$ and the scalar variable $u > 0$,

$$\int_0^\infty v^{\alpha-1} e^{-v-uv^{-1}} \mathrm{d}v = G_{0,2}^{2,0}\left(u\big|_{\alpha,0}\right)$$

$$= 2u^{\frac{\alpha}{2}} K_\alpha\left(2u^{\frac{1}{2}}\right) \tag{2.9.9}$$

where $K_\alpha(\cdot)$ is a Bessel function. The second part of the result in (2.9.9) is available from Mathai and Saxena (1978, p.145) and the first part can be seen from standard tables or by taking two independent positive scalar variables X and Y with the densities $f_1(x) = c_1 x^\alpha e^{-x}$ and $f_2(y) = c_2 e^{-y}$ where c_1 and c_2 are normalizing constants, and then evaluating the density of $U = XY$ by using transformation of variables to get the integral representation and by using the Mellin transform technique to get the G-function. The integral in (2.9.9) is a basic integral in many problems in astrophysics, see for example, Mathai and Haubold (1988).

Note also that when $p = 2,\ \Sigma_{11} = \sigma_{11} = \sigma_1^2,\ \Sigma_{22} = \sigma_{22} = \sigma_2^2,\ \Sigma_{12} = \rho\sigma_1\sigma_2$ where

ρ is the linear correlation , one has

$$\frac{\Gamma_r\left(\alpha-\frac{q}{2}\right)}{\Gamma_p(\alpha)}=\frac{\Gamma\left(\alpha-\frac{1}{2}\right)}{\pi^{\frac{1}{2}}\Gamma(\alpha)\Gamma\left(\alpha-\frac{1}{2}\right)}=\frac{1}{\pi^{\frac{1}{2}}\Gamma(\alpha)},$$

$$\frac{|\Sigma^{11}|^{-\alpha+\frac{q}{2}}|\Sigma^{22}|^{-\alpha+\frac{r}{2}}}{|\Sigma|^{\alpha}}=\frac{(1-\rho^2)^{\alpha-1}}{\sigma_1\sigma_2},$$

and

$$\Sigma^{12}S_{21}=-\frac{\rho s_{21}}{\sigma_1\sigma_2(1-\rho^2)}.$$

Hence from (2.9.8) one has

$$\begin{aligned}g(s_{21})&=\frac{1}{\pi^{\frac{1}{2}}\Gamma(\alpha)}\frac{(1-\rho^2)^{\alpha-1}}{\sigma_1\sigma_2}G^{2,0}_{0,2}\left(u\big|_{\alpha,0}\right)\\&\quad\times\exp[2\rho s_{21}/(\sigma_1\sigma_2(1-\rho^2))]\\&=\frac{2}{\pi^{\frac{1}{2}}\Gamma(\alpha)}\frac{(1-\rho^2)^{\alpha-1}}{\sigma_1\sigma_2}u^{\frac{\alpha}{2}}K_\alpha\left(2u^{\frac{1}{2}}\right)\\&\quad\times\exp\left[2\rho s_{21}/(\sigma_1\sigma_2(1-\rho^2))\right],\end{aligned}$$

$$u=s_{21}^2/\left((1-\rho^2)^2\sigma_1^2\sigma_2^2\right),\ -\infty<s_{21}<\infty.$$

When $\rho=0$ the density reduces to

$$g(s_{21})=\frac{2}{\sigma_1\sigma_2\pi^{\frac{1}{2}}\Gamma(\alpha)}G^{2,0}_{0,2}\left(u^*\big|_{\alpha,0}\right),\ u^*=\frac{s_{21}^2}{\sigma_1^2\sigma_2^2},\ -\infty<s_{21}<\infty.$$

This type of a structure also comes from the density of the product of two independent real random variables where one is the square root of a constant multiple of a gamma variable and the other is a constant multiple of a standard normal variable.

When $S\sim W_p(2\alpha,I)$ and when S is written as $S=ZZ'$ where $Z=(z_{ij})$ with $z_{ii}>0,\ i=1,\ldots,p,\ z_{ij}=0,\ i<j$ then it is known that the z_{ij}'s $i\geq j$ are mutually independently distributed with the z_{ij}'s $i>j$ as standard normals and the z_{ii}'s as square root of chi-square variables. Thus if

$$Z=\begin{bmatrix}Z_1 & O\\ Z_2 & Z_3\end{bmatrix}$$

where Z_1,Z_2,Z_3 are submatrices then the submatrix $S_{12}=Z_1Z_2'$. Thus the general density worked out for S_{12} also serves for the configuration of normal and chi-square variables in Z_1Z_2' or Z_2Z_1'.

Some more results on Laplacianness, independence and the distributions of bilinear forms in normal variables may be seen from Mathai (1992) and Mathai (1993a,b,c,d).

EXERCISES

2.1 Prove Lemma 2.5.1.

2.2 Prove that $Q(\mathbf{Z}) = \mathbf{Z}'A\mathbf{Z}+\mathbf{a}'\mathbf{Z}+d$ cannot be a NGL or a gamma difference if $A \geq 0$.

2.3 For Corollary 2.5.5 and Corollary 2.6.1 write down the conditions explicitly.

2.4 By using the Notation 2.4.2 write down the explicit forms of the joint cumulants $K_{2,3}$ and $K_{3,2}$ of a quadratic form and a bilinear form in the nonsingular normal case under (a): independence; (b): equicorrelation.

2.5 Write down the explicit form of $K_{2,2}$ in equation (2.4.17) for the independence and equicorrelated cases.

2.6 Let $\mathbf{X}' = (X_1, X_2)$ and $\mathbf{Y}' = (Y_1, Y_2, Y_3)$ be jointly normally distributed with mean value null and the following covariance matrices.

$$\text{Cov}(\mathbf{X}) = \begin{bmatrix} 2 & 1 \\ 1 & 2 \end{bmatrix}, \quad \text{Cov}(\mathbf{Y}) = \begin{bmatrix} 2 & 0 & 1 \\ 0 & 1 & -1 \\ 1 & -1 & 2 \end{bmatrix},$$

$$\text{Cov}(\mathbf{X},\ \mathbf{Y}) = \begin{bmatrix} 1 & 1 & 1 \\ 1 & 1 & 1 \end{bmatrix}.$$

Let

$$Q_1 = 2X_1^2 + X_2^2 + 2X_1X_2,$$

$$Q_2 = X_1Y_1 + X_1Y_2 + X_1Y_3 + X_2Y_1 + X_2Y_2 + X_2Y_3$$

and

$$Q_3 = 2X_1Y_1 + X_1Y_2 - X_1Y_3 + X_2Y_1 - X_2Y_2 + 2X_2Y_3.$$

Then show that

$$\text{Var}(Q_1) = 116, \quad \text{Var}(Q_2) = 66, \quad \text{Cov}(Q_1,\ Q_2) = 60.$$

2.7 For the same Q_1 and Q_2 of Problem 2.6 but with $\text{Cov}(\mathbf{X}, \mathbf{Y}) = 0$ show that $K_{2,2}$ (between Q_1 and Q_2) $= 1710$ and $K_{2,2}$ (between Q_2 and Q_3) $= 9288$.

2.8 Let $\mathbf{X}$ and $\mathbf{Y}$ be distributed as in Problem 2.6 but with $E(\mathbf{X}') = (0,0)$, and $E(\mathbf{Y}') = (0,0,-1)$. Let

$$Q = X_1Y_1 + X_1Y_2 + X_1Y_3 + X_2Y_1 + X_2Y_2 + X_2Y_3$$

$$- 2X_1 + 3Y_2 + 280.$$

Then show that the second cumulant of Q is 77.

2.9 Let $\mathbf{X}' = (X_1, X_2, X_3)$ and $\mathbf{X} \sim N_3(\mu,\ \Sigma)$ with

$$\mu = \begin{pmatrix} 0 \\ -1 \\ 1 \end{pmatrix}, \quad \Sigma = \begin{pmatrix} 2 & 1 & 0 \\ 1 & 2 & 0 \\ 0 & 0 & 6 \end{pmatrix}.$$

Let $Q = X_1^2 + X_2^2 - X_3^2 + 2X_1X_2$. Then show that Q is a NGL with the parameters $\alpha = \frac{1}{2}$, $\beta = 12$, $\lambda = \frac{1}{12}$.

2.10 Verify that U_1, U_2, U_3 and U_4 in Example 2.8.3 are distributed as gamma differences.

CHAPTER 3

Quadratic and Bilinear Forms in Elliptically Contoured Distributions

3.0 Introduction

The multivariate normal distribution is the most widely used multivariate distribution in statistical theory today. A p-variate vector $\boldsymbol{X}$ is said to have a *real nonsingular normal distribution* if its density is given by

$$f(\boldsymbol{x}) = \frac{e^{-\frac{1}{2}(\boldsymbol{x}-\boldsymbol{\mu})'\Sigma^{-1}(\boldsymbol{x}-\boldsymbol{\mu})}}{(2\pi)^{\frac{p}{2}}|\Sigma|^{\frac{1}{2}}} \tag{3.0.1}$$

where Σ is a real symmetric positive definite matrix of parameters, that is, $\Sigma = \Sigma' > 0$, and $\boldsymbol{\mu}$ is a p-vector of real parameters. It can be shown that $\boldsymbol{\mu} = E(\boldsymbol{X})$ = expected value of $\boldsymbol{X}$ and Σ is the covariance matrix of $\boldsymbol{X}$, that is,

$$\Sigma = \text{Cov}(\boldsymbol{X}) = E(\boldsymbol{X} - E\boldsymbol{X})(\boldsymbol{X} - E\boldsymbol{X})' .$$

If $\boldsymbol{\mu} = \boldsymbol{0}$ (null vector) and $\Sigma = I$ (identity matrix), then $\boldsymbol{X}$ is said to have a *standard normal distribution.* A standard notation used in the literature is $\boldsymbol{X} \sim N_p(\boldsymbol{\mu}, \Sigma)$ meaning that the p-vector $\boldsymbol{X}$ is normally distributed with the parameters $\boldsymbol{\mu}$ and Σ. Thus the standard normal will be denoted as $\boldsymbol{X} \sim N_p(\boldsymbol{0}, I)$.

If the covariance matrix of $\boldsymbol{X}$ is singular then one can write $\Sigma = BB'$ since Σ is at least positive semi-definite where B is $p \times r$ and r is the rank of Σ. Whether Σ is singular or nonsingular we can always find a vector $\boldsymbol{Y}$ such that

$$\boldsymbol{X} = \boldsymbol{\mu} + B\boldsymbol{Y} \tag{3.0.2}$$

where $\boldsymbol{Y} \sim N_r(\boldsymbol{0}, I)$. By using the representation in (3.0.2) in the *real singular normal* case, that is Σ is singular, and from (3.0.1) in the nonsingular case, one can evaluate the characteristic function of the p-vector $\boldsymbol{X}$. In both the real singular and nonsingular normal cases it can be shown that the characteristic function reduces to the following:

$$E(e^{it'X}) = e^{it'\mu - \frac{1}{2}t'\Sigma t} \tag{3.0.3}$$

where t is a $p \times 1$ vector of parameters and $i = \sqrt{-1}$. Note that in the singular normal case the density of $\boldsymbol{X}$ will not exist but all the properties of $\boldsymbol{X}$ can be studied through the vector $\boldsymbol{Y}$ of (3.0.2). In the nonsingular case B of (3.0.2) is $p \times p$ of rank p.

Some basic properties of the normal density in (3.0.1) are the following. Consider the equation

$$f(\boldsymbol{x}) = c_1 \tag{3.0.4}$$

where c_1 is a constant. Then

$$-\ln f(\boldsymbol{x}) = -\ln c_1 \Rightarrow (\boldsymbol{x} - \boldsymbol{\mu})'\Sigma^{-1}(\boldsymbol{x} - \boldsymbol{\mu}) = c \tag{3.0.5}$$

where c is a constant. Thus the contours of constant density are ellipsoids where $\boldsymbol{X} \sim N_p(\boldsymbol{\mu}, \Sigma)$, and when $\Sigma = I$, the identity matrix, these ellipsoids are spheres with the centre at $\boldsymbol{\mu}$. We may call $\boldsymbol{X} \sim N_p(\boldsymbol{0}, I)$ a spherical distribution. General definitions of elliptically contoured and spherical distributions will be given in the next section.

Another interesting property of the normal distribution is that if $\boldsymbol{Y} \sim N_p(\boldsymbol{0}, I)$ then $V = \boldsymbol{Y}'\boldsymbol{Y} = ||\boldsymbol{Y}||^2$ has a chi-square distribution with p degrees of freedom. Also note that the characteristic function of $\boldsymbol{Y} \sim N_p(\boldsymbol{0}, I)$ as well as that of $\boldsymbol{Z} = P\boldsymbol{Y}$, where P is an orthogonal matrix, are one and the same which means that there is invariance under orthogonal transformations.

If the factor $e^{-\frac{1}{2}t'\Sigma t}$ of (3.0.3) is replaced by a general positive function of $\boldsymbol{t}'\Sigma\boldsymbol{t}$, what will be the distribution of $\boldsymbol{X}$? Distributions of this type are said to belong to the class of elliptically contoured distributions.

Various distributional results on quadratic forms in elliptically contoured vectors and matrices are presented in this chapter. By virtue of the relationship (2.1.4), these results also apply to bilinear forms.

The main references used in this chapter are Anderson and Fang (1987) for basic results, Cochran's theorem and extensions to quadratic forms in elliptically contoured random matrices [Sections 3.1 and 3.5]; Gang (1987) in connection with the moments of quadratic forms in elliptically contoured vectors [Section 3.2]; Fan (1986) for distributional results on quadratic forms and noncentral versions of Cochran's theorem [Sections 3.3 and 3.7]; Hsu (1990) for the distribution of quadratic forms in noncentral vectors [Section 3.4]; and Fang, Fan and Xu (1987) for the distribution of quadratic forms of random idempotent matrices [Section 3.6]. Section 3.8 which is devoted to test statistics for elliptically contoured distributions contains several additional references. More details and additional results on spherically symmetric and elliptically contoured distributions may be found in the books Fang and Anderson (1990) and Fang, Kotz and Ng (1990). Related results and additional references are available in Cacoullos and Koutraz (1984), Chemielewst (1981), Das Gupta, Eaton, Olkin, Perlman, Savage and Sobel (1972), Fang and Wu (1984), Fraser and Ng (1980), Graybill and Milliken (1969), and Kelker (1970).

3.1 Definitions and Basic Results

Definition 3.1.1 *Elliptically contoured distributions* Let $\boldsymbol{\mu}$ be a p-dimensional real vector, Σ be a $p \times p$ nonnegative definite matrix and $\xi(\cdot)$ be a nonnegative function; then the p-dimensional vector $\boldsymbol{X}$ is said to have an elliptically contoured distribution if its characteristic function $\phi(\boldsymbol{t})$ can be written as

$$\phi(\boldsymbol{t}) = e^{i\boldsymbol{t}'\boldsymbol{\mu}}\xi(\boldsymbol{t}'\Sigma\boldsymbol{t}) . \tag{3.1.1}$$

This will be denoted by

$$\boldsymbol{X} \sim \mathcal{C}_p(\xi; \boldsymbol{\mu}, \Sigma) . \tag{3.1.2}$$

Definition 3.1.2 *Spherical distributions* When $\boldsymbol{\mu}$ is the null vector and Σ is the identity matrix of order p, the notation $\boldsymbol{X} \sim \mathcal{C}_p(\xi; \boldsymbol{0}, I)$ is shortened to

$$\boldsymbol{X} \sim \mathcal{S}_p(\xi) \tag{3.1.3}$$

and $\boldsymbol{X}$ is said to have a spherical distribution or spherically symmetric distribution.

A spherical distribution can also be defined by a general density of the form $g(\boldsymbol{y}'\boldsymbol{y})$ where g is an arbitrary density. Note that $g(\cdot)$ is invariant under orthogonal transformations. The density of $\boldsymbol{Y} \sim N_p(\boldsymbol{0}, I)$ is a particular case of $g(\cdot)$.

If $\boldsymbol{Y}$ has the density $g(\boldsymbol{y}'\boldsymbol{y})$ then the density of $\mathcal{W} = ||\boldsymbol{Y}||$ can be evaluated by changing to polar coordinates. Consider the transformation

$$\begin{aligned} y_1 &= w\ \sin\theta_1 , \\ y_j &= w\ \cos\theta_1\ \cos\theta_2 \cdots \cos\theta_{j-1}\ \sin\theta_j,\ j = 2, 3, \ldots, p-1 \\ y_p &= w\ \cos\theta_1\ \cos\theta_2 \cdots \cos\theta_{p-1}, \\ &\qquad 0 < \theta_j \le \pi,\ \ j = 1, \ldots, p-2,\ 0 < \theta_{p-1} \le 2\pi . \end{aligned} \tag{3.1.4}$$

Then from the Jacobian we have

$$\mathrm{d}y_1 \ldots \mathrm{d}y_p = w^{p-1}\Big\{\prod_{j=1}^{p-1} |\cos\theta_j|^{p-j-1}\Big\}\mathrm{d}w\ \mathrm{d}\theta_1 \ldots \mathrm{d}\theta_{p-1} . \tag{3.1.5}$$

The joint density of $w, \theta_1, \ldots, \theta_{p-1}$ is given by

$$g(w^2)w^{p-1}\Big\{\prod_{j=1}^{p-1} |\cos\theta_j|^{p-j-1}\Big\}\mathrm{d}w\ \mathrm{d}\theta_1 \ldots \mathrm{d}\theta_{p-1} . \tag{3.1.6}$$

Note that $\mathcal{W}$ and $(\theta_1, \ldots, \theta_{p-1})$ are independently distributed; w represents an observed value of the random variable $\mathcal{W}$. Integrating out the θ_j's, $0 < \theta_j \le \pi,\ j = 1, \ldots, p-2$ and $0 < \theta_{p-1} \le 2\pi$, one has the density of $\mathcal{W} = V^{\frac{1}{2}} = ||\boldsymbol{Y}||$ given by

$$f_{\mathcal{W}}(w) = \begin{cases} \frac{2\pi^{\frac{p}{2}}}{\Gamma(\frac{p}{2})} w^{p-1} g(w^2), & 0 < w < \infty \\ 0, & \text{elsewhere} . \end{cases} \tag{3.1.7}$$

The density given in (3.1.7) can also be obtained from Theorem 3.1.2 by letting $n = p$, $q = 1$, and $F = w^2$. The transformation in (3.1.4) can be written as

$$Y = \mathcal{W}U^{(p)} \tag{3.1.8}$$

where $U^{(p)}$ is the p-vector involving $\theta_1, \ldots, \theta_{p-1}$. Note that $U^{(p)}$ has a uniform distribution on the p-dimensional sphere. Further, $U^{(p)}$ and $\mathcal{W}$ are independently distributed. The characteristic function of Y is then given by

$$E(e^{it'Y}) = E(e^{it'\mathcal{W}U^{(p)}}) . \tag{3.1.9}$$

Integrating out $U^{(p)}$ at $\mathcal{W} = w$, over the joint density of $\theta_1, \ldots, \theta_{p-1}$ which is available from (3.1.6), one can show that the conditional expectation of $e^{it'Y}$ at $\mathcal{W} = w$ is a function of the form $\varphi(w^2 t't)$; this can also be seen by noticing that the conditional expectation is in fact the characteristic function of $U^{(p)}$ evaluated at the point wt which has the form $\varphi(w^2 t't)$ in view of (3.1.1) since the distribution of $U^{(p)}$ belongs to the class of elliptically contoured distributions. Now integrating out over the density of $\mathcal{W}$, which is available form (3.1.7), it is evident that the characteristic function is of the form $\xi(t't)$ where $\xi(\cdot)$ is a positive function. That is,

$$E(e^{it'Y}) = \xi(t't) . \tag{3.1.10}$$

If $Y \sim \mathcal{S}_p(\xi)$ with the characteristic function $\xi(t't)$ and if X is written as

$$X = \mu + BY$$

where μ is a $p \times 1$ constant vector, then letting $\Sigma = BB'$ the characteristic function of X is given by

$$E(e^{it'X}) = e^{it'\mu}\xi(t'\Sigma t) . \tag{3.1.11}$$

If Y has the density $g(y'y)$ and if B is nonsingular, then X has the density

$$|\Sigma|^{-\frac{1}{2}} g[(x - \mu)'\Sigma^{-1}(x - \mu)] . \tag{3.1.12}$$

Note that the contours of constant density are still ellipsoids:

$$(x - \mu)'\Sigma^{-1}(x - \mu) = \text{ constant} . \tag{3.1.13}$$

From the characteristic function in (3.1.11), note that if some components of t are set as zeros the structure still remains intact which means that if X has an elliptically contoured distribution then all the marginal distributions are also elliptical.

Note that if the $p \times 1$ random vector X has an elliptically contoured distribution with the characteristic function as in (3.1.11), and if $U = AX$ where A is a $q \times p$ constant matrix of rank $q \leq p$, then the characteristic function of U is given by

$$E(e^{is'U}) = E(e^{is'AX}) = E(e^{i(s'A)X})$$

where s is a $q \times 1$ vector of parameters. Now proceeding as in the derivation of (3.1.11) one has

$$E(e^{is'U}) = e^{is'A\mu}\xi(s'A\Sigma A's) \tag{3.1.14}$$

which means that $U = AX$ also has an elliptically contoured distribution.

Notation 3.1.1 The notation $X \simeq Y$ will be used to indicate that the random vectors X and Y are identically distributed.

Theorem 3.1.1 (Cambanis, Huang, and Simmons (1981)) Let $\Sigma = L'L$ where L is a $q \times p$ matrix of rank q and let the rank of Σ be $q \le p$. Let the random variable $\mathcal{W}$ represent the distance from the point μ in every direction and $U^{(\alpha)}$ denote a random vector which is uniformly distributed on the unit sphere in $\Re^\alpha$, then $X \sim \mathcal{C}_p(\xi; \mu, \Sigma)$ if and only if

$$X \simeq \mu + L'\mathcal{W}U^{(q)} \tag{3.1.15}$$

where $\mathcal{W}$ is a nonnegative random variable which is distributed independently of $U^{(q)}$ and whose distribution function, $F(w)$, is such that

$$\xi(t't) = E_{\mathcal{W}}\left[E_{U^{(q)}|\mathcal{W}=w}(e^{it'wU^{(q)}})\right] = \int_0^\infty E(e^{iwt'U^{(q)}})\, dF(w) \tag{3.1.16}$$

where $\xi(t't)$ is the characteristic function of $\mathcal{W}U^{(q)} \sim S_q(\xi)$.

Definition 3.1.3 *Stieffel manifold* Let A be a $q\times p$, $q \ge p$ matrix such that $A'A = I_p$, that is, the p columns of A are orthonormal vectors. The set of all such matrices A is known as the Stieffel manifold, denoted by $V_{p,q}$. That is, for all $q \times p$ matrices A, $q \ge p$,

$$V_{p,q} = \{A : A'A = I_p\} .$$

When $p = q$, $V_{p,q}$ is the full orthogonal group denoted by $O(p)$, that is,

$$O(p) = \{B : B'B = I_p\} .$$

Theorem 3.1.2 Let X be a $n \times q$ random matrix of rank q, $n \ge q$, whose p.d.f. can be written as $g(X'X)$; then the p.d.f. of $F = X'X$ is given by

$$\frac{\pi^{\frac{nq}{2}}|F|^{\frac{n}{2}-\frac{q+1}{2}}g(F)}{\Gamma_q(\frac{n}{2})} \tag{3.1.17}$$

where $\Gamma_q(\cdot)$ denotes the matrix-variate gamma function; for example,

$$\Gamma_q(\alpha) = \pi^{\frac{q(q-1)}{4}}\Gamma(\alpha)\Gamma(\alpha-\tfrac{1}{2})\cdots\Gamma(\alpha-\tfrac{q-1}{2})\,,\quad \mathrm{Re}(\alpha) > \tfrac{q-1}{2}\ . \tag{3.1.18}$$

PROOF Let $X = UT$ where T is an upper triangular matrix with positive diagonal elements and U is a semiorthogonal matrix. Then from the Jacobian of this transformation

$$dX = \Big\{ \prod_{j=1}^{q} t_{jj}^{n-j} \Big\} \, dT \Big\{ \prod_{j=1}^{q} |U_{(j)}| \Big\} \, dU \tag{3.1.19}$$

where t_{jj}'s are the diagonal elements of T, $U_{(j)}$ is the j-th leading submatrix of U and $|U_{(j)}|$, its determinant. Also, we have (see Theorems 1.1a.5, 1.1a.11 or Muirhead (1982)),

$$F = X'X = T'T \Rightarrow dF = 2^q \Big\{ \prod_{j=1}^{p} t_{jj}^{q+1-j} \Big\} \, dT \, . \tag{3.1.20}$$

Substituting for dT and noting that $|F| = \prod_{j=1}^{q} t_{jj}^2$, we have

$$\begin{aligned} dX &= 2^{-q} \Big\{ \prod_{j=1}^{q} t_{jj}^{n-(q+1)} \Big\} \, dF \Big\{ \prod_{j=1}^{q} |U_{(j)}| \Big\} \, dU \\ &= 2^{-q} |F|^{\frac{n}{2} - \frac{q+1}{2}} \, dF \Big\{ \prod_{j=1}^{q} |U_{(j)}| \Big\} \, dU \, . \end{aligned} \tag{3.1.21}$$

Integrating out U over the Stieffel manifold $V_{q,n}$, we have

$$\int_{V_{q,n}} \Big\{ \prod_{j=1}^{q} |U_{(j)}| \Big\} \, dU = 2^q \frac{\pi^{\frac{qn}{2}}}{\Gamma_q(\frac{n}{2})} \, . \tag{3.1.22}$$

Thus

$$g(X'X) \, dX = \frac{\pi^{\frac{qn}{2}}}{\Gamma_q(\frac{n}{2})} |F|^{\frac{n}{2} - \frac{q+1}{2}} g(F) \, dF \, .$$

This completes the proof.

Definition 3.1.4 *Dirichlet distribution (type 1)* A random vector $V = (V_1, \ldots, V_\ell)'$ is said to have a Dirichlet distribution of the first type with parameters $a_1, \ldots, a_\ell$ if its p.d.f. is

$$\frac{\Gamma(\sum_{i=1}^{\ell} a_i)}{\prod_{i=1}^{\ell} \Gamma(a_i)} v_1^{a_1 - 1} \cdots v_\ell^{a_\ell - 1} \, , \quad v_\ell = 1 - \sum_{i=1}^{\ell-1} v_i \, , \tag{3.1.23}$$

for $0 \le v_i \le 1$, $a_i > 0$, $i = 1, \ldots, \ell$, and 0 elsewhere. This is denoted by

$$(V_1, \ldots, V_{\ell-1})' \sim \mathcal{D}_\ell(a_1, \ldots, a_\ell)$$

or

$$(V_1, \ldots, V_\ell)' \sim \mathcal{D}_\ell(a_1, \ldots, a_\ell) \, .$$

The Dirichlet distribution of the first type is a generalization of the *beta (type 1) distribution* denoted Beta(a_1, a_2) for which $\ell = 2$ and $V_2 = 1 - V_1$. The probability density function of a type-1 real beta random variable is given in (1.4.9).

Definition 3.1.5 *Real gamma distribution* A scalar random variable X is said to have a real gamma distribution, denoted by $X \sim$ Gamma (α, β) if its density is of the form

$$f(x) = \begin{cases} \frac{x^{\alpha-1}e^{-x/\beta}}{\beta^{\alpha}\Gamma(\alpha)}, & x > 0,\ \alpha > 0,\ \beta > 0 \\ 0, & \text{elsewhere.} \end{cases}$$

Theorem 3.1.3 Let $Y_j \sim \text{Gamma}(\alpha_j, \beta)$, $j = 1, \ldots, \ell$ and mutually independently distributed and let $Y = \sum_{j=1}^{\ell} Y_j$ then $(\frac{Y_1}{Y}, \ldots, \frac{Y_\ell}{Y})' \sim D_\ell(\alpha_1, \ldots, \alpha_{\ell-1}, \alpha_\ell)$.

PROOF Consider the transformation $z_j = \frac{y_j}{y}, j = 1, \ldots, \ell - 1$ and $z = y$. Then the Jacobian is $y^{\ell-1}$ and the joint density $g(z_1, \ldots z_{\ell-1}, z)$ of $Z_1, \ldots, Z_{\ell-1}, Z$ is available from the joint density of $Y_1, \ldots, Y_\ell$ which is the product of the marginal densities due to independence. That is,

$$\begin{aligned} g(z_1, \ldots, z_{\ell-1}, z) &= \left\{ \prod_{j=1}^{\ell} \frac{1}{\beta^{\alpha_j}\Gamma(\alpha_j)} y_j^{\alpha_j - 1} e^{-y_j/\beta} \right\} \frac{1}{y^{\ell-1}} \\ &= \frac{1}{\beta^{\alpha_1+\cdots+\alpha_\ell}\Gamma(\alpha_1)\cdots\Gamma(\alpha_\ell)} z_1^{\alpha_1-1} \cdots z_{\ell-1}^{\alpha_{\ell-1}-1} \Big(1 - \sum_{j=1}^{\ell-1} z_j\Big)^{\alpha_\ell - 1} \\ &\quad \times z^{\alpha_1+\cdots+\alpha_\ell-1} e^{-z/\beta} . \end{aligned}$$

Integrating out z one has the result.

It follows that $(Y_1/Y, \ldots, Y_{\ell-1}/Y)$ is distributed independently of $Y \sim \text{Gamma}(\alpha_1 + \cdots + \alpha_\ell, \beta)$ and that if $\boldsymbol{X} \sim N_p(0, I)$ is partitioned into ℓ subvectors , that is, $\boldsymbol{X}' = (\boldsymbol{X}_1', \ldots, \boldsymbol{X}_\ell')$ where $\boldsymbol{X}_i$ has p_i components, $i = 1, \ldots, \ell$, then

$$\Big(\frac{\boldsymbol{X}_1'\boldsymbol{X}_1}{\boldsymbol{X}'\boldsymbol{X}}, \ldots, \frac{\boldsymbol{X}_\ell'\boldsymbol{X}_\ell}{\boldsymbol{X}'\boldsymbol{X}}\Big) \sim D_\ell\Big(\frac{p_1}{2}, \ldots, \frac{p_\ell}{2}\Big) \tag{3.1.24}$$

is distributed independently of $\boldsymbol{X}'\boldsymbol{X}$.

Theorem 3.1.4 When $\boldsymbol{X} \sim N_p(0, I)$, where $\boldsymbol{X} \simeq \mathcal{W}U^{(p)}$, then $\mathcal{W}^2 \simeq \boldsymbol{X}'\boldsymbol{X}$ is distributed as a chi-square variable having p degrees of freedom; letting p_j denote the number of components of $\boldsymbol{X}_j$, $j = 1, \ldots, \ell$, $\boldsymbol{X}_i'\boldsymbol{X}_i$ and $\boldsymbol{X}_j'\boldsymbol{X}_j$ are independently distributed for $1 \le i \ne j \le \ell$ and

$$\boldsymbol{X}_j'\boldsymbol{X}_j \sim \chi^2_{p_j} \ , \quad j = 1, \ \ldots, \ \ell \ ; \tag{3.1.25}$$

furthermore when $\mathcal{W}^2 \sim \chi^2_p$ and $U^{(p)}$ are independently distributed, then

$$\boldsymbol{X} = (\boldsymbol{X}_1', \ldots, \boldsymbol{X}_\ell')' \simeq \mathcal{W}U^{(p)} \sim N_p(0, I) \ . \tag{3.1.26}$$

PROOF The first part is evident and hence we shall prove the second part. Let $\mathcal{W}^2 \sim \chi^2_p$ and $\boldsymbol{U}^{(p)}$ be uniform on the unit sphere and let $\mathcal{W}^2$ and $\boldsymbol{U}^{(p)}$ be independently distributed. Writing $\boldsymbol{U}^{(p)}$ in polar coordinates the joint density of $\mathcal{W}^2$ and $\boldsymbol{U}^{(p)}$ is given by

$$f(w^2, \boldsymbol{u}^{(p)}) = c\frac{(w^2)^{\frac{p}{2}-1}e^{-w^2/2}}{2^{p/2}\Gamma(p/2)}\left\{\prod_{j=1}^{p-1}|\cos\theta_j|^{p-j-1}\right\}$$

where c is a constant (see also (3.1.5)). Thus the density of $\mathcal{W}$ and $\boldsymbol{U}^{(p)}$ is given by

$$f(w, \boldsymbol{u}^{(p)}) = c_1\frac{w^{p-1}e^{-w^2/2}}{2^{p/2}\Gamma(p/2)}\left\{\prod_{j=1}^{p-1}|\cos\theta_j|^{p-j-1}\right\}.$$

Writing $\boldsymbol{x} = w\boldsymbol{u}^{(p)}$ and using the same transformations as in (3.1.5), one sees that the density of $\boldsymbol{x}$ is of the form $c_2e^{-w^2/2} = c_2e^{-\boldsymbol{x}'\boldsymbol{x}/2}$ where c_2 is a normalizing constant. Then $c_2 = \frac{1}{(2\pi)^{p/2}}$ and this establishes the result.

Definition 3.1.6 *Generalized chi-square distribution* Let $\boldsymbol{X} \sim \mathcal{S}_p(\xi)$ as defined in (3.1.3) with p.d.f. $g(\cdot)$, $\Pr(\boldsymbol{X} = \boldsymbol{0}) = 0$, and let $\boldsymbol{X} = (\boldsymbol{X}_1', \ldots, \boldsymbol{X}_\ell')'$ where $\boldsymbol{X}_i$ is a p_i-dimensional subvector of $\boldsymbol{X}$, $i = 1, \ldots, \ell$. Then the random vector $(\boldsymbol{X}_1'\boldsymbol{X}_1, \ldots, \boldsymbol{X}_\ell'\boldsymbol{X}_\ell)'$ is said to follow a generalized chi-square distribution with pdf

$$\frac{\pi^{p/2}}{\Gamma(p_1/2)\cdots\Gamma(p_\ell/2)}w^{\frac{p}{2}-1}g(w)\, v_1^{\frac{p_1}{2}-1}\cdots v_{\ell-1}^{\frac{p_{\ell-1}}{2}-1}\left(1-\sum_{j=1}^{\ell-1}v_j\right)^{\frac{p_\ell}{2}-1}, \tag{3.1.27}$$

where $v_i = (\boldsymbol{x}_i'\boldsymbol{x}_i)/(\boldsymbol{x}'\boldsymbol{x})$, $i = 1, \ldots, \ell$ and $w = \boldsymbol{x}'\boldsymbol{x}$, and we write

$$(\boldsymbol{X}_1'\boldsymbol{X}_1, \ldots, \boldsymbol{X}_\ell'\boldsymbol{X}_\ell)' \sim \chi^2(\xi; p_1, \ldots, p_\ell).$$

Note that (3.1.27) which can also be obtained directly from (3.1.33), reduces to (3.1.32).

Theorem 3.1.5 Let $\boldsymbol{X} \simeq \mathcal{W}\boldsymbol{U}^{(p)} \sim \mathcal{S}_p(\xi)$, $F(\cdot)$ denote the distribution function of $\mathcal{W}$ which satisfies the relationship (3.1.16), and $\boldsymbol{X}' = (\boldsymbol{X}_1', \ldots, \boldsymbol{X}_\ell')$ where $\boldsymbol{X}_i$ is a p_i-dimensional subvector of $\boldsymbol{X}$, $i = 1, \ldots, \ell$. Then

$$\boldsymbol{X} = \begin{pmatrix}\boldsymbol{X}_1\\ \vdots\\ \boldsymbol{X}_\ell\end{pmatrix} \simeq \begin{pmatrix}\mathcal{W}\sqrt{V_1}\,\boldsymbol{U}_1\\ \vdots\\ \mathcal{W}\sqrt{V_\ell}\,\boldsymbol{U}_\ell\end{pmatrix} \tag{3.1.28}$$

where $(V_1, \ldots, V_{\ell-1})'$ has the Dirichlet distribution $\mathcal{D}_\ell(\frac{p_1}{2}, \ldots, \frac{p_{\ell-1}}{2}, \frac{p_\ell}{2})$, $V_1 + \cdots + V_\ell = 1$, $\boldsymbol{U}_i \sim \boldsymbol{U}^{(p_i)}, i = 1, \ldots, \ell$, and $\mathcal{W}$, $(V_1, \ldots, V_{\ell-1})$, $\boldsymbol{U}_1, \ldots, \boldsymbol{U}_{\ell-1}$, and $\boldsymbol{U}_\ell$ are independently distributed.

PROOF

$$\boldsymbol{X}'\boldsymbol{X} = \boldsymbol{X}_1'\boldsymbol{X}_1 + \cdots + \boldsymbol{X}_\ell'\boldsymbol{X}_\ell;\ \boldsymbol{X}_j'\boldsymbol{X}_j \simeq \mathcal{W}^2V_j,\ j = 1, \ldots, \ell \tag{3.1.29}$$

since $U^{j\prime}U^j = 1$. Also $X'X = \mathcal{W}^2$ since $V_1 + \cdots + V_\ell = 1$. Note that $d\boldsymbol{x} = d\boldsymbol{x}_1 d\boldsymbol{x}_2 \ldots d\boldsymbol{x}_\ell$. From (3.1.21) for $q = 1$ we have

$$d\boldsymbol{x}_j = \{2^{-1}|X_j'X_j|^{\frac{p_j}{2}-1}d(\boldsymbol{x}_j'\boldsymbol{x}_j)\}\{|\boldsymbol{u}_{(1)}^j|\}d\boldsymbol{u}^j \,.$$

Integrating out $\boldsymbol{u}^j$ over the Stieffel manifold one has the factor $\frac{2\pi^{\frac{p_j}{2}}}{\Gamma(\frac{p_j}{2})}$. After integrating out all the $\boldsymbol{u}^j$'s one has

$$d\boldsymbol{x}_1 \ldots d\boldsymbol{x}_\ell = (\boldsymbol{x}_1'\boldsymbol{x}_1)^{\frac{p_1}{2}-1} \cdots (\boldsymbol{x}_\ell'\boldsymbol{x}_\ell)^{\frac{p_\ell}{2}-1} d(\boldsymbol{x}_1'\boldsymbol{x}_1) \ldots d(\boldsymbol{x}_\ell'\boldsymbol{x}_\ell) \qquad (3.1.30)$$

$$\times \frac{\pi^{p/2}}{\Gamma(\frac{p_1}{2}) \cdots \Gamma(\frac{p_\ell}{2})}$$

$$= (w^2)^{\frac{p}{2}-\ell} v_1^{\frac{p_1}{2}-1} \cdots v_{\ell-1}^{\frac{p_{\ell-1}}{2}-1} \Big(1 - \sum_{j=1}^{\ell-1} v_j\Big)^{\frac{p_\ell}{2}-1} \qquad (3.1.31)$$

$$\times \frac{\pi^{p/2}}{\Gamma(\frac{p_1}{2}) \cdots \Gamma(\frac{p_\ell}{2})} \quad dw dv_1 \ldots dv_{\ell-1}\ 2(w^2)^{\ell-\frac{1}{2}} \,.$$

The density of X is assumed to be of the form $g(\boldsymbol{x}'\boldsymbol{x}) = g(w^2)$ and hence from the above structure we note that $\mathcal{W}, (V_1, \ldots, V_{\ell-1})'$ and $U^1, \ldots, U^\ell$ are mutually independently distributed. Further, the joint density of $(V_1, \ldots, V_{\ell-1})'$ is of the form

$$c v_1^{\frac{p_1}{2}-1} \cdots v_{\ell-1}^{\frac{p_{\ell-1}}{2}-1} \Big(1 - \sum_{j=1}^{\ell-1} v_j\Big)^{\frac{p_\ell}{2}-1}$$

where c is a normalizing constant and thus $(V_1, \ldots, V_{\ell-1})'$ has the type–1 Dirichlet distribution $D_\ell(\frac{p_1}{2}, \ldots, \frac{p_\ell}{2})$.

From (3.1.30) and (3.1.31) note that the joint density of $X_1'X, \ldots, X_\ell'X_\ell$ is given by

$$\frac{\pi^{p/2}}{\Gamma(\frac{p_1}{2}) \cdots \Gamma(\frac{p_\ell}{2})} (\boldsymbol{x}_1'\boldsymbol{x}_1)^{\frac{p_1}{2}-1} \cdots (\boldsymbol{x}_\ell'\boldsymbol{x}_\ell)^{\frac{p_\ell}{2}-1} g(\boldsymbol{x}_1'\boldsymbol{x}_1 + \cdots + \boldsymbol{x}_\ell'\boldsymbol{x}_\ell) \qquad (3.1.32)$$

where $g(\boldsymbol{x}'\boldsymbol{x})$ is the density of X.

It follows that

$$(X_1'X_1, \ldots, X_\ell'X_\ell)' \simeq \mathcal{W}^2(V_1, \ldots, V_\ell)' \sim \chi^2(\xi; p_1, \ldots, p_\ell) \,. \qquad (3.1.33)$$

Furthermore, whenever $(Z_1, \ldots, Z_\ell)' \sim \chi^2(\xi; p_1, \ldots, p_\ell)$, $(Z_1, \ldots, Z_j)' \sim \chi^2(\xi; p_1, \ldots, p_j, p_{j+1} + \cdots + p_\ell)$, $1 \le j < \ell$. This implies that

$$(X_1'X_1, \ldots, X_j'X_j)' \sim \mathcal{W}^2 \mathcal{D}_{j+1}(\tfrac{p_1}{2}, \ldots, \tfrac{p_j}{2}, \tfrac{p_{j+1}+\cdots+p_\ell}{2}) \qquad (3.1.34)$$

where X_i, $i = 1, \ldots, \ell$, is defined in Theorem 3.1.5.

Theorem 3.1.6 Let $X = (X_1', \ldots, X_\ell')' \simeq \mathcal{W}U^{(p)}$, p_i be the number of components of X_j, $i = 1, \ldots, \ell$, $p_\ell \ge 1$ and $\Pr(X = 0) = 0$, then the p.d.f. of $(Z_1, \ldots, Z_j) =$

$(X_1'X_1, \ldots, X_j'X_j)$ is

$$\frac{\Gamma(\frac{p}{2})}{[\prod_{i=1}^{j}\Gamma(\frac{p_i}{2})]\Gamma(\frac{p^c}{2})}\left[\prod_{i=1}^{j} z_i^{\frac{p_i}{2}-1}\right] \tag{3.1.35}$$
$$\times \int_{(\sum_{i=1}^{j} z_i)^{\frac{1}{2}}}^{\infty} t^{2-p}\left(t^2 - \sum_{i=1}^{j} z_i\right)^{\frac{p^c}{2}-1} \mathrm{d}F(t),$$

for $z_i > 0$, $i = 1, \ldots, j$ where $F(\cdot)$ denotes the distribution function of $\mathcal{W}$ and $p^c = p_{j+1} + \cdots + p_\ell$.

PROOF From (3.1.32) the joint density of $Z_1, \ldots, Z_\ell$ is given by

$$\frac{\pi^{p/2}}{\Gamma(\frac{p_1}{2})\cdots\Gamma(\frac{p_\ell}{2})} z_1^{\frac{p_1}{2}-1} \cdots z_\ell^{\frac{p_\ell}{2}-1} g(w^2) \tag{3.1.36}$$

since $z_1 + \cdots + z_\ell = w^2$. Note that $z_\ell = w^2 - z_1 - \cdots - z_{\ell-1}$. Also note that

$$\mathrm{d}z_1 \ldots \mathrm{d}z_\ell = \mathrm{d}z_1 \ldots \mathrm{d}z_{\ell-1}(2w\mathrm{d}w) .$$

From (3.1.36) integrate out $z_{\ell-1}$ by observing that $0 < z_{\ell-1} < w^2 - z_1 - \cdots - z_{\ell-2}$.

$$\int_{z_{\ell-1}} z_{\ell-1}^{\frac{p_{\ell-1}}{2}-1}(w^2 - z_1 - \cdots - z_{\ell-1})^{\frac{p_\ell}{2}-1}\mathrm{d}z_{\ell-1}$$
$$= (w^2 - z_1 - \cdots - z_{\ell-2})^{\frac{p_\ell}{2}-1}$$
$$\times \int_{z_{\ell-1}} z_{\ell-1}^{\frac{p_{\ell-1}}{2}-1}\left(1 - \frac{z_{\ell-1}}{w^2 - z_1 - \cdots - z_{\ell-2}}\right)^{\frac{p_\ell}{2}-1} \mathrm{d}z_{\ell-1}$$
$$= (w^2 - z_1 - \cdots - z_{\ell-2})^{\frac{p_\ell+p_{\ell-1}}{2}-1}\frac{\Gamma(\frac{p_{\ell-1}}{2})\Gamma(\frac{p_\ell}{2})}{\Gamma(\frac{p_\ell+p_{\ell-1}}{2})} .$$

Successive integrations of $z_{\ell-1}, z_{\ell-2}, \ldots, z_{j+1}$ leave the factors

$$\frac{2\pi^{p/2}w}{[\Gamma(\frac{p_1}{2})\cdots\Gamma(\frac{p_j}{2})]\Gamma(\frac{p_c}{2})} z_1^{\frac{p_1}{2}-1} \cdots z_j^{\frac{p_j}{2}-1}\left(w^2 - \sum_{i=1}^{j} z_i\right)^{\frac{p_c}{2}-1} g(w^2)\mathrm{d}w\mathrm{d}z_1 \ldots \mathrm{d}z_j$$

where $p_c = p_{j+1} + \cdots + p_\ell$. Now integrate out w noting that $(\sum_{i=1}^{j} z_i)^{\frac{1}{2}} < w < \infty$. We may either evaluate

$$2\pi^{p/2}\int_{(\sum_{i=1}^{j} z_j)^{\frac{1}{2}}}^{\infty} w\left(w^2 - \sum_{i=1}^{j} z_i\right)^{\frac{p_c}{2}-1} g(w^2)\mathrm{d}w$$

or replace $2\pi^{p/2}wg(w^2)$ by $\Gamma(\frac{p}{2})w^{2-p}\mathrm{d}F(w)$, where $\mathrm{d}F(w)$ is the density of $\mathcal{W}$, see also

(3.1.7), to establish the result.

Definition 3.1.7 *Generalized noncentral chi-square distribution* Let $X \simeq \mu + \mathcal{W}U^{(p)} \sim C_p(\xi; \mu, I)$ as defined in (3.1.15) with $\mu \neq 0$ and $\Pr(X = \mu) = 0$. Then $Q = X'X$ is said to follow a generalized non-central chi-square distribution with p.d.f.

$$\frac{\Gamma(p/2)}{2\delta\Gamma(\frac{1}{2})\Gamma((p-1)/2)} \int_{|\sqrt{q}-\delta|}^{\sqrt{q}+\delta} w^{-1}\left[1 - \left(\frac{q-\delta^2-w^2}{2w\delta}\right)^2\right]^{\frac{(p-3)}{2}} dG(w)$$

for $q > 0$, where $\delta^2 = \mu'\mu$ where $G(w)$ denotes the distribution function of $\mathcal{W}$ and we write

$$X'X \sim \chi^2(\xi; p; \delta^2) .$$

Definition 3.1.8 *Generalized F-distribution* Let $X' = (X_1', X_2') \sim C_{m+n}(\xi; \mu, I)$ as defined in (3.1.2) where X_1 is $m\times 1$, $\mu' = (\mu_1', 0')$ where μ_1 is $m\times 1$ and $\Pr(X = \mu) = 0$ and let $g((x-\mu)'(x-\mu))$ denote the density of X. Then

$$F = \frac{n}{m}\frac{X_1'X_1}{X_2'X_2}$$

is said to follow a generalized F-distribution (central when $\mu_1 = 0$ and noncentral when $\mu_1 \neq 0$) with p.d.f.

$$\frac{2\pi^{\frac{1}{2}(m+n-1)}}{\Gamma(\frac{1}{2}(m-1))\Gamma(\frac{n}{2})}\frac{m}{n}\left(\frac{mf}{n}\right)^{\frac{(m-2)}{2}}\left(1+\frac{mf}{n}\right)^{-\frac{(m+n)}{2}}$$

$$\times \int_0^\pi \int_0^\infty (\sin\theta)^{m-2} y^{m+n-1} g(z^2 - 2\delta' z \cos\theta + \delta^2) d\theta dz$$

where $\delta^2 = \mu_1'\mu_1$ and $\delta' = (mF/(n+mF))^{\frac{1}{2}}\delta$ and we write $F \sim F(\xi; m, n; \delta^2)$.

Definition 3.1.9 *Class* Ξ Let $\phi(t)$ be a characteristic function and let $\psi(t)$ be any characteristic function satisfying $\phi(t) = \psi(t)$ in the neighbourhood of zero. If this implies that the characteristic functions are equal for all real number t, then $\phi(t)$ is said to belong to the class Ξ.

Lemma 3.1.1 If U, V, W and X are random variables $W \simeq X, U$ and W are independently distributed, V and X are also independently distributed, and $U \simeq V$, then $U + W \simeq V + X$. Furthermore, if the characteristic function of U belongs to the class Ξ or if the characteristic function of W is not equal to zero almost everywhere, then whenever $U + W \simeq V + X$, we have that $U \simeq V$.

Lemma 3.1.2 If U and V are independently distributed, U and W are also independently distributed, and $UV \simeq UW$, then $V \simeq W$ provided one of the following conditions is satisfied:

(i) $\Pr(U > 0) = \Pr(V > 0) = \Pr(W > 0) = 1$ and the characteristic function of $\log V$ belongs to the class Ξ or that of $\log U$ is not equal to zero almost everywhere;
(ii) $\Pr(U > 0) = 1$ and the characteristic function of $\log U$ is not equal to 0 almost everywhere;
(iii) $\Pr(U > 0) = 1$, the characteristic function of $\{\log V | V > 0\}$ and the characteristic function of $\{\log(-V) | V < 0\}$ belong to the class Ξ .

Lemma 3.1.3 Let the joint p.d.f. of $\boldsymbol{X}_1$ and $\boldsymbol{X}_2$ be

$$f(\boldsymbol{x}_1, \boldsymbol{x}_2) = c_p|\Sigma|^{-\frac{1}{2}} g((\boldsymbol{x}_1' - \boldsymbol{\mu}_1', \boldsymbol{x}_2' - \boldsymbol{\mu}_2')\Sigma^{-1}(\boldsymbol{x}_1' - \boldsymbol{\mu}_1', \boldsymbol{x}_2' - \boldsymbol{\mu}_2')')$$

where c_p is a constant and $\boldsymbol{X}_1$ and $\boldsymbol{X}_2$ have respectively q and r components, $q + r = p$, and let

$$\Sigma = \begin{pmatrix} \Sigma_{11} & \Sigma_{12} \\ \Sigma_{21} & \Sigma_{22} \end{pmatrix}$$

where Σ_{11} and Σ_{22} are respectively $q \times q$ and $r \times r$ submatrices. Then there exists a function $g_r(\cdot)$ such that the marginal p.d.f. of $\boldsymbol{X}_2$ is

$$f_{\boldsymbol{X}_2}(\boldsymbol{x}_2) = c_q|\Sigma_{22}|^{-\frac{1}{2}} g_r((\boldsymbol{x}_2 - \boldsymbol{\mu}_2)'\Sigma_{22}^{-1}(\boldsymbol{x}_2 - \boldsymbol{\mu}_2)) \; .$$

The proof is similar to that used for the multivariate normal. Two characterizations of the multivariate normal distribution are given below.

Lemma 3.1.4 Let $\boldsymbol{X} = (\boldsymbol{X}_1', \boldsymbol{X}_2')' \sim C_p(\xi; \boldsymbol{\mu}, \Sigma)$ and $\boldsymbol{\mu} = (\boldsymbol{\mu}_1', \boldsymbol{\mu}_2')'$. Then under the notation of Lemma 3.1.3, the conditional mean of $\boldsymbol{X}_1$ given $\boldsymbol{X}_2 = \boldsymbol{x}_2$, if it exists, is given by

$$E(\boldsymbol{X}_1 | \boldsymbol{X}_2 = \boldsymbol{x}_2) = \boldsymbol{\mu}_1 + \Sigma_{12}\Sigma_{22}^{-1}(\boldsymbol{x}_2 - \boldsymbol{\mu}_2)$$

and the conditional covariance parameter matrix is $\Sigma_{11} - \Sigma_{12}\Sigma_{22}^{-1}\Sigma_{21}$. The proof is left as an exercise.

Lemma 3.1.5 Let $\boldsymbol{X} \sim C_p(\xi; \boldsymbol{\mu}, \Sigma)$. If any marginal p.d.f. is a multivariate normal density, then $\boldsymbol{X}$ has a multivariate normal distribution.

This follows from the fact that the characteristic functions of the marginal and parent distributions have the same functional form.

Lemma 3.1.6 Let $\boldsymbol{X} \sim C_p(\xi; \boldsymbol{\mu}, \Sigma)$, $\boldsymbol{X} = (\boldsymbol{X}_1', \boldsymbol{X}_2')'$ where $\boldsymbol{X}_2$ is an r-dimensional vector. If $\boldsymbol{X}_1$ given $\boldsymbol{X}_2 = \boldsymbol{x}_2$ has a multivariate normal distribution for $r = 1, 2, \ldots, p-1$, then $\boldsymbol{X}$ also has a multivariate normal distribution.

PROOF It is assumed without loss of generality that $\boldsymbol{\mu} = \boldsymbol{0}$ and $\Sigma = I$. We can write the conditional p.d.f. as

$$f_{\boldsymbol{X}_1|\boldsymbol{x}_2}(\boldsymbol{x}_1) = \frac{c_p g(\boldsymbol{x}_1'\boldsymbol{x}_1 + \boldsymbol{x}_2'\boldsymbol{x}_2)}{f_{\boldsymbol{X}_2}(\boldsymbol{x}_2)}$$

which is a function of $\boldsymbol{x}_1'\boldsymbol{x}_1$ for a given $\boldsymbol{x}_2$. According to Lemma 3.1.4, the conditional mean is $\mathbf{0}$ and the conditional covariance matrix is $\alpha(\boldsymbol{x}_2)I_{p-r}$ (see also Lemma 3.8.1). Assuming normality, we have

$$f_{\boldsymbol{X}_1|\boldsymbol{x}_2}(\boldsymbol{x}_1) = \frac{e^{-\frac{1}{2}\left(\frac{\boldsymbol{x}_1'\boldsymbol{x}_1}{\alpha(\boldsymbol{x}_2)}\right)}}{(2\pi\alpha(\boldsymbol{x}_2))^{\frac{p-r}{2}}}$$

and then

$$c_p g(\boldsymbol{x}_1'\boldsymbol{x}_1 + \boldsymbol{x}_2'\boldsymbol{x}_2) = \frac{f_{\boldsymbol{X}_2}(\boldsymbol{x}_2)}{(2\pi\alpha(\boldsymbol{x}_2))^{\frac{p-r}{2}}} e^{-\frac{1}{2}\left(\frac{\boldsymbol{x}_1'\boldsymbol{x}_1}{\alpha(\boldsymbol{x}_2)}\right)}$$

or equivalently for fixed $\boldsymbol{x}_2$

$$c_p g(\boldsymbol{x}_1'\boldsymbol{x}_1) = \frac{f_{\boldsymbol{X}_2}(\boldsymbol{x}_2)}{(2\pi\alpha(\boldsymbol{x}_2))^{\frac{p-k}{2}}} e^{\frac{1}{2}\left(\frac{\boldsymbol{x}_2'\boldsymbol{x}_2}{\alpha(\boldsymbol{x}_2)}\right)} e^{-\frac{1}{2}\left(\frac{\boldsymbol{x}_1'\boldsymbol{x}_1}{\alpha(\boldsymbol{x}_1)}\right)} .$$

Hence $f_{\boldsymbol{X}}(\boldsymbol{x}) = (2\pi\alpha)^{p/2} e^{-\frac{1}{2}(\frac{\boldsymbol{x}'\boldsymbol{x}}{\alpha})}$.

Theorem 3.1.7 Let $\boldsymbol{X} \simeq \mathcal{W}\boldsymbol{U}^{(p)} \sim \mathcal{S}_p(\xi)$ where $\Pr(\boldsymbol{X} = \mathbf{0}) = 0$. Let B be a symmetric idempotent matrix which does not have full rank that is, $1 \le \rho(B) < p$, and let $A = I - B$. Then $\boldsymbol{X}'A\boldsymbol{X}$ and $\boldsymbol{X}'B\boldsymbol{X}$ are independently distributed if and only if $\boldsymbol{X}$ is normally distributed with mean $\mathbf{0}$ and covariance matrix $\sigma^2 I$, where $\sigma > 0$.

PROOF $A = I - B$ and $B = B' = B^2 \Rightarrow AB = B - B^2 = B - B = O$. If $\boldsymbol{X} \sim N_p(\mathbf{0}, \sigma^2 I_p)$ and $AB = O$, then it follows from Craig's theorem that $\boldsymbol{X}'A\boldsymbol{X}$ and $\boldsymbol{X}'B\boldsymbol{X}$ are independently distributed. Now, let $\boldsymbol{X}'A\boldsymbol{X}$ and $\boldsymbol{X}'B\boldsymbol{X}$ be independently distributed. The matrix A can be diagonalized by means of an orthonormal matrix P as follows

$$A = P' \begin{bmatrix} I_r & O \\ O & O \end{bmatrix} P, \quad 1 \le r < p \tag{3.1.37}$$

where r is the rank of A. Since spherically symmetric distributions are invariant under orthogonal transformations, $\boldsymbol{Y} = P\boldsymbol{X} \sim \mathcal{S}_p(\xi)$. On partitioning the vector $\boldsymbol{Y}$ into two subvectors $\boldsymbol{Y}_1$ and $\boldsymbol{Y}_2$ where $\boldsymbol{Y}_1$ has r components, we have that

$$Z_1 \simeq \boldsymbol{X}'A\boldsymbol{X} = \boldsymbol{Y}_1'\boldsymbol{Y}_1$$

and

$$Z_2 \simeq \boldsymbol{X}'B\boldsymbol{X} = \boldsymbol{Y}_2'\boldsymbol{Y}_2 \tag{3.1.38}$$

are independently distributed and, in view of Theorem 3.1.5,

$$\begin{pmatrix} Z_1 \\ Z_2 \end{pmatrix} \sim \begin{pmatrix} \mathcal{W}^2 V_1 \\ \mathcal{W}^2 V_2 \end{pmatrix} . \tag{3.1.39}$$

Noting that $Z_1/Z_2 \equiv \mathcal{W}^2 V_1/(\mathcal{W}^2 V_2) = V_1/V_2$ which is free of $\mathcal{W}^2$ and that $Z_1 + Z_2 = \boldsymbol{Y}'\boldsymbol{Y} \equiv \mathcal{W}^2$. Hence we have that Z_1/Z_2 and $Z_1 + Z_2$ are independently distributed. Then, from Lukacs (1956, p. 208), Z_1 and Z_2 must have a gamma distribution with the same scale parameter; hence we can write

$$Z_1 \sim \sigma^2 \chi^2_{d_1} \text{ and } Z_2 \sim \sigma^2 \chi^2_{d_2},\ \sigma > 0\,.$$

It follows that $\mathcal{W}^2 \sim \sigma^2 \chi^2_d$ where $d = d_1 + d_2$ and $V_1 \sim D_2(\frac{d_1}{2}, \frac{d_2}{2})$. However, in view of Theorem 3.1.3, $V_1 \sim D_2(\frac{r}{2}, \frac{p-r}{2})$. Hence $d_1 = r$, $d_2 = p - r$, and from Theorem 3.1.4, we conclude that $\boldsymbol{X} \simeq \boldsymbol{Y} \sim N_p(\boldsymbol{0}, \sigma^2 I_p)$.

Example 3.1.1 Let $\bar{X}^2 = (\sum_{i=1}^{p} X_i/p)^2 = \boldsymbol{X}'A\boldsymbol{X}$ and $S = \sum_{i=1}^{p}(X_i - \bar{X})^2 = \boldsymbol{X}'B\boldsymbol{X}$ where $\boldsymbol{X} = (X_1, \ldots, X_p)' \sim \mathcal{S}_p(\xi)$.

Let $A = \boldsymbol{1}_p\boldsymbol{1}_p'/p$ and $B = I - A$ where $\boldsymbol{1}_p$ is a p-dimensional vector whose components are all equal to one; then we can write $\bar{X}^2 = \boldsymbol{X}'A\boldsymbol{X}$ and $S = \boldsymbol{X}'B\boldsymbol{X}$ and by Theorem 3.1.7 we have that $\bar{X}^2$ and S are independently distributed if and only if $\boldsymbol{X} \sim N_p(\boldsymbol{0}, \sigma^2 I)$ for $\sigma > 0$.

Theorem 3.1.8 Let $\boldsymbol{X} = (X_1, \ldots, X_p)' \sim N_p(\boldsymbol{0}, I)$, $(V_1, \ldots, V_{p-1}) \sim D_p(\frac{1}{2}, \ldots, \frac{1}{2}, \frac{1}{2})$ where $\sum_{i=1}^{p} V_i = 1$, and let $A_1, \ldots, A_\ell$ be $p \times p$ symmetric matrices with $\ell \le p$. Then

$$(\boldsymbol{X}'A_1\boldsymbol{X}, \ldots, \boldsymbol{X}'A_\ell\boldsymbol{X}) \simeq \Big(\sum_{i=1}^{p} c_{1i}X_i^2, \ldots, \sum_{i=1}^{p} c_{\ell i}X_i^2\Big) \tag{3.1.40}$$

if and only if

$$\Big(\frac{\boldsymbol{X}'A_1\boldsymbol{X}}{||\boldsymbol{X}||^2}, \ldots, \frac{\boldsymbol{X}'A_\ell\boldsymbol{X}}{||\boldsymbol{X}||^2}\Big) \simeq \Big(\sum_{i=1}^{p} c_{1i}V_i, \ldots, \sum_{i=1}^{p} c_{\ell i}V_i\Big) \tag{3.1.41}$$

PROOF We assume that $\ell = 1$. (The proof in the general case is similar). If (3.1.40) is true, it follows that

$$\boldsymbol{X}'A_1\boldsymbol{X} \simeq \sum_{i=1}^{p} c_{1i}X_i^2\,.$$

Since $\boldsymbol{X} \sim N_p(\boldsymbol{0}, I)$ we have $X_i^2 \sim \chi^2_1$, $i = 1, \ldots, p$ and mutually independently distributed, $||\boldsymbol{X}||^2 \sim \chi^2_p$ and further $||\boldsymbol{X}||^2$ and $\frac{X_i^2}{||\boldsymbol{X}||^2}$ are independently distributed. Furthermore, $\Big(\frac{X_1^2}{||\boldsymbol{X}||^2}, \ldots, \frac{X_p^2}{||\boldsymbol{X}||^2}\Big)$ has a type-1 Dirichlet distribution. Hence,

$$\frac{\boldsymbol{X}'A_1\boldsymbol{X}}{||\boldsymbol{X}||^2} \simeq \sum_{i=1}^{p} c_{1i}\frac{X_i^2}{||\boldsymbol{X}||^2} = \sum_{i=1}^{p} c_{1i}V_i\,.$$

The proof for the converse is similar and hence omitted.

The previous result provides a link between the generalized chi-square distribution (see Definition 3.1.6) and the Dirichlet distribution as well as between quadratic forms in normal vectors and quadratic forms in elliptically contoured vectors.

Corollary 3.1.1 If $X \sim N_p(0, I)$ and $A = A'$ then $\boldsymbol{X}'A\boldsymbol{X}/\|\boldsymbol{X}\|^2 \sim \text{Beta}(k/2, (p-k)/2)$ (see Definition 3.1.4) if and only if A is an idempotent matrix of rank k.

Corollary 3.1.2 If $\boldsymbol{X} \sim N_p(0, I), A = A'$ and $B = B'$. Then

$$(\boldsymbol{X}'A\boldsymbol{X}/\|\boldsymbol{X}\|^2,\ \boldsymbol{X}'B\boldsymbol{X}/\|\boldsymbol{X}\|^2) \sim D_3(k/2, m/2, (p-k-m)/2)$$

if and only if $AB = O$ and A and B are idempotent matrices of rank k and m respectively.

Theorem 3.1.9 Let $\boldsymbol{X} \simeq \mathcal{W}\boldsymbol{U}^{(p)} \sim S_p(\xi)$, $A = A'$ and $\Pr(\boldsymbol{X} = \boldsymbol{0}) = 0$; then $\boldsymbol{X}'A\boldsymbol{X} \sim \chi^2(\xi; k, (p-k))$ if and only if A is an idempotent matrix of rank k.

PROOF Let $\boldsymbol{X}'A\boldsymbol{X} \sim \chi^2(\xi; k, (p-k))$, then $\|\boldsymbol{X}\|^2(\boldsymbol{X}'A\boldsymbol{X}/\|\boldsymbol{X}\|^2) = \boldsymbol{X}'A\boldsymbol{X} \simeq \mathcal{W}^2 V$ where V is distributed independently of $\mathcal{W}^2$ and $V \sim D_2(k/2, (n-k)/2)$. Clearly $\Pr(V > 0) = 1$ and therefore $\Pr(\boldsymbol{X}'A\boldsymbol{X}/\|\boldsymbol{X}\|^2 > 0) = \Pr(\mathcal{W}^2 > 0) = \Pr(\boldsymbol{X} \neq \boldsymbol{0}) = 1$. Note that the characteristic function of $\log V$ belongs to the class Ξ since $\Pr(0 < V < 1) = 1$. Using Lemma 3.1.1, we have $\boldsymbol{X}'A\boldsymbol{X}/\|\boldsymbol{X}\|^2 \simeq V$. Noting that $\boldsymbol{X} \simeq \mathcal{W}\boldsymbol{U}^{(p)}$, we have $\boldsymbol{X}'A\boldsymbol{X}/\|\boldsymbol{X}\|^2 \simeq \boldsymbol{U}^{(p)\prime}A\boldsymbol{U}^{(p)} \simeq \boldsymbol{Y}'A\boldsymbol{Y}/\|\boldsymbol{Y}\|^2$ where $\boldsymbol{Y} \sim N_p(\boldsymbol{0}, I)$. The result is obtained from Corollary 3.1.1. The other implication is easy to prove.

Corollary 3.1.3 If $\boldsymbol{X} \sim \mathcal{C}_p(\xi; 0, \Sigma)$, $\Pr(\boldsymbol{X} = \boldsymbol{0}) = 0$, $\Sigma \geq 0$, rank $(\Sigma) = s \geq k \geq 1$, and $A = A'$. Then $\boldsymbol{X}'A\boldsymbol{X} \sim \chi^2(\xi; k, (p-k))$ if and only if $\Sigma A \Sigma A \Sigma = \Sigma A \Sigma$ and rank $(\Sigma A \Sigma) = k$.

Theorem 3.1.10 Let $\boldsymbol{X} \simeq \mathcal{W}\boldsymbol{U}^{(p)} \sim S_p(\xi)$, $A_1, A_2, \ldots, A_\ell$ be symmetric matrices and $\Pr(\boldsymbol{X} = \boldsymbol{0}) = 0$. Then $(\boldsymbol{X}'A_1\boldsymbol{X}, \ldots, \boldsymbol{X}'A_\ell\boldsymbol{X}) \sim \chi^2(\xi; p_1, \ldots, p_{\ell+1})$ with $p = \sum_{i=1}^{\ell+1} p_i$ if and only if $A_iA_j = \delta_{ij}A_i$ and the rank of A_i is p_i, $i, j = 1, 2, \ldots, \ell$ with $\delta_{ii} = 1$ and $\delta_{ij} = 0$ when $i \neq j$.

PROOF We will prove the "only if" part for $\ell = 2$. Let $(\boldsymbol{X}'A_1\boldsymbol{X}, \boldsymbol{X}'A_2\boldsymbol{X}) \sim \chi^2(\xi; p_1, p_2, p_3)$. Then we can write

$$(\boldsymbol{X}'A_1\boldsymbol{X}, \boldsymbol{X}'A_2\boldsymbol{X}) \simeq \mathcal{W}^2(V_1, V_2)$$

with $\mathcal{W}$ distributed independently of (V_1, V_2) and $(V_1, V_2) \sim D_3(p_1/2, p_2/2, p_3/2)$. It follows that $\boldsymbol{X}'A_1\boldsymbol{X} \sim \chi^2(\xi; p_1, (p_2 + p_3))$, $\boldsymbol{X}'A_2\boldsymbol{X} \sim \chi^2(\xi; p_2, (p_1 + p_3))$ and $\boldsymbol{X}'(A_1 + A_2)\boldsymbol{X} \sim \chi^2(\xi; (p_1 + p_2), p_3)$, and in view of Theorem 3.1.9, A_1 and A_2 are idempotent matrices of rank p_1 and p_2 respectively and $(A_1 + A_2)^2 = (A_1 + A_2)$; hence $A_1A_2 = O$.

Corollary 3.1.4 Let $\boldsymbol{X} \sim C_p(\xi; \mathbf{0}, \Sigma), \Sigma > 0$, $\Pr(\boldsymbol{X} = \mathbf{0}) = 0$, $A_i = A_i'$, $i = 1, \ldots, \ell$. Then $\boldsymbol{X}'A_1\boldsymbol{X}, \ldots, \boldsymbol{X}'A_\ell\boldsymbol{X} \sim \chi^2(\xi; p_1, \ldots, p_{\ell+1})$ if and only if $A_i\Sigma A_j = A_i\delta_{ij}$ and the rank of A_i is p_i, $i, j = 1, \ldots, \ell$.

3.2 Moments of Quadratic Forms

Some representations of the first four moments of elliptically contoured vectors are given in this section. The mean and variance of quadratic forms in elliptically contoured vectors are also obtained.

Definition 3.2.1 The Kronecker product of the matrices $A_{q\times p} = (a_{ij})$ and $B_{n\times m}$ is denoted by $A \otimes B$ where

$$A \otimes B = (a_{ij})B = \begin{pmatrix} a_{11}B & \ldots & a_{1p}B \\ \vdots & & \vdots \\ a_{q1}B & \ldots & a_{qp}B \end{pmatrix}.$$

Properties of the Kronecker Product The following properties hold for any vectors $\boldsymbol{v}$ and $\boldsymbol{w}$ and conformable matrices $A, B, C, E, A_1, \ldots, A_n, B_1, \ldots, B_n$:

$$(i)\ (B \otimes C)' = B' \otimes C' \tag{3.2.1}$$

$$(ii)\ \boldsymbol{v} \otimes \boldsymbol{w}' = \boldsymbol{v}\boldsymbol{w}' \tag{3.2.2}$$

$$(iii)\ (A \otimes B)(C \otimes E) = (AC) \otimes (BE) \tag{3.2.3}$$

$$(iv)\ (A_1 \otimes B_1)(A_2 \otimes B_2)\cdots(A_n \otimes B_n) = (A_1A_2\ldots A_n) \otimes (B_1B_2\ldots B_n) \tag{3.2.4}$$

$$(v)\ (A_1B_1) \otimes \cdots \otimes (A_nB_n) = (A_1 \otimes \cdots \otimes A_n)(B_1 \otimes \cdots \otimes B_n) \tag{3.2.5}$$

$$(vi)\ \mathrm{tr}\,(B \otimes C) = \mathrm{tr}\,(B)\mathrm{tr}\,(C) \tag{3.2.6}$$

$$(vii)\ \mathrm{tr}\,(K_{pp}(B'_{p\times p} \otimes C_{p\times p})) = \mathrm{tr}\,(B'C) = [\mathrm{vec}(B)]'\mathrm{vec}(C) \tag{3.2.7}$$

where K_{pp} is a $p^2 \times p^2$ orthogonal matrix whose (i,j)-th $p \times p$ submatrix is $\boldsymbol{e}_j\boldsymbol{e}_i', i, j = 1, \ldots, p$ and $\boldsymbol{e}_\ell$ is a p-dimensional vector whose ℓ-th component is one, the other components being zero. K_{pp} is called a permutation matrix as $\mathrm{vec}(B'_{p\times p}) = K_{pp}\ \mathrm{vec}(B)$.

$$(viii)\ \boldsymbol{v} \otimes \boldsymbol{w}' = \boldsymbol{w}' \otimes \boldsymbol{v} \tag{3.2.8}$$

$$(ix)\ (B \otimes C)^{-1} = B^{-1} \otimes C^{-1} \tag{3.2.9}$$

$$(x)\ \mathrm{tr}\,[(C \otimes C)\ \mathrm{vec}(I)\ (\mathrm{vec}(I))'] = \mathrm{tr}\,(C^2),\ C = C'\ . \tag{3.2.10}$$

Definition 3.2.2 Let $B_{q\times p} = (\boldsymbol{b}_1, \ldots \boldsymbol{b}_p)$. Then $\mathrm{vec}(B)$ is the qp-dimensional vector

$(b'_1, \ldots, b'_p)'$, that is,

$$\text{vec}(B) = \begin{pmatrix} b_1 \\ \vdots \\ b_p \end{pmatrix}.$$

Definition 3.2.3 The k-th moment of a p-dimensional random vector X is

$$M_k(X) = \begin{cases} E[\overset{k/2}{\otimes}(XX')] & \text{when } k \text{ is even,} \\ E[\{\overset{(k-1)/2}{\otimes}(XX')\} \otimes X] & \text{when } k \text{ is odd,} \end{cases}$$

where the notation $\overset{\ell}{\otimes} A$ means $A \otimes A \otimes \cdots \otimes A$.

Lemma 3.2.1 Let $X = (X_1, \ldots, X_p)'$ be a p-dimensional random vector. Let $M_1(X) = E(X), M_2(X) = E(X \otimes X'), M_3(X) = E(X \otimes X' \otimes X), \ldots$. Let $M_k(X) = (m_{rs}^{(k)})$ where $(m_{rs}^{(k)})$ denotes the $(r, s) - th$ element in $M_k(X)$. Then

$$r = 1 + \sum_{j=1}^{[(k+1)/2]} (i_{2j-1} - 1)p^{[(k+1)/2]-j}$$

and

$$s = 1 + \sum_{j=1}^{[k/2]} (i_{2j} - 1)p^{[k/2]-j}$$

where $[b]$ denotes the largest integer which is less than or equal to b.

PROOF

$$M_2(X) = E(XX') \text{ and } E(X_{i_1} X_{i_2}) = m_{i_1 i_2}^{(2)}, \; i_1, i_2 = 1, \ldots, p;$$
$$M_3(X) = E(X \otimes X' \otimes X) = E(X \otimes X \otimes X')$$

and

$$E(X_{i_1} X_{i_2} X_{i_3}) = E(X_{i_1} X_{i_3} X_{i_2}) = m_{(i_1-1)p+i_3, i_2}^{(3)}, \; i_1, i_2, i_3 = 1, \ldots, p;$$

$$\begin{aligned} M_4(X) =& E(X \otimes X' \otimes X \otimes X') \\ =& E(X \otimes X \otimes X' \otimes X') \\ =& E((X \otimes X)(X \otimes X)') \end{aligned}$$

and

$$\begin{aligned} E(X_{i_1} X_{i_2} X_{i_3} X_{i_4}) =& E(X_{i_1} X_{i_3} X_{i_2} X_{i_4}) \\ =& m_{(i_1-1)p+i_3,\, (i_2-1)p+i_4}^{(4)}, \; i_1, i_2, i_3, i_4 = 1, \ldots, p\,. \end{aligned}$$

The general result can be obtained by induction.

Lemma 3.2.2 Let $M_k(\boldsymbol{X})$ be the k-th moment matrix of $\boldsymbol{X}$ and $\xi_{\boldsymbol{X}}(\boldsymbol{t})$ denote the characteristic function of $\boldsymbol{X}$; then

$$M_k(\boldsymbol{X}) = \begin{cases} \frac{1}{i^k} \frac{\partial^k \xi_{\boldsymbol{X}}(\boldsymbol{t})}{\partial \boldsymbol{t} \partial \boldsymbol{t}' \dots \partial \boldsymbol{t} \partial \boldsymbol{t}'} |_{\boldsymbol{t}=\boldsymbol{0}} & \text{when } k \text{ is even,} \\ \frac{1}{i^k} \frac{\partial^k \xi_{\boldsymbol{X}}(\boldsymbol{t})}{\partial \boldsymbol{t} \partial \boldsymbol{t}' \dots \partial \boldsymbol{t} \partial \boldsymbol{t}' \partial \boldsymbol{t}} |_{\boldsymbol{t}=\boldsymbol{0}} & \text{when } k \text{ is odd.} \end{cases}$$

PROOF Noting that

$$\frac{\partial \xi_{\boldsymbol{X}}(\boldsymbol{t})}{\partial \boldsymbol{t}} = E\Big[\frac{\partial e^{(i\boldsymbol{t}'\boldsymbol{X})}}{\partial \boldsymbol{t}}\Big] = iE[e^{(i\boldsymbol{t}'\boldsymbol{X})}\boldsymbol{X}]$$

and

$$\begin{aligned} \frac{\partial^2 \xi_{\boldsymbol{X}}(\boldsymbol{t})}{\partial \boldsymbol{t} \partial \boldsymbol{t}'} &= iE\Big[\frac{\partial}{\partial \boldsymbol{t}'}(e^{i\boldsymbol{t}'\boldsymbol{X}}\boldsymbol{X})\Big] = iE\Big[\frac{\partial e^{(i\boldsymbol{t}'\boldsymbol{X})}}{\partial \boldsymbol{t}'} \otimes \boldsymbol{X}\Big] \\ &= i^2 E[e^{i\boldsymbol{t}'\boldsymbol{X}}(\boldsymbol{X}' \otimes \boldsymbol{X})] = i^2 E[e^{i\boldsymbol{t}'\boldsymbol{X}}(\boldsymbol{X} \otimes \boldsymbol{X}')] , \end{aligned}$$

we have

$$M_1(\boldsymbol{X}) = \frac{1}{i} \frac{\partial \xi_{\boldsymbol{X}}(\boldsymbol{t})}{\partial \boldsymbol{t}} |_{\boldsymbol{t}=\boldsymbol{0}}$$

and

$$M_2(\boldsymbol{X}) = \frac{1}{i^2} \frac{\partial^2 \xi_{\boldsymbol{X}}(\boldsymbol{t})}{\partial \boldsymbol{t} \partial \boldsymbol{t}'} |_{\boldsymbol{t}=\boldsymbol{0}} .$$

Induction may be used to prove the general result.

Lemma 3.2.3 Let $\boldsymbol{X} = \boldsymbol{\mu} + B\boldsymbol{Y}$ with $B = B'$, then

$$\begin{aligned} (i)\ & M_1(\boldsymbol{X}) = \boldsymbol{\mu} + BM_1(\boldsymbol{Y}) \\ (ii)\ & M_2(\boldsymbol{X}) = M_1(\boldsymbol{X}) \otimes \boldsymbol{\mu}' + \boldsymbol{\mu} M_1'(\boldsymbol{Y})B + BM_2(\boldsymbol{Y})B \\ (iii)\ & M_3(\boldsymbol{X}) = M_2(\boldsymbol{X}) \otimes \boldsymbol{\mu} + \boldsymbol{\mu}\boldsymbol{\mu}' \otimes BM_1(\boldsymbol{Y}) + (\boldsymbol{\mu} \otimes B)M_2(\boldsymbol{Y})B \\ & \quad + \mathrm{vec}[BM_2(\boldsymbol{Y})B]\boldsymbol{\mu}' + (B \otimes B)M_3(\boldsymbol{Y})B \\ (iv)\ & M_4(\boldsymbol{X}) = M_3(\boldsymbol{X}) \otimes \boldsymbol{\mu}' + \boldsymbol{\mu}\boldsymbol{\mu}' \otimes \boldsymbol{\mu} M_1'(\boldsymbol{Y})B \\ & \quad + (\boldsymbol{\mu} \otimes \boldsymbol{\mu})[\mathrm{vec}(BM_2(\boldsymbol{Y})B)]' \\ & \quad + f(B, \boldsymbol{\mu})M_2(\boldsymbol{Y})(\boldsymbol{\mu}' \otimes B) + f(B, \boldsymbol{\mu})M_3'(\boldsymbol{Y})(B \otimes B) \\ & \quad + (B \otimes B)M_3(\boldsymbol{Y})(\boldsymbol{\mu}' \otimes B) + (B \otimes B)M_4(\boldsymbol{Y})(B \otimes B) , \end{aligned}$$

where

$$f(B_1, B_2) = B_1 \otimes B_2 + B_2 \otimes B_1 . \tag{3.2.11}$$

The proof follows from property (3.2.4).

Theorem 3.2.1 Let $A_1, A_2, \dots, A_\ell$ be $p \times p$ symmetric matrices. Consider the quadratic forms $\boldsymbol{X}'A_j\boldsymbol{X}, j = 1, \dots, \ell$; then their mixed moments are

$$E[(\boldsymbol{X}'A_1\boldsymbol{X})^{k_1}(\boldsymbol{X}'A_2\boldsymbol{X})^{k_2} \cdots (\boldsymbol{X}'A_\ell\boldsymbol{X})^{k_\ell}]$$

$$= \mathrm{tr}\,[(A_1 \otimes \cdots \otimes A_1 \otimes \cdots \otimes A_\ell \otimes \cdots \otimes A_\ell) M_k(X)]$$

with $k = 2(k_1 + k_2 + \cdots + k_\ell)$.

PROOF

$$\begin{aligned}
E[(X'A_1X)^{k_1} \cdots (X'A_\ell X)^{k_\ell}]
&= E[(X'A_1X) \otimes \cdots \otimes (X'A_1X) \otimes \cdots \otimes (X'A_\ell X) \otimes \cdots \otimes (X'A_\ell X)] \\
&= E[(X' \otimes \cdots \otimes X')(A_1 \otimes \cdots \otimes A_1 \otimes \cdots \otimes A_\ell \otimes \\
&\qquad \cdots \otimes A_\ell)(X \otimes \cdots \otimes X)] \\
&= E[\mathrm{tr}\,\{(A_1 \otimes \cdots \otimes A_1 \otimes \cdots \otimes A_\ell \otimes \cdots \otimes A_\ell) \\
&\qquad (X \otimes \cdots \otimes X)(X' \otimes \cdots \otimes X')\}] \\
&= \mathrm{tr}\,\{E[(A_1 \otimes \cdots \otimes A_1 \otimes \cdots \otimes A_\ell \otimes \cdots \otimes A_\ell)(XX' \otimes \cdots \otimes XX')]\} \\
&= \mathrm{tr}\,[(A_1 \otimes \cdots \otimes A_1 \otimes \cdots \otimes A_\ell \otimes \cdots \otimes A_\ell) M_k(X)]\,.
\end{aligned}$$

We will now use the characteristic function of an elliptically contoured random vector X to derive its moments where $X \sim \mathcal{C}_p(\xi; \mu, \Sigma)$ with $\xi = \xi(t'\Sigma t)$ such that $\xi_X(t) = e^{it'\mu}\xi(t'\Sigma t)$.

Theorem 3.2.2 Let $X \sim \mathcal{C}_p(\xi; \mu, \Sigma)$, then

(*i*) $M_1(X) = \mu$

(*ii*) $M_2(X) = \mu\mu' - 2\xi'(0)\Sigma$; $\mathrm{Cov}\,(X, X) = -2\xi'(0)\Sigma$

(*iii*) $M_3(X) = \mu \otimes \mu' \otimes \mu - 2\xi'(0)[f(\mu, \Sigma) + \mathrm{vec}(\Sigma)\mu']$
where f(,) is defined in (3.2.11)

(*iv*) $$\begin{aligned} M_4(X) = \mu \otimes \mu' \otimes \mu \otimes \mu' &- 2\xi'(0)[(I_{p^2} + K_{pp}) f(\mu\mu', \Sigma) \\ &+ (\mu \otimes \mu')(\mathrm{vec}(\Sigma))' + \mathrm{vec}(\Sigma)(\mu \otimes \mu')] \\ &+ 4\xi''(0)[(I_{p^2} + K_{pp})(\Sigma \otimes \Sigma) + \mathrm{vec}(\Sigma)(\mathrm{vec}\,(\Sigma))']\,. \end{aligned}$$

PROOF

(*i*) $$\begin{aligned} \frac{\partial \xi_X(t)}{\partial t} &= \frac{\partial [e^{it'\mu}\xi(t'\Sigma t)]}{\partial t} \\ &= i e^{it'\mu}\xi(t'\Sigma t)\mu + 2e^{it'\mu}\xi'(t'\Sigma t)\Sigma t \\ &= i\mu \text{ at } t = 0 \Rightarrow M_1(X) = \mu\,. \end{aligned}$$

(*ii*) $$\begin{aligned} \frac{\partial^2 \xi_X(t)}{\partial t \partial t'} &= i^2 e^{it'\mu}\xi(t'\Sigma t)\mu\mu' + 2ie^{it'\mu}\xi'(t'\Sigma t)\,[\mu \otimes t'\Sigma \\ &\qquad + \mu' \otimes \Sigma t - i\Sigma] + 4e^{it'\mu}\xi''(t'\Sigma t)[\Sigma t \otimes t'\Sigma]. \\ &= i^2\mu\mu' - i^2 2\xi'(0)\Sigma \text{ at } t = 0 \\ &\Rightarrow M_2(X) = \mu\mu' - 2\xi'(0)\Sigma\,. \end{aligned}$$

(iii) and (iv) are obtained similarly by differentiation.

Using Lemma 3.2.1 and Theorem 3.2.2 we have

Corollary 3.2.1. Let $\boldsymbol{X} = (X_1, \ldots, X_p) \sim C_p(\xi; \boldsymbol{\mu}, \Sigma)$, $\boldsymbol{\mu}' = (\mu_1, \ldots, \mu_p)$ and $\Sigma = (\sigma_{ij})$, $i, j = 1, \ldots, p$; then

$$
\begin{aligned}
&(i)\ E(X_i) = \mu_i \\
&(ii)\ E(X_iX_j) = \mu_i\mu_j - 2\xi'(0)\sigma_{ij} \text{ and Var } (X_i) = -2\xi'(0)\sigma_{ii} \\
&(iii)\ E(X_iX_jX_k) = \mu_i\mu_j\mu_k - 2\xi'(0)[\mu_i\sigma_{kj} + \mu_j\sigma_{ki} + \mu_k\sigma_{ij}] \\
&(iv)\ E(X_iX_jX_kX_\ell) = \mu_i\mu_j\mu_k\mu_\ell - 2\xi'(0)[\mu_i\mu_j\sigma_{k\ell} + \mu_k\mu_\ell\sigma_{ij} \\
&\qquad + \mu_k\mu_j\sigma_{i\ell} + \mu_i\mu_\ell\sigma_{kj} + \mu_i\mu_k\sigma_{\ell i} + \mu_\ell\mu_j\sigma_{ki}] \\
&\qquad + 4\xi''(0)[\sigma_{ij}\sigma_{k\ell} + \sigma_{kj}\sigma_{i\ell} + \sigma_{ki}\sigma_{\ell j}] \,.
\end{aligned}
$$

Letting $i = j = k = \ell$, we have

Corollary 3.2.2. Let $X \sim C_p(\xi; \boldsymbol{\mu}, \Sigma)$; then for $i = 1, \ldots, p$, we have

$$
\begin{aligned}
&(i)\ E(X_i) = \mu_i \\
&(ii)\ E(X_i^2) = \mu_i^2 - 2\xi'(0)\sigma_{ii} \\
&(iii)\ E(X_i^3) = \mu_i^3 - 6\xi'(0)\mu_i\sigma_{ii} \\
&(iv)\ E(X_i^4) = \mu_i^4 - 12\xi'(0)\mu_i^2\sigma_{ii} + 12\xi''(0)\sigma_{ii}^2,\ i = 1, \ldots, p \,.
\end{aligned}
$$

Theorem 3.2.3 Let $\boldsymbol{X} \sim C_p(\xi; \boldsymbol{\mu}, \Sigma)$ and A, A_1, A_2 be symmetric matrices; then

$$
\begin{aligned}
&(i)\ E(\boldsymbol{X}'A\boldsymbol{X}) = \boldsymbol{\mu}'A\boldsymbol{\mu} - 2\xi'(0)\text{tr}\,(\text{A}\Sigma) \\
&(ii)\ E[(\boldsymbol{X}'A\boldsymbol{X})^2] = (\boldsymbol{\mu}'A\boldsymbol{\mu})^2 - 4\xi'(0)[\boldsymbol{\mu}'A\boldsymbol{\mu}\ \text{tr}\,(\text{A}\Sigma) + 2\boldsymbol{\mu}'\text{A}\Sigma\text{A}\boldsymbol{\mu}] \\
&\qquad + 4\xi''(0)\{[\text{tr}\,(\text{A}\Sigma)]^2 + 2\text{tr}\,(\text{A}\Sigma\text{A}\Sigma)\} \\
&(iii)\ \text{Var}\,(\boldsymbol{X}'A\boldsymbol{X}) = -8\xi'(0)[\boldsymbol{\mu}'A\Sigma\boldsymbol{\mu}] + 4[\xi''(0) - \xi'^2(0)] \\
&\qquad \times [\text{tr}\,(\text{A}\Sigma)]^2 + 8\xi''(0)\text{tr}\,(\text{A}\Sigma\text{A}\Sigma) \\
&(iv)\ E[(\boldsymbol{X}'A_1\boldsymbol{X})(\boldsymbol{X}'A_2\boldsymbol{X})] = \boldsymbol{\mu}'A_1\boldsymbol{\mu}\boldsymbol{\mu}'A_2\mu - 2\xi'(0)[\boldsymbol{\mu}'A_1\boldsymbol{\mu}\ \text{tr}\,(\text{A}_2\Sigma) \\
&\qquad + \boldsymbol{\mu}'A_2\boldsymbol{\mu}\ \text{tr}\,(\text{A}_1\Sigma) + \boldsymbol{\mu}'(\text{A}_1\Sigma\text{A}_2 + 3\text{A}_2\Sigma\text{A}_1)\boldsymbol{\mu}] \\
&\qquad + 4\xi''(0)\,[\text{tr}\,(\text{A}_1\Sigma)\text{tr}\,(\text{A}_2\Sigma) + 2\text{tr}\,(\text{A}_1\Sigma\text{A}_2\Sigma)] \\
&(v)\ \text{Cov}\,(\boldsymbol{X}'A_1\boldsymbol{X}, \boldsymbol{X}'A_2\boldsymbol{X}) = -2\xi''(0)\boldsymbol{\mu}'(A_1\Sigma A_2 + 3A_2\Sigma A_1)\boldsymbol{\mu} \\
&\qquad + 4[\xi''(0) - \xi'^2(0)]\text{tr}\,(\text{A}_1\Sigma)\text{tr}\,(\text{A}_2\Sigma) \\
&\qquad + 8\xi''(0)\text{tr}\,(\text{A}_1\Sigma\text{A}_2\Sigma) \,.
\end{aligned}
$$

The results follow from Theorem 3.2.1 by noting that

$$E(\boldsymbol{X}'A\boldsymbol{X}) = \text{tr}\,[\text{AM}_2(\boldsymbol{X})] \tag{3.2.12}$$

and that

$$E(\boldsymbol{X}'A_1\boldsymbol{X}\boldsymbol{X}'A_2\boldsymbol{X}) = \text{tr}\,[(\text{A}_1 \otimes \text{A}_2)\text{M}_4(\boldsymbol{X})] \,, \tag{3.2.13}$$

and by using the fact that $\text{tr}\,(\text{A}_1\text{A}_2) = \text{tr}\,(\text{A}_2\text{A}_1)$ along with the properties (3.2.6) and (3.2.7) of the Kronecker product.

We may also use the decomposition $\boldsymbol{X} = \boldsymbol{\mu} + \mathcal{W}BU^{(p)}$, where $\mathcal{W}$ are $U^{(p)}$ are independently distributed, in order to obtain the moments of elliptically contoured distributions.

Lemma 3.2.4 Let $\boldsymbol{Y} \sim N_p(\mathbf{0}, I)$; then

$$M_1(\boldsymbol{Y}) = \mathbf{0},\ M_2(\boldsymbol{Y}) = I_p,\ M_3(\boldsymbol{Y}) = \mathbf{0}$$

and

$$M_4(\boldsymbol{Y}) = I_{p^2} + K_{pp} + \text{ vec}(I_p)(\text{vec}(I_p))' \,.$$

where K_{pp} is defined in (3.2.7).

The moments of a uniform distribution on a unit sphere in $\Re^p$ are given in the next lemma.

Lemma 3.2.5 Let $U^{(p)}$ denote the uniform distribution on the p-dimensional unit sphere; then

$$M_1(U^{(p)}) = \mathbf{0},\ M_2(U^{(p)}) = (I_p)/p,\ M_3(U^{(p)}) = O$$
$$M_4(U^{(p)}) = [I_{p^2} + K_{pp} + \text{ vec }(I_p)(\text{vec }(I_p))']/[p(p+2)] \,.$$

where K_{pp} is defined in (3.2.7).

PROOF When $\boldsymbol{Y} \sim N_p(\mathbf{0}, I)$, $Y = \mathcal{W}U^{(p)}$ where $\mathcal{W}$ and $U^{(p)}$ are independently distributed and $\mathcal{W} \sim \sqrt{\chi^2_p}$. Then

$$M_k(\boldsymbol{Y}) = E(\mathcal{W}^k)M_k(U^{(p)})$$

and since

$$E(\mathcal{W}^k) = 2^{k/2}\frac{\Gamma((p+k)/2)}{\Gamma(p/2)} \,,$$

$M_k(U^{(p)})$ is obtained from Lemma 3.2.4 as the ratio $M_k(\boldsymbol{Y})/E(\mathcal{W}^k)$.

Now let $\boldsymbol{X} \sim \mathcal{C}_p(\xi; \boldsymbol{\mu}, \Sigma)$ where $\Sigma = B'B$ is positive semi-definite. Then $\boldsymbol{X}$ is distributed as $\boldsymbol{\mu} + \mathcal{W}B'U^{(p)}$ where $\mathcal{W}$ is a nonnegative random variable whose distribution is determined by $\xi(\ )$, *and* $\mathcal{W}$ and $U^{(p)}$ are independently distributed. Let $m_2 = E(\mathcal{W}^2)$ and $m_4 = E(\mathcal{W}^4)$. The first four moments of $\boldsymbol{X}$ are given in the next theorem.

Theorem 3.2.4 Let $\boldsymbol{X} \sim \mathcal{C}_p(\xi; \boldsymbol{\mu}, \Sigma)$ where Σ is positive semi-definite. Then

$$(i)\ M_1(\boldsymbol{X}) = \boldsymbol{\mu};\ \ (ii)\ M_2(\boldsymbol{X}) = \boldsymbol{\mu}'\boldsymbol{\mu} + (m_2/p)\Sigma;$$
$$(iii)\ M_3(\boldsymbol{X}) = \boldsymbol{\mu} \otimes \boldsymbol{\mu}' \otimes \boldsymbol{\mu} + (m_2/p)[f(\boldsymbol{\mu}, \Sigma) + \text{vec}(\Sigma)\boldsymbol{\mu}'];$$
$$\begin{aligned}(iv)\ M_4(\boldsymbol{X}) = {} & \boldsymbol{\mu} \otimes \boldsymbol{\mu}' \otimes \boldsymbol{\mu} \otimes \boldsymbol{\mu}' + (m_2/p)[(I_{p^2} + K_{pp})f(\boldsymbol{\mu}\boldsymbol{\mu}',\ \Sigma) \\ & + \text{ vec}(\Sigma)(\boldsymbol{\mu} \otimes \boldsymbol{\mu})' + (\boldsymbol{\mu} \otimes \boldsymbol{\mu})(\text{vec}(\Sigma))'] \\ & + [m_4/(p(p+2))](I_{p^2} + K_{pp})(\Sigma \otimes \Sigma) + \text{ vec}(\Sigma)(\text{vec}(\Sigma))'] \,.\end{aligned}$$

PROOF The results follow from Lemma 3.2.5 and the fact that $\boldsymbol{X} \sim \boldsymbol{\mu} + \mathcal{W}B'\boldsymbol{U}^{(p)}$.

Remark 3.2.1 We obtain the following relationships from Theorem 3.2.4 and Theorem 3.2.2:

$$m_2 = -2p\xi'(0) \tag{3.2.14}$$

$$m_4 = 4p(p+2)\xi''(0)\ . \tag{3.2.15}$$

Example 3.2.1 If $\boldsymbol{X}$ is a uniformly distributed random vector in the p-dimensional unit ball, then

$$(i)\ \ E(\boldsymbol{X}'A\boldsymbol{X}) = \mathrm{tr}\,(\mathrm{A})/(\mathrm{p}+2) \tag{3.2.16}$$

$$(ii)\ \ \mathrm{Var}(\boldsymbol{X}'A\boldsymbol{X}) = \frac{2[\mathrm{tr}\,(\mathrm{A}^2) - (\mathrm{tr}\,(\mathrm{A}))^2/(\mathrm{p}+2)]}{(p+2)(p+4)}\ . \tag{3.2.17}$$

PROOF Consider $\boldsymbol{X} \equiv ||\boldsymbol{X}||(\boldsymbol{X}/||\boldsymbol{X}||) \simeq \mathcal{W}\boldsymbol{U}^{(p)}$ where $\mathcal{W}$ and $\boldsymbol{U}^{(p)}$ are independently distributed and the radius $\mathcal{W}$ has p.d.f.

$$f(w) = \begin{cases} pw^{p-1} & \text{for } 0 \le w \le 1, \\ 0 & \text{elsewhere}\ . \end{cases}$$

Note that the density function of $\mathcal{W}$ has to be proportional to w^{p-1} in order to achieve a uniform distribution since the surface areas covering concentric spheres in $\Re^p$ are proportional to w^{p-1}. The second and fourth moment of $\mathcal{W}$ can be easily evaluated; they are respectively

$$m_2 = E(\mathcal{W}^2) = p/(p+2) \tag{3.2.18}$$

and

$$m_4 = E(\mathcal{W}^4) = p/(p+4)\ . \tag{3.2.19}$$

Letting $\boldsymbol{\mu} = \mathbf{0}$, $m_2 = p/(p+2)$ and $m_4 = p/(p+4)$ in Theorem 3.2.4, one obtains the following first four moments of $\boldsymbol{X}$:

$$M_1(\boldsymbol{X}) = \mathbf{0};\ \ M_2(\boldsymbol{X}) = I_p/(p+2);\ \ M_3(\boldsymbol{X}) = O\,;$$

$$M_4(\boldsymbol{X}) = \frac{I_{p^2} + K_{pp} + \ \mathrm{vec}\ (I_p)(\ \mathrm{vec}\ (I_p))'}{(m+2)(m+4)}\ .$$

The expected value and the variance of $\boldsymbol{X}'A\boldsymbol{X}$ are therefore

$$E(\boldsymbol{X}'A\boldsymbol{X}) = \ \mathrm{tr}\,(\mathrm{AM}_2(\boldsymbol{X})) = \mathrm{tr}\,(\mathrm{A})/(\mathrm{p}+2)$$

and

$$\mathrm{Var}(\boldsymbol{X}'A\boldsymbol{X}) = \ E[(\boldsymbol{X}'A\boldsymbol{X})^2] - [E(\boldsymbol{X}'A\boldsymbol{X})]^2$$

$$= \mathrm{tr}\,[(\mathrm{A}\otimes\mathrm{A})\mathrm{M}_4(\boldsymbol{X})] - [\mathrm{tr}\,(\mathrm{A})/(\mathrm{p}+2)]^2$$
$$= \frac{\mathrm{tr}\,[(\mathrm{A}\otimes\mathrm{A})(\mathrm{I}_{\mathrm{p}^2} + \mathrm{K}_{\mathrm{pp}} + \mathrm{vec}(\mathrm{I_p})(\mathrm{vec}(\mathrm{I_p}))']}{(p+2)(p+4)}$$
$$-\,(\frac{\mathrm{tr}\,(\mathrm{A})}{(p+2)})^2$$
$$= \frac{(\mathrm{tr}\,(\mathrm{A}))^2 + \mathrm{tr}\,(\mathrm{A}^2) + \mathrm{tr}\,(\mathrm{A}^2)}{(p+2)(p+4)} - \frac{[\mathrm{tr}\,(\mathrm{A})]^2}{(p+2)^2}$$
$$= \frac{2[\mathrm{tr}\,(\mathrm{A}^2) - (\mathrm{tr}\,(\mathrm{A}))^2/(\mathrm{p}+2)]}{(p+2)(p+4)}\,.$$

3.3 The Distribution of Quadratic Forms

Representations of the probability density function of $\boldsymbol{X}'\boldsymbol{X}$ and of the characteristic function of a quadratic expression are given in this section.

Let $\boldsymbol{X} \sim \mathcal{C}_p(\xi;\boldsymbol{\mu}, I)$; then, in view of (3.1.15), $\boldsymbol{X} \simeq \boldsymbol{\mu} + \mathcal{W}\boldsymbol{U}^{(p)}$. When $\boldsymbol{\mu} = \mathbf{0}$, one may use Theorem 3.1.2 with $q = 1$ to obtain the probability density function of $\mathcal{W}^2$ as

$$\frac{\pi^{p/2} w^{p/2-1}}{\Gamma(\frac{p}{2})} f(w)\,, \tag{3.3.1}$$

where $f(\;)$ denotes the probability density function of $\boldsymbol{X}$, and then that of $\mathcal{W}$ as

$$g(w) = \frac{2\pi^{p/2} w^{p-1}}{\Gamma(p/2)} f(w^2)\,. \tag{3.3.2}$$

We denote the distribution of $Q = \boldsymbol{X}'\boldsymbol{X}$ by $\chi^2(\xi;p;\delta^2)$ where $\delta^2 = \boldsymbol{\mu}'\boldsymbol{\mu}$ (cf Definition 3.1.7). This distribution is referred to as the generalized noncentral chi-square distribution. Whenever the p.d.f. of $\boldsymbol{X}$ exists, that of Q can be obtained as follows.

Theorem 3.3.1 Let $\boldsymbol{X} \sim \mathcal{C}_p(\xi;\boldsymbol{\mu}, I)$ with p.d.f. $f((\boldsymbol{x}-\boldsymbol{\mu})'(\boldsymbol{x}-\boldsymbol{\mu})), \boldsymbol{X}'\boldsymbol{X} = Q$, $\delta = \sqrt{\boldsymbol{\mu}'\boldsymbol{\mu}} > 0$, and let the distribution function of $\mathcal{W}$ specified by ξ be denoted by $G(w)$; then the probability density function of Q can be expressed as follows:

$$\frac{1}{2\delta B(\frac{1}{2},\frac{p-1}{2})}\int_{|\sqrt{q}-\delta|}^{\sqrt{q}+\delta} w^{-1}\left[1-\left(\frac{q-\delta^2-w^2}{2w\delta}\right)^2\right]^{\frac{p-3}{2}} \mathrm{d}G(w)$$

where $B(a,b) = \Gamma(a)\Gamma(b)/\Gamma(a+b)$.

PROOF From the representation $\boldsymbol{X} \simeq \boldsymbol{\mu} + \mathcal{W}\boldsymbol{U}^{(p)}$, and noting that there exists an orthogonal transformation such that $\boldsymbol{\mu}'\boldsymbol{U}^{(p)} \simeq U_1\delta$, U_1 scalar, one may write

$$Q \simeq \delta^2 + 2\delta\mathcal{W}V + \mathcal{W}^2 \tag{3.3.3}$$

where $V = U^1$, $U^{(p)'} = (U^1, (U^2)')'$, $\mathcal{W}$ and V are independently distributed, and V^2 follows a type-1 beta distribution with parameters $\frac{1}{2}$ and $\frac{p-1}{2}$. The probability density function of V is then

$$\frac{1}{B(\frac{1}{2}, \frac{p-1}{2})}(1 - v^2)^{\frac{p-3}{2}} \text{ for } |v| < 1$$

and for any Borel function $h(Q)$ such that $E(|h(Q)|) < \infty$,

$$E(h(Q)) = E(h(\delta^2 + 2\delta\mathcal{W}V + \mathcal{W}^2))$$

$$= \frac{1}{B(\frac{1}{2}, \frac{p-1}{2})} \int_0^\infty \int_{-1}^1 h(\delta^2 + 2\delta wv + w^2)(1 - v^2)^{\frac{p-3}{2}} \mathrm{d}v\mathrm{d}G(w)$$

$$= \frac{1}{B(\frac{1}{2}, \frac{p-1}{2})} \int_0^\infty h(q) \int_{|\sqrt{q}-\delta|}^{\sqrt{q}+\delta} \frac{1}{2\delta w}\left[1 - \left(\frac{q - \delta^2 - w^2}{2w\delta}\right)^2\right]^{\frac{p-3}{2}} \mathrm{d}G(w)\mathrm{d}q \,. \tag{3.3.4}$$

Corollary 3.3.1 Let $\boldsymbol{X} \sim \mathcal{C}_p(\xi; \boldsymbol{\mu}, I)$ and $f((\boldsymbol{x} - \boldsymbol{\mu})'(\boldsymbol{x} - \boldsymbol{\mu}))$ be the p.d.f. of $\boldsymbol{X}$, then the p.d.f. of $Q = \boldsymbol{X}'\boldsymbol{X}$ is

$$\frac{\pi^{\frac{p-1}{2}}}{\delta\Gamma(\frac{p-1}{2})} \int_{|\sqrt{q}-\delta|}^{\sqrt{q}+\delta} w^{p-2} \left[1 - \left(\frac{q - \delta^2 - w^2}{2\delta w}\right)^2\right]^{\frac{p-3}{2}} f(w^2)\mathrm{d}w \,.$$

Corollary 3.3.2 Let $\boldsymbol{X} = (\boldsymbol{X}_1', \ldots, \boldsymbol{X}_\ell')' \sim \mathcal{C}_p(\xi; \boldsymbol{\mu}, I)$ where $\boldsymbol{X}_i$ is a p_i-dimensional vector, let $\boldsymbol{\mu} = (\boldsymbol{\mu}_1', \ldots, \boldsymbol{\mu}_\ell')'$ be partitioned similarly, and let $Y_i = (\boldsymbol{X}_i + \boldsymbol{\mu}_i)'(\boldsymbol{X}_i + \boldsymbol{\mu}_i)$, $i = 1, \ldots, \ell$, then

$$(Y_1, \ldots, Y_\ell)' \sim \chi^2(\xi; p_1, \ldots, p_\ell; \delta_1^2, \ldots, \delta_\ell^2) \tag{3.3.5}$$

where for $i = 1, \ldots, \ell$, $\delta_i^2 = \boldsymbol{\mu}_i'\boldsymbol{\mu}_i$. It follows that

$$\sum_{i=1}^m Y_i \sim \chi^2(\xi; p_1 + \cdots + p_m; \delta_1^2 + \cdots + \delta_m^2) \,.$$

Theorem 3.3.2 Let $f((\boldsymbol{x} - \boldsymbol{\mu})'(\boldsymbol{x} - \boldsymbol{\mu}))$ be the p.d.f. of $\boldsymbol{X}$ then the p.d.f. of $\boldsymbol{Y} = (Y_1, \ldots, Y_\ell)'$ as defined in (3.3.5) is

$$\left(\prod_{j=1}^\ell \frac{\pi^{\frac{p_j-1}{2}}}{\Gamma(\frac{p_j-1}{2})} y_j^{\frac{p_j}{2}-1}\right) \int_0^\pi \cdots \int_0^\pi f\Big(\sum_{j=1}^\ell (y_j + \delta_j^2) - 2\sum_{j=1}^\ell \delta_j\sqrt{y_j}\, \cos\theta_j\Big)$$
$$\times \prod_{i=1}^\ell \sin^{p_i-2}\theta_i \mathrm{d}\theta_i$$

for $p_i \geq 2$ and $j = 1, \ldots, \ell$.

PROOF Letting $h(\ )$ be a Borel function such that $E(h(Y_1,\ldots,Y_\ell))$ exists, then by making polar coordinates transformations on each of the subvectors of $\boldsymbol{X} = (\boldsymbol{X}_1',\ldots,\boldsymbol{X}_\ell')' = (X_1,\ldots,X_p)'$, the Jacobian being $\prod_{j=1}^{\ell} w_j^{p_j-1} \sin^{p_j-2}\theta_j$ where $w_j^2 = \sum_{i=1}^{p_j} x_{i+p_1+\cdots+p_{j-1}}^2$ and integrating out all the angular coordinates except the first one of each transformed subvector, we have

$$E(h(Y_1,\ldots,Y_\ell)) = \int_{-\infty}^{\infty}\cdots\int_{-\infty}^{\infty} h\Big(\sum_{j=1}^{p_1} x_j^2,\ldots,\sum_{j=1}^{p_\ell} x_{j+p_1+\cdots+p_{\ell-1}}^2\Big)$$

$$\times f\Big((x_1-\delta_1)^2+\sum_{j=2}^{p_1}x_j^2+\cdots+(x_{1+p_1+\cdots+p_{\ell-1}}-\delta_\ell)^2$$

$$+\sum_{j=2}^{p_\ell} x_{j+p_1+\cdots+p_{\ell-1}}^2\Big)dx_1\ldots dx_p$$

$$= 2^\ell\Big(\prod_{j=1}^{\ell}\frac{\pi^{\frac{p_j-1}{2}}}{\Gamma(\frac{p_j-1}{2})}\Big)\int_0^\pi\cdots\int_0^\pi\int_0^\infty\cdots\int_0^\infty h(w_1^2,\ldots,w_\ell^2)$$

$$\times f\Big(\sum_{j=1}^{\ell}(w_j^2-2w_j\delta_j\cos\theta_j+\delta_j^2)\Big)$$

$$\times\prod_{j=1}^{\ell} w_j^{p_j-1}\sin^{p_j-2}\theta_j d\theta_j dw_j$$

(the first component of each subvector of $\boldsymbol{X}$ being equal to $w_j\cos\theta_j$, $j=1,\ldots,\ell$)

$$=\Big(\prod_{j=1}^{\ell}\frac{\pi^{\frac{p_j-1}{2}}}{\Gamma(\frac{p_j-1}{2})}\Big)\int_0^\infty\cdots\int_0^\infty h(y_1,\ldots,y_\ell)$$

$$\int_0^\pi\cdots\int_0^\pi g\Big(\sum_{j=1}^{\ell}(y_j-2\delta_j\sqrt{y_j}\cos\theta_j+\delta_j^2)\Big)$$

$$\times\prod_{j=1}^{\ell} y_j^{(p_j/2)-1}\sin^{p_j-2}\theta_j d\theta_j dy_j \quad (\text{letting } y_j = w_j^2)\,,$$

hence the result.

Corollary 3.3.3 If $p_m = 1$, $p_{m+1} = 1,\ldots,p_\ell = 1$, then noting that $\cos\theta_j$ can only be equal to 1 or -1 for $j = m,\ldots,\ell$, the density of $(Y_1,\ldots,Y_\ell)' \sim \chi^2(\xi;p_1,\ldots,p_{m-1},1,\ldots,1;\delta_1^2,\ldots,\delta_\ell^2)$ is

$$\Big(\prod_{j=1}^{m-1}\frac{\pi^{\frac{p_j-1}{2}}}{\Gamma(\frac{p_j-1}{2})}y_j^{(p_j/2)-1}\Big)\Big(\prod_{j=m}^{\ell} y_j^{-1/2}\Big)\sum_{k_m=-1,1}\cdots\sum_{k_\ell=-1,1}$$

$$\int_0^\pi \cdots \int_0^\pi f\Big(\sum_{j=1}^{\ell}(y_j+\delta_j^2) - 2\sum_{j=1}^{m-1}\delta_j\sqrt{y_i}\cos\theta_j - 2\sum_{j=m}^{\ell}k_j\delta_j\sqrt{y_j}\Big)$$
$$\times\Big(\prod_{j=1}^{m-1}\sin^{p_j-2}\theta_j\Big)\mathrm{d}\theta_1\ldots\mathrm{d}\theta_{m-1}\,. \tag{3.3.6}$$

Corollary 3.3.4 For the central case, that is, $(Y_1,\ldots,Y_\ell)' \sim \chi^2(\xi;p_1,\ldots,p_\ell;0,\ldots,0)$, the p.d.f. of $(Y_1,\ldots,Y_\ell)'$ is

$$\Big(\prod_{j=1}^{\ell}\frac{\pi^{p_j/2}}{\Gamma(p_j/2)}y_j^{(p_j/2)-1}\Big)f\Big(\sum_{j=1}^{\ell}y_j\Big)\,,$$

which is obtained by integrating out the θ_j's in (3.3.6).

We now consider the characteristic function of a quadratic expression in spherical vectors. Let $\boldsymbol{X}\sim S_p(\xi)$ with $Pr(\boldsymbol{X}=\boldsymbol{0})=0$. Then in view of (3.1.8), $\boldsymbol{X}\simeq \mathcal{W}\boldsymbol{U}^{(p)}$ with $\boldsymbol{U}^{(p)}\simeq \boldsymbol{Z}/||\boldsymbol{Z}||$ where $\boldsymbol{Z}\sim N_p(\boldsymbol{0},I)$. The characteristic function of $\boldsymbol{X}'A\boldsymbol{X}+2\boldsymbol{b}'\boldsymbol{X}+c$ can be expressed as

$$\psi(t)=\int_0^\infty h(w)\mathrm{d}F(w) \tag{3.3.7}$$

where $F(w)$ denotes the distribution function of $\mathcal{W}$ and

$$h(w)=\int_{S_p} e^{it(w^2\boldsymbol{u}^{(p)\prime}A\boldsymbol{u}^{(p)}+2w\boldsymbol{b}'\boldsymbol{u}^{(p)}+c)}\mathrm{d}\boldsymbol{u}^{(p)} \tag{3.3.8}$$

where S_n is the unit sphere $||\boldsymbol{u}^{(p)}||^2=1$. Letting $w=s/v$ where v is a constant and multiplying both sides of (3.3.8) by $k(s)=\frac{s^{p-1}e^{-s^2/2}}{2^{(p-2)/2}\Gamma(p/2)}$, the p.d.f. of $||\boldsymbol{Z}||$ which is distributed as the square root of a chi-square random variable with p degrees of freedom, we have

$$\int_0^\infty \frac{s^{p-1}e^{-s^2/2}}{2^{(p-2)/2}\Gamma(p/2)}h(s/v)\mathrm{d}s \tag{3.3.9}$$
$$=\int_{-\infty}^{\infty}\cdots\int_{-\infty}^{\infty} e^{it(\boldsymbol{y}'A\boldsymbol{y}/v^2+2\boldsymbol{b}'\boldsymbol{y}/v+c)}\frac{e^{-\boldsymbol{y}'\boldsymbol{y}/2}}{(2\pi)^{p/2}}\mathrm{d}\boldsymbol{y}$$

where $\boldsymbol{Y}\simeq ||\boldsymbol{Z}||\boldsymbol{U}^{(p)}\simeq S\boldsymbol{U}^{(p)}\sim N_p(\boldsymbol{0},I)$

$$=v^p|v^2I_p-2itA|^{-1/2}e^{itc-2t^2\boldsymbol{b}'(v^2I_n-2itA)^{-1}\boldsymbol{b}}$$
$$=v^pg(v^2)\,. \tag{3.3.10}$$

Letting $z=s^2/(2v^2)$, (3.3.9) can be written as

$$\frac{1}{\Gamma(p/2)}v^p\int_0^\infty e^{-v^2z}z^{(p/2)-1}h(\sqrt{2z})\mathrm{d}z\,. \tag{3.3.11}$$

Noting that the integral which is, in view of (3.3.10), equal to $\Gamma(p/2)g(v^2)$ is the Laplace transform of $z^{(p/2)-1}h(\sqrt{2z})$, the inverse Laplace transform is given by

$$\frac{1}{2\pi i}\int_{a-i\infty}^{a+i\infty} e^{vz}\Gamma(p/2)g(v)dv \quad \text{(changing } v^2 \text{to } v\text{)}$$

where the integration is taken along the Bromwitch path $(a - i\infty,\ a + i\infty)$.

Hence

$$z^{(p/2)-1}h(\sqrt{2z}) = \frac{\Gamma(p/2)}{2\pi i}\int_{a-i\infty}^{a+i\infty} e^{vz}|vI_p - 2itA|^{-\frac{1}{2}} \times e^{(itc-2t^2b'(vI_n-2itA)^{-1}b)}dv \tag{3.3.12}$$

and letting $w = \sqrt{2z}$, we have

$$h(w) = \frac{2^{(p/2)-1}\Gamma(p/2)}{2\pi i}\int_{a-i\infty}^{a+i\infty} w^{2-p}e^{w^2v/2}|vI_n - 2itA|^{-\frac{1}{2}} \times e^{itc-2t^2b'(vI_n-2itA)^{-1}b}dv\,. \tag{3.3.13}$$

Theorem 3.3.3 Let $X \simeq \mathcal{W}U^{(p)}$ with $\Pr(X = 0) = 0$, then the characteristic function of $X'AX + 2b'X + c$ is

$$\psi(t) = \frac{2^{(p/2)-1}\Gamma(p/2)}{2\pi i}\int_0^{\infty}\int_{a-i\infty}^{a+i\infty} e^{y/2}|yI_n - 2itw^2A|^{-\frac{1}{2}} \times e^{(itc-2t^2w^2b'(yI_n-2itw^2A)^{-1}b)}dydF(w)$$

where $F(w)$ denotes the distribution function of $\mathcal{W}$.

PROOF The result follows from (3.3.7) and the change of variables $y = w^2v$ in (3.3.13).

Corollary 3.3.5 Let $X \sim C_p(\xi; \mu, I)$ and $Pr(X = 0) = 0$, $A = A'$, and T be an orthogonal matrix such that $T'AT = \operatorname{diag}(\lambda_1, \ldots, \lambda_p)$, then the moment-generating function of $X'AX$ is

$$E(e^{tX'AX}) = \frac{2^{(p/2)-1}\Gamma(p/2)}{2\pi i}\int_0^{\infty}\int_{a-i\infty}^{a+i\infty} e^{y/2} \times \prod_{j=1}^{p}(y - 2t\lambda_j w^2)^{-\frac{1}{2}}e^{t\lambda_j\delta_j^2 y/(y-2\lambda_j w^2 t)}dydF(w)$$

where $\delta = T'\mu$.

This is easily seen by observing the following. When $X \sim C_p(\xi; \mu, I), Y = X - \mu \sim C_p(\xi; 0, I)$ and

$$\begin{aligned} X'AX &= Y'AY + 2\mu'AY + \mu'A\mu, Y = X - \mu \\ &= Z'DZ + 2\mu'ATZ + \mu'A\mu,\ Z = T'Y,\ D = \operatorname{diag}(\lambda_1, \ldots, \lambda_p)\,. \end{aligned}$$

Comparing with the quadratic expression in Theorem 3.3.3 one has

$$c = \boldsymbol{\mu}' A \boldsymbol{\mu} = (T'\boldsymbol{\mu})' D (T'\boldsymbol{\mu}) = \boldsymbol{\delta}' D \boldsymbol{\delta},$$

$$\boldsymbol{b}' = \boldsymbol{\mu}' A T = (T'\boldsymbol{\mu})' D = \boldsymbol{\delta}' D\ .$$

Then replacing it by t, $itc - 2t^2 w^2 \boldsymbol{b}'(yI_n - 2itw^2 A)^{-1}\boldsymbol{b}$ becomes

$$\begin{aligned} & tc + 2t^2 w^2 \boldsymbol{b}'(yI_n - 2tw^2 A)^{-1}\boldsymbol{b} \\ &= t\boldsymbol{\delta}' D \boldsymbol{\delta} + 2t^2 w^2 \sum_{j=1}^{p} \delta_j^2 (y - 2tw^2 \lambda_j)^{-1} \qquad (3.3.14) \\ &= \sum_{j=1}^{p} [t\lambda_j \delta_j^2 y (y - 2\lambda_j w^2 t)^{-1}] \qquad (3.3.15) \end{aligned}$$

and

$$|yI_p - 2tw^2 A| = \prod_{j=1}^{p} (y - 2t\lambda_j w^2)$$

which establishes the result.

Provost and Kawczak (1995) obtained an integral representation of the probability density function of quadratic forms in spherically symmetric random vectors for $p = 3$. Some generalizations are discussed in Provost (1995). The probability density function of quadratic forms in correlated t variables is given in Menzefricke (1981); it is expressed in terms of a representation of the density of a quadratic form in a normal vector by means of a transformation. Several results on the distribution of quadratic forms (positive semi-definite or indefinite) in possibly singular normal vectors are derived in Provost (1988,1989,1994) and Provost and Rudiuk (1992,1993,1994).

3.4 Noncentral Distribution

We consider in this section the distribution of $\boldsymbol{X}_1'\boldsymbol{X}_1$ where $\boldsymbol{X}_1$ is a subvector of a noncentral elliptically contoured random vector.

Let $\boldsymbol{X} \sim \mathcal{C}_p(\xi; \boldsymbol{\mu}, I)$ that is, $\boldsymbol{X} \simeq \boldsymbol{\mu} + \mathcal{W}\boldsymbol{U}^{(p)}$ where $\boldsymbol{U}^{(p)}$ is a random vector which is uniformly distributed on the p-dimensional unit sphere and $\mathcal{W}$ is a nonnegative random variable distributed independently of $\boldsymbol{U}^{(p)}$. Note also that $\boldsymbol{X} - \boldsymbol{\mu} \sim \mathcal{C}_p(\xi; \boldsymbol{0}, I)$.

Recall the representation in Theorem 3.1.5, that is,

$$\boldsymbol{X} - \boldsymbol{\mu} \simeq \begin{pmatrix} \mathcal{W}\sqrt{V_1}U_1 \\ \vdots \\ \mathcal{W}\sqrt{V_\ell}U_\ell \end{pmatrix}.$$

Then for $\ell = 2$, $(V_1, V_2)'$ is a $\mathcal{D}_2(q/2, (p-q)/2)$. But $\mathcal{D}(\cdot,\cdot)$ is a type-1 beta. Thus we may write

$$\begin{pmatrix} X_1 \\ X_2 \end{pmatrix} \simeq \begin{pmatrix} \mu_1 + WBU_1 \\ \mu_2 + W(1-B^2)^{\frac{1}{2}} U_2 \end{pmatrix} . \tag{3.4.1}$$

where B is a type-1 beta random variable independently distributed of W and $(U_1', U_2')'$, $\mu' = (\mu_1', \mu_2')$, U_1 is $q \times 1$, U_2 is $(p-q) \times 2$.

We are investigating the distribution of $X_1'X_1$ in the noncentral case. Letting $Y = X_1'X_1$, the distribution function of Y at the point $y \geq 0$ is

$$\begin{aligned} F_Y(y) &= \Pr\{(\mu_1 + WBU_1)'(\mu_1 + WBU_1) \leq y\} \\ &= \Pr\{\mu_1'\mu_1 + 2\mu_1'WBU_1 + W^2B^2 \leq y\} \\ &= \Pr\{\mu_1'\mu_1 + 2Z^{\frac{1}{2}}\mu_1'U_1 + Z \leq y\} \end{aligned} \tag{3.4.2}$$

where $Z = W^2B^2$. Noting that $TU_1 \simeq U_1$ for any orthogonal matrix T and letting the orthogonal matrix

$$T_1 = \begin{pmatrix} \frac{\mu_1}{\sqrt{\mu_1'\mu_1}}, & \frac{\mu_2}{\sqrt{\mu_1'\mu_1}}, & \dots, & \frac{\mu_q}{\sqrt{\mu_1'\mu_1}} \\ t_{21} & t_{22} & \dots & t_{2q} \\ \vdots & \vdots & & \vdots \\ t_{q1} & t_{q2} & \dots & t_{qq} \end{pmatrix}$$

where $\mu_1' = (\mu_1, \dots, \mu_q)$, we have $\mu_1^{*\prime}U_1 \simeq \mu_1'T_1'U_1 \simeq (U\sqrt{\mu_1'\mu_1}, U_2^{*\prime})$ where U is a scalar and U_2^* is a null vector since $\mu_1'T_1' = (\sqrt{\mu_1'\mu_1}, 0, \dots, 0)$. Then in view of the independence of Z and U, and denoting by $I_{\{A\}}$ the indicator function of the set A which is equal to one for points belonging to the set A and, to zero for points outside the set A, we have

$$\begin{aligned} F_Y(y) &= \Pr\{\mu_1'\mu_1 + 2\sqrt{Z}U\sqrt{\mu_1'\mu_1} + Z \leq y\} \\ &= E_Z E_U (I_{\{\mu_1'\mu_1 + 2\sqrt{Z}U\sqrt{\mu_1'\mu_1} + Z \leq y\}}) \\ &= \int_0^\infty \Pr\left\{U \leq \frac{y - \mu_1'\mu_1 - Z}{2\sqrt{Z}\sqrt{\mu_1'\mu_1}}\right\} f(z) dz \\ &= \int_0^\infty \left[\int_{-\infty}^a \frac{(1-u^2)^{\frac{q-1}{2}-1}}{B(1/2,\ (q-1)/2)} I_{\{|u| \leq 1\}} du\right] f(z) dz \end{aligned} \tag{3.4.3}$$

where

$$B(\delta, \epsilon) = \frac{\Gamma(\delta)\Gamma(\epsilon)}{\Gamma(\delta + \epsilon)},$$

$$a = \frac{y - \mu_1'\mu_1 - z}{2\sqrt{z}\sqrt{\mu_1'\mu_1}},$$

$$f(z) = \frac{z^{\frac{q}{2}-1}}{B(q/2,\ (p-q)/2)} \int_{\sqrt{z}}^\infty w^{-(p-2)}(w^2 - z)^{\frac{p-q}{2}-1} dG(w)$$

and $G(w)$ is the distribution function of W.

Assume without any loss of generality that z and y are nonnegative real numbers and define the sets

$$
\begin{aligned}
S_1 &= \left\{\frac{y - \mu_1'\mu_1 - z}{2\sqrt{z}\sqrt{\mu_1'\mu_1}} < 1\right\} \\
&\equiv \{y < (\sqrt{z} + \sqrt{\mu_1'\mu_1})^2\} \\
&\equiv \{\sqrt{y} - \sqrt{\mu_1'\mu_1} < \sqrt{z}\} \\
&\equiv \{(\sqrt{y} - \sqrt{\mu_1'\mu_1})^2 < z\} \cup \{y < \mu_1'\mu_1\}
\end{aligned}
$$

and

$$
\begin{aligned}
S_2 &= \left\{\frac{y - \mu_1'\mu_1 - z}{2\sqrt{z}\sqrt{\mu_1'\mu_1}} > -1\right\} \\
&\equiv \{y > (\sqrt{z} - \sqrt{\mu_1'\mu_1})^2\} \\
&\equiv \{\sqrt{y} + \sqrt{\mu_1'\mu_1} > \sqrt{z}\} \cap \{\sqrt{z} > \mu_1'\mu_1 - \sqrt{y}\} \\
&\equiv \{(\sqrt{y} + \sqrt{\mu_1'\mu_1})^2 > z\} \cap [\{z > (\sqrt{\mu_1'\mu_1} - \sqrt{y})^2\} \cup \{y > \mu_1'\mu_1\}] \,.
\end{aligned}
$$

In view of (3.4.3), the set $S_1 \cap S_2$ corresponds to the case $a < 1$ and the set S_1^c corresponds to the case $a \geq 1$. By inspection we find that $S_1 \cap S_2 = \{(\sqrt{y} - \sqrt{\mu_1'\mu_1})^2 < z\} \cap \{(\sqrt{y} + \sqrt{\mu_1'\mu_1})^2 > z\}$ and then (3.4.3) becomes

$$
\begin{aligned}
F_Y(y) &= \int_0^\infty \left[\int_{-\infty}^a \frac{(1-u^2)^{\frac{q-1}{2}-1}}{B(1/2,\ (q-1)/2)} (I_{S_1 \cap S_2} + I_{S_1^c}) \mathrm{d}u\right] f(z) \mathrm{d}z \\
&= \int_{(\sqrt{y}-\sqrt{\mu_1'\mu_1})^2}^{(\sqrt{y}+\sqrt{\mu_1'\mu_1})^2} \left[\int_{-1}^a \frac{(1-u^2)^{\frac{q-1}{2}-1}}{B(1/2,\ (q-1)/2)} \mathrm{d}u\right] f(z) \mathrm{d}z \\
&\quad + I_{\{y \geq \mu_1'\mu_1\}} \int_0^{(\sqrt{y}-\sqrt{\mu_1'\mu_1})^2} f(z) \mathrm{d}z \,.
\end{aligned}
$$

Using Leibnitz's differentiation rule for an integral the probability density function of $Y = \boldsymbol{X}_1'\boldsymbol{X}_1$ is found to be

$$
\begin{aligned}
h(y) &= \int_{(\sqrt{y}-\sqrt{\mu_1'\mu_1})^2}^{(\sqrt{y}+\sqrt{\mu_1'\mu_1})^2} \frac{(2\sqrt{z}\sqrt{\mu_1'\mu_1})^{-1}(1-a^2)^{\frac{q-3}{2}}}{B(1/2,\ (q-1)/2)} f(z) \mathrm{d}z \qquad (3.4.4) \\
&= \int_{(\sqrt{y}-\sqrt{\mu_1'\mu_1})^2}^{(\sqrt{y}+\sqrt{\mu_1'\mu_1})^2} \frac{(2\sqrt{z}\sqrt{\mu_1'\mu})^{-(q-2)}}{B(1/2,\ (q-1)/2)} \\
&\quad \times \{[z - (\sqrt{y} - \sqrt{\mu_1'\mu_1})^2][(\sqrt{y} + \sqrt{\mu_1'\mu_1})^2 - z]\}^{\frac{q-3}{2}} f(z) \mathrm{d}z \\
&= \int_{(\sqrt{y}-\sqrt{\mu_1'\mu_1})^2}^{(\sqrt{y}+\sqrt{\mu_1'\mu_1})^2} \frac{1}{\Gamma((q-1)/2)\Gamma(\frac{1}{2})(2\sqrt{\mu_1'\mu_1})^{q-2}2^{q/2}} \\
&\quad \times \{[z - (\sqrt{y} - \sqrt{\mu_1'\mu_1})^2][(\sqrt{y} + \sqrt{\mu_1'\mu_1})^2 - z]^{\frac{q-3}{2}} e^{-z/2} \phi(z) \mathrm{d}z
\end{aligned}
$$

where

$$\phi(z) = \frac{\Gamma(q/2)2^{q/2}}{z^{(q/2)-1}e^{-z/2}} f(z)\,. \tag{3.4.5}$$

Now letting $\delta = y + \mu_1'\mu_1$, $\sigma = 2\sqrt{y}\sqrt{\mu_1'\mu_1}$, and $\nu = (z-\delta)/\sigma$, the p.d.f. of $Y = X_1'X_1$ is

$$\begin{aligned} h(y) &= \frac{\sigma^{q-2}e^{-\delta/2}}{\Gamma((q-1)/2)\Gamma(1/2)(2\sqrt{\mu_1'\mu_1})^{q-2}2^{q/2}} \\ &\quad \times \int_{-1}^{1} (1-\nu^2)^{\frac{q-3}{2}} e^{-\sigma\nu/2}\phi(\sigma\nu+\delta)\mathrm{d}\nu \\ &= \frac{y^{\frac{q}{2}-1}e^{-\frac{1}{2}(y+\mu_1'\mu_1)}}{\Gamma((q-1)/2)\Gamma(1/2)2^{q/2}} I \end{aligned} \tag{3.4.6}$$

where

$$\begin{aligned} I &= \int_{-1}^{1} (1-\nu^2)^{\frac{(q-3)}{2}} e^{-\sigma\nu/2}\phi(\sigma\nu+\delta)\mathrm{d}\nu \\ &= \int_{0}^{1} (1-\nu^2)^{\frac{(q-3)}{2}} e^{-\sigma\nu/2}\phi(\sigma\nu+\delta)\mathrm{d}\nu \\ &\quad + \int_{0}^{1} (1-\nu^2)^{(q-3)/2} e^{\sigma\nu/2}\phi(\delta-\sigma\nu)\mathrm{d}\nu \\ &= \int_{0}^{1} (1-\nu^2)^{(q-3)/2}[e^{-\sigma\nu/2}\phi(\sigma\nu+\delta) + e^{\sigma\nu/2}\phi(\delta-\sigma\nu)]\mathrm{d}\nu\,. \end{aligned}$$

Letting $\nu = \sqrt{t}$,

$$\begin{aligned} I &= \int_{0}^{1} (1-t)^{\frac{(q-3)}{2}} \Big[e^{-\sigma\sqrt{t}/2}\phi(\sigma\sqrt{t}+\delta) \\ &\quad + e^{\sigma\sqrt{t}/2}\phi(\delta-\sigma\sqrt{t})\Big]\left(\frac{1}{2\sqrt{t}}\right)\mathrm{d}t \\ &= \frac{1}{2}\int_{0}^{1} (1-t)^{\frac{(q-3)}{2}} \left[\sum_{i=0}^{\infty}\frac{(-\sigma\sqrt{t}/2)^i}{i!}\phi(\sigma\sqrt{t}+\delta)\right. \\ &\quad \left. + \sum_{1=0}^{\infty}\frac{(\sigma\sqrt{t}/2)^i}{i!}\phi(\delta-\sigma\sqrt{t})\right] t^{-1/2}\mathrm{d}t \\ &= \frac{1}{2}\int_{0}^{1} (1-t)^{\frac{(q-3)}{2}} \left[[\phi(\sigma\sqrt{t}+\delta)+\phi(\delta-\sigma\sqrt{t})]\sum_{j=0}^{\infty}\frac{(\sigma\sqrt{t}/2)^{2j}}{(2j)!}\right. \\ &\quad \left. + [\phi(\sigma\sqrt{t}+\delta)-\phi(\delta-\sigma\sqrt{t})]\sum_{j=0}^{\infty}\frac{(\sigma\sqrt{t}/2)^{2j+1}}{(2j+1)!}\right] t^{-1/2}\mathrm{d}t \end{aligned}$$

$$= \frac{1}{2}\sqrt{\pi}\int_0^1 (1-t)^{\frac{(q-3)}{2}}$$

$$\times \Bigg[[\phi(\sigma t^{-\frac{1}{2}}+\delta)+\phi(\delta-\sigma t^{-\frac{1}{2}})]\Big(\sum_{j=0}^{\infty}\frac{(\sigma\sqrt{t}/2)^{2j}}{2^{2j}j!\Gamma(j+\frac{1}{2})}\Big)\sqrt{t}$$

$$+[\phi(\sigma\sqrt{t}+\delta)-\phi(\delta-\sigma\sqrt{t})\frac{\sigma}{2}\sum_{j=0}^{\infty}\frac{(\sigma\sqrt{t}/2)^{2j}}{2^{2j+1}j!\Gamma(j+\frac{3}{2})}\Bigg] \,\mathrm{d}t$$

(using the doubling formula $\frac{2^{2x-1}}{\sqrt{\pi}}\Gamma(x)\Gamma(x+\frac{1}{2})=\Gamma(2x)$ with $x=j+\frac{1}{2}$)

$$= \frac{\Gamma(1/2)\sqrt{\sigma}}{4}\int_0^1 (1-t)^{\frac{(q-3)}{2}}t^{-1/4}\Big\{[\phi(\sigma\sqrt{t}+\delta)+\phi(\delta-\sigma\sqrt{t})]$$

$$\times I_{-\frac{1}{2}}(\sigma\sqrt{t}/2)+[\phi(\sigma\sqrt{t}+\delta)-\phi(\delta-\sigma\sqrt{t})]I_{\frac{1}{2}}(\sigma\sqrt{t}/2)\Big\}\,\mathrm{d}t$$

$$= \frac{\Gamma(\frac{1}{2})y^{1/4}(\mu_1'\mu_1)^{1/4}}{2^{3/2}}\int_0^1 (1-t)^{\frac{(q-3)}{2}}t^{-\frac{1}{4}}\Big\{[\phi(\sigma\sqrt{t}+\delta)+\phi(\delta-\sigma\sqrt{t})]$$

$$\times I_{-\frac{1}{2}}(y^{1/2}\sqrt{\mu_1'\mu_1}\sqrt{t})+[\phi(\sigma\sqrt{t}+\delta)-\phi(\delta-\sigma\sqrt{t})]I_{\frac{1}{2}}(y^{1/2}\sqrt{\mu_1'\mu_1}\sqrt{t})\Big\}\,\mathrm{d}t$$

where $I_\nu(z)$ denotes the modified Bessel Function of the first kind and of order ν:

$$I_\nu(z)=\sum_{k=0}^{\infty}\frac{(z/2)^{\nu+2k}}{k!\Gamma(\nu+k+1)}\,.$$

Theorem 3.4.1 The probability density function of $Y = \boldsymbol{X}_1'\boldsymbol{X}_1$ where $\boldsymbol{X}_1$ is a q-dimensional subvector of $\boldsymbol{X}=(\boldsymbol{X}_1',\ \boldsymbol{X}_2')'\sim C_p(\xi;(\boldsymbol{\mu}_1',\ \boldsymbol{\mu}_2')',\ I)$ is

$$\frac{y^{\frac{2q-3}{4}}e^{-\frac{1}{2}(y+\mu_1'\mu_1)}\sqrt{\mu_1'\mu_1}}{\Gamma((q-1)/2)2^{(q+3)/2}}\int_0^1 (1-t)^{\frac{(q-3)}{2}}t^{-\frac{1}{4}}\Big\{[\phi(\sigma\sqrt{t}+\delta)+\phi(\delta-\sigma\sqrt{t})]$$

$$\times I_{-\frac{1}{2}}(y^{\frac{1}{2}}\sqrt{\mu_1'\mu_1}\sqrt{t})+[\phi(\sigma\sqrt{t}+\delta)-\phi(\delta-\sigma\sqrt{t})]I_{\frac{1}{2}}(y^{\frac{1}{2}}\sqrt{\mu_1'\mu_1}\sqrt{t})\Big\}\,\mathrm{d}t$$

where $\phi(\ )$ is given in (3.4.5), $\delta=y+\boldsymbol{\mu}_1'\boldsymbol{\mu}_1$, $\sigma=2y^{1/2}\sqrt{\boldsymbol{\mu}_1'\boldsymbol{\mu}_1}$ and $I_\nu(z)$ denotes the modified Bessel function of the first kind and of order ν.

Example 3.4.1 Let $\boldsymbol{X}=(\boldsymbol{X}_1',\ \boldsymbol{X}_2')'\sim N_p((\boldsymbol{\mu}_1',\ \boldsymbol{\mu}_2')',\ I)$ and $\boldsymbol{X}_1$ be a q-dimensional subvector of $\boldsymbol{X}$. It is well known that $\boldsymbol{X}_1'\boldsymbol{X}_1\sim\chi_q^2$ when $\boldsymbol{\mu}_1=\boldsymbol{0}$. When $\boldsymbol{\mu}_1\neq\boldsymbol{0}$, in view of (3.4.5), $\phi(z)=1$ and the probability density function of $Y=\boldsymbol{X}_1'\boldsymbol{X}$ given by Theorem 3.4.1 is

$$h_1(y)=\frac{2e^{-\frac{1}{2}(y+\mu_1'\mu_1)}y^{\frac{2q-3}{4}}\sqrt{\mu_1'\mu_1}}{\Gamma((q-1)/2)2^{(q+3)/2}}$$

$$\times \int_0^1 (1-t)^{\frac{(q-3)}{2}} t^{-\frac{1}{4}} I_{-\frac{1}{2}}(\sqrt{y}\sqrt{\mu_1'\mu_1}\sqrt{t})\mathrm{d}t$$

where the integral is

$$\int_0^1 (1-t)^{\frac{q}{2}-\frac{1}{2}-1} \sum_{j=0}^{\infty} (t^{j+\frac{1}{2}-1} \frac{\sqrt{y}\sqrt{\mu_1'\mu_1}/2)^{2j-\frac{1}{2}}}{j!\Gamma(-\frac{1}{2}+j+1)})\mathrm{d}t$$

$$= \sum_{j=0}^{\infty} \frac{\Gamma(\frac{q}{2}-\frac{1}{2})\Gamma(j+\frac{1}{2})}{\Gamma(\frac{q}{2}+j)} \frac{(\sqrt{y}\sqrt{\mu_1'\mu_1}/2)^{2j-\frac{1}{2}}}{j!\Gamma(j+1/2)} ;$$

hence,

$$h_1(y) = \frac{1}{2}\Big(\frac{y}{\mu_1'\mu_1}\Big)^{(q-2)/4} e^{-\frac{1}{2}(y+\mu_1'\mu_1)} I_{(q-2)/2}(\sqrt{y}\sqrt{\mu_1'\mu_1}) \tag{3.4.7}$$

where $I_\nu(z)$ denotes the modified Bessel function of the first kind and of order ν.

Example 3.4.2 Let $\boldsymbol{X} \sim (\boldsymbol{X}_1', \boldsymbol{X}_2')'$ be distributed as a p-variate t-distribution with m degrees of freedom where $\boldsymbol{X}_1$ is a q-dimensional subvector of $\boldsymbol{X}$. When $\boldsymbol{\mu}_1 = \boldsymbol{0}$, then $\boldsymbol{X}_1'\boldsymbol{X}_1/m \sim F_{q,m}$ and the probability density function of $\boldsymbol{X}_1'\boldsymbol{X}_1$ is therefore

$$f(z) = \frac{\Gamma((q+m)/2)}{\Gamma(q/2)\Gamma(m/2)} m^{-q/2} z^{(q-2)/2} (1+\frac{z}{m})^{-\frac{q+m}{2}} .$$

When $\boldsymbol{\mu}_1 \neq \boldsymbol{0}$, the probability density function of $Y = \boldsymbol{X}_1'\boldsymbol{X}_1$ given by Theorem 3.4.1 with $\delta = y + \boldsymbol{\mu}_1'\boldsymbol{\mu}_1$, $\sigma = 2\sqrt{y}\sqrt{\boldsymbol{\mu}_1'\boldsymbol{\mu}_1}$ and $\nu = (z-\delta)/\sigma$ is

$$h_1^*(y) = \frac{\sigma^{q-2}}{\Gamma(\frac{(q-1)}{2})\Gamma(\frac{1}{2})(2\sqrt{\mu_1'\mu_1})^{q-2}}$$

$$\times \int_{-1}^1 (1-\nu^2)^{\frac{(q-3)}{2}} \frac{\Gamma(\frac{q+m}{2})m^{-q/2}}{\Gamma(\frac{m}{2})(1+\frac{\sigma\nu+\delta}{m})^{(q+m)/2}}\mathrm{d}\nu$$

$$= \frac{y^{\frac{q}{2}-1}\Gamma(\frac{q+m}{2})m^{m/2}}{\Gamma(\frac{q-1}{2})\Gamma(\frac{1}{2})\Gamma(\frac{m}{2})} \int_{-1}^1 (1-\nu^2)^{\frac{q-3}{2}} (\sigma\nu+\delta+m)^{-\frac{(q+m)}{2}}\mathrm{d}\nu$$

where the integral can be expressed as follows by splitting $\int_{-1}^1$ into $\int_{-1}^0$ and $\int_0^1$, changing v to $-v$ in $\int_{-1}^0$ and then letting $\nu = \sqrt{t}$:

$$\int_0^1 (1-t)^{\frac{q-3}{2}} \Big[(\sigma\sqrt{t}+(\delta+m))^{-\frac{(q+m)}{2}} + (-\sigma\sqrt{t}+(\delta+m))^{\frac{(q+m)}{2}}\Big] \frac{1}{2\sqrt{t}}\mathrm{d}t$$

$$= \sum_{j=0}^{\infty} \frac{\Gamma(\frac{q+m}{2}+2j)}{\Gamma(\frac{q+m}{2})(2j)!} (\delta+m)^{-\frac{(q+m)}{2}-2j} \int_0^1 (1-t)^{\frac{q-3}{2}} \frac{(\sigma\sqrt{t})^{2j}}{\sqrt{t}}\mathrm{d}t$$

$$= \sum_{j=0}^{\infty} \frac{\Gamma(\frac{q+m}{2}+2j)\Gamma(1/2)}{\Gamma(\frac{q+m}{2})2^{2j}j!\Gamma(j+\frac{1}{2})}(\delta+m)^{-\frac{(q+m)}{2}-2j}\sigma^{2j}\frac{\Gamma(j+\frac{1}{2})\Gamma(\frac{q-1}{2})}{\Gamma(\frac{q}{2}+j)}$$

using the doubling formula on $(2j)! = \Gamma(2j+1)$; hence

$$h_1^*(y) = \frac{y^{\frac{q}{2}-1}m^{m/2}}{\Gamma(m/2)} \sum_{j=0}^{\infty} \frac{\Gamma(\frac{q+m}{2}+2j)y^j(\mu_1'\mu_1)^j}{j!(y+\mu_1'\mu_1+m)^{2j+(q+m)/2}\Gamma(\frac{q}{2}+j)} . \tag{3.4.8}$$

3.5 Quadratic Forms in Random Matrices

This section examines the conditions under which some quadratic forms in random matrices have a generalized Wishart distribution. Let P be a matrix belonging to $O(n)$, the set of $n \times n$ orthonormal real matrices, and X be an $n \times p$ random matrix. X is said to be left-spherical (see Dawid (1977)) if $X \simeq PX$ for every $P \in O(n)$. Then the characteristic function of X has the form $\xi(T'T)$ where T is an $n \times p$ matrix and the p.d.f. of X if it exists has the form $f(X'X)$. We write $X \sim LS_{n\times p}(\xi)$.

Definition 3.5.1 *Generalized Wishart distribution* Let $X = (X_1', \ldots, X_\ell')'$ where X_i is an $n_i \times p$ matrix, $i = 1, \ldots, \ell$, and let $Y_i = X_i'X_i$, $i = 1, \ldots, \ell$. Then assuming that $X \sim LS_{n\times p}(\xi)$ and that $n_i \geq p$, $i = 1, \ldots, \ell$, the generalized Wishart distribution denoted by $W_\ell(\xi; n_1, \ldots, n_\ell)$ is defined as the joint distribution of $(Y_1, \ldots, Y_\ell)$ and its p.d.f. is given by

$$\Big\{\prod_{i=1}^{\ell} b_{p,n_i}|Y_i|^{(n_i-p-1)/2}\Big\} f\Big(\sum_{i=1}^{\ell} Y_i\Big)$$

with

$$b_{p,n} = \frac{\pi^{(pn/2)-p(p-1)/4}}{\prod_{j=1}^{p}\Gamma(\frac{n-j+1}{2})} \quad , \; n \geq p,$$

Furthermore, letting $n^c = n_{k+1} + \cdots + n_\ell$, $1 \leq k < \ell$, the marginal distribution of $Y_1, \ldots, Y_k$, $1 < k \leq \ell$, denoted by $W_k(\xi; n_1, \ldots, n_k, n^c)$ has the following p.d.f.

$$\Big\{b_{p,n^c}\prod_{i=1}^{k} b_{p,n_i}|Y_i|^{(n_i-p-1)/2}\Big\} \int_{Y>0} f\Big(\sum_{i=1}^{k} Y_i + Y\Big)|Y|^{(n^c-p-1)/2}dY$$

where

$$Y = \sum_{i=k+1}^{\ell} Y_i \, .$$

Lemma 3.5.1 Let the $p \times p$ matrix $W \sim W_2(\xi; k, (n-k))$; then $\boldsymbol{u}'W\boldsymbol{u} \sim \chi^2(\psi; k, (n-k))$ where $\boldsymbol{u}$ is a p-dimensional non-null vector and $\psi(\boldsymbol{t}'\boldsymbol{t}) = \xi(\boldsymbol{u}\boldsymbol{t}'\boldsymbol{t}\boldsymbol{u}')$ where $\boldsymbol{t}$ is an n-dimensional vector.

PROOF Let $W = X'X$ and the parameters of W_2 indicate that we have the following partition: $X = \binom{X_1}{X_2} \sim LS_{n\times p}(\xi)$ where X_1 is $k \times p$. The characteristic function of $X\boldsymbol{u}$ is $\psi_{X\boldsymbol{u}}(\boldsymbol{t}'\boldsymbol{t}) = \xi_X(\boldsymbol{u}\boldsymbol{t}'\boldsymbol{t}\boldsymbol{u}')$; hence $X\boldsymbol{u} \sim S_n(\psi)$ and $\boldsymbol{u}'W\boldsymbol{u} = (X\boldsymbol{u})'X\boldsymbol{u} \sim \chi^2(\psi; k, (n-k))$.

Theorem 3.5.1 Let $X \sim LS_{n\times p}(\xi), A = A'$ and $\Pr(X\boldsymbol{u} = \boldsymbol{0}) = 0$ for some $\boldsymbol{u} \neq \boldsymbol{0}$. Then $X'AX \sim W_2(\xi; k, (n-k))$ if and only if A is an idempotent matrix of rank k.

PROOF We prove only the "only if" part. Using the notation of Lemma 3.5.1, we have that $X\boldsymbol{u} \sim S_n(\psi)$. Now, since $X'AX \sim W_2(\xi; k, (n-k))$, by Lemma 3.5.1, we have $\boldsymbol{u}'X'AX\boldsymbol{u} \sim \chi^2(\psi; k, (n-k))$. The result follows from Theorem 3.1.9.

Theorem 3.5.2 Let $X \sim LS_{n\times p}(\xi)$, $A_i = A_i'$, $i = 1, \ldots, \ell$ and $\Pr(X\boldsymbol{u} = \boldsymbol{0}) = 0$ for some $\boldsymbol{u} \neq \boldsymbol{0}$; then $(X'A_1X, \ldots, X'A_\ell X) \sim W_{\ell+1}(\xi; n_1, \ldots, n_\ell, n^c)$ where $n^c = n - \sum_{i=1}^{\ell} n_i$, if and only if $A_iA_j = \delta_{ij}A_i$ and the rank of A_i is n_i $i, j = 1, \ldots, \ell$.

Theorem 3.5.3 Let $X = (X_1', X_2')' \sim LS_{n\times p}(\xi)$ and $\Pr(X\boldsymbol{u} = \boldsymbol{0}) = 0$ for some $\boldsymbol{u} \neq \boldsymbol{0}$. Then $X_1'X_1$ and $X_2'X_2$ are independently distributed if and only if X is normally distributed or equivalently the $\boldsymbol{X}_i$'s are i.i.d. $N_p(\boldsymbol{0}, \Sigma)$, $i = 1, \ldots, n$ with $X = (\boldsymbol{X}_1, \ldots, \boldsymbol{X}_n)'$.

3.6 Quadratic Forms of Random Idempotent Matrices

The conditions for some quadratic forms of random idempotent matrices to have generalized chi-square or Wishart distributions are discussed in this section. The generalized Hotelling's T^2 distribution is also defined.

Notation 3.6.1 Let ξ be the characteristic function of a k-dimensional real random vector then Φ_k denotes the class of characteristic functions which can be expressed as $\xi(t_1^2 + \cdots + t_k^2)$.

Lemma 3.6.1 Let $\boldsymbol{U}_1$, $\boldsymbol{U}_2$ and $\boldsymbol{T}$ be random vectors defined on the probability triple $(\Omega, \mathcal{F}, \mathcal{P})$ and let $\boldsymbol{U}_1|\boldsymbol{T} \simeq \boldsymbol{U}_2|\boldsymbol{T}$ (except possibly on a set A with $\Pr(A) = 0$ where the density of $\boldsymbol{T}$ is zero) then $\boldsymbol{U}_1 \simeq \boldsymbol{U}_2$. ($\boldsymbol{U}_i|\boldsymbol{T}$ denotes the conditional distribution of $\boldsymbol{U}_i$ given $\boldsymbol{T}$, $i = 1, 2$.)

Property 3.6.1 Let $\boldsymbol{X} \sim C_p(\xi; \boldsymbol{\mu}, \Sigma)$, rank $(\Sigma) = q$, B be a $p \times m$ constant matrix of full rank, and $\boldsymbol{\lambda}$ be a constant m-dimensional vector. Then $\boldsymbol{\lambda} + B'\boldsymbol{X} \sim C_p(\xi; B'\boldsymbol{\mu} + \boldsymbol{\lambda}, B'\Sigma B)$.

The proof is left as an exercise.

Theorem 3.6.1 Let $\boldsymbol{X} \sim \mathcal{C}_p(\xi, \boldsymbol{\mu}, I)$, K be an $r \times p$ constant matrix whose rank is $k < p$, A be a $p \times p$ random matrix whose elements $a_{ij} = g_{ij}(K\boldsymbol{X})$ where $g_{ij}(\cdot)$ is a Borel measurable function, and let $\chi^2(\cdot;\cdot,\cdot)$ denote the generalized noncentral chi-square distribution, see Definition 3.1.8. Then

$$\boldsymbol{X}'A\boldsymbol{X} \sim \chi^2(\xi; m; \delta^2)$$

provided all the following conditions are satisfied with probability one:

(i) A is a symmetric idempotent matrix
(ii) the rank of A is m
(iii) $A = M'AM$ and $MK' = O$ for some $p \times p$ matrix M
(iv) $\delta^2 = \boldsymbol{\mu}'A\boldsymbol{\mu}$ is constant.

PROOF Let

$$\boldsymbol{Y}_{(r+p)\times 1} = \begin{pmatrix} K \\ M \end{pmatrix} \boldsymbol{X} = \begin{pmatrix} \boldsymbol{Y}_1 \\ \boldsymbol{Y}_2 \end{pmatrix} ;$$

Then, in view of (3.1.14) and (iii),

$$\boldsymbol{Y} \sim \mathcal{C}_{r+p}\left(\xi; \begin{pmatrix} K\boldsymbol{\mu} \\ M\boldsymbol{\mu} \end{pmatrix}, \begin{pmatrix} KK' & O \\ O & MM' \end{pmatrix}\right).$$

If the rank of M is m, then clearly $k + m \le p$, and we can write

$$KK' = K_1'K_1 \text{ where } K_1 \text{ is a } k \times r \text{ matrix of rank } k ,$$

$$MM' = M_1'M_1 \text{ where } M_1 \text{ is an } m \times p \text{ matrix of rank } m ,$$

and

$$\begin{pmatrix} KK' & O \\ O & MM' \end{pmatrix} = \begin{pmatrix} K_1 & O \\ O & M_1 \end{pmatrix}' \begin{pmatrix} K_1 & O \\ O & M_1 \end{pmatrix} .$$

Then we can represent the random vector $\boldsymbol{Y}$ as follows:

$$\boldsymbol{Y} = \begin{pmatrix} \boldsymbol{Y}_1 \\ \boldsymbol{Y}_2 \end{pmatrix} \equiv \begin{pmatrix} K\boldsymbol{\mu} \\ M\boldsymbol{\mu} \end{pmatrix} + \begin{pmatrix} K_1 & O \\ O & M_1 \end{pmatrix}' \begin{pmatrix} \boldsymbol{Z}_1 \\ \boldsymbol{Z}_2 \end{pmatrix} = \begin{pmatrix} K\boldsymbol{\mu} + K_1'\boldsymbol{Z}_1 \\ M\boldsymbol{\mu} + M_1'\boldsymbol{Z}_2 \end{pmatrix} \tag{3.6.1}$$

where

$$\begin{pmatrix} \boldsymbol{Z}_1 \\ \boldsymbol{Z}_2 \end{pmatrix} \simeq \mathcal{W}\boldsymbol{U}^{(k+m)} \sim \mathcal{C}_{k+m}(\xi; \boldsymbol{0}, I)$$

and $\boldsymbol{Z}_1$ is a k-dimensional vector. It follows that

$$\begin{aligned} Q &= \boldsymbol{X}'A\boldsymbol{X} = \boldsymbol{X}'M'AM\boldsymbol{X} \\ &= \boldsymbol{Y}_2'A\boldsymbol{Y}_2' \simeq (M\boldsymbol{\mu} + M_1'\boldsymbol{Z}_2)'A^*(M\boldsymbol{\mu} + M_1'\boldsymbol{Z}_2) = Q^* \end{aligned} \tag{3.6.2}$$

where the elements of A^* are $a_{ij}^* = g_{ij}(K\boldsymbol{\mu} + K_1'\boldsymbol{Z}_1)$.

The conditional distribution of Z_2 given z_1 is $\mathcal{C}_m(\xi|\boldsymbol{z}_1; \boldsymbol{0}, I)$ where $\xi|\boldsymbol{z}_1 \in \Phi_m$ and

$$\mathcal{W}|\boldsymbol{z}_1 \simeq ((\mathcal{W}^2 - \|\boldsymbol{z}_1\|^2)^{\frac{1}{2}}|\boldsymbol{Z}_1 = \boldsymbol{z}_1) = (\|\boldsymbol{Z}_2\| \; |\boldsymbol{z}_1) . \tag{3.6.3}$$

The conditional distribution of Q^* is

$$Q^*|z_1 \sim \chi^2(\xi|z_1; m; \delta^{*2}) \tag{3.6.4}$$

where in view of (3.6.1) and (3.6.2)

$$\delta^{*2} = (M\mu)'A^*(M\mu) = \mu'A\mu = \delta^2; \tag{3.6.5}$$

then

$$Q^*|z_1 \simeq \delta^2 + 2\delta\sqrt{V}(\mathcal{W}|z_1) + (\mathcal{W}|z_1)^2 \simeq (\delta^2 + 2\delta||Z_2||\sqrt{V} + ||Z_2||^2|z_1)$$

where V is independently distributed of $(Z_1', Z_2')'$ as a Beta$(\frac{1}{2}, \frac{(m-1)}{2})$ random variable; it follows that the p.d.f. of $T = \sqrt{V}$ is

$$g(t) = \frac{\Gamma(m/2)}{\Gamma(\frac{1}{2})\Gamma((m-1)/2)}(1-t^2)^{(m-3)/2}, \ -1 < t < 1, \ m = 2, 3, \ldots$$

Then, by (3.6.2) and Lemma 3.6.1

$$Q \simeq Q^* \simeq ||Z_2||^2 + 2\delta\sqrt{V}||Z_2|| + \delta^2 \sim \chi^2(\xi; m; \delta^2)\ .$$

Definition 3.6.1 *Multivariate generalized noncentral chi-square* Let $X \sim \mathcal{C}_p(\xi; \mu, I)$, $X = (X_1', \ldots, X_\ell')'$, $\mu = (\mu_1', \ldots, \mu_\ell')'$, and $\delta_i^2 = \mu_i'\mu_i$, where X_i and μ_i have p_i components, $i = 1, \ldots, \ell$. Then $(X_1'X_1, \ldots, X_\ell'X_\ell)' \sim \chi^2(\xi; p_1, \ldots, p_\ell; \delta_1^2, \ldots, \delta_\ell^2)$ which denotes the multivariate generalized noncentral chi-square distribution.

Theorem 3.6.2 Let $(Q_1, Q_2, \ldots, Q_\ell)' \sim \chi^2(\xi; p_1, \ldots, p_\ell; \delta_1^2, \ldots, \delta_\ell^2)$. Then

$$\begin{pmatrix} Q_1 \\ Q_2 \\ \vdots \\ Q_\ell \end{pmatrix} \simeq \begin{pmatrix} \mathcal{W}^2 V_1 + 2\delta_1 \mathcal{W}\sqrt{V_1} Z_1 + \delta_1^2 \\ \vdots \\ \mathcal{W}^2 V_\ell + 2\delta_\ell \mathcal{W}\sqrt{V_\ell} Z_\ell + \delta_\ell^2 \end{pmatrix}$$

where $\sqrt{V_i}$ is positive, $i = 1, \ldots, \ell$, $(V_1, \ldots, V_\ell)'$, $\mathcal{W}$, $Z_1, \ldots, Z_\ell$ are independently distributed, $(V_1, \ldots, V_\ell)' \sim D_\ell(p_1/2, \ldots, p_\ell/2)$ and the p.d.f. of Z_i is

$$g_i(z_i) = \frac{\Gamma(p_i/2)}{\Gamma(\frac{1}{2})\Gamma((p_i-1)/2)}(1-z_i^2)^{(p_i-3)/2}, \ -1 < z_i < 1, \ p_i = 2, 3, \ldots, \ i = 1, \ldots, \ell.$$

For p_i=1, $\Pr(Z_i = -1) = \Pr(Z_i = 1) = \frac{1}{2}$.

PROOF We can write

$$\begin{pmatrix} Q_1 \\ \vdots \\ Q_\ell \end{pmatrix} \sim \begin{pmatrix} (Y_1 + \mu_1)'(Y_1 + \mu_1) \\ \vdots \\ (Y_\ell + \mu_\ell)'(Y_\ell + \mu_\ell) \end{pmatrix}$$

where $(Y_1',\ldots,Y_\ell')' \sim \mathcal{C}_p(\xi;0,I)$. By Theorem 3.1.5, we have that

$$\begin{pmatrix} Y_1 \\ \vdots \\ Y_\ell \end{pmatrix} \simeq \mathcal{W} \begin{pmatrix} \sqrt{V_1}U_1 \\ \vdots \\ \sqrt{V_\ell}U_\ell \end{pmatrix}$$

where $(V_1,\ldots,V_\ell)' \sim D_\ell(p_1/2,\ldots,p_\ell/2)$, $U_i \sim U^{(p_i)}, i = 1,\ldots,\ell$, and $(V_1,\ldots,V_\ell)'$, $\mathcal{W}$, $U_1,\ldots, U_\ell$ are independently distributed. Let the first component of U_i be Z_i, $i = 1,\ldots,\ell$; we shall prove the result for $i = 1$. Let G_1 be a $p_1 \times p_1$ orthogonal matrix whose first row is $(\frac{\mu_1}{\|\mu_1\|},\ldots,\frac{\mu_{p_1}}{\|\mu_1\|})$ where $\mu_1' = (\mu_1,\mu_2,\ldots,\mu_{p_1})$; then

$$(Y_1 + \mu_1)'(Y_1 + \mu_1) = (Y_1 + \mu_1)'G_1'G_1(Y_1 + \mu_1)$$

where

$$G_1(Y_1 + \mu_1) \simeq G_1(\mathcal{W}\sqrt{V_1}U_1 + \mu_1) \simeq \mathcal{W}\sqrt{V_1}U_1 + (\delta_1,0,\ldots,0)'$$

which shows that

$$(Y_1 + \mu_1)'(Y_1 + \mu_1) \simeq \mathcal{W}^2V_1 + 2Z_1\mathcal{W}\sqrt{V_1}\delta_1 + \delta_1^2 \, .$$

It follows from Definition 3.1.8 that if $(Q_1,\ldots,Q_\ell)' \sim \chi^2(\xi;p_1,\ldots, p_\ell;\delta_1^2,0,\ldots,0)$ where $\mathcal{W}$ (specified by ξ) is such that $\Pr(\mathcal{W} = 0) = 0$, then

$$\frac{Q_1/p_1}{Q_i/p_i} \sim F(\xi;n_1,n_i;\delta^2) \tag{3.6.6}$$

which denotes a generalized noncentral F distribution. If $\delta^2 = 0$, then

$$\frac{Q_1/p_1}{Q_i/p_i} \sim F_{n_1,n_i} \text{ (the central } F \text{ distribution).}$$

Theorem 3.6.3 Let $X \sim \mathcal{C}_p(\xi;\mu,I)$, K be an $r \times p$ constant matrix whose rank is $k < p$, A_i be a $p \times p$ random matrix whose elements $a_{mj}^{(i)} = g_{mj}(KX)$ where $g_{mj}(KX)$ is a Borel measurable function. Then

$$(X'A_1X,\ldots,XA_\ell X)' \sim \chi^2(\xi;p_1,\ldots,p_\ell;\delta_1^2,\ldots,\delta_\ell^2)$$

provided all the following conditions are satisfied with probability one:

(i) the matrices A_i, $i = 1,\ldots,\ell$ are symmetric and idempotent
(ii) the rank of A_i is p_i, $i = 1,\ldots,\ell$
(iii) $A_i = M'A_iM$ and $MK' = O$ for some $p \times p$ matrix M, $i = 1,\ldots,\ell$
(iv) $\delta_i^2 = \mu'A\mu$ are constants, $i = 1,\ldots,\ell$
(v) $A_iA_j = O$ for $i \neq j$, i, $j = 1,\ldots,\ell$.

PROOF Using the notation of Theorem 3.6.1, we can write

$$\boldsymbol{Q} = \begin{pmatrix} Q_1 \\ \vdots \\ Q_\ell \end{pmatrix} = \begin{pmatrix} \boldsymbol{X}'A_i\boldsymbol{X} \\ \vdots \\ \boldsymbol{X}'A_\ell\boldsymbol{X} \end{pmatrix}$$

$$\simeq \begin{pmatrix} (M\boldsymbol{\mu} + M_1'\boldsymbol{Z}_2)'A_1^*(M\boldsymbol{\mu} + M_1'\boldsymbol{Z}_2) \\ \vdots \\ (M\boldsymbol{\mu} + M_\ell'\boldsymbol{Z}_2)A_\ell^*(M_\mu + M_\ell'\boldsymbol{Z}_2) \end{pmatrix} = \boldsymbol{Q}^*$$

where $A_k^* = (a_{ij}^{(k)*})$, $a_{ij}^{(k)*} = g_{ij}^{(k)}(K\boldsymbol{\mu} + K_1'Z_1)$. Then for a given $\boldsymbol{z}_1$

$$\boldsymbol{Z}_2|\boldsymbol{z}_1 \sim \mathcal{C}_m(\xi_{[\boldsymbol{Z}_2|\boldsymbol{z}_1]}; \boldsymbol{0}, I)$$

if the rank of M is m, and

$$\boldsymbol{Q}^*|\boldsymbol{z}_1 \sim \chi^2(\xi_{[\boldsymbol{Z}_2|\boldsymbol{z}_1]},\ p_1^2, \ldots, p_\ell^2; \delta_1^{*2}, \ldots, \delta_\ell^{*2})$$

where $\delta_i^{*2} = (M\boldsymbol{\mu})'A_i^*(M\boldsymbol{\mu}) = \boldsymbol{\mu}'A_i\boldsymbol{\mu} = \delta_i^2$, $i = 1, \ldots, \ell$.

In view of Theorem 3.6.2, we have

$$\boldsymbol{Q}|\boldsymbol{z}_1 \simeq \begin{pmatrix} \delta_1^2 + 2\delta_1||\boldsymbol{Z}_2||\sqrt{V_1}Z_1 + ||\boldsymbol{Z}_2||^2V_1 \\ \vdots \\ \delta_\ell^2 + 2\delta_\ell||\boldsymbol{Z}_2||\sqrt{V_\ell}Z_\ell + ||\boldsymbol{Z}_2||^2V_\ell \end{pmatrix} \text{ given } \boldsymbol{z}_1.$$

The result follows from Lemma 3.6.1.

Theorem 3.6.4 Let $X_{n\times p} \sim N_{n\times p}(M, I_n \otimes \Sigma)$ and $A_{n\times n}$ be a random matrix where $a_{ij} = g_{ij}(X_2)$, $X = (X_1, X_2)$, X_1 is an $n \times p_1$ matrix, $M = (M_1, M_2)$ is similarly partitioned and g_{ij} is a Borel measurable function. Then, letting $\Sigma_{11.2} = \Sigma_{11} - \Sigma_{12}\Sigma_{22}^{-1}\Sigma_{21}$ with Σ_{11} $p_1 \times p_1$,

$$X_1'AX_1 \sim W(p_1, k, \Sigma_{11.2}, \Omega) \tag{3.6.7}$$

which denotes a noncentral Wishart distribution if the following conditions are satisfied with probability one:

(i) A is idempotent and symmetric
(ii) the rank of A is k
(iii) $(B + C)'A(B + C) = \Omega$, a constant matrix where $B = M_1 - M_2\Sigma_{22}^{-1}\Sigma_{12}'$, $C = X_2\Sigma_{22}^{-1}\Sigma_{12}'$ and

$$\Sigma = \begin{pmatrix} \Sigma_{11} & \Sigma_{12} \\ \Sigma_{21} & \Sigma_{22} \end{pmatrix}, \quad \Sigma_{11} \text{ being of dimension } p_1 \times p_1 .$$

PROOF We know that

$$(X_1|X_2) \sim N_{n\times p_1}(M_1 + (X_2 - M_2)\Sigma_{22}^{-1}\Sigma_{21}, \Sigma_{11.2} \otimes I)$$

that is,

$$(X_1|X_2) \sim N_{n\times p_1}(B+C, \Sigma_{11.2}\otimes I)\ .$$

From (i), (ii) and (iii), we have

$$X_1'AX_1|X_2 \sim W(p_1, k, \Sigma_{11.2}, \Omega)$$

which does not depend on X_2, hence the result.

Generalized Hotelling's T^2 Distribution Let $X_{n\times p} \sim S_{n\times p}(\xi; O,\ I\otimes I)$ and assume that the p.d.f. of X can be expressed as $f(\mathrm{tr}\,(X'X))$. Let $Y_{n\times p} = M + X\Sigma^{\frac{1}{2}}$ where $\Sigma^{\frac{1}{2}}$ denote the $p\times p$ symmetric square root of the symmetric positive definite matrix Σ and

$$M = \mathbf{1}\boldsymbol{\mu}' = \begin{pmatrix} \boldsymbol{\mu}' \\ \boldsymbol{\mu}' \\ \vdots \\ \boldsymbol{\mu}' \end{pmatrix},\quad \mathbf{1}' = (1,1,\ldots,1,1)\ .$$

The rows of Y denoted by $\mathbf{Y}_{(i)}'$ represents observation vectors from a random sample of size n. In order to test $H_0 : \boldsymbol{\mu} = \mathbf{0}$ vs $H_1 : \boldsymbol{\mu} \neq \mathbf{0}$, Anderson, Fang and Hsu (1986) proposed the following likelihood ratio statistic

$$T^2 = n\bar{\mathbf{Y}}'S^{-1}\bar{\mathbf{Y}} \tag{3.6.8}$$

where

$$\bar{\mathbf{Y}} = \frac{1}{n}\sum_{i=1}^{n}\mathbf{Y}_{(i)} = \frac{1}{n}Y'\mathbf{1}\ ,$$

$$S = \sum_{i=1}^{n}(\mathbf{Y}^{(i)} - \bar{\mathbf{Y}})(\mathbf{Y}^{(i)} - \bar{\mathbf{Y}})' = Y'D_nY, \tag{3.6.9}$$

and

$$D_n = I_n - \frac{1}{n}\mathbf{1}\mathbf{1}',\ D_n = D_n^2,\ D_n = D_n'\ . \tag{3.6.10}$$

We will show that

$$\frac{(n-p)}{p}T^2 \sim F(\xi; p, n-p; \delta^2)$$

where

$$\delta^2 = n\boldsymbol{\mu}'\Sigma^{-1}\boldsymbol{\mu}\ .$$

Let Ψ be an orthogonal matrix such that

$$\Psi'\bar{\mathbf{Y}} = (||\bar{\mathbf{Y}}||, 0, \ldots, 0)'\ .$$

Then $T^2 = n(||\bar{\mathbf{Y}}||, 0, \ldots, 0)(\Psi'S\Psi)^{-1}(||\mathbf{Y}||, 0\ldots, 0)'$. Note that in view of (3.6.9) and (3.6.10)

$$\Psi'S\Psi = \Psi'Y'D_n'D_nY\Psi = (D_nY)'(D_nY)\ .$$

Let

$$D_n Y = (\boldsymbol{h}_1, H_2) \tag{3.6.11}$$

where $\boldsymbol{h}_1$ is $n \times 1$ and H_2 is $n \times (p-1)$;

$$\begin{aligned} T^2 &= n(\|\bar{Y}\|, 0, \ldots, 0) \begin{pmatrix} \boldsymbol{h}_1'\boldsymbol{h}_1 & \boldsymbol{h}_1'H_2 \\ H_2'\boldsymbol{h}_1 & H_2'H_2 \end{pmatrix}^{-1} (\|\bar{Y}\|, 0, \ldots, 0)' \\ &= n\|\bar{Y}\|(\boldsymbol{h}_1'\boldsymbol{h}_1 - \boldsymbol{h}_1'H_2(H_2'H_2)^{-1}H_2'\boldsymbol{h}_1)^{-1}\|\bar{Y}\| \\ &= n\frac{\|\bar{Y}\|^2}{\boldsymbol{h}_1'(I_n - H_2(H_2'H_2)^{-1}H_2')\boldsymbol{h}_1} \\ &= \frac{(\mathrm{vec}(Y))'(I_p \otimes \frac{1}{n}\mathbf{1}\mathbf{1}')(\mathrm{vec}(Y))}{(\mathrm{vec}(Y))'(B \otimes C)(\mathrm{vec}(Y))} \\ &= \frac{(\mathrm{vec}(Y))'A_1(\mathrm{vec}(Y))}{(\mathrm{vec}(Y))'A_2(\mathrm{vec}(Y))} = \frac{Q_1}{Q_2} \end{aligned} \tag{3.6.12}$$

where

$$\begin{aligned} A_1 &= I_p \otimes \frac{1}{n}\mathbf{1}\mathbf{1}' , \\ A_2 &= B \otimes C , \\ B_{p\times p} &= \boldsymbol{e}_1\boldsymbol{e}_1', \ \boldsymbol{e}_1' = (1, 0, \ldots, 0) , \end{aligned}$$

and

$$C_{n\times n} = D_n'(I_n - H_2(H_2'H_2)^{-1}H_2')D_n .$$

The denominator of (3.6.12) is obtained by noticing that $\boldsymbol{h}_1 = D_n^*$ vec(Y) where $D_n^* = (D_n, O, \ldots, O)_{n\times np}$. Letting

$$Z = Y\Sigma^{-\frac{1}{2}} ,$$

$$\mathrm{vec}(Z) = (\Sigma^{-\frac{1}{2}} \otimes I_n)(\mathrm{vec}(Y)) \sim C_{np}(\xi; (\Sigma^{-\frac{1}{2}} \otimes I_n)\mathrm{vec}(M), I_{np}) ,$$

and we may now apply Theorem 3.6.1 to the numerator and the denominator of T^2 noting that the substitution of Z to Y in (3.6.12) does not affect the distribution of T^2. Defining

$$K = \begin{pmatrix} 0 & \mathbf{0}' \\ \mathbf{0} & I_{p-1} \end{pmatrix} \otimes D_n$$

and

$$L = \begin{pmatrix} I_n & O \\ O & I_{p-1} \otimes \frac{1}{n}\mathbf{1}_n\mathbf{1}_n' \end{pmatrix} ,$$

we have that

$$LK' = O, \ L'A_1L = A_1 \text{ and } L'A_2L = A_2 \tag{3.6.13}$$

with

$$\begin{aligned} A_1 &= A_1^2, \ A_2 = A_2^2 , \\ \mathrm{tr}\,(A_1) &= p \\ \mathrm{tr}\,(A_2) &= \mathrm{tr}\,(B)\mathrm{tr}\,(C) \end{aligned} \tag{3.6.14}$$

$$
\begin{aligned}
&= \operatorname{tr}(C)\\
&= \operatorname{tr}(D_n'D_n - D_n'H_2(H_2'H_2)^{-1}H_2'D_n)\\
&= \operatorname{tr}(D_n'D_n) - \operatorname{tr}(D_nD_n'H_2(H_2'H_2)^{-1}H_2')\\
&= \operatorname{tr}(D_n) - \operatorname{tr}(D_nH_2(H_2'H_2)^{-1}H_2')\\
&= \operatorname{tr}(D_n) - [\operatorname{tr}(H_2(H_2'H_2)^{-1}H_2') - \operatorname{tr}(\frac{1}{n}\mathbf{1}\mathbf{1}'H_2(H_2'H_2)^{-1}H_2')]\\
&= \operatorname{tr}(D_n - H_2(H_2'H_2)^{-1}H_2')
\end{aligned}
$$

noting that $\mathbf{1}'H_2 = \mathbf{0}$ since H_2 is of the form $\boldsymbol{Y}_i - \bar{\boldsymbol{Y}}$. Hence

$$
\begin{aligned}
\operatorname{tr}(A_2) &= \operatorname{tr}(D_n) - \operatorname{tr}((H_2'H_2)^{-1}H_2'H_2)\\
&= (n-1) - (p-1) = n-p\,.
\end{aligned}
$$

Furthermore

$$A_1A_2 = O \tag{3.6.15}$$

$$(\operatorname{vec}(M))'(\Sigma^{-\frac{1}{2}} \otimes I_n)A_1(\Sigma^{-\frac{1}{2}} \otimes I_n)\operatorname{vec}(M) = n\boldsymbol{\mu}'\Sigma^{-1}\boldsymbol{\mu} \tag{3.6.16}$$

and

$$(\operatorname{vec}(M))'(\Sigma^{-\frac{1}{2}} \otimes I_n)A_2(\Sigma^{-\frac{1}{2}} \otimes I_n)\operatorname{vec}(M) = 0\,. \tag{3.6.17}$$

Therefore, by Theorem 3.6.3 we have that

$$\frac{n-p}{p}\frac{Q_1}{Q_2} = \frac{n-p}{p}T^2 \sim F(\xi; p, n-p; \delta^2)$$

where $\delta^2 = n\boldsymbol{\mu}'\Sigma^{-1}\boldsymbol{\mu}$.

Example 3.6.1 Let

$$x_{ij} = \mu + \alpha_i + \beta_j + \delta\alpha_i\beta_j + \epsilon_{ij}, \tag{3.6.18}$$

$i = 1, \ldots, n$; $j = 1, \ldots, p$; where $\sum_i \alpha_i = \sum_j \beta_j = 0$ and μ, $\{\alpha_i\}$, $\{\beta_j\}$ and δ are unknown. Let $X_{n\times p} = (x_{ij})$ and $E_{n\times p} = (\epsilon_{ij})$ where $E \sim C_{n\times p}(\xi; O, I_n \otimes I_p)$. In order to test $H_0 : \delta = 0$ vs $H_1 : \delta \neq 0$, Tukey (1949) proposed the statistic

$$\frac{S_1^2}{S_2^2 - S_1^2}$$

where

$$S_1^2 = \frac{\left\{\sum_{i=1}^{n}\sum_{j=1}^{p}(x_{ij} - x_{i.} - x_{.j} + x_{..})(x_{i.} - x_{..})(x_{.j} - x_{..})\right\}^2}{\sum_{i=1}^{n}(x_{i.} - x_{..})^2\sum_{j=1}^{p}(x_{.j} - x_{..})^2}\,;$$

$$S_2^2 = \sum_{i=1}^{n}\sum_{j=1}^{p}(x_{ij} - x_{i.} - x_{j.} + x_{..})^2\,;$$

$$x_{i.} = \sum_{j=1}^{p} x_{ij}/p \ ; \ x_{.j} = \sum_{i=1}^{n} x_{ij}/n \ ; \ x_{..} = \sum_{i=1}^{n}\sum_{j=1}^{p} x_{ij}/(np) \, .$$

It is left as exercises to verify that

$$S_1^2 = (\text{vec}\,(X))' A_1 (\text{vec}\,(X)) \tag{3.6.19}$$

where

$$A_1 = \frac{(D_p X' \mathbf{1}_n \mathbf{1}_n' X D_p) \otimes (D_n X \mathbf{1}_p \mathbf{1}_p' X' D_n)}{(\mathbf{1}_p' X' D_n X \mathbf{1}_p)(\mathbf{1}_n' X D_p X' \mathbf{1}_n)} \, ,$$

with

$$D_k = I_k - \frac{1}{k} \mathbf{1}_k \mathbf{1}_k' \, ,$$

that

$$S_2^2 = (\text{vec}\,(X))' A_2 (\text{vec}\,(X)) \tag{3.6.20}$$

where

$$A_2 = D_p \otimes D_n \ , \ A_2 = A_2^2 \, ,$$

and

$$\text{rank}(A_2) = \text{tr}\,(D_p \otimes D_n) = [\text{tr}\,(\text{D}_\text{p})][\text{tr}\,(\text{D}_\text{n})] = (\text{p}-1)(\text{n}-1)$$

and that

$$B = B^2, \ \text{rank}(B) = (n-p)(p-1) - 1 \ \text{where} \ B = A_2 - A_1, \tag{3.6.21}$$

and

$$BA_2 = O \, . \tag{3.6.22}$$

Then under the null hypothesis, it follows from Theorem 3.6.3 that

$$[(n-1)(p-1)-1] \frac{S_1^2}{S_2^2 - S_1^2} \sim F_{1,(n-1)(p-1)-1} \, .$$

3.7 Cochran's Theorem

Some generalizations of Cochran's theorem involving elliptically contoured distributions are discussed in this section.

Lemma 3.7.1 Let $X \equiv \mu + \mathcal{W}U^{(p)} \sim C_p(\xi; \mu; I)$ and $\Pr(\mathcal{W}^2 > \ ||\mu||^2) > 0$; then $X'AX \geq 0$ (almost everywhere) if and only if $A \geq 0$.

This result is proved by contradiction, using the Heine-Borel theorem and the fact that a matrix is positive semi-definite if and only if there is a closed surface Θ containing some neighborhood of the origin such that $x'Ax \geq 0$ for every vector $x \in \Theta$.

Corollary 3.7.1 Let $\boldsymbol{X} \sim C_p(\xi; \boldsymbol{\mu}, I)$, $\Pr(||\boldsymbol{X} - \boldsymbol{\mu}||^2 > ||\boldsymbol{\mu}||^2) > 0$, and $A = A'$; then if $\boldsymbol{X}'A\boldsymbol{X} \sim \chi^2_\ell(\xi; \delta^2)$, A must be positive semi-definite.

Lemma 3.7.2 Let $f((\boldsymbol{x} - \boldsymbol{\mu})'(\boldsymbol{x} - \boldsymbol{\mu})) > 0$ denote the continuous density function of $\boldsymbol{X} \sim C_p(\xi; \boldsymbol{\mu}, I)$, $\boldsymbol{X}'A\boldsymbol{X} \sim \chi^2(\xi; \ell; \delta^2)$ and $A = A'$; then the rank or A is ℓ.

PROOF It follows from Corollary 3.7.1 that A is positive semi-definite. If the rank of A is q then there exists an orthogonal matrix T that diagonalizes A as follows: $T'AT = \text{diag}\,(\theta_1, \theta_2, \ldots, \theta_q, 0, \ldots, 0)$ where $\theta_i \neq 0$ for $i = 1, \ldots, q$. Let

$$\boldsymbol{\delta} = (\delta_1, \ldots, \delta_p)' = T'\boldsymbol{\mu}; \boldsymbol{Y}^* \sim C_p(\xi; \boldsymbol{0}, I);$$

$$\boldsymbol{Y} = (y_1, \ldots, y_p)' = T'\boldsymbol{X} \sim C_p(\xi; \boldsymbol{\delta}; I)\ .$$

then we can write

$$\boldsymbol{Y} \simeq \boldsymbol{Y}^* + \boldsymbol{\delta}\ .$$

But it is given that

$$\boldsymbol{X}'A\boldsymbol{X} = \sum_{i=1}^{q} \theta_i Y_i^2 \sim \chi^2(\xi; \ell; \delta^2)\ . \tag{3.7.1}$$

From (3.7.1), we have

$$\theta_1(Y_1^* + \delta_1)^2 + \cdots + \theta_q(Y_q^* + \delta_q)^2 \simeq (Y_1 + \delta^*)^2 + (Y_2^*)^2 + \cdots + (Y_\ell^*)^2 \tag{3.7.2}$$

for some δ^*. Let $f(\boldsymbol{y}^{*\prime}\boldsymbol{y}^*)$ denote the density function of $\boldsymbol{Y}^*$; then the density function of $(Y_1^*, \ldots Y_q^*)'$ and $(Y_1, Y_2^* \ldots Y_\ell^*)'$ are respectively $f_q(y_1^2 + y_2^{*^2} + \cdots + y_q^{*^2})$ and $f_\ell(y_1^2 + y_2^{*^2} + \cdots + y_\ell^{*^2})$. Since $f(\ )$ defined in Lemma 3.7.2 is a positive continuous function, so are the marginal density functions $f_q(\cdot)$ and $f_\ell(\cdot)$.

In view of (3.7.2), the distribution function of $\boldsymbol{X}'A\boldsymbol{X}$ at the point x^2 is

$$\begin{aligned} F(x^2) &= \int \cdots \int_{\theta_1(z_1+\delta_1)^2 + \cdots + \theta_q(z_q+\delta_q)^2 \leq x^2} f_q(z_1^2 + \cdots + z_q^2)\mathrm{d}z_1 \ldots \mathrm{d}z_q \\ &= \int \cdots \int_{(z_1+\delta^*)^2 + z_2^2 + \cdots + z_\ell^2 \leq x^2} f_\ell(z_1^2 + \cdots + z_\ell^2)\mathrm{d}z_1 \ldots \mathrm{d}z_\ell\ . \end{aligned} \tag{3.7.3}$$

Since A is positive semi-definite, the θ_i's are all positive. Letting $V(L)$ denote the volume of the set L, one has

$$\lim_{x^2 \to 0} \frac{F(x^2)}{V(\theta_1(z_1 + \delta_1)^2 + \cdots + \theta_q(z_q + \delta_q)^2 \leq x^2)} = f_q(\delta_1^2 + \cdots + \delta_q^2) \neq 0$$

and

$$\lim_{x^2 \to 0} \frac{F(x^2)}{V((z_1 + \delta^*)^2 + z_2^2 + \cdots + z_\ell^2 \leq x^2)} = f_\ell(\delta_1^2 + \cdots + \delta_\ell^2) = f_\ell(\delta^{*2}) \neq 0.$$

Hence, $q = \ell$.

Lemma 3.7.3 Let $X \sim \mathcal{C}_p(\xi;\mu;I)$, $p \geq 3$, $M_4(X) < \infty$, that is, the elements of the fourth moment of X as defined in Section 3.2 are finite, $\Pr(X = \mu) = 0$, and let A be a symmetric matrix of rank ℓ. Furthermore, if $q < p$ for some q or $\ell < p$ and $X'AX \sim \chi^2(\xi;q;\delta^2)$ then $\sum_{j=1}^{\ell} \theta_j = q$ and $\sum_{j=1}^{\ell} \theta_j^2 = q$ where the θ_j's, $j = 1,\ldots,\ell$, are the non-zero characteristic roots of A.

PROOF Using the notation of Lemma 3.7.3, the following equality holds in view of Corollary 3.3.5

$$\frac{2^{\frac{p}{2}-1}\Gamma(p/2)}{2\pi i}\int_0^\infty g(w)\int_{a-i\infty}^{a+i\infty} e^{y/2}y^{-(p-\ell)/2} \times\Big\{\prod_{j=1}^{\ell}(y+2t\theta_j w^2)^{-\frac{1}{2}}e^{-((t\theta_j\delta_j^2 y)/(y+2\theta_j w^2 t))}\Big\}dydw \tag{3.7.4}$$

$$= \frac{2^{\frac{p}{2}-1}\Gamma(p/2)}{2\pi i}\int_0^\infty g(w)\int_{a-i\infty}^{a+i\infty} e^{y/2}y^{-(p-\ell)/2}(y+2tw^2)^{-\ell/2} \times e^{-((t\delta^2 y)/(y+2w^2 t))}dydw$$

where $g(w)$ denotes the density function of $\mathcal{W}$ which is determined by $\xi(\cdot)$, the characteristic function of $X - \mu$.

Let the expectation with respect to $\mathcal{W}$ be denoted by $E_{\mathcal{W}}$; then noting that the integrals are independent of the value of a provided that it is positive and letting $y = tz/c$, we have

$$E_{\mathcal{W}}\Big[\frac{1}{2\pi i}\int_{a-i\infty}^{a+i\infty} z^{-(p-\ell)/2}e^{-tz/(2c)}\Big\{\prod_{j=1}^{\ell}(z+2\theta_j w^2 c)^{-\frac{1}{2}}e^{-\frac{(\theta_j\delta_j^2 z)}{(z+2\theta_j w^2 c)}}\Big\}dz\Big] \tag{3.7.5}$$

$$= E_{\mathcal{W}}\Big[\frac{1}{2\pi i}\int_{a-i\infty}^{a+i\infty} z^{-(p-q)/2}e^{-tz/(2c)}(z+2w^2c)^{-\frac{q}{2}}e^{-\frac{(\delta^2 z)}{(z+2w^2c)}}dz\Big].$$

Now fixing c and letting b tend to zero, we have

$$E_{\mathcal{W}}\Big[\frac{1}{2\pi i}\int_{a-i\infty}^{a+i\infty} z^{-(p-\ell)/2}\prod_{j=1}^{\ell}(z+2\theta_j w^2 c)^{-\frac{1}{2}}dz\Big] \tag{3.7.6}$$

$$= E_{\mathcal{W}}\Big[\frac{1}{2\pi i}\int_{a-i\infty}^{a+i\infty} z^{-(p-q)/2}(z+2w^2c)^{-\frac{q}{2}}dz\Big].$$

Taking the first and second derivatives on both sides of (3.7.6) with respect to c and letting c tend to zero, one obtains the following relationships:

$$E(\mathcal{W}^2)\sum_{j=1}^{\ell}\theta_j = qE(\mathcal{W}^2)$$

and

$$E(\mathcal{W}^4)\sum_{j=1}^{\ell}\theta_j^2 = qE(\mathcal{W}^4)\ ;$$

hence the result.

Theorem 3.7.1 Let $\boldsymbol{X} \sim \mathcal{C}_p(\xi;\boldsymbol{\mu}, I)$, $M_4(\boldsymbol{X}) < \infty$, that is, the elements of the fourth moment of $\boldsymbol{X}$ as defined in Section 3.2 are finite, let the density function of $\boldsymbol{X}$ be continuous and positive, $A = A'$, and let the rank of A be less than p or ℓ be less than p. Then $\boldsymbol{X}'A\boldsymbol{X} \sim \chi^2(\xi;\ell;\delta^2)$ if and only if A is an idempotent matrix of rank ℓ and $\boldsymbol{\mu}'A\boldsymbol{\mu} = \delta^2$.

PROOF We will only prove the necessity, the sufficiency being trivial. Let $\boldsymbol{X}'A\boldsymbol{X} \sim \chi^2(\xi;\ell;\delta^2)$, then in view of Lemmas 3.7.2 and 3.7.3, we have $\sum_{j=1}^{p}\theta_j = \ell$ and $\sum_{j=1}^{p}\theta_j^2 = \ell$ which implies that $\theta_1 = \ldots = \theta_\ell = 1$ and hence that A is idempotent. The relationship (3.7.2) in Lemma 3.7.2 then becomes

$$(Y_1^* + \delta_1)^2 + \cdots + (Y_\ell^* + \delta_\ell)^2 \simeq (Y_1 + \delta^*)^2 + (Y_2^*)^2 + \cdots + (Y_\ell^*)^2 \tag{3.7.7}$$

where the left-hand side of (3.7.7) is distributed as $\chi^2(\xi;\ell;\sum_{j=1}^{\ell}\delta_j^2)$. On taking the expected values of both sides of (3.7.7), we have

$$\sum_{j=1}^{\ell}\delta_j^2 = \delta^{*2}$$

where

$$\sum_{j=1}^{\ell}\delta_j^2 = \boldsymbol{\mu}'T(T'AT)T\boldsymbol{\mu} = \boldsymbol{\mu}'A\boldsymbol{\mu}\ .$$

This result is generalized in the next theorem.

Theorem 3.7.2 Let $\boldsymbol{X} \sim \mathcal{C}_\ell(\xi;\boldsymbol{\mu}, I)$, $M_4(\boldsymbol{X}) < \infty$, that is, the elements of the fourth moment of $\boldsymbol{X}$ as defined in Section 3.2 are finite, let the density function of $\boldsymbol{X}$ be continuous and positive and $A_i = A_i'$, $i = 1,\ldots,\ell$, then

$$\begin{pmatrix} \boldsymbol{X}'A_1\boldsymbol{X} \\ \vdots \\ \boldsymbol{X}'A_\ell\boldsymbol{X} \end{pmatrix} \sim \chi^2(\xi;p_1,\ldots,p_\ell;\delta_1^2,\ldots,\delta_\ell^2)$$

if and only if A_j is an idempotent matrix of rank p_j,

$$\delta_j^2 = \mu' A_j \mu,\ j = 1,\ldots,\ell, \text{ and } A_i A_j = O,\ i \neq j,\ i,j = 1,\ldots,\ell\,.$$

PROOF Sufficiency.

Since the matrices are mutually orthogonal, they commute and can be diagonalized with the same orthogonal matrix T; so without loss of generality let

$$T' A_i T = \text{ diag } (0,\ldots,0,1,\ldots,1,0,\ldots,0),\ i = 1,\ldots,\ell\,,$$

where p_i diagonal elements are equal to 1 (since A_i is idempotent of rank p_i).

Letting $Y = T'X$, we have $Y \sim C_p(\xi; T'\mu, I)$. Now let

$$Y = \begin{pmatrix} Y_1 \\ \vdots \\ Y_\ell \\ Y_{\ell+1} \end{pmatrix} \text{ and } v = \begin{pmatrix} v_1 \\ \vdots \\ v_\ell \\ v_{\ell+1} \end{pmatrix}$$

where the first ℓ subvectors have respectively $p_1, p_2, \ldots, p_\ell$ components and the last one has $p - \Sigma_{j=1}^{\ell} p_j$ components. Then

$$\begin{pmatrix} X'A_1X \\ \vdots \\ X'A_\ell X \end{pmatrix} \simeq \begin{pmatrix} Y_1'Y_1 \\ \vdots \\ Y_\ell'Y_\ell \end{pmatrix} \sim \chi^2(\xi; p_1,\ldots,p_\ell; v_1'v_1,\ldots,v_\ell'v_\ell)$$

and

$$\delta_j^2 = \mu' A_j \mu = v_j' v_j,\ j = 1,\ldots,\ell\,.$$

Necessity.

In view of Corollary 3.3.2,

$$X' A_j X \sim \chi^2(\xi; p_j; \delta_j^2)$$

and

$$X'(A_i + A_j)X \sim \chi^2(\xi; p_i + p_j; \delta_i^2 + \delta_j^2),\ i,j = 1,\ldots,\ell\,.$$

From Theorem 3.7.1, we have that A_j is an idempotent matrix of rank $p_j, \delta_j^2 = \mu' A_j \mu,\ j = 1,\ldots \ell$, and $A_i + A_j$ is an idempotent matrix of rank $p_i + p_j,\ i \neq j$. This implies that $A_i A_j = O$ for $i \neq j$.

Theorem 3.7.3 Assuming that $\mu = 0$ in Theorem 3.7.1, then

$$X'AX + 2b'X + c \sim \chi^2(\xi; \ell; \delta^2)$$

if and only if A is an idempotent matrix of rank ℓ, $Ab = b$, and $\delta^2 = b'Ab = c$.

PROOF Sufficiency

Let T be an orthogonal matrix which diagonalizes A as follows:

$$T'AT = \begin{pmatrix} I_\ell & O \\ O & O \end{pmatrix} .$$

Let $(\beta_1, \ldots, \beta_p)' = T'\boldsymbol{b} = T'A\boldsymbol{b} = T'ATT'\boldsymbol{b} = (\beta_1, \ldots, \beta_\ell, 0, \ldots, 0)$. Then $\delta^2 = c = \boldsymbol{b}'A\boldsymbol{b} = \boldsymbol{b}'ATT'A\boldsymbol{b} = \sum_{i=1}^{\ell} \beta_i^2$ and $\boldsymbol{X}'A\boldsymbol{X} + 2\boldsymbol{b}'\boldsymbol{X} + c = \sum_{j=1}^{\ell} (Y_j + \beta_j)^2 \sim \chi^2(\xi; \ell; \delta^2)$ where $\boldsymbol{Y} = (Y_1, \ldots, Y_p)' = T'\boldsymbol{X}$.

Necessity

Let $f(\boldsymbol{x}'\boldsymbol{x})$ denote the density function of $\boldsymbol{X}$ and ℓ be the rank of A. Let T be an orthogonal matrix such that

$$T'AT = \text{ diag } (\theta_1, \ldots, \theta_\ell, 0, \ldots, 0) \ , \theta_i \neq 0, \ i = 1, \ldots, \ell \ .$$

Let $\boldsymbol{\beta} = (\beta_1, \ldots, \beta_p)' = T'\boldsymbol{b}$ and $\boldsymbol{Y} = T'\boldsymbol{X} = (Y_1, \ldots, Y_p)'$ whose density function is $f(\boldsymbol{y}'\boldsymbol{y})$. Then

$$\sum_{i=1}^{\ell} \theta_i (Y_i + \beta_i/\theta_i)^2 + 2 \sum_{i=\ell+1}^{p} \beta_i Y_i + c - \sum_{i=1}^{\ell} \beta_i^2/\theta_i^2$$

$$\simeq (Y_1 - \delta)^2 + Y_2^2 + \cdots + Y_\ell^2 > 0 \tag{3.7.8}$$

(almost surely).

If some of the β_i's, $i = \ell + 1, \ldots, p$ are not equal to zero, then there will be a positive probability that the left-hand side of (3.7.8) be negative. Hence all the β_i's, $i = \ell + 1, \ldots, p$ must be zero. Similarly it can be shown that the last two terms on the left-hand side of (3.7.8) cancel out. Then we are left with only the first term on the left-hand side of (3.7.8), and by Theorem 3.7.1, $\theta_i = 1, \ i = 1, \ldots, \ell$ and $\sum_{i=1}^{\ell} \beta_i^2 = \delta^2$.

This result can be generalized as follows.

Theorem 3.7.4 Let $\boldsymbol{X} \sim C_p(\xi; \boldsymbol{0}; I)$, $M_4(\boldsymbol{X}) < \infty$, the density function of $\boldsymbol{X}$ be positive and continuous, and $A_i = A_i'$, $i = 1, \ldots, \ell$; then

$$\begin{pmatrix} \boldsymbol{X}'A_1\boldsymbol{X} + 2\boldsymbol{b}_1'\boldsymbol{X} + c \\ \vdots \\ \boldsymbol{X}'A_\ell\boldsymbol{X} + 2\boldsymbol{b}_\ell'\boldsymbol{X} + c \end{pmatrix} \sim \chi^2(\xi; p_1, \ldots, p_\ell; \delta_1^2, \ldots, \delta_\ell^2)$$

if and only if A_i is an indempotent matrix of rank p_i, $A_i\boldsymbol{b}_i = \boldsymbol{b}_i$, $\delta_i^2 = \boldsymbol{b}_i'A_i\boldsymbol{b}_i = c_i$, $i = 1, \ldots, \ell$ and $A_iA_j = O, \ i \neq j, \ i, j = 1, \ldots, \ell$.

Let X be an $n \times p$ random matrix whose rows are denoted by $\boldsymbol{X}_{(i)}'$, $i = 1, \ldots, n$, and whose columns are denoted by $\boldsymbol{X}_j$, $j = 1, \ldots, p$. Let $\mathcal{F}$ be the class of random matrices X such that $TX \simeq X$ for every $n \times n$ orthogonal matrix T. Whenever X belongs to $\mathcal{F}$, its characteristic function can be expressed as $\xi(S'S)$ where S is an $n \times p$ matrix. It follows that the characteristic function of $(X_{(1)},\ldots,X_{(\ell)})'$, $\ell \le n$, is $\xi(S_1'S_1)$ where S_1 is an $\ell \times p$ matrix.

If G is $p\times q$ with $q \le p$ then XG belongs to $\mathcal{F}$ since $TXG \simeq XG$ and the characteristic function of XG is $\xi_G = \xi(GS'SG')$ as $E(e^{i\mathrm{tr}(S'XG)}) = E(e^{i\mathrm{tr}(GS'X)})$. Furthermore, whenever X belong to the class $\mathcal{F}$, its density function can be expressed as $f(X'X)$; the density function of $Y = M + XA$, where A is positive definite, is

$$|A|^{-n} f(A'^{-1}(Y-M)'(Y-M)A^{-1}) \,. \tag{3.7.9}$$

Let

$$Y_{n\times p} = \begin{pmatrix} Y_{(1)} \\ \vdots \\ Y_{(\ell)} \end{pmatrix}, \; X = \begin{pmatrix} X_{(1)} \\ \vdots \\ X_{(\ell)} \end{pmatrix} \text{ and } M = \begin{pmatrix} M_{(1)} \\ \vdots \\ M_{(\ell)} \end{pmatrix} \tag{3.7.10}$$

where $Y_{(i)}$, $X_{(i)}$ and $M_{(i)}$ are $n_i \times p$ submatrices of Y, X and M respectively, with $n_i \ge p$, $i = 1,\ldots,\ell$ and $\sum_{i=1}^{\ell} n_j = n$. Let

$$W_i = A'(X_{(i)}A + M_{(i)})'(X_{(i)}A + M_{(i)})A, \quad i = 1,\ldots,\ell, \tag{3.7.11}$$

and let the distribution of $W_1, \ldots, W_\ell$ be denoted by $W_{p,\ell}(\xi; n_1, \ldots, n_\ell; M_{(1)}'M_{(1)}, \ldots, M_{(\ell)}'M_{(\ell)}; A)$. Note that the distribution of $W_1,\ldots,W_\ell$ depends on M only through $M_{(1)}'M_{(1)}, \ldots, M_{(\ell)}'M_{(\ell)}$.

Corollary 3.7.2 Let $M_i = M_{(i)}'M_{(i)}$, $i = 1,\ldots,\ell$ and $(W_1,\ldots,W_\ell) \sim W_{p,\ell}(\xi; n_1, \ldots, n_\ell; M_1,\ldots,M_\ell; A)$, then, for $k \le \ell$, $\sum_{i=1}^{k} W_i \sim W_{p,1}(\xi; n_1 + \cdots + n_k; M_1,\ldots,M_k; A)$.

Corollary 3.7.3 Let G be a $p \times q$ matrix, $q \le p$, and $Z \sim W_{p,1}(\xi; M; I)$; then

$$G'ZG \sim W_{p,1}(\xi_G; G'MG; I)$$

where $\xi_G = \xi(GS'SG')$.

Corollary 3.7.4 Let g be a non-null p-dimensional vector, then $g'Zg \sim \chi^2(\xi; g'Mg; I)$.

Theorem 3.7.5 Assume that $X \in \mathcal{F}$, that X possesses a positive and continuous density function and that the fourth moment is finite, and let A be an $n \times n$ symmetric matrix where $n > p$ and V be a constant matrix of dimension $n \times p$, then

$$(X+V)'A(X+V) \sim W_{p,1}(\xi; k; V^*; I)$$

if and only if A is an idempotent matrix of rank ℓ and $V^* = V'AV$.

PROOF We only prove the necessity, the sufficiency being obvious. Let g be a non-null p-dimensional vector, then $Xg \sim \mathcal{C}_n(\boldsymbol{\xi}; 0, I)$ and by Corollary 3.7.4, $((X+V)g)'A((X+V)g) \sim \chi^2(\xi_g; \ell; g'V^*g)$. In view of Theorem 3.7.1, $A = A^2$, the rank of A is ℓ, $g'V'AVg = g'V^*g$ and hence $V^* = V'AV$. Theorem 3.7.5 can be generalized as follows.

Theorem 3.7.6 Assume that $X \in \mathcal{F}$, that X possesses a positive continuous density function, that the fourth moment is finite and that $n > p$, then

$$\begin{pmatrix} (X+V_1)'A(X+V_1) \\ \vdots \\ (X+V_\ell)'A(X+V_\ell) \end{pmatrix} \sim W_{p,\ell}(\xi; n_1, \ldots, n_\ell; V_1^*, \ldots, V_\ell^*; I)$$

if and only if A_i is an idempotent matrix of rank n_i,

$$V_i^* = V_i A V_i,\ i = 1, \ldots, \ell \text{ and } A_i A_j = O,\ i \neq j,\ i,j = 1, \ldots, \ell\,.$$

The following result is a generalization of Theorem 3.7.3.

Theorem 3.7.7 Assume $X \in \mathcal{F}$, that X possesses a positive continuous density function, that the fourth moment is finite and A is an $n \times n$ symmetric matrix. Then

$$X'AX + B'X + X'B + C \sim W_{p,1}(\xi; \ell; V^*; I)$$

if and only if A is an idempotent matrix of rank ℓ, $AB = B$, $C = V^* = B'AB$, where A, B, and C are respectively $n \times n$, $n \times p$ and $p \times p$ matrices.

3.8 Test Statistics for Elliptically Contoured Distributions

The robust properties of certain statistics based on the normal population are studied for elliptically contoured samples in this section. The asymptotic expansions are given and some limiting distributions are considered. It will be shown that some statistics have similar distributional properties for the finite sample case and in a certain limiting case. We begin this section with an alternate definition of elliptically contoured distributions.

Definition 3.8.1 *Elliptically Contoured Distribution* Let $\boldsymbol{X}$ be a real m-dimensional random vector with the probability density function

$$c_m |V|^{-\frac{1}{2}} g_m((\boldsymbol{x}-\boldsymbol{\mu})'V^{-1}(\boldsymbol{x}-\boldsymbol{\mu})) \tag{3.8.1}$$

for some nonnegative function g_m, where c_m is a positive constant and V is a symmetric positive definite matrix, then $\boldsymbol{X}$ is said to have an elliptically contoured distribution with parameters $\boldsymbol{\mu}$ and V and we denote it by

$$\boldsymbol{X} \sim \mathcal{C}_m(\boldsymbol{\mu}, V)\,.$$

Lemma 3.8.1 Let $X \sim C_m(\mu, V)$, then the characteristic function $\varphi(t) = E[\exp\{it'X\}]$ has the form

$$\varphi(t) = \exp\{it'\mu\}\psi(t'Vt) \tag{3.8.2}$$

where

$$\psi(t'Vt) = c_m \frac{\pi^{\frac{m}{2}}}{\Gamma(\frac{m}{2})} \int_0^\infty g_m(z) z^{\frac{m}{2}-1} \, {}_0F_1\left(\frac{m}{2}; -\frac{1}{4} t'Vtz\right) \mathrm{d}z$$

and

$$\mathrm{E}[X] = \mu, \quad \mathrm{Cov}\,[X] = \alpha V \text{ with } \alpha = -2\psi'(0)\,.$$

PROOF The characteristic function $\varphi(t)$ can be expressed as

$$\begin{aligned}
\varphi(t) &= \exp(it'\mu) c_m \int_{\Re^m} \exp(it'V^{\frac{1}{2}}y) g_m(y'y) \mathrm{d}y \\
&= \exp(it'\mu) c_m \int_{\Re^m} \left[\int_{O(m)} \mathrm{etr}(iyt'V^{\frac{1}{2}}H)\mathrm{d}(H)\right] g_m(y'y)\mathrm{d}y \\
&= \exp(it'\mu) c_m \int_{\Re^m} g_m(y'y)\, {}_0F_1\left(\frac{m}{2}; -\frac{1}{4}t'Vty'y\right)\mathrm{d}y \\
&= \exp(it'\mu) c_m \frac{\pi^{\frac{m}{2}}}{\Gamma(\frac{m}{2})} \int_0^\infty g_m(z) z^{\frac{m}{2}-1}\, {}_0F_1\left(\frac{m}{2}; -\frac{1}{4}t'Vtz\right)\mathrm{d}z \\
&= \exp\,(it'\mu)\psi(t'Vt)\,.
\end{aligned}$$

Lemma 3.8.2 If the function g_m does not depend on m, then the characteristic function $\varphi(t)$ can be expressed as

$$\varphi(t) = \exp(it'\mu) \sum_{k=0}^{\infty} \frac{1}{k!} \left(-\frac{t'Vt}{4\pi}\right)^k \frac{c_m}{c_{m+2k}}\,. \tag{3.8.3}$$

PROOF With the help of the equality

$$\int_0^\infty g(z) z^{\frac{m}{2}-1} dz = \frac{\Gamma(\frac{m}{2})}{\pi^{\frac{m}{2}} c_m}\,, \tag{3.8.4}$$

term by term integration in $\varphi(t)$ gives (3.8.3).

3.8.1 Sample Correlation Coefficient

Let

$$x_\alpha = \begin{bmatrix} x_{1\alpha} \\ x_{2\alpha} \end{bmatrix}, \quad \alpha = 1, 2, \ldots, N$$

be random vectors from the elliptically contoured population $C_2(\boldsymbol{\mu}, \Sigma)$ with $\Pr\{\boldsymbol{x}_\alpha = \mathbf{0}\} = 0$, $\alpha = 1, 2, \ldots, N$. The sample correlation coefficient r is defined as

$$r = a_{12}/\sqrt{a_{11}a_{22}} = s_{12}/\sqrt{s_{11}s_{22}} \,,$$

where

$$a_{ij} = \sum_{\alpha=1}^{N}(x_{i\alpha} - \bar{x}_i)(x_{j\alpha} - \bar{x}_j), \quad \bar{x}_i = \frac{1}{N}\sum_{\alpha=1}^{N} x_{i\alpha} \,,$$

$$s_{ij} = a_{ij}/n, \quad n = N - 1, \quad i, j = 1, 2 \,.$$

Note that a_{12} can be expressed as a bilinear form and that a_{11} and a_{22} can be expressed as quadratic forms.

Theorem 3.8.1 For an elliptically contoured population with $\rho = \sigma_{12}/\sqrt{\sigma_{11}\sigma_{22}} = 0$ and $\Pr\{\boldsymbol{x}_\alpha = \mathbf{0}\} = 0$,

$$(N-2)^{\frac{1}{2}} r/\sqrt{1 - r^2}$$

has the t-distribution with $N - 2$ degrees of freedom.

PROOF See Kariya and Eaton (1977) and Muirhead (1982, Theorem 5.1.1).

The asymptotic expansion of the distribution of a certain function $f(r)$ of the sample correlation coefficient r is considered. The sample correlation coefficient r is expanded asymptotically as follows:

$$\begin{aligned} r = \rho\Big[1 &- \frac{1}{2}\Big\{\frac{s_{11}-\sigma_{11}}{\sigma_{11}} + \frac{s_{22}-\sigma_{22}}{\sigma_{22}} - 2\frac{s_{12}-\sigma_{12}}{\sigma_{12}}\Big\} \\ &+ \frac{3}{8}\Big(\frac{s_{11}-\sigma_{11}}{\sigma_{11}}\Big)^2 + \frac{1}{4}\Big(\frac{s_{11}-\sigma_{11}}{\sigma_{11}}\Big)\Big(\frac{s_{22}-\sigma_{22}}{\sigma_{22}}\Big) + \frac{3}{8}\Big(\frac{s_{22}-\sigma_{22}}{\sigma_{22}}\Big)^2 \\ &- \frac{1}{2}\Big(\frac{s_{11}-\sigma_{11}}{\sigma_{11}}\Big)\Big(\frac{s_{12}-\sigma_{12}}{\sigma_{12}}\Big) - \frac{1}{2}\Big(\frac{s_{22}-\sigma_{22}}{\sigma_{22}}\Big)\Big(\frac{s_{12}-\sigma_{12}}{\sigma_{12}}\Big) \\ &- \Big\{\frac{5}{16}\Big(\frac{s_{11}-\sigma_{11}}{\sigma_{11}}\Big)^3 + \frac{5}{16}\Big(\frac{s_{22}-\sigma_{22}}{\sigma_{22}}\Big)^3 + \frac{3}{16}\Big(\frac{s_{11}-\sigma_{11}}{\sigma_{11}}\Big)^2\Big(\frac{s_{22}-\sigma_{22}}{\sigma_{22}}\Big) \\ &+ \frac{3}{16}\Big(\frac{s_{11}-\sigma_{11}}{\sigma_{11}}\Big)\Big(\frac{s_{22}-\sigma_{22}}{\sigma_{22}}\Big)^2 - \frac{3}{8}\Big(\frac{s_{11}-\sigma_{11}}{\sigma_{11}}\Big)^2\Big(\frac{s_{12}-\sigma_{12}}{\sigma_{12}}\Big) \\ &- \frac{3}{8}\Big(\frac{s_{22}-\sigma_{22}}{\sigma_{22}}\Big)^2\Big(\frac{s_{12}-\sigma_{12}}{\sigma_{12}}\Big) - \frac{1}{4}\Big(\frac{s_{11}-\sigma_{11}}{\sigma_{11}}\Big)\Big(\frac{s_{22}-\sigma_{22}}{\sigma_{22}}\Big)\Big(\frac{s_{12}-\sigma_{12}}{\sigma_{12}}\Big)\Big\}\Big] \\ &+ \text{remainder term} \,. \end{aligned} \tag{3.8.5}$$

To evaluate $E[(r-\rho)^k]$, $k = 2, 3, 4$, we need the expectations of the terms $(s_{ij} - \sigma_{ij})/\sigma_{ij}$. They are given in the next lemma in terms of the parameter κ which is defined as follows:

$$\kappa = \{\psi''(0) - (\psi'(0))^2\}/(\psi'(0))^2 \,.$$

Lemma 3.8.3 For $\sigma_{12} \neq 0$,

$$
\begin{aligned}
E\Big[\Big(\frac{s_{11}-\sigma_{11}}{\sigma_{11}}\Big)^2\Big] &= \frac{1}{n}(2+3\kappa)+\frac{2}{n^2}+\frac{2}{n^3}+o\Big(\frac{1}{n^3}\Big)\\
E\Big[\Big(\frac{s_{12}-\sigma_{12}}{\sigma_{12}}\Big)^2\Big] &= \frac{1}{n}\Big\{\Big(2+\frac{1}{\rho^2}\Big)(1+\kappa)-1\Big\}+\frac{2}{n^2}+\frac{2}{n^3}+o\Big(\frac{1}{n^3}\Big)\\
E\Big[\Big(\frac{s_{11}-\sigma_{11}}{\sigma_{11}}\Big)\Big(\frac{s_{22}-\sigma_{22}}{\sigma_{22}}\Big)\Big] &= \frac{1}{n}\{(1+2\rho^2)(1+\kappa)-1\}+\frac{2\rho^2}{n^2}+\frac{2\rho^2}{n^3}+o\Big(\frac{1}{n^3}\Big)\\
E\Big[\Big(\frac{s_{11}-\sigma_{11}}{\sigma_{11}}\Big)\Big(\frac{s_{12}-\sigma_{12}}{\sigma_{12}}\Big)\Big] &= \frac{1}{n}(2+3\kappa)+\frac{2}{n^2}+\frac{2}{n^3}+o\Big(\frac{1}{n^3}\Big).
\end{aligned}
$$

If we set

$$
\ell_1 = -\frac{1}{6}\frac{\psi^{(3)}(0)}{(\psi'(0))^3}, \quad \ell_2 = \frac{\psi''(0)}{2\{\psi'(0)\}^2} = \frac{1+\kappa}{2},
$$

then

$$
\begin{aligned}
E\Big[\Big(\frac{s_{11}-\sigma_{11}}{\sigma_{11}}\Big)^3\Big] &= \frac{1}{n^2}(-90\ell_1-18\ell_2+2)+O\Big(\frac{1}{n^3}\Big)\\
E\Big[\Big(\frac{s_{12}-\sigma_{12}}{\sigma_{12}}\Big)^3\Big] &= \frac{1}{n^2}\Big\{-(36\ell_1+12\ell_2)-\frac{1}{\rho^2}(54\ell_1+6\ell_2)\\
&\qquad +2\Big\}+O\Big(\frac{1}{n^3}\Big)\\
E\Big[\Big(\frac{s_{11}-\sigma_{11}}{\sigma_{11}}\Big)^2\Big(\frac{s_{22}-\sigma_{22}}{\sigma_{22}}\Big)\Big] &= \frac{1}{n^2}\{-(72\rho^2+18)\ell_1-(10+8\rho^2)\ell_2\\
&\qquad +2\}+O\Big(\frac{1}{n^3}\Big)\\
E\Big[\Big(\frac{s_{11}-\sigma_{11}}{\sigma_{11}}\Big)^2\Big(\frac{s_{12}-\sigma_{12}}{\sigma_{12}}\Big)\Big] &= \frac{1}{n^2}(-90\ell_1-18\ell_2+2)+O\Big(\frac{1}{n^3}\Big)\\
E\Big[\Big(\frac{s_{11}-\sigma_{11}}{\sigma_{11}}\Big)\Big(\frac{s_{12}-\sigma_{12}}{\sigma_{12}}\Big)^2\Big] &= \frac{1}{n^2}\Big\{-(72\ell_1+16\ell_2)-\frac{1}{\rho^2}(18\ell_1+2\ell_2)\\
&\qquad +2\Big\}+O\Big(\frac{1}{n^3}\Big)\\
E\Big[\Big(\frac{s_{11}-\sigma_{11}}{\sigma_{11}}\Big)\Big(\frac{s_{22}-\sigma_{22}}{\sigma_{22}}\Big)\Big(\frac{s_{12}-\sigma_{12}}{\sigma_{12}}\Big)\Big] &= \frac{1}{n^2}\{-(54+36\rho^2)\ell_1\\
&\qquad -(14+4\rho^2)\ell_2+2\}+O\Big(\frac{1}{n^3}\Big)\\
E\Big[\Big(\frac{s_{11}-\sigma_{11}}{\sigma_{11}}\Big)^4\Big] &= \frac{3}{n^2}(2+3\kappa)^2+O\Big(\frac{1}{n^3}\Big)\\
E\Big[\Big(\frac{s_{12}-\sigma_{12}}{\sigma_{12}}\Big)^4\Big] &= \frac{1}{n^2}\Big[12\ell_2^2\Big(2+\frac{1}{\rho^2}\Big)^2\\
&\qquad -12\ell_2\Big(2+\frac{1}{\rho^2}\Big)+3\Big]+O\Big(\frac{1}{n^3}\Big)
\end{aligned}
$$

$$E\left[\left(\frac{s_{11}-\sigma_{11}}{\sigma_{11}}\right)^3\left(\frac{s_{22}-\sigma_{22}}{\sigma_{22}}\right)\right] = \frac{1}{n^2}\,[36\ell_2^2(1+2\rho^2) \\ -12\ell_2(2+\rho^2)+3] + O\left(\frac{1}{n^3}\right)$$

$$E\left[\left(\frac{s_{11}-\sigma_{11}}{\sigma_{11}}\right)^3\left(\frac{s_{12}-\sigma_{12}}{\sigma_{12}}\right)\right] = \frac{3}{n^2}(2+3\kappa)^2 + O\left(\frac{1}{n^3}\right)$$

$$E\left[\left(\frac{s_{11}-\sigma_{11}}{\sigma_{11}}\right)^2\left(\frac{s_{22}-\sigma_{22}}{\sigma_{22}}\right)^2\right] = \frac{1}{n^2}\,[\ell_2^2\{8(1+2\rho^2)^2+36\} \\ -\ell_2\{8(1+2\rho^2)+12\}+3] + O\left(\frac{1}{n^3}\right)$$

$$E\left[\left(\frac{s_{11}-\sigma_{11}}{\sigma_{11}}\right)^2\left(\frac{s_{12}-\sigma_{12}}{\sigma_{12}}\right)^2\right] = \frac{1}{n^2}\left[\ell_2^2\left\{72+12\left(2+\frac{1}{\rho^2}\right)\right\} \\ -\ell_2\left\{30+2\left(2+\frac{1}{\rho^2}\right)\right\}+3\right] \\ + O\left(\frac{1}{n^3}\right)$$

$$E\left[\left(\frac{s_{11}-\sigma_{11}}{\sigma_{11}}\right)\left(\frac{s_{12}-\sigma_{12}}{\sigma_{12}}\right)^3\right] = \frac{1}{n^2}\left[36\ell_2^2\left(2+\frac{1}{\rho^2}\right) \\ -6\ell_2\left(5+\frac{1}{\rho^2}\right)+1\right] + O\left(\frac{1}{n^3}\right)$$

$$E\left[\left(\frac{s_{11}-\sigma_{11}}{\sigma_{11}}\right)^2\left(\frac{s_{22}-\sigma_{22}}{\sigma_{22}}\right)\left(\frac{s_{12}-\sigma_{12}}{\sigma_{12}}\right)\right] = \frac{1}{n^2}\,[\ell_2^2(60+48\rho^2) \\ -\ell_2(28+8\rho^2)+3] + O\left(\frac{1}{n^3}\right)$$

$$E\left[\left(\frac{s_{11}-\sigma_{11}}{\sigma_{11}}\right)\left(\frac{s_{22}-\sigma_{22}}{\sigma_{22}}\right)\left(\frac{s_{12}-\sigma_{12}}{\sigma_{12}}\right)^2\right] = \frac{1}{n^2}\left[\ell_2^2\left\{72+\frac{4}{\rho^2}(1+2\rho^2)^2\right\} \\ -\ell_2\left\{2\left(2+\frac{1}{\rho^2}\right)+24+2(1+2\rho^2)\right\} \\ +3\right] + O\left(\frac{1}{n^3}\right).$$

The proof is left to the reader.

Lemma 3.8.4 The following relations hold for elliptically contoured distributions.

$$E[(r-\rho)] = -\frac{1}{2n}\rho(1-\rho^2)(1+\kappa) + O\left(\frac{1}{n^2}\right)$$
$$E[(r-\rho)^2] = \frac{1}{n}(1-\rho^2)^2(1+\kappa) + O\left(\frac{1}{n^2}\right)$$
$$E[(r-\rho)^3] = -\frac{1}{n^2}\frac{15}{2}\rho(1-\rho^2)^3(1+\kappa)^2 + O\left(\frac{1}{n^3}\right)$$
$$E[(r-\rho)^4] = \frac{1}{n^2}3(1-\rho^2)^4(1+\kappa)^2 + O\left(\frac{1}{n^3}\right).$$

PROOF The expectations are obtained with the help of (3.8.5) and Lemma 3.8.1.

Theorem 3.8.2 Let $f(r)$ be a one-to-one and twice continuously differentiable function in the neighbourhood of the population correlation coefficient $\rho \neq 0$. The Edgeworth expansion of the distribution of the normalized statistic of $f(r)$ under the elliptical population is given by

$$\Pr[\sqrt{n}\{f(r) - f(\rho) - c/n\}/[(1+\kappa)^{\frac{1}{2}}(1-\rho^2)f'(\rho)] \leq x]$$

$$= \Phi(x) + \sqrt{\frac{1+\kappa}{n}}\Big[-\frac{\rho}{2} + \frac{c}{(1+\kappa)f'(\rho)(1-\rho^2)} + x^2\Big\{\rho - \frac{1}{2}(1-\rho^2)\frac{f''(\rho)}{f'(\rho)}\Big\}\Big]\phi(x) + O\Big(\frac{1}{n}\Big) \tag{3.8.6}$$

where $\Phi(x)$ and $\phi(x)$ are respectively the standard normal distribution and density functions.

PROOF With the help of the Taylor expansion of $f(r)$ at $r = \rho$, the normalized statistic T is expressed as

$$\begin{aligned} T &= \sqrt{n}\{f(r) - f(\rho) - c/n\}/\{(1+\kappa)^{\frac{1}{2}}(1-\rho^2)f'(\rho)\} \\ &= -\frac{1}{\sqrt{n}}\frac{c}{(1+\kappa)^{\frac{1}{2}}(1-\rho^2)f'(\rho)} \\ &\quad + \sqrt{n}(r-\rho)/[(1+\kappa)^{\frac{1}{2}}(1-\rho^2)] \\ &\quad + \frac{\sqrt{n}}{2}(r-\rho)^2 f''(\rho)/[(1+\kappa)^{\frac{1}{2}}(1-\rho^2)f'(\rho)] \\ &\quad + \frac{\sqrt{n}}{6}(r-\rho)^3 f'''(\rho)/[(1+\kappa)^{\frac{1}{2}}(1-\rho^2)f'(\rho)] + \cdots . \end{aligned}$$

By Lemma 3.8.4, the first few moments of T are given by

$$\begin{aligned} E[T] &= \frac{1}{\sqrt{n}}\Big[-\frac{\rho}{2} + \frac{1}{2}(1-\rho^2)\frac{f''(\rho)}{f'(\rho)} \\ &\qquad - \frac{c}{(1+\kappa)(1-\rho^2)f'(\rho)}\Big](1+\kappa)^{\frac{1}{2}} + O\Big(\frac{1}{n}\Big) \\ E[T^2] &= 1 + O\Big(\frac{1}{n}\Big) \\ E[T^3] &= \frac{1}{\sqrt{n}}\Big[-\frac{15}{2}\rho + \frac{9}{2}(1-\rho^2)\frac{f''(\rho)}{f'(\rho)} \\ &\qquad - \frac{3c}{(1+\kappa)(1-\rho^2)f'(\rho)}\Big](1+\kappa)^{\frac{1}{2}} + O\Big(\frac{1}{n}\Big), \end{aligned}$$

and the first cumulants are obtained as

$$\kappa_1 = E[T], \quad \kappa_2 = 1 + O\Big(\frac{1}{n}\Big),$$

$$\kappa_3 = \frac{1}{\sqrt{n}}\Big\{ -6\rho + 3(1-\rho^2)\frac{f''(\rho)}{f'(\rho)}\Big\}(1+\kappa)^{\frac{1}{2}} + O\Big(\frac{1}{n}\Big) .$$

Thus the Edgeworth expansion of T is given by (3.8.6)

Remark 3.8.1 The function f which makes the coefficient of x^2 in (3.8.6) vanish such that

$$\rho - \frac{1}{2}(1-\rho^2)\frac{f''(\rho)}{f'(\rho)} = 0 ,$$

is given by

$$f(\rho) = \frac{1}{2}\log\left[\frac{1+\rho}{1-\rho}\right] ,$$

which is Fisher's z-transformation.

Letting $c = \rho(1+\kappa)/2$, we have

$$\Pr\Big[\Big(\frac{n}{1+\kappa}\Big)^{\frac{1}{2}}\Big\{\frac{1}{2}\log\Big(\frac{1+r}{1-r}\Big) - \log\Big(\frac{1+\rho}{1-\rho}\Big) - \frac{\rho}{2}(1+\kappa)\Big\} \le x\Big]$$

$$= \Phi(x) + O\Big(\frac{1}{n}\Big) . \tag{3.8.7}$$

Remark 3.8.2 Let $\boldsymbol{x}' = (x_1, \ldots, x_p)$ and $\boldsymbol{y}' = (y_1, \ldots, y_q)$, $p < q$ have the covariance structure

$$\Sigma = \begin{bmatrix} I_p & P & O \\ P & I_p & O \\ O & O & I_{q-p} \end{bmatrix}$$

where $P =$ diag $(\rho_1, \rho_2, \ldots, \rho_p)$, and $1 > \rho_1 > \rho_2 > \cdots > \rho_p > 0$ are the population canonical correlation coefficients. Let S be a sample variance-covariance matrix based on $(n+1)$ observations from a $(p+q)$-variate elliptically contoured distribution. The following asymptotic expansion of the function $f(r_i^2)$ of the i-th sample canonical correlation coefficient r_i, $i = 1, 2, \ldots, p$ is obtained by using steps similar to those used in the proof of Theorem 3.8.2.

$$\Pr[\sqrt{n}\{f(r_i^2) - f(\rho_i^2) - c/n\}/\{2\rho_i(1-\rho_i^2)f'(\rho_i^2)(1+\kappa)^{\frac{1}{2}}\} \le x]$$

$$= \Phi(x) - \frac{(1+\kappa)^{\frac{1}{2}}}{2\rho_i\sqrt{n}}\Big[p + q - 2 + \rho_i^2 + 2(1-\rho_i^2)\sum_{j\neq i}\frac{\rho_j^2}{\rho_i^2 - \rho_j^2}$$

$$- \frac{c}{(1-\rho_i^2)(1+\kappa)f'(\rho_i^2)}$$

$$+\left\{1-3\rho_i^2+2\rho_i^2(1-\rho_i^2)\frac{f''(\rho_i^2)}{f'(\rho_i^2)}\right\}x^2\Big]\phi(x)+o(1/\sqrt{n})\,. \tag{3.8.8}$$

The function f which makes the coefficient of x^2 vanish is

$$f(\rho_i^2)=\frac{1}{2}\log\left(\frac{1+\rho_i}{1-\rho_i}\right).$$

3.8.2 Likelihood Ratio Criteria

The limiting distribution of a certain likelihood ratio criterion for the covariance matrix of a normal population can be shown to have two types of representations, that of a noncentral chi-square type and that of a noncentral quadratic form under the elliptical population.

Let $\boldsymbol{x}_\alpha$, $\alpha=1,2,\ldots,n$ be a random sample from $N(\mathbf{0},\Sigma)$. The m.l.e. of Σ is given by $\hat{\Sigma}=S/n=\sum_{\alpha=1}^{n}\boldsymbol{x}_\alpha\boldsymbol{x}'_\alpha/n$.

Lemma 3.8.5 Let $\boldsymbol{x}_\beta$, $\beta=1,2,\ldots,n$ be a random sample from $C_m(\mathbf{0},V)$, with $\mathrm{Cov}(\boldsymbol{X})=\Sigma=\alpha V$ and let $S=\sum_{\beta=1}^{n}\boldsymbol{x}_\beta\boldsymbol{x}'_\beta$. The asymptotic expansion of the probability density function of

$$W=\sqrt{n}\Sigma^{-\frac{1}{2}}(S/n-\Sigma)\Sigma^{-\frac{1}{2}}$$

is given by

$$\begin{aligned} f(W)&=(2\pi)^{-m(m+1)/4}|\Sigma_1|^{-\frac{1}{2}}(1+\kappa)^{-m(m-1)/4}\\ &\quad\times\exp\left\{-\frac{1}{2}\boldsymbol{w}_1'\Sigma_1^{-1}\boldsymbol{w}_1-\frac{\boldsymbol{w}_2'\boldsymbol{w}_2}{2(1+\kappa)}\right\}\\ &\quad\times\Big[1+\frac{1}{\sqrt{n}}\{a_0\mathrm{tr}\,(W)-a_1(\mathrm{tr}\,(W))^3-a_2\mathrm{tr}\,(W)\mathrm{tr}\,(W^2)-a_3\mathrm{tr}\,(W^3)\}\\ &\qquad+o\Big(\frac{1}{\sqrt{n}}\Big)\Big]\end{aligned} \tag{3.8.9}$$

where $\boldsymbol{w}_1'=(w_{11},\ldots,w_{mm})$, $\boldsymbol{w}_2'=(w_{12},w_{13},\ldots,w_{m-1,m})$,

$$\Sigma_1^{-1}=uI_m+v\mathbf{1}\mathbf{1}',\quad \mathbf{1}'=(1,1,\ldots,1)\,,$$

$$u=\frac{1}{2(1+\kappa)},\quad v=\frac{1}{m}\left\{\frac{1}{2+(m+2)\kappa}-\frac{1}{2(1+\kappa)}\right\},$$

$$\ell_1=\frac{4}{3}\frac{\psi^{(3)}(0)}{\alpha^3},\ \ell_2=\frac{2\psi''(0)}{\alpha^2},\ \ell_3=-\frac{1}{3}\,,$$

$$\begin{aligned}
a_0 &= (u+mv)[3(m+2)(m+4)(u+v)\ell_1 + \{(m+2)^2u + 3m(m+2)v\}\ell_2 \\
&\quad + 3m(u+mv)\ell_3] \\
a_1 &= \{u^3 + 3(m+4)u^2v + 3(m+2)(m+4)uv^2 + m(m+2)(m+4)v^3\}\ell_1 \\
&\quad + \{u^3 + (3m+4)u^2v + 3m(m+2)uv^2 + m^2(m+2)v^3\}\ell_2 \\
&\quad + (u+mv)^3\ell_3 \\
a_2 &= 6\{u^3 + (m+4)u^2v\}\ell_1 + 2u^2(u+mv)\ell_2 \\
a_3 &= 8u^3\ell_1 \ .
\end{aligned}$$

The domain of W is $W > -\sqrt{n}I_m$.

Let

$$W/\sqrt{n} = \exp(Z/\sqrt{n}) - I_m \ ,$$

and, for $Z^2 = (z_{ij}^{(2)})$, set

$$\boldsymbol{z}_1' = (z_{11}, \ldots, z_{mm}), \quad \boldsymbol{z}_2' = (z_{12}, \ldots, z_{m-1,m}) \ ,$$

$$\boldsymbol{z}_1^{(2)\prime} = (z_{11}^{(2)}, \ldots, z_{mm}^{(2)}), \quad \boldsymbol{z}_2^{(2)\prime} = (z_{12}^{(2)}, \ldots, z_{m-1,m}^{(2)}) \ .$$

The asymptotic expansion of the probability density function of Z for an elliptically contoured distribution is

$$\begin{aligned}
f(Z) =& (2\pi)^{-m(m+1)/4}|\Sigma_1|^{-\frac{1}{2}}(1+\kappa)^{-m(m-1)/4} \\
&\times \exp\Big\{-\frac{1}{2}\boldsymbol{z}_1'\Sigma_1^{-1}\boldsymbol{z}_1 - \frac{1}{2(1+\kappa)}\boldsymbol{z}_2'\boldsymbol{z}_2\Big\} \qquad (3.8.10) \\
&\times \Big[1 + \frac{1}{\sqrt{n}}\Big\{-\frac{1}{2}\boldsymbol{z}_1^{(2)\prime}\Sigma_1^{-1}\boldsymbol{z}_1 - \frac{\boldsymbol{z}_2^{(2)\prime}\boldsymbol{z}_2}{2(1+\kappa)} + \Big(\frac{m+1}{2} + a_0\Big)\mathrm{tr}\,(Z) \\
&\quad - a_1(\mathrm{tr}\,Z)^3 - a_2\mathrm{tr}\,(Z)\mathrm{tr}\,(Z^2) - a_3\mathrm{tr}\,(Z^3)\Big\} + o\Big(\frac{1}{\sqrt{n}}\Big)\Big] \ .
\end{aligned}$$

The range of the z_{ij}'s is the whole real line.

PROOF The characteristic function of W is defined for a symmetric matrix $T = (\frac{1}{2}(1+\delta_{ij})t_{ij})$ as

$$\begin{aligned}
\phi(T) =& E[\ \mathrm{etr}\,(iTW)] \\
=& \mathrm{etr}\,(-i\sqrt{n}T)[E\{\ \mathrm{etr}\,(in^{-\frac{1}{2}}\Sigma^{-\frac{1}{2}}T\Sigma^{-\frac{1}{2}}\boldsymbol{x}\boldsymbol{x}')\}]^n \ .
\end{aligned}$$

The probability density function of $\boldsymbol{y} = V^{-\frac{1}{2}}\boldsymbol{x}$ is $c_m g(\boldsymbol{y}'\boldsymbol{y})$ and the characteristic function is $\psi(\boldsymbol{t}'\boldsymbol{t})$. Thus

$$E[\ \mathrm{etr}\,(in^{-\frac{1}{2}}\Sigma^{-\frac{1}{2}}T\Sigma^{-\frac{1}{2}}\boldsymbol{x}\boldsymbol{x}')] = \sum_{k=0}^{\infty}\frac{1}{k!}E[\{in^{-\frac{1}{2}}\alpha^{-1}(\mathrm{tr}\,(\mathrm{T}\boldsymbol{y}\boldsymbol{y}'))\}^k]$$

$$= \sum_{k=0}^{\infty} \frac{1}{k!}\left(\frac{i}{\sqrt{n}\alpha}\right)^k c_m \int_{\Re^m} \{\mathrm{tr}\,(\mathrm{T}\boldsymbol{y}\boldsymbol{y}')\}^k g(\boldsymbol{y}'\boldsymbol{y})d\boldsymbol{y}$$

$$= \sum_{k=0}^{\infty} \frac{1}{k!}\left(\frac{i}{\sqrt{n}\alpha}\right)^k c_m \int_{\Re^m} \left\{\int_{O(m)} \{\mathrm{tr}\,(\mathrm{T}\mathrm{H}\boldsymbol{y}\boldsymbol{y}'\mathrm{H}')\}^k d(\mathrm{H})\right\} g(\boldsymbol{y}'\boldsymbol{y})d\boldsymbol{y}$$

$$= \sum_{k=0}^{\infty} \frac{1}{k!}\left(\frac{1}{\sqrt{n}\alpha}\right)^k c_m \frac{Z_{(k)}(T)}{Z_{(k)}(I_m)} \int_{\Re_m} (\boldsymbol{y}'\boldsymbol{y})^k g(\boldsymbol{y}'\boldsymbol{y})d\boldsymbol{y} .$$

On the other hand, using (3.8.2) we have

$$\begin{aligned} E[\exp(i\boldsymbol{t}'\boldsymbol{y})] &= \psi(\boldsymbol{t}'\boldsymbol{t}) = \sum_{k=0}^{\infty} \frac{\psi^{(k)}(0)}{k!}(\boldsymbol{t}'\boldsymbol{t})^k \\ &= \sum_{k=0}^{\infty} \frac{1}{k!(\frac{m}{2})_k}\left(-\frac{1}{4}\right)^k E[(\boldsymbol{y}'\boldsymbol{y})^k](\boldsymbol{t}'\boldsymbol{t})^k . \end{aligned}$$

Therefore

$$E[(\boldsymbol{y}'\boldsymbol{y})^k] = (-4)^k \left(\frac{m}{2}\right)_k \psi^{(k)}(0) .$$

Summarizing these, the characteristic function of W is expanded asymptotically as

$$\begin{aligned} \phi(T) = {} & \exp\left[\frac{1}{2}(\mathrm{tr}\,\mathrm{T})^2 - \ell_2 \mathrm{Z}_{(2)}(\mathrm{T})\right] \\ & \times \exp\left[\frac{i}{\sqrt{n}}\{\ell_1 Z_{(3)}(T) + \ell_2 \mathrm{tr}\,(\mathrm{T})\mathrm{Z}_{(2)}(\mathrm{T}) - \frac{1}{3}(\mathrm{tr}\,(\mathrm{T}))^3\}\right] \\ & + o\left(\frac{1}{\sqrt{n}}\right) . \end{aligned}$$

The exponent of the first exponential is

$$-\frac{1}{2}\{2\alpha^2 Z_{(2)}(T) - (\mathrm{tr}\,(\mathrm{T}))^2\} = -\frac{1}{2}\{\boldsymbol{t}_1'\Sigma_1\boldsymbol{t}_1 + (1+\kappa)\boldsymbol{t}_2'\boldsymbol{t}_2\} ,$$

where $\boldsymbol{t}_1' = (t_{11}, \ldots, t_{mm})$ and $\boldsymbol{t}_2' = (t_{12}, \ldots, t_{m-1\ m})$, which implies that the limiting distributions of $\boldsymbol{W}_1$ and $\boldsymbol{W}_2$ are $N(\boldsymbol{0}, \Sigma_1)$ and $N(\boldsymbol{0}, (1+\kappa)I_f), f = \frac{1}{2}m(m-1)$, respectively. Inversion of $\phi(T)$ gives (3.8.9).

Lemma 3.8.6 Let Y_α, $\alpha = 1, 2, \ldots, n$ be distributed independently as $N(0,1)$ and let $\lambda_1 \geq \lambda_2 \geq \cdots \geq \lambda_n > 0$. Then the characteristic function of

$$Q = \sum_{i=1}^{n} \lambda_i (Y_i - w_i)^2 \tag{3.8.11}$$

is expressed as

$$\phi(t) = \exp\left\{-\frac{1}{2}\sum_{i=1}^{n} w_i^2\right\} \exp\left\{\frac{1}{2}\sum_{j=1}^{n} \frac{w_j^2}{1-2it\lambda_j}\right\} \prod_{j=1}^{n}(1-2it\lambda_j)^{-\frac{1}{2}} . \quad (3.8.12)$$

The distribution function of Q is given by

$$F(y) = \sum_{j=0}^{\infty} e_j \Pr\{\chi^2_{n+2j} \leq y/w\} , \quad (3.8.13)$$

where w is an appropriate constant which makes the series converge rapidly, and

$$e_0 = \exp\left\{-\frac{1}{2}\sum_{j=1}^{n} w_j^2\right\} \prod_{j=1}^{n}\left(\frac{w}{\lambda_j}\right)^{\frac{1}{2}}$$

$$e_r = \frac{1}{2r}\sum_{j=0}^{r-1} G_{r-j} e_j \quad (r \geq 1)$$

$$G_r = \sum_{j=1}^{n}\left(1-\frac{w}{\lambda_j}\right)^r + rw\sum_{j=1}^{n}\left(\frac{w_j^2}{\lambda j}\right)\left(1-\frac{w}{\lambda j}\right)^{r-1} .$$

PROOF See Johnson and Kotz (1970, Ch. 29.4) and Mathai and Provost (1992, Section 4.2d).

Remark 3.8.3. If the characteristic function is expressed as

$$\exp\left\{-\frac{1}{2}(a\delta_1^2 + b\delta_2^2)\right\} \exp\left\{\frac{1}{2}\left(\frac{a\delta_1^2}{1-2it\nu_1} + \frac{b\delta_2^2}{1-2it\nu_2}\right)\right\}$$

$$\times(1-2it\nu_1)^{-\frac{a}{2}}(1-2it\nu_2)^{-\frac{b}{2}} , \quad (3.8.14)$$

then the distribution function is given by

$$\sum_{j=0}^{\infty} e'_j \Pr\{\chi^2_{a+b+2j} \leq x/w\} , \quad (3.8.15)$$

where

$$e'_0 = \exp\left\{-\frac{1}{2}(a\delta_1^2 + b\delta_2^2)\right\}\left(\frac{w}{\nu_1}\right)^{\frac{a}{2}}\left(\frac{w}{\nu_2}\right)^{\frac{b}{2}}$$

$$e'_r = \frac{1}{2r}\sum_{j=0}^{r-1} G'_{r-j} e'_j$$

$$G_r' = a\left(1-\frac{w}{\nu_1}\right)^r + b\left(1-\frac{w}{\nu_2}\right)^r$$
$$+ rw\left\{\frac{a\delta_1^2}{\nu_1}\left(1-\frac{w}{\nu_1}\right)^{r-1} + \frac{b\delta_2^2}{\nu_2}\left(1-\frac{w}{\nu_2}\right)^{r-1}\right\}.$$

Here we consider the asymptotic behavior of some likelihood ratio criteria (LRC's) for testing structures of the covariance matrix of a normal population and elliptically contoured populations.

Let $\boldsymbol{x}_\alpha$, $\alpha = 1,\ldots,n$ be a random sample from $N(\mathbf{0},\Sigma)$. The m.l.e. of Σ is $\hat{\Sigma} = S/n = \sum_{\alpha=1}^{n} \boldsymbol{x}_\alpha \boldsymbol{x}_\alpha'/n = XX'/n$ where $X = (\boldsymbol{x}_1,\ldots,\boldsymbol{x}_n)$.

(i) The LRC for testing the hypothesis $H_1 : \Sigma = I_m$ is given by

$$L_1 = \left(\frac{e}{n}\right)^{\frac{1}{2}mn} |S|^{\frac{n}{2}} \operatorname{etr}\left(-\frac{S}{2}\right). \tag{3.8.16}$$

(ii) The LRC for testing the hypothesis $H_2 : \Sigma = \sigma^2 I_m$ is given by

$$L_2 = |S|^{\frac{n}{2}}/\{\operatorname{tr}(S/m)\}^{\frac{1}{2}mn}. \tag{3.8.17}$$

(iii) Let the covariance matrix Σ be partitioned into submatrices as

$$\Sigma = \begin{bmatrix} \Sigma_{11} & \ldots & \Sigma_{1q} \\ \vdots & & \vdots \\ \Sigma_{q1} & \ldots & \Sigma_{qq} \end{bmatrix},$$

where $\Sigma_{\alpha\beta}$ is an $m_\alpha \times m_\beta$ matrix.

The LRC for testing the hypothesis $H_3 : \Sigma_{\alpha\beta} = O$ $(\alpha \neq \beta)$ is given by

$$L_3 = \left\{|S|/\prod_{\alpha=1}^{q} |S_{\alpha\alpha}|\right\}^{\frac{1}{2}n}. \tag{3.8.18}$$

(iv) Let $\Lambda = \operatorname{diag}(\lambda_1,\ldots,\lambda_m)$ where the diagonal elements are the characteristic roots of Σ with multiplicity one. Let Γ be an $m\times m$ orthogonal matrix whose α-th column vector corresponds to the α-th characteristic root, which implies that $\Sigma = \Gamma\Lambda\Gamma'$. The LRC for testing the hypothesis $H_4 : \Lambda = \Lambda_0$ (a given matrix) is given by

$$L_4 = \left(\frac{e}{n}\right)^{\frac{1}{2}mn} \left(\prod_{i=1}^{m} \frac{d_i}{\lambda_{i0}}\right)^{\frac{n}{2}} \exp\left(-\frac{1}{2}\sum_{i=1}^{m} \frac{d_i}{\lambda_{i0}}\right) \tag{3.8.19}$$

where $d_1 \geq \cdots \geq d_m$ are the characteristic roots of S/n. The LRC for testing the hypothesis $H_5 : \Gamma_1 = \Gamma_{1o}$ (a given matrix), where Γ_{1o} is a specified $m \times k$ matrix whose columns are orthonormal vectors, is given by

$$L_5 = \left[|S|/\left\{\left(\prod_{i=1}^{k} (\Gamma_{1o}' S \Gamma_{1o})_{ii}\right)|\Gamma_2' S \Gamma_2|\right\}\right]^{\frac{n}{2}}, \tag{3.8.20}$$

where $(A)_{ii}$ denotes the i-th diagonal element of the matrix A. The $m \times (m-k)$ matrix Γ_2 is such that $\Gamma = [\Gamma_{1o}, \Gamma_2]$ is an orthogonal matrix.

Theorem 3.8.3 Under the elliptically contoured distribution $C_m(0, V)$ and $V = I_m + \Theta/\sqrt{n}$ the distribution of L_1 is expanded as follows

$$\Pr\{-2\log L_1 \le x\} = F_{0,0}^{(1)}(x) + \frac{1}{\sqrt{n}} \sum_{\alpha=0}^{3} \sum_{\beta=0}^{3} d_{\alpha\beta}^{(1)}(x) F_{\alpha,\beta}^{(1)} + o\left(\frac{1}{\sqrt{n}}\right) \tag{3.8.21}$$

where

$$\begin{aligned}
d_{3,0}^{(1)} &= -\left(\frac{1}{3}u + a_3\right)\left(s_3 - \frac{3s_1s_2}{m} + \frac{2}{m^2}s_1^3\right)\\
d_{2,0}^{(1)} &= \left(\frac{4}{3}u + 3a_3\right)s_3 - \left(\frac{5}{2}u - \frac{1}{2}mv - a_2m + 6a_3\right)\frac{s_1s_2}{m}\\
&\quad + \left(\frac{7}{6}u - \frac{1}{2}mv - a_2m + 3a_3\right)\frac{s_1^3}{m^2}\\
d_{1,0}^{(1)} &= -(2u + 3a_3)s_3 + (2u - mv - 2a_2m + 3a_3)\frac{s_1s_2}{m}\\
&\quad + (mv + 2a_2m)\frac{s_1^3}{m^2} + \frac{1}{4u}(m-1)(m+2)\\
&\quad \times (3u + mu + 2a_2m + 6a_3)\frac{s_1}{m}\\
d_{0,0}^{(1)} &= (u + a_3)s_3 + (v + a_2)s_1s_2 + a_1s_1^3 - \left(\frac{m+1}{2} + a_0\right)s_1
\end{aligned}$$

$$\begin{aligned}
d_{2,1}^{(1)} &= -\frac{1}{2}(2u + mv + 2a_2m + 6a_3)\left(\frac{s_1s_2}{m} - \frac{s_1^3}{m^2}\right)\\
d_{1,1}^{(1)} &= \left(\frac{5}{2}u + mv + 2a_2m + 6a_3\right)\left(\frac{s_1s_2}{m} - \frac{s_1^3}{m^2}\right)\\
&\quad - \frac{1}{4u}(m-1)(m+2)(2u + mv + 2a_2m + 6a_3)\frac{s_1}{m}\\
d_{0,1}^{(1)} &= -(2u + mv + a_2m + 3a_3)\frac{s_1s_2}{m} - (mv + 3a_1m^2 + 2a_2m)\frac{s_1^3}{m^2}\\
&\quad + \left\{\frac{m^2+m+6}{4} + a_0m + \frac{3}{u+mv}(a_1m^2 + a_2m + a_3)\right\}\frac{s_1}{m}\\
d_{0,2}^{(1)} &= \left\{\frac{4}{3}(u + mv) + 3(a_1m^2 + a_2m + a_3)\right\}\frac{s_1^3}{m^2}\\
&\quad - \left\{1 + \frac{3}{u+mv}(a_1m^2 + a_2m + a_3)\right\}\frac{s_1}{m}\\
d_{0,3}^{(1)} &= -\left\{\frac{1}{3}(u + mv) + a_1m^2 + a_2m + a_3\right\}\frac{s_1^3}{m^2},
\end{aligned}$$

the rest of the $d_{\alpha,\beta}^{(1)}$'s being all zero,

$$
\begin{aligned}
s_\alpha =& \operatorname{tr}(\Theta^\alpha),\ \alpha = 1,2,3\ ,\\
F_{\alpha,\beta}(x) =& \sum_{k=0}^{\infty} e_k \Pr\{\chi^2_{f+2\alpha+2\beta+2k} \le x/w\}\ ,\\
e_0 =& \exp\Big\{-\frac{1}{2}(us_2+vs_1^2)\Big\}(2w)^{\frac{f}{2}+\alpha+\beta}u^{\frac{1}{4}(m-1)(m+2)+\alpha}(u+mv)^{\frac{1}{2}+\beta},\\
e_k =& \frac{1}{2k}\sum_{j=0}^{k-1} G_{k-j}e_j,\ k \ge 1\ ,\\
G_k =& \Big\{\frac{1}{4}(m-1)(m+2)+2\alpha\Big\}(1-2uw)^k + (1+2\beta)\\
&\times\{1-2(u+mv)w\}^k + kw\Big[2(u+mv)^2\frac{s_1^2}{m}\{1-2(u+mv)w\}^{k-1}\\
&+2u^2\Big(s_2-\frac{s_1^2}{m}\Big)(1-2uw)^{k-1}\Big]
\end{aligned}
$$

where $f = \frac{1}{2}m(m+1)$, and u, v, a_1, a_2, and a_3 are given in Lemma 3.8.5.

PROOF Expanding L_1 by using

$$
\frac{S}{n} = \Big(I+\frac{\Theta}{\sqrt{n}}\Big)^{\frac{1}{2}} \exp\Big(\frac{Z}{\sqrt{n}}\Big)\Big(I+\frac{\Theta}{\sqrt{n}}\Big)^{\frac{1}{2}}\ ,
$$

we have

$$
-2\log L_1 = \frac{1}{2}\operatorname{tr}(Z+\Theta)^2 + \frac{1}{\sqrt{n}}\Big\{\frac{1}{6}\operatorname{tr}(Z^3) + \frac{1}{2}\operatorname{tr}(Z^2\Theta) - \frac{1}{3}\operatorname{tr}(\Theta^3)\Big\} + o_p\Big(\frac{1}{\sqrt{n}}\Big)\ .
$$

The characteristic function of $-2\log L_1$ is obtained using (3.8.10) as

$$
\begin{aligned}
&\exp\Big\{-\frac{1}{2}(us_2+vs_1^2)\Big\}\exp\Big[\frac{1}{2}\Big\{c_1u\Big(s_2-\frac{s_1^2}{m}\Big)+c_2(u+mv)\frac{s_1^2}{m}\Big\}\Big]\\
&\times c_1^{\frac{1}{4}(m-1)(m+2)}c_2^{\frac{1}{2}}\\
&\times\Big[1+\frac{1}{\sqrt{n}}\sum_{\alpha=0}^{3}\sum_{\beta=0}^{3} d^{(1)}_{\alpha,\beta}c_1^\alpha c_2^\beta + o\Big(\frac{1}{\sqrt{n}}\Big)\Big]\ ,
\end{aligned}
$$

where

$$
c_1 = \Big(1-\frac{it}{u}\Big)^{-1},\quad c_2 = \Big(1-\frac{it}{u+mv}\Big)^{-1}
$$

and the $d^{(1)}_{\alpha,\beta}$'s are defined in (3.8.21). We obtain (3.8.21) with the help of Remark 3.8.3.

Theorem 3.8.4 Under the elliptically contoured distribution $C_m(0, V)$ with $V =$

$\sigma^2(I_m + \Theta/\sqrt{n})$, the distribution of L_2 can be expanded as follows:

$$\Pr\{-2\log L_2 \le x\} = F_0(x) + \frac{1}{\sqrt{n}}\sum_{\alpha=0}^{3} d_\alpha F_\alpha(x) + o\Big(\frac{1}{\sqrt{n}}\Big) , \tag{3.8.22}$$

where

$$\begin{aligned}
d_3 &= -\Big(\frac{1}{3}u + a_3\Big)\Big(s_3 - \frac{3s_1s_2}{m} + \frac{2s_1^3}{m^2}\Big),\\
d_2 &= \Big(\frac{4}{3}u + 3a_3\Big)s_3 - (4u + 9a_3)\frac{s_1s_2}{m} + \Big(\frac{8}{3}u + 6a_3\Big)\frac{s_1^3}{m^2},\\
d_1 &= -(2u + 3a_3)s_3 + (5u + 9a_3)\frac{s_1s_2}{m} - (3u + 6a_3)\frac{s_1^3}{m^2},\\
d_0 &= (u + a_3)s_3 - (2u + 3a_3)\frac{s_1s_2}{m} + (u + 2a_3)\frac{s_1^3}{m^2},
\end{aligned}$$

and

$$\begin{aligned}
F_\alpha(x) &= \sum_{k=0}^{\infty}\bar{e}_k \Pr\{\chi^2_{f+2\alpha+2k} \le x/w\} ,\\
f &= \frac{1}{2}(m-1)(m+2) ,
\end{aligned}$$

$$\begin{aligned}
\bar{e}_0 &= \exp\Big\{-\frac{u}{2}\Big(s_2 - \frac{s_1^2}{m}\Big)\Big\}(2uw)^{f+\alpha} ,\\
\bar{e}_k &= \frac{1}{2k}\sum_{j=0}^{k-1}\bar{G}_{k-j}\bar{e}_j ,
\end{aligned}$$

$$\begin{aligned}
\bar{G}_k = &\Big\{\frac{(m-1)(m+2)}{2} + 2\alpha\Big\}(1-2uw)^k\\
&+ 2kwu^2(1-2uw)^{k-1}\Big(s_2 - \frac{s_1^2}{m}\Big) .
\end{aligned}$$

PROOF (3.8.22) is obtained by steps similar to those used in the proof of Theorem 3.8.3. The proof is left to the reader.

Theorem 3.8.5 Under the elliptically contoured distribution $C_m(0, V)$ and $V = I_m + \Theta/\sqrt{n}$, where $\Theta = (\Theta_{\alpha\beta})$ with $\Theta_{\alpha\alpha} = O$ and $\Theta'_{\alpha\beta} = \Theta_{\beta\alpha}$, the distribution of L_3 is expressed as

$$\Pr\{-2\log L_3/(1+\kappa) \le x\} = \bar{P}_f(\delta^2) + \frac{s_3}{\sqrt{n}}\sum_{\alpha=0}^{3} b_\alpha \bar{P}_{f+2\alpha}(\delta^2) + o\Big(\frac{1}{\sqrt{n}}\Big) \tag{3.8.23}$$

where

$$b_3 = -\left(\frac{1}{3}u + a_3\right), \quad b_2 = \frac{4}{3}u + 3a_3$$
$$b_1 = -(2u + 3a_3), \quad b_0 = u + a_3 ,$$

and $\bar{P}_f(\delta^2) = \Pr\{\chi_f^2(\delta^2) \le x\}$, $\chi_f^2(\delta^2)$ is a noncentral chi-square random variable with $f = \frac{1}{2}(m^2 - \sum_{\alpha=1}^{q} m_\alpha^2)$ degrees of freedom and noncentrality parameter $\delta^2 = \frac{1}{2}\frac{s_2}{1+\kappa}$.

PROOF Since L_3 is invariant for the transformation $S \to \Sigma_D^{\frac{1}{2}} S \Sigma_D^{\frac{1}{2}}$, where $\Sigma_D^{\frac{1}{2}} = \operatorname{diag}(\Sigma_{11}^{\frac{1}{2}}, \ldots, \Sigma_{qq}^{\frac{1}{2}})$, we assume without loss of generality that $\Sigma_{\alpha\alpha} = I_{m_\alpha}$, $\alpha = 1, \ldots, q$. We expand $-2 \log L_3$ as follows:

$$\begin{aligned} -2 \log L_3 = & \sum_{\alpha<\beta} tr((Z_{\alpha\beta} + \Theta_{\alpha\beta})(Z_{\beta\alpha} + \Theta_{\beta\alpha})) \\ & + \frac{1}{\sqrt{n}} \left\{ \frac{1}{\delta}(tr(Z^3) + 3tr(Z^2\Theta) - 2tr(\Theta^3)) \right. \\ & - \frac{1}{4} \sum_{\alpha=1}^{q} tr(Z_{\alpha\alpha}\{Z_{\alpha\alpha}^{(2)} + \sum_{\beta} \Theta_{\alpha\beta} Z_{\beta\alpha} + \sum_{\beta} Z_{\alpha\beta}\Theta_{\beta\alpha}\}) \\ & \left. + \frac{1}{3} \sum_{\alpha=1}^{q} tr(Z_{\alpha\alpha}^3) \right\} + o_p\left(\frac{1}{\sqrt{n}}\right) , \end{aligned}$$

that is, the leading term is expressed in terms of off-diagonal elements of Z. Steps similar to those used in the proof of Theorem 3.8.3 give (3.8.23).

Theorem 3.8.6 Under the elliptically contoured distribution $C_m(\mathbf{0}, V)$ and letting $\Lambda = \Lambda_0 + \Theta/\sqrt{n}$ where $\Theta = \operatorname{diag}(\theta_1, \ldots, \theta_m)$, the distribution of L_4 as defined in (3.8.19) is expanded as

$$\Pr\{-2\log L_4 \le x\} = F_{0,0}^{(4)}(x) + \frac{1}{\sqrt{n}} \sum_{\alpha=0}^{3} \sum_{\beta=0}^{3} d_{\alpha\beta}^{(4)} F_{\alpha,\beta}^{(4)}(x) + o(1/\sqrt{n}), \tag{3.8.24}$$

where

$$\begin{aligned} d_{3,0}^{(4)} = & -\left(\frac{1}{3}u + a_3\right)s_3 , \\ d_{2,0}^{(4)} = & \left(\frac{4}{3}u + 3a_3\right)s_3 + (u + a_2 m + 3a_3)\bar{\theta}s_2 , \\ d_{1,0}^{(4)} = & -(2u + 3a_3)s_3 - (3u + 2a_2 m + 6a_3)\bar{\theta}s_2 \\ & + \frac{1}{2}(m-1)\left\{m(1+\kappa)u + \frac{1}{u}(3u + mv + 2a_2 m + 6a_3)\right\}\bar{\theta} \\ & + (1+\kappa)u \sum_{i=1}^{m} \theta_i \sum_{j \ne i} \lambda_{ijo}\lambda_{jo} , \end{aligned}$$

$$d_{0,0}^{(4)} = (u+a_3)s_3 + (3u+mv+a_2m+3a_3)\bar{\theta}s_2$$
$$+(u+mv+a_1m^2+a_2m+a_3)m\bar{\theta}^3$$
$$+\left\{\frac{1}{2}m(m-1)(1+\kappa)(mv+2a_2m+6a_3)-\frac{1}{2}m(m+1)-a_0m\right\}\bar{\theta}$$
$$-(1+\kappa)u\sum_{i=1}^{m}\theta_i\sum_{j\neq i}\lambda_{ijo}\lambda_{jo}\ ,$$

$$d_{2,1}^{(4)} = -(u+a_2m+3a_3)\bar{\theta}s_2\ ,$$
$$d_{1,1}^{(4)} = (3u+2a_2m+6a_3)\bar{\theta}s_2-\frac{1}{2}(m-1)\frac{1}{u}(2u+mv+2a_2m+6a_3)\bar{\theta}\ ,$$
$$d_{0,1}^{(4)} = -(3u+mv+a_2m+3a_3)\bar{\theta}s_2$$
$$-(2u+2mv+3a_1m^2+3a_2m+3a_3)m\bar{\theta}^3$$
$$+\left\{\frac{3}{u+mv}(a_1m^2+a_2m+a_3)+\frac{1}{2}(m^2+3)+a_0m\right.$$
$$\left.-\frac{1}{2}m(m-1)(1+\kappa)(u+mv+2a_2m+6a_3)\right\}\bar{\theta}\ ,$$
$$d_{0,2}^{(4)} = \left\{\frac{4}{3}(u+mv)+3(a_1m^2+a_2m+a_3)\right\}m\bar{\theta}^3$$
$$-\left\{1+\frac{3}{u+mv}(a_1m^2+a_2m+a_3)\right\}\bar{\theta}\ ,$$
$$d_{0,3}^{(4)} = -\left\{\frac{1}{3}(u+mv)+(a_1m^2+a_2m+a_3)\right\}m\bar{\theta}^3\ ,$$

the rest of the $d_{\alpha,\beta}^{(4)}$'s being zero,

$$F_{\alpha,\beta}(x) = \sum_{k=0}^{\infty} e_k \Pr\{\chi^2_{m+2\alpha+2\beta+2k} \le x/w\}\ ,$$
$$e_0 = \exp\left[-\frac{1}{2}\{us_2+(u+mv)m\bar{\theta}^2\}\right](2w)^{\frac{1}{2}m+\alpha+\beta}u^{\frac{1}{2}(m-1)+\alpha}$$
$$\times(u+pv)^{\frac{1}{2}+\beta}\ ,$$
$$e_k = \frac{1}{2k}\sum_{j=0}^{k-1}G_{k-j}e_j,\ (k\ge 1)\ ,$$

$$G_k = (m-1+2\alpha)(1-2uw)^k+(1+2\beta)\{1-2(u+mv)w\}^k$$
$$+kw\left[2(u+mv)^2m\bar{\theta}^2\{1-2(u+mv)w\}^{k-1}\right.$$
$$\left.+2u^2s_2(1-2uw)^{k-1}\right]\ ,$$

$$s_k=\sum_{j=1}^{m}(\theta_j-\bar{\theta})^k,\ \ k=2,3,\ \ \bar{\theta}=\sum_{j=1}^{m}\theta_j/m,\ \text{and}\ \lambda_{ijo}=(\lambda_{io}-\lambda_{jo})^{-1}\ .$$

Theorem 3.8.7 Under the elliptically contoured distribution $C_m(0, V)$ with $\Gamma_1 = \Gamma_{1o} \exp(\Theta_{11}/\sqrt{n})$ where Θ_{11} is a $k \times k$ skew symmetric matrix, the distribution of L_5 as defined in (3.8.20) is expanded as

$$\Pr\{-2 \log L_5/(1+\kappa) \le x\} = P_f(\xi^2) - \frac{m_3}{\sqrt{n}} \sum_{\alpha=0}^{3} b_\alpha P_{f+2\alpha}(\xi^2) + o(1/\sqrt{n})$$

where

$$b_3 = \frac{1}{1+\kappa} + \frac{6\ell_1}{(1+\kappa)^3}, \quad b_2 = -\frac{4}{1+\kappa} - \frac{18\ell_1}{(1+\kappa)^3},$$
$$b_1 = \frac{9}{2(1+\kappa)} + \frac{18\ell_1}{(1+\kappa)^3}, \quad b_0 = -\frac{3}{2(1+\kappa)} - \frac{6\ell_1}{(1+\kappa)^3},$$
$$\xi^2 = \frac{1}{2(1+\kappa)} \operatorname{tr}(\Xi_{11}^2), \quad \Xi_{11} = (\xi_{ij}),$$

$$\xi_{ij} = \begin{cases} [(\lambda_j/\lambda_i)^{\frac{1}{2}} - (\lambda_i/\lambda_j)^{\frac{1}{2}}]\theta_{ij}, & 1 \le i < j \le k \\ 0, & \text{otherwise} \end{cases}$$

$$f = \frac{1}{2}k(k-1) + k(m-k), \quad m_3 = \operatorname{tr}(\Lambda^{-1}\Theta\Lambda\Theta^2)$$

and

$$P_f(\xi^2) = \Pr\{\chi_f^2(\xi^2) \le x\}.$$

Additional results can be found in Hayakawa (1972d,1987,1989) and Hayakawa and Puri (1985).

EXERCISES

3.1 Prove Theorem 3.1.1.

3.2 Obtain (3.1.27) using Theorem 3.1.2 and (3.1.24).

3.3 Referring to the proof of Theorem 3.1.10, show that if A_1 and A_2 are two symmetric idempotent $p \times p$ matrices, then $(A_1 + A_2)^2 = (A_1 + A_2)$ implies that $A_1 A_2 = O$.

3.4 Prove Lemma 3.1.4.

3.5 Show that (3.3.14) is equal to (3.3.15).

3.6 Show that Leibnitz' differentiation rule leads to (3.4.4).

3.7 (i) Prove Property 3.6.1;
(ii) Verify (3.6.19), (3.6.20), (3.6.21) and (3.6.22).

3.8 Let $\boldsymbol{x}_\alpha, \alpha = 1, 2, \ldots, N$ be a random sample from $C_m(\boldsymbol{\mu}, \Sigma)$, $\bar{\boldsymbol{x}} = \frac{1}{N}\sum_{\alpha=1}^{N} \boldsymbol{x}_\alpha$, and

$$W = \sum_{\alpha=1}^{N} (\boldsymbol{x}_\alpha - \boldsymbol{\mu})(\boldsymbol{x}_\alpha - \boldsymbol{\mu})'.$$

(i) Proceeding as in Lemma 3.8.5 find the asymptotic expansion of the joint probability density function of $z = \sqrt{N}\Sigma^{-\frac{1}{2}}(\bar{x} - \mu)$ and $U = \sqrt{N}\Sigma^{-\frac{1}{2}}(W/N - \Sigma)\Sigma^{-\frac{1}{2}}$, $\Sigma = (2\psi'(0))V$ where $\psi(x)$ is defined in Lemma 3.8.1.

(ii) Let $S = \sum_{\alpha=1}^{N}(x_\alpha - \bar{x})(x_\alpha - \bar{x})'$. Noting that $S = \frac{N}{n}(W - \bar{y}\bar{y}')$ where $\bar{y} = \bar{x} - \mu$ and $n = N - 1$, show that the asymptotic expansion of the joint probability density function of z and $Z = \sqrt{n}\Sigma^{-\frac{1}{2}}(\frac{S}{n} - \Sigma)\Sigma^{-\frac{1}{2}}$ is given by

$$\begin{aligned} f(z, Z) =&(2\pi)^{-\frac{1}{4}m(m+3)}|\Sigma_1|^{-\frac{1}{2}}(1+\kappa)^{-\frac{1}{4}m(m-1)} \\ &\times \exp\Big(-\frac{1}{2}z'z\Big)\exp\Big(-\frac{1}{2}z_1'\Sigma_1^{-1}z_1\Big)\exp\Big(-\frac{1}{2(1+k)}z_2'z_2\Big) \\ &\times \Big[1 + \frac{1}{\sqrt{N}}\Big\{\Big(a_0 - \frac{1}{2} + (u + mv)\Big)\mathrm{tr}(Z) - a_1(\mathrm{tr}(Z))^3 \\ &\qquad - a_2\mathrm{tr}(Z)\mathrm{tr}(Z^2) - a_3\mathrm{tr}(Z^3) \\ &\qquad + \Big(\frac{1}{2} - u\Big)z'Zz - vz'z\mathrm{tr}(Z)\Big\} + o\Big(\frac{1}{\sqrt{N}}\Big)\Big] . \end{aligned}$$

(iii) Show that the probability density function of z can be expanded up to the order $\frac{1}{N}$ as

$$\begin{aligned} f(z) =&(2\pi)^{-\frac{1}{2}m}\exp\Big\{-\frac{1}{2}z'z\Big\} \\ &\times\Big[1 + \frac{\kappa}{8N}\{m(m+2) - 2(m+2)z'z + (z'z)^2\} + o\Big(\frac{1}{N}\Big)\Big] \end{aligned}$$

Wakaki (1994).

3.9 Let $r = S_{12}/\sqrt{S_{11}S_{22}}$ be a sample correlation coefficient centered at $(\bar{x}_1, \bar{x}_2)$ under $C_2(\mu, V)$. Show that $E[(r-\rho)^k]$, $k = 1, 2, 3, 4$ are given as in Lemma 3.8.4 by using Exercise 3.8 (ii).

3.10 Let x_α, $\alpha = 1, 2, \ldots, N$ be a random sample from $N(\mu, \Sigma)$. The modified likelihood ratio criteria for testing a covariance structure H_i, $i = 1, 2, 3, 4$ are obtained by replacing n by $N - 1$ and S by $\sum_{\alpha=1}^{N}(x_\alpha - \bar{x})(x_\alpha - \bar{x})'$. Show that the asymptotic expansions of the distribution functions of $-2\log L_i$ under the alternative hypotheses are respectively the same as those obtained in the Section 3.8.

3.11 Let x_α, $\alpha = 1, 2, \ldots, N$ be a random sample from $N(\mu, \Sigma)$. Nagao (1973) proposed the following statistics for testing some structures of the covariance matrix:

$$\begin{aligned}
T_1 &= \frac{n}{2}\mathrm{tr}\left(\frac{S}{n}-I\right)^2 \quad \text{for} \quad H_1:\Sigma = I_m\\
T_2 &= \frac{1}{2}m^2 n\ \mathrm{tr}\,(S/\mathrm{tr}\,(S) - I/m)^2 \quad \text{for} \quad H_2:\Sigma=\sigma^2 I_m\\
T_3 &= \frac{n}{2}\mathrm{tr}\,(SS_0^{-1}-I)^2 \quad \text{for} \quad H_3:\Sigma_{\alpha\beta}=O,\ \alpha\neq\beta\\
S_0 &= \mathrm{diag}(S_{11},\ldots,S_{qq})
\end{aligned}$$

where $S=\sum_{\alpha=1}^{N}(\boldsymbol{x}_\alpha-\bar{\boldsymbol{x}})(\boldsymbol{x}_\alpha-\bar{\boldsymbol{x}})'$, $n=N-1$,

$$\Sigma=\begin{bmatrix}\Sigma_{11} & \ldots & \Sigma_{1q}\\ \vdots & & \vdots\\ \Sigma_{q1} & \ldots & \Sigma_{qq}\end{bmatrix},$$

and $\Sigma_{\alpha\beta}$ is an $m_\alpha\times m_\beta$ matrix.

(i) Under the elliptically contoured distribution $C_m(\boldsymbol{\mu},V)$ and $\Sigma=I+\Theta/\sqrt{n}$, show that the distribution of T_1 can be expanded as follows:

$$\Pr\{T_1\le x\}=F_{00}^{(1)}(x)+\frac{1}{\sqrt{n}}\sum_{\alpha=0}^{3}\sum_{\beta=0}^{3}d_{\alpha\beta}^{(1)}F_{\alpha\beta}^{(1)}(x)+o(1/\sqrt{n})$$

where

$$\begin{aligned}
d_{3,0}^{(1)} &= -a_3\left(s_3-\frac{3}{m}s_1s_2+\frac{2}{m^2}s_1^3\right)\\
d_{2,1}^{(1)} &= -\left(\frac{mv}{2}+a_2m+3a_3\right)\left(\frac{s_1s_2}{m}-\frac{s_1^3}{m^2}\right)\\
d_{1,2}^{(1)} &= 0,\ d_{0,3}^{(1)}=-(a_1m^2+a_2m+a_3)\frac{s_1^3}{m^2}\\
d_{2,0}^{(1)} &= (u+3a_3)s_3-\left(\frac{3}{2}u-\frac{mv}{2}-a_2m+6a_3\right)\frac{s_1s_2}{m}\\
&\quad+\left(\frac{u}{2}-\frac{mv}{2}-a_2m+3a_3\right)\frac{s_1^3}{m^2}\\
d_{1,1}^{(1)} &= \left(\frac{3}{2}u+mv+2a_2m+6a_3\right)\left(\frac{s_1s_2}{m}-\frac{s_1^3}{m^2}\right)\\
&\quad-\frac{1}{4u}(m-1)(m+2)(mv+2a_2m+6a_3)\frac{s_1}{m}\\
d_{0,2}^{(1)} &= \{u+mv+3(a_1m^2+a_2m+a_3)\}\frac{s_1^3}{m^2}\\
&\quad-\frac{3}{u+mv}(a_1m^2+a_2m+a_3)\frac{s_1}{m}
\end{aligned}$$

$$d_{1,0}^{(1)} = -(2u + 3a_3)s_3 + (2u - mv - 2a_2m + 3a_3)\frac{s_1s_2}{m}$$
$$+ (mv + 2a_2m)\frac{s_1^3}{m^2} + \frac{1}{4u}(m-1)(m+2)(3u + mv + 2a_2m + 6a_3)\frac{s_1}{m}$$
$$d_{0,1}^{(1)} = -(2u + mv + a_2m + 3a_3)\frac{s_1s_2}{m} - (mv + 3a_1m^2 + 2a_2m)\frac{s_1^3}{m^2}$$
$$+ \left\{a_0m - \frac{1}{4}(m-2)(m+3) + \frac{3}{u+mv}(a_1m^2 + a_2m + a_3)\right\}\frac{s_1}{m}$$
$$d_{0,0}^{(1)} = (u + a_3)s_3 + (v + a_2)s_1s_2 + a_1s_1^3 - \left(\frac{m+1}{2} + a_0\right)s_1$$

$$s_\alpha = \mathrm{tr}(\Theta^\alpha),\ \alpha = 1, 2, 3\ .$$

$$F_{\alpha,\beta}^{(1)}(x) = \sum_{k=0}^{\infty} e_k \Pr\{\chi^2_{f+2\alpha+2\beta+2k} \le x/w\}$$
$$e_0 = \exp\left\{-\frac{1}{2}(us_2 + vs_1^2)\right\}(2w)^{\frac{1}{2}f+\alpha+\beta}u^{\frac{1}{4}(m-1)(m+2)}(u+mv)^{\frac{1}{2}+\beta}$$
$$e_k = \frac{1}{2k}\sum_{j=0}^{k-1} G_{k-j}e_j,\ k \ge 1$$
$$G_k = \left\{\frac{1}{4}(m-1)(m+2) + 2\alpha\right\}(1-2uw)^k + (1+2\beta)\{1-2(u+mv)w\}^k$$
$$+ kw\Big[2(u+mv)^2\frac{s_1^2}{m}\{1-2(u+mv)w\}^{k-1}$$
$$+ 2u^2\left(s_2 - \frac{s_1^2}{m}\right)(1-2uw)^{k-1}\Big]$$
$$f = \frac{1}{2}m(m+1)\ .$$

(ii) Under the elliptically contoured distribution $C_m(\boldsymbol{\mu}, V)$ show that when $\Sigma = \sigma^2 I + \Theta/\sqrt{n}$ the distribution of T_2 can be expanded as follows:

$$\Pr\{T_2 \le x\} = F_0(x) + \frac{1}{\sqrt{n}}\sum_{\alpha=0}^{3} d_\alpha F_\alpha(x) + o(1/\sqrt{n})$$

where

$$d_3 = -a_3\left(s_3 - \frac{3s_1s_2}{m} + \frac{2s_1^3}{m^2}\right)$$
$$d_2 = -(u + 3a_3)\left(s_3 - \frac{3s_1s_2}{m} + \frac{2s_1^2}{m^2}\right)$$
$$d_1 = -(2u + 3a_3)s_3 + (5u + 9a_3)\frac{s_1s_2}{m} - 3(u + 2a_3)\frac{s_1^3}{m^2}$$
$$d_0 = (u + a_3)s_3 - (2u + 3a_3)\frac{s_1s_2}{m} + (u + 2a_3)\frac{s_1^3}{m^2}$$

$$F_\alpha(x) = \sum_{k=0}^{\infty} e_k \Pr\{\chi^2_{f+2\alpha+2k} \le x/w\}$$

$$e_0 = \exp\Big\{-\frac{u}{2}(s_2 - \frac{s_1^2}{m})\Big\}(2u)^{\frac{1}{4}(m-1)(m+2)+\alpha} w^{\frac{1}{4}(m-1)(m+2)+\alpha}$$

$$e_k = \frac{1}{2k}\sum_{j=0}^{k-1} G_{k-j} e_j$$

$$G_k = \Big\{\frac{1}{2}(m-1)(m+2) + 2\alpha\Big\}(1-2uw)^k + 2kwu^2(1-2uw)^{k-1}(s_2 - \frac{s_1^2}{m})$$

$$f = \frac{1}{2}(m-1)(m+2)\,.$$

(iii) **Under the elliptically contoured distribution $C_m(\boldsymbol{\mu}, V)$ with $\Sigma = I + \Theta/\sqrt{n}$, $\Theta = (\Theta_{\alpha\beta})$, $\Theta_{\alpha\alpha} = 0$, and $\Theta_{\alpha\beta} = \Theta_{\beta\alpha}$, $\alpha \neq \beta$, show that the distribution of T_3 can be expanded as follows:**

$$\Pr\{T_3/(1+\kappa) \le x\} = P_f(\delta^2) + \frac{s_3}{\sqrt{n}}\sum_{\alpha=0}^{3} b_\alpha P_{f+2\alpha}(\delta^2) + o\Big(\frac{1}{\sqrt{n}}\Big)$$

$$b_3 = -a_3, \quad b_2 = u + 3a_3, \quad b_1 = -(2u + 3a_3)$$

$$b_0 = u + a_3, \quad f = \frac{1}{2}(m^2 - \sum_{\alpha=1}^{q} m_\alpha^2), \; \delta^2 = \frac{s_2}{2(1+\kappa)}\,,$$

and $P_f(\delta^2) = \Pr\{\chi_f^2(\delta^2) \le x\}$, $\chi_f^2(\delta^2)$ is a noncentral chi-square random variable with f degrees of freedom and noncentrality parameter δ^2. Hayakawa (1986).

3.12 Consider the discrimination problem of an observation $\boldsymbol{x}$ falling into one of two populations $\Pi_1 : C_m(\boldsymbol{\mu}_1, \Lambda)$ and $\Pi_2 : C_m(\boldsymbol{\mu}_2, \Lambda)$ with probability density function

$$|\Lambda|^{-\frac{1}{2}} h((\boldsymbol{x} - \boldsymbol{\mu})'\Lambda^{-1}(\boldsymbol{x} - \boldsymbol{\mu})).$$

Consider the parameters $\boldsymbol{\theta} = (\boldsymbol{\mu}_1, \boldsymbol{\mu}_2, \Delta)$ and $\boldsymbol{\tau} = (\boldsymbol{\eta}_1, \boldsymbol{\eta}_2, \Xi)$. Let the discrimination rule be: assign $\boldsymbol{x}$ to Π_i if $(\boldsymbol{x} - \bar{\boldsymbol{\eta}})'\Xi^{-1}(\boldsymbol{\eta}_i - \boldsymbol{\eta}_j) > 0$ where $\bar{\boldsymbol{\eta}} = (\boldsymbol{\eta}_1 + \boldsymbol{\eta}_2)/2$ and $i + j = 3$, $i = 1, 2$. Let $P_i(\boldsymbol{\tau}, \boldsymbol{\theta})$ denote the probability of misclassifying $\boldsymbol{x}$ belonging to Π_i as belonging to Π_j and Q denote a distribution function with corresponding probability density function

$$q(x) = \frac{\pi^{\frac{1}{2}(m-1)}}{\Gamma(\frac{m-1}{2})}\int_0^\infty u^{\frac{1}{2}(m-3)} h(x^2 + u)\,du\,.$$

Let

$$c_i(\boldsymbol{\tau}, \boldsymbol{\theta}) = (\bar{\boldsymbol{\eta}} - \boldsymbol{\mu}_i)'\Xi^{-1}(\boldsymbol{\eta}_i - \boldsymbol{\eta}_j)/||\Lambda^{\frac{1}{2}}\Xi^{-1}(\boldsymbol{\eta}_i - \boldsymbol{\eta}_j)||\,,$$
$$i + j = 3\,.$$

(i) Show that the misclassification probability can be expressed as

$$\Pr(\boldsymbol{\tau},\boldsymbol{\theta}) = Q(c_i(\boldsymbol{\tau},\boldsymbol{\theta})), \quad i = 1,2 .$$

(Hint) Let $\boldsymbol{X} = \Lambda^{\frac{1}{2}}\boldsymbol{Y} + \boldsymbol{\mu}_i$ and show that

$$P_i(\boldsymbol{\tau},\boldsymbol{\theta}) = \Pr\{U < c_i(\boldsymbol{\tau},\boldsymbol{\theta})\},$$

where

$$U = \boldsymbol{Y}'\Lambda^{\frac{1}{2}}\Xi^{-1}(\boldsymbol{\eta}_i - \boldsymbol{\eta}_j)/||\Lambda^{-\frac{1}{2}}\Xi^{-1}(\boldsymbol{\eta}_i - \boldsymbol{\eta}_j)||.$$

Note that U is distributed as Y_1 of $\boldsymbol{Y}' = (Y_1, Y_2, \ldots, Y_m)$.

(ii) Suppose that the training samples of sizes n_1 and n_2 are drawn from populations Π_1 and Π_2, respectively. Let $\bar{\boldsymbol{x}}_i$, $i = 1,2$ denote the sample means and S_i, $i = 1,2$ denote the sample covariance matrices. The pooled sample covariance matrix is given by

$$S = \frac{1}{N-2}\{(n_1 - 1)S_1 + (n_2 - 1)S_2\}$$

where $N = n_1 + n_2$.

Let $\hat{\boldsymbol{\theta}} = (\bar{\boldsymbol{x}}_1, \bar{\boldsymbol{x}}_2, S/\alpha)$, $\alpha = -2\psi'(0)$ and $\rho_i = n_i/N$, $i = 1,2$.

(a) Show that the following equalities holds:

$$c_i(\hat{\boldsymbol{\theta}},\boldsymbol{\theta}) = -\frac{1}{2}\Delta + \frac{1}{\sqrt{N}}c_i^{(1)} + \frac{1}{N}c_i^{(2)} + O_p\Big(\frac{1}{N\sqrt{N}}\Big),$$

where

$$\Delta^2 = (\boldsymbol{\mu}_1 - \boldsymbol{\mu}_2)'\Lambda^{-1}(\boldsymbol{\mu}_1 - \boldsymbol{\mu}_2) = \boldsymbol{\xi}_1'\boldsymbol{\xi}_1, \ \boldsymbol{\xi}_1 = \Lambda^{-\frac{1}{2}}(\boldsymbol{\mu}_1 - \boldsymbol{\mu}_2)$$

$$\begin{aligned}
c_i^{(1)} &= \frac{1}{2}\Delta^{-1}\alpha^{\frac{1}{2}}\boldsymbol{\xi}_i'(Y_i/\sqrt{\rho_i} + Y_j/\sqrt{\rho_j}) \\
c_i^{(2)} &= \frac{1}{4}\Delta^{-1}\alpha(3\boldsymbol{Y}_i/\sqrt{\rho_i} + \boldsymbol{Y}_j/\sqrt{\rho_j})'(\boldsymbol{Y}_i/\sqrt{\rho_i} - \boldsymbol{Y}_j/\sqrt{\rho_j}) \\
&\quad - \Delta^{-1}\alpha^{\frac{1}{2}}\boldsymbol{\xi}_i'(\sqrt{\rho_i}Z_i + \sqrt{\rho_j}Z_j)\boldsymbol{Y}_i/\sqrt{\rho_i} \\
&\quad + \frac{1}{4}\Delta^{-1}\boldsymbol{\xi}_i'(\sqrt{\rho_i}Z_i + \sqrt{\rho_j}Z_j)^2\boldsymbol{\xi}_i \\
&\quad - \frac{1}{4}\Delta^{-3}\{\alpha^{\frac{1}{2}}\boldsymbol{\xi}_i'(\boldsymbol{Y}_i/\sqrt{\rho_i} - \boldsymbol{Y}_j/\sqrt{\rho_j}) - \boldsymbol{\xi}_i'(\sqrt{\rho_i}Z_i + \sqrt{\rho_j}Z_j)\boldsymbol{\xi}_j\} \\
&\qquad \times \{\alpha^{\frac{1}{2}}\boldsymbol{\xi}_i'(3\boldsymbol{Y}_i/\sqrt{\rho_i} + \boldsymbol{Y}_j\sqrt{\rho_j}) - \boldsymbol{\xi}_i'(\sqrt{\rho_i}Z_i + \sqrt{\rho_j}Z_j)\boldsymbol{\xi}_j\}
\end{aligned}$$

and

$$\boldsymbol{Y}_i = \sqrt{n_i}\Sigma^{-\frac{1}{2}}(\bar{\boldsymbol{X}}_i - \boldsymbol{\mu}_i)\ ,$$

$$Z_i = \sqrt{n_i}\Sigma^{-\frac{1}{2}}(S_i - \Sigma)\Sigma^{-\frac{1}{2}}\ .$$

(b) Let $[PN]_i = \sqrt{N}(P_i(\hat{\boldsymbol{\theta}}, \boldsymbol{\theta}) - P(\boldsymbol{\theta}, \boldsymbol{\theta}))$; show that

$$[PN]_i = q_1 c_i^{(1)} + \frac{1}{\sqrt{N}}\{q_1 c_i^{(2)} + \frac{1}{2} q_2 (c_i^{(1)})^2\} + O_p\left(\frac{1}{N}\right),$$

where

$$q_1 = q(-\frac{1}{2}\Delta), \quad q_2 = q'(-\frac{1}{2}\Delta).$$

(c) Show that the marginal distribution function of $[PN]_i$, $i = 1, 2$ can be expanded as

$$\Pr\{[PN]_i \le x\} = \Phi(x/v) + \frac{1}{\sqrt{N}}\phi(x/v)/v\{(x^2/v^2 - 1)b_0 - b_i\} + O\left(\frac{1}{N}\right),$$

where

$$\begin{aligned} v^2 &= \frac{1}{4}\{q(-\frac{1}{2}\Delta)\}^2 \alpha \rho^{(1)}, \quad \rho^{(1)} = 1/\rho_1 + 1/\rho_2, \\ b_0 &= -\frac{1}{8} q_2 \alpha \rho^{(2)}/\rho^{(1)}, \quad \rho^{(2)} = 1/\rho_1^2 + 1/\rho_2^2, \\ b_i &= \frac{1}{8} q_2 \alpha \rho^{(1)} + \frac{1}{4} q_1 w(3/\rho_i - 1/\rho_j)(m-1)\Delta^{-1} \\ &\quad + \frac{1}{4} q_1 (m-1)(\kappa+1)\Delta, \quad i + j = 3 \end{aligned}$$

and $\Phi(\cdot)$ and $\phi(\cdot)$ denote respectively the distribution function and the probability density function of a standard normal variable.

(d) Show that the expected misclassification probabilities can be expanded as

$$E[P_i(\hat{\boldsymbol{\theta}}, \boldsymbol{\theta})] = Q(-\frac{1}{2}\Delta) + b_i/N + O(1/N^{\frac{3}{2}})$$
$$i = 1, 2.$$

(iii) Let

$$\hat{\Delta}^2 = (\bar{\boldsymbol{X}}_1 - \bar{\boldsymbol{X}}_2)'(S/\alpha)^{-1}(\bar{\boldsymbol{X}}_1 - \bar{\boldsymbol{X}}_2).$$

(a) Show that the following equalities hold:

$$\hat{\Delta}^2 = \Delta^2 + \frac{1}{\sqrt{N}}\delta_1 + \frac{1}{N}\delta_2 + O_p(1/N^{\frac{3}{2}}),$$

$$\begin{aligned} \delta_1 &= 2\alpha^{\frac{1}{2}} \boldsymbol{\xi}_1'(\boldsymbol{Y}_1/\sqrt{\rho_1} - \boldsymbol{Y}_2/\sqrt{\rho_2}) - \boldsymbol{\xi}_1'(\sqrt{\rho_1} Z_1 + \sqrt{\rho_2} Z_2)\boldsymbol{\xi}_1, \\ \delta_2 &= \alpha(\boldsymbol{Y}_1/\sqrt{\rho_1} - \boldsymbol{Y}_2/\sqrt{\rho_2})'(\boldsymbol{Y}_1/\sqrt{\rho_1} - \boldsymbol{Y}_2/\sqrt{\rho_2}) \\ &\quad - 2\alpha^{\frac{1}{2}} \boldsymbol{\xi}_1'(\sqrt{\rho_1} Z_1 + \sqrt{\rho_2} Z_2)(\boldsymbol{Y}_1/\sqrt{\rho_1} - \boldsymbol{Y}_2/\sqrt{\rho_2}) \\ &\quad + \boldsymbol{\xi}_1'(\sqrt{\rho_1} Z_1 + \sqrt{\rho_2} Z_2)^2 \boldsymbol{\xi}_1. \end{aligned}$$

(b) An estimator of the misclassification probability $Q(-\Delta/2)$ is $Q(-\hat{\Delta}/2)$.

Show that the bias is given by

$$E[Q(-\hat{\Delta}/2)] - Q(-\Delta/2)$$

$$= \frac{1}{N}\left[q_2\left(\frac{1}{8}\rho^{(1)}\alpha + \frac{1}{32}(3\kappa+2)\Delta^2\right) - \frac{1}{4}q_1(m-1)\rho^{(1)}\alpha\Delta^{-1}\right.$$

$$\left. + q_1\left(-\frac{1}{4}(m+2)\kappa - \frac{1}{4}(m+1) + \frac{1}{16}(3\kappa+2)\right)\Delta\right] + O(1/N^{\frac{3}{2}})$$

Wakaki (1994).

CHAPTER 4

Zonal Polynomials

4.0 Introduction

The first section of this chapter is devoted to some properties of Wishart distribution. Section 4.2 gives some relations between certain symmetric functions and elementary symmetric functions. Zonal polynomials of a matrix argument are constructed in Section 4.3. Section 4.4 deals with the Laplace transform of a zonal polynomial and hypergeometric functions of matrix argument. Some weighted sums of zonal polynomials are also given in this section. Section 4.5 introduces two types of binomial coefficients which are extensions of a usual binomial coefficient. Section 4.6 constructs some special functions, Laguerre polynomials, Hermite polynomials and P-polynomials, and gives generating functions of these which are used for the derivation of the probability density function of a generalized quadratic form.

4.1 Wishart Distribution

This section deals with some basic properties of the Wishart distribution in connection with the orthogonal invariant measure and Bartlett decomposition of a Wishart matrix.

In the distribution theory of the multivariate statistical analysis the integration over the orthogonal group plays a fundamental role and the determination of the volume of the orthogonal group is important.

Let X be an $m \times n$ matrix whose elements x_{ij}, $i = 1, \ldots, m$, $j = 1, \ldots, n$ are independent real standard normal random variables. Then

$$2^{\frac{1}{2}mn}\pi^{\frac{1}{2}mn} = \prod_{i=1}^{m}\prod_{j=1}^{n}\int_{-\infty}^{\infty}\exp\left\{-\frac{1}{2}x_{ij}^2\right\} \mathrm{d}x_{ij}$$
$$= \int_X \operatorname{etr}\left(-\frac{1}{2}XX'\right) \mathrm{d}X\,, \tag{4.1.1}$$

where $\operatorname{etr}(A) = \exp(\operatorname{tr}(A))$.

Let $T = (t_{ij})$ be an $m \times m$ lower triangular matrix with positive diagonal elements and L be an $n \times m$ matrix satisfying $L'L = I_m$. If $X = TL'$, then by Theorem 1.1a.11, (4.1.1) can be expressed as

$$2^{\frac{1}{2}mn}\pi^{\frac{1}{2}mn}$$
$$= \int_T \operatorname{etr}\left(-\frac{1}{2}TT'\right)\left\{\prod_{i=1}^{m} t_{ii}^{n-i}\right\}\mathrm{d}T \int_{L'L=I_m} \mathrm{d}(\tilde{L})$$
$$= \prod_{i=1}^{m}\int_0^{\infty}\exp\left(-\frac{1}{2}t_{ii}^2\right)t_{ii}^{n-i}\mathrm{d}t_{ii}$$
$$\times\left\{\prod_{i>j}\int_{-\infty}^{\infty}\exp\left(-\frac{1}{2}t_{ij}^2\right)\mathrm{d}t_{ij}\right\}\int_{L'L=I_m}\mathrm{d}(\tilde{L})$$
$$= 2^{\frac{1}{2}mn-m}\Gamma_m\left(\frac{n}{2}\right)\int_{L'L=I_m}\mathrm{d}(\tilde{L}), \tag{4.1.2}$$

where

$$\Gamma_m(a) = \pi^{\frac{1}{4}m(m-1)}\prod_{j=1}^{m}\Gamma\left(a - \frac{1}{2}(j-1)\right). \tag{4.1.3}$$

This yields

$$\int_{L'L=I_m}\mathrm{d}(\tilde{L}) = \frac{2^m\pi^{\frac{1}{2}mn}}{\Gamma_m\left(\frac{n}{2}\right)}. \tag{4.1.4}$$

If $m = n$, L is then an orthogonal matrix H of order m and (4.1.4) becomes

$$\int_{O(m)}\mathrm{d}(\tilde{H}) = \frac{2^m\pi^{\frac{1}{2}m^2}}{\Gamma_m\left(\frac{m}{2}\right)}. \tag{4.1.5}$$

Hence we define the normalized orthogonal measures as

$$\mathrm{d}(H) = \frac{\Gamma_m\left(\frac{m}{2}\right)}{2^m\pi^{\frac{1}{2}m^2}}\mathrm{d}(\tilde{H}). \tag{4.1.6}$$

$$\mathrm{d}(L) = \frac{\Gamma_m\left(\frac{n}{2}\right)}{2^m\pi^{\frac{1}{2}mn}}\mathrm{d}(\tilde{L}). \tag{4.1.7}$$

Lemma 4.1.1 If the density function of an $m \times n$ matrix X is $f(XX')$, then the density function of $S = XX'$ is expressed as

$$\frac{\pi^{\frac{1}{2}mn}}{\Gamma_m\left(\frac{n}{2}\right)} |S|^{\frac{1}{2}(n-m-1)} f(S) \tag{4.1.8}$$

PROOF By Theorem 1.1a.12 the joint pdf of S and L is expressed as

$$f(XX')\mathrm{d}(X) = 2^{-m} |S|^{\frac{1}{2}(n-m-1)} f(S)\mathrm{d}S\mathrm{d}(\tilde{L}) \tag{4.1.9}$$

and integration of the right-hand side over $L'L = I_m$ yields, by using (4.1.4),

$$\frac{\pi^{\frac{1}{2}mn}}{\Gamma_m\left(\frac{n}{2}\right)} |S|^{\frac{1}{2}(n-m-1)} f(S)\mathrm{d}S.$$

Remark 4.1.1 If the density function of an $m \times n$ matrix X is expressed as

$$\frac{1}{(2\pi)^{\frac{1}{2}mn}|\Sigma|^{\frac{n}{2}}} \operatorname{etr}\left(-\frac{1}{2}\Sigma^{-1}XX'\right),$$

then the density function of S is given by

$$\frac{1}{\Gamma_m\left(\frac{n}{2}\right)|2\Sigma|^{\frac{n}{2}}} |S|^{\frac{1}{2}(n-m-1)} \operatorname{etr}\left(-\frac{1}{2}\Sigma^{-1}S\right). \tag{4.1.10}$$

The probability density function of S is known as a Wishart density function with n degrees of freedom and parameter matrix Σ, and is denoted as $w_m(n, \Sigma)$. The associated distribution is represented by $W_m(n, \Sigma)$.

If the pdf of an $m \times n$ matrix X has the form

$$\frac{1}{(2\pi)^{mn/2}|\Sigma|^{\frac{n}{2}}|B|^{\frac{m}{2}}} \operatorname{etr}\left(-\frac{1}{2}\Sigma^{-1}(X-M)B^{-1}(X-M)'\right), \tag{4.1.11}$$

then the associated distribution is denoted by $X \sim N_{m,n}(M, \Sigma, B)$.

Lemma 4.1.2 (Bartlett decomposition) Let S be distributed as $W_m(n, I_m)$ and put $S = TT'$ for T a lower triangular matrix with positive diagonal elements, then t_{ij} $(1 \le j < i \le n)$ is $N(0,1)$, and t_{ii}^2 $(1 \le i \le n)$ is χ^2_{n-i+1}, and all t_{ij}'s are mutually independently distributed.

PROOF From Theorem 1.1a.5, (4.1.10) is decomposed as

$$\left\{\prod_{i>j}^{m} \frac{1}{(2\pi)^{1/2}} \exp\left(-\frac{1}{2}t_{ij}^2\right)\right\} \prod_{i=1}^{m} \left\{\frac{(t_{ii}^2)^{\frac{1}{2}(n-i-1)}}{\Gamma\left(\frac{n-i+1}{2}\right)2^{\frac{1}{2}(n-i+1)}} \exp\left(-\frac{1}{2}t_{ii}^2\right)\right\}, \tag{4.1.12}$$

which establishes Lemma 4.1.2.

The detailed discussion on the invariant measures over the orthogonal group and the Stieffel manifold can be seen in James (1954). For the fundamental properties of Wishart distribution and Jacobians of matrix transformations, see Anderson (1984), Deemer and Olkin (1951), Jack (1966), Muirhead (1982), Olkin and Sampson (1972), Roy (1957), Siotani, Hayakawa and Fujikoshi (1985) and Srivastava and Khatri (1979).

4.2 Symmetric Polynomials

In multivariate statistical analysis some test statistics are expressed by using certain symmetric functions of the latent roots of symmetric matrices. To handle the distribution problem James (1961a) introduced zonal polynomials associated with a given matrix, which are certain homogeneous symmetric functions of the latent roots of the matrix. Those polynomials are constructed by certain symmetric functions such as monomial symmetric functions, elementary symmetric functions etc. This section deals with some fundamental properties of these symmetric functions. For a detailed discussion, see David, Kendall and Barton (1966) and Macdonald (1979).

Let $\kappa = (k_1, k_2, \ldots, k_m)$ be a partition of k into not more than m parts and $k = k_1 + k_2 + \cdots + k_m$, $k_1 \geq k_2 \geq \cdots \geq k_m \geq 0$.

Definition 4.2.1 *Height, weight, multiplicity and lexicographic ordering of partitions.*

$$k_1 = \text{ height of } \kappa, \qquad k = |\kappa| = \text{ weight of } \kappa. \tag{4.2.1}$$

The multiplicity ν_i of i in κ is defined as

$$\nu_i = \text{ number of } j \text{ such that } \kappa_j = i. \tag{4.2.2}$$

Then κ is written as

$$\kappa = (1^{\nu_1} 2^{\nu_2} 3^{\nu_3} \cdots), \qquad k = \nu_1 + 2\nu_2 + 3\nu_3 + \cdots + k\nu_k. \tag{4.2.3}$$

The lexicographic ordering of two partitions $\kappa = (k_1, \ldots, k_m)$ and $\tau = (t_1, \ldots, t_m)$ of k is defined as $\kappa > \tau$ if $k_1 = t_1, \ldots, k_i = t_i, k_{i+1} > t_{i+1}$ for some i. This is a total ordering.

Example 4.2.1

$$(5) > (4,1) > (3,2) > (3,1,1) > (2,2,1) > (2,1,1,1) > (1,1,1,1,1)$$

or

$$(5) > (41) > (32) > (31^2) > (2^2 1) > (21^3) > (1^5).$$

Definition 4.2.2 *Monomial symmetric function* The monomial symmetric function $m_\kappa(\alpha)$ of m elements $\alpha_1, \ldots, \alpha_m$ associated with the partition κ of k into not more than

m parts is defined as

$$m_\kappa(\alpha) = \sum \alpha_{i_1}^{k_1} \alpha_{i_2}^{k_2} \cdots \alpha_{i_m}^{k_m} \tag{4.2.4}$$

where the summation takes place over all $i_1, i_2, \ldots, i_m$ which are different.

Definition 4.2.3 *Elementary symmetric functions* The r-th elementary symmetric function is defined as

$$a_r = \sum_{1 \le i_1 < \cdots < i_r \le m} \alpha_{i_1} \alpha_{i_2} \cdots \alpha_{i_r}, \qquad r = 1, 2, \ldots, m \tag{4.2.5}$$

and for the partition κ, it is defined as

$$a_\kappa(\alpha) = a_1^{k_1 - k_2} a_2^{k_2 - k_3} \cdots a_{m-1}^{k_{m-1} - k_m} a_m^{k_m}. \tag{4.2.6}$$

The degree of $a_\kappa(\alpha)$ is

$$\begin{aligned}(k_1 - k_2) + 2(k_2 - k_3) + \cdots + (m-1)(k_{m-1} - k_m) + mk_m \\ = k_1 + k_2 + \cdots + k_m = k.\end{aligned}$$

For the partition κ define

$$s_\kappa(\alpha) = s_1^{k_1 - k_2} s_2^{k_2 - k_3} \cdots s_{m-1}^{k_{m-1} - k_m} s_m^{k_m} \tag{4.2.7}$$

where

$$s_r = \sum_{i=1}^{m} \alpha_i^r \qquad 1 \le r \le m. \tag{4.2.8}$$

Similarly, for the partition κ define

$$t_\kappa(\alpha) = s_1^{\pi_1} s_2^{\pi_2} \cdots s_m^{\pi_m} = s_{k_1} s_{k_2} \cdots s_{k_m} \tag{4.2.9}$$

where π_i is the number of i's in $k_1, k_2, \ldots, k_m$.

We order $m_\kappa(\alpha)$, $a_\kappa(\alpha)$, $s_\kappa(\alpha)$ and $t_\kappa(\alpha)$ according to the lexicographic ordering and define

$$\boldsymbol{m}^{(k)} = \begin{bmatrix} m_{(k)}(\alpha) \\ m_{(k-1,1)}(\alpha) \\ \vdots \\ m_{(1^k)}(\alpha) \end{bmatrix}, \qquad \boldsymbol{a}^{(k)} = \begin{bmatrix} a_{(k)}(\alpha) \\ a_{(k-1,1)}(\alpha) \\ \vdots \\ a_{(1^k)}(\alpha) \end{bmatrix}, \tag{4.2.10}$$

$$\boldsymbol{s}^{(k)} = \begin{bmatrix} s_{(k)}(\alpha) \\ s_{(k-1,1)}(\alpha) \\ \vdots \\ s_{(1^k)}(\alpha) \end{bmatrix} = \begin{bmatrix} s_1^k \\ s_1^{k-2} s_2 \\ \vdots \\ s_k \end{bmatrix},$$

$$t^{(k)} = \begin{bmatrix} t_{(k)}(\alpha) \\ t_{(k-1,1)}(\alpha) \\ \vdots \\ t_{(1^k)}(\alpha) \end{bmatrix} = \begin{bmatrix} s_k \\ s_{k-1}s_1 \\ \vdots \\ s_1^k \end{bmatrix}.$$

It should be noted that if $\kappa = (k_1, k_2, \ldots, k_m, k_{m+1}, \ldots, k_r)$, $r > m$ and $k_{m+1} > 0$, then

$$a_\kappa(\alpha_1, \ldots, \alpha_m) = m_\kappa(\alpha_1, \ldots, \alpha_m) = 0. \tag{4.2.11}$$

Lemma 4.2.1 There exists a non-singular upper triangular matrix $T_A^{(k)}$ with integer entries and diagonal entries equal to one such that

$$a^{(k)} = T_A^{(k)} m^{(k)}. \tag{4.2.12}$$

PROOF For a partition $\kappa = (k_1, k_2, \ldots, k_m)$

$$\begin{aligned} a_\kappa(\alpha) &= a_1^{k_1-k_2} a_2^{k_2-k_3} \cdots a_m^{k_m} \\ &= (\alpha_1 + \cdots + \alpha_m)^{k_1-k_2} (\alpha_1\alpha_2 + \cdots + \alpha_{m-1}\alpha_m)^{k_2-k_m} \\ &\qquad \cdots (\alpha_1 \cdots \alpha_m)^{k_m}. \end{aligned} \tag{4.2.13}$$

The highest order term of $a_\kappa(\alpha)$ is

$$\alpha_1^{k_1-k_2} (\alpha_1\alpha_2)^{k_2-k_3} \cdots (\alpha_1\alpha_2 \cdots \alpha_m)^{k_m} = \alpha_1^{k_1} \alpha_2^{k_2} \cdots \alpha_m^{k_m}, \tag{4.2.14}$$

which has coefficient 1. The other terms in (4.2.13) are lower in the lexicographic ordering and have integer coefficients.

Lemma 4.2.2 There exists a nonsingular lower triangular matrix $T_s^{(k)}$ such that

$$s^{(k)} = T_s^{(k)} a^{(k)} \tag{4.2.15}$$

where the diagonal element $t_{\kappa\kappa}^{(s)}$ of $T_s^{(k)}$ is expressed as

$$t_{\kappa\kappa}^{(s)} = (-1)^{k-k_1} \prod_{\ell=1}^{m} \ell^{k_\ell - k_{\ell+1}}. \tag{4.2.16}$$

PROOF The generating function of a_ℓ's is defined as

$$A(x) = \prod_{i=1}^{m} (1 + \alpha_i x) = 1 + \sum_{r=1}^{m} a_r x^r \tag{4.2.17}$$

and

$$\ln A(x) = \sum_{\ell=1}^{\infty} (-1)^{\ell-1} \frac{s_\ell}{\ell} x^\ell. \tag{4.2.18}$$

This implies

$$\ln A(x) = \sum_{\ell=1}^{\infty} \frac{(-1)^{\ell-1}}{\ell} \left\{ \sum_{r=1}^{m} a_r x^r \right\}^{\ell} . \tag{4.2.19}$$

Comparing the coefficient of x^ℓ in (4.2.17) and (4.2.18) and noting that a_ℓ is the lowest element $a_{(1^\ell)}$ in $\boldsymbol{a}^{(\ell)}$, we have for $\tau = (t_1, t_2, \ldots, t_m) > (1^\ell)$

$$(-1)^{\ell-1} \frac{s_\ell}{\ell} = a_{(1^\ell)} + \sum_{\tau > (1^\ell)} \frac{(-1)^{t_1-1}(t_1-1)!}{(t_1-t_2)! \cdots t_m!} a_\tau . \tag{4.2.20}$$

For the partition κ

$$\begin{aligned} s_\kappa &= \prod_{\ell=1}^{m} s_\ell^{k_\ell - k_{\ell+1}} \\ &= \prod_{\ell=1}^{m} \left[(-1)^{\ell-1} \ell \left\{ a_{(1^\ell)} + \sum_{\tau > (1^\ell)} \frac{(-1)^{t_1-1}(t_1-1)!}{(t_1-t_2)! \cdots t_m!} a_\tau \right\} \right]^{k_\ell - k_{\ell+1}} . \end{aligned}$$

The lowest term in s_κ is

$$\begin{aligned} \prod_{\ell=1}^{m} \left[(-1)^{\ell-1} \ell a_{(1^\ell)} \right]^{k_\ell - k_{\ell+1}} &= \left\{ \prod_{\ell=1}^{m} \left[(-1)^{\ell-1} \ell \right]^{k_\ell - k_{\ell+1}} \right\} \prod_{\ell=1}^{m} a_\ell^{k_\ell - k_{\ell+1}} \\ &= (-1)^{k-k_1} \prod_{\ell=1}^{m} \ell^{k_\ell - k_{\ell+1}} a_\kappa \end{aligned}$$

and the highest term in s_κ is simply $a_{(k)}$. The remaining terms are expressed by the terms between $a_{(k)}$ and a_κ in $\boldsymbol{a}^{(k)}$.

The following lemma can be easily proved.

Lemma 4.2.3 There exists a nonsingular lower triangular matrix T_t with diagonal entries equal one such that

$$\boldsymbol{t}^{(k)} = T_t^{(k)} \boldsymbol{m}^{(k)} . \tag{4.2.21}$$

Table 4.2.1 $T_A^{(k)}$ and $T_s^{(k)}$.

Based on the Table 2 and Table 5 of David, Kendall and Barton (1966), $T_A^{(k)}$ and $T_s^{(k)}$ are given as follows. Omitted entries in the matrices are equal to zero.

$$\boldsymbol{a}^{(k)} = T_A^{(k)} \boldsymbol{m}^{(k)} \tag{4.2.22}$$

$$T_A^{(2)} = \begin{bmatrix} 1 & 2 \\ 0 & 1 \end{bmatrix}, \qquad T_A^{(2)^{-1}} = \begin{bmatrix} 1 & -2 \\ 0 & 1 \end{bmatrix}$$

$$T_A^{(3)} = \begin{bmatrix} 1 & 3 & 6 \\ 0 & 1 & 3 \\ 0 & 0 & 1 \end{bmatrix}, \qquad T_A^{(3)^{-1}} = \begin{bmatrix} 1 & -3 & 3 \\ 0 & 1 & -3 \\ 0 & 0 & 1 \end{bmatrix}$$

$$T_A^{(4)} = \begin{bmatrix} 1 & 4 & 6 & 12 & 24 \\ & 1 & 2 & 5 & 12 \\ & & 1 & 2 & 6 \\ & & & 1 & 4 \\ & & & & 1 \end{bmatrix}, \quad T_A^{(4)^{-1}} = \begin{bmatrix} 1 & -4 & 2 & 4 & -4 \\ & 1 & -2 & -1 & 4 \\ & & 1 & -2 & 2 \\ & & & 1 & -4 \\ & & & & 1 \end{bmatrix}$$

$$T_A^{(5)} = \begin{bmatrix} 1 & 5 & 10 & 20 & 30 & 60 & 120 \\ & 1 & 3 & 7 & 12 & 27 & 60 \\ & & 1 & 2 & 5 & 12 & 30 \\ & & & 1 & 2 & 7 & 20 \\ & & & & 1 & 3 & 10 \\ & & & & & 1 & 5 \\ & & & & & & 1 \end{bmatrix}$$

$$T_A^{(5)^{-1}} = \begin{bmatrix} 1 & -5 & 5 & 5 & -5 & -5 & 5 \\ & 1 & -3 & -1 & 5 & 1 & -5 \\ & & 1 & -2 & -1 & 5 & -5 \\ & & & 1 & -2 & -1 & 5 \\ & & & & 1 & -3 & 5 \\ & & & & & 1 & -5 \\ & & & & & & 1 \end{bmatrix}.$$

$$\boldsymbol{s}^{(k)} = T_s^{(k)} \boldsymbol{a}^{(k)} \tag{4.2.23}$$

$$T_s^{(2)} = \begin{bmatrix} 1 & 0 \\ 1 & -2 \end{bmatrix}, \qquad T_s^{(2)^{-1}} = \frac{1}{2} \begin{bmatrix} 2 & 0 \\ 1 & -1 \end{bmatrix}$$

$$T_s^{(3)} = \begin{bmatrix} 1 & 0 & 0 \\ 1 & -2 & 0 \\ 1 & -3 & 3 \end{bmatrix}, \qquad T_s^{(3)^{-1}} = \frac{1}{6} \begin{bmatrix} 6 & 0 & 0 \\ 3 & -3 & 0 \\ 1 & -3 & 2 \end{bmatrix}$$

$$T_s^{(4)} = \begin{bmatrix} 1 & & & & \\ 1 & -2 & & & \\ 1 & -4 & 4 & & \\ 1 & -3 & 0 & 3 & \\ 1 & -4 & 2 & 4 & -4 \end{bmatrix}, \quad T_s^{(4)^{-1}} = \frac{1}{24} \begin{bmatrix} 24 & & & & \\ 12 & -12 & & & \\ 6 & -12 & 6 & & \\ 4 & -12 & 0 & 8 & \\ 1 & -6 & 3 & 8 & -6 \end{bmatrix}$$

$$T_s^{(5)} = \begin{bmatrix} 1 & & & & & & \\ 1 & -2 & & & & & \\ 1 & -4 & 4 & & & & \\ 1 & -3 & 0 & 3 & & & \\ 1 & -5 & 6 & 3 & -6 & & \\ 1 & -4 & 2 & 4 & 0 & -4 & \\ 1 & -5 & 5 & 5 & -5 & -5 & 5 \end{bmatrix}$$

$$T_s^{(5)^{-1}} = \frac{1}{120} \begin{bmatrix} 120 & & & & & & \\ 60 & -60 & & & & & \\ 30 & -60 & 30 & & & & \\ 20 & -60 & 0 & 40 & & & \\ 10 & -40 & 30 & 20 & -20 & & \\ 5 & -30 & 15 & 40 & 0 & -30 & \\ 1 & -10 & 15 & 20 & -20 & -30 & 24 \end{bmatrix}.$$

4.3 Zonal Polynomials

James (1961a) introduced zonal polynomials associated with a given matrix by using group representation theory. Hua (1959) has expressed the zonal polynomials as linear functions of the characters of the generalized linear group. But the coefficients involve integrals which seems to be difficult to evaluate. Recently there have been several methods proposed for the derivation of these polynomials. Saw (1977) gave many properties and an algorithm for the determination of the coefficients of zonal polynomials. Takemura (1982, 1984) gave a simple method with the help of the eigenvectors of a certain matrix and derived some fundamental properties of the polynomials. However, there is only one point where the character of the representation of a symmetric group is used. In this section we will use the approach due to Takemura to introduce zonal polynomials.

Let A be an $m \times m$ symmetric matrix and α_i, the i-th latent root of A in the descending order, $\alpha_1 \geq \alpha_2 \geq \cdots \geq \alpha_m$. Let $A_{i_1,\ldots,i_k}$ be the $k \times k$ matrix formed from A by deleting all but the $i_1, i_2, \ldots, i_k$-th rows and columns. It is well known that

$$a_{(1^k)}(A) = a_k(A) = \sum_{1 \leq i_1 < \cdots < i_k \leq m} |A_{i_1,\ldots,i_k}| \tag{4.3.1}$$

and

$$a_\kappa(A) = a_1^{k_1-k_2}(A)\, a_2^{k_2-k_3}(A) \cdots a_m^{k_m}(A). \tag{4.3.2}$$

Lemma 4.3.1 Let W be an $m \times m$ random matrix which has a Wishart distribution

$W_m(n, I_m)$, then

$$E[a_\kappa(AW)] = \sum_{\kappa \geq \tau} d_{\kappa\tau}^{(n)} a_\tau(A) \tag{4.3.3}$$

and

$$d_{\kappa\kappa}^{(n)} = 2^k \frac{\Gamma_m\left(\frac{n}{2}; \kappa\right)}{\Gamma_m\left(\frac{n}{2}\right)} = 2^k \left(\frac{n}{2}\right)_\kappa \tag{4.3.4}$$

where

$$\Gamma_m(a; \kappa) = \pi^{\frac{1}{4}m(m-1)} \prod_{\alpha=1}^{m} \Gamma\left(a + k_\alpha - \frac{1}{2}(\alpha - 1)\right) \tag{4.3.5}$$

$$(a)_\kappa = \frac{\Gamma_m(a; \kappa)}{\Gamma_m(a)} = \prod_{\alpha=1}^{m} \left(a - \frac{1}{2}(\alpha - 1)\right)_{k_\alpha} \tag{4.3.6}$$

and

$$(a)_k = a(a+1)\cdots(a+k-1).$$

PROOF Let A be a diagonal matrix denoted by $\text{diag}(\alpha_1, \ldots, \alpha_m)$, then

$$|(AW)_{i_1,i_2,\ldots,i_k}| = (\alpha_{i_1}\alpha_{i_2}\cdots\alpha_{i_k})|W_{i_1,i_2,\ldots,i_k}|. \tag{4.3.7}$$

This yields

$$a_\kappa(AW) = \left\{\sum_{i_1=1}^{m} \alpha_{i_1}|W_{i_1}|\right\}^{k_1-k_2} \left\{\sum_{i_1<i_2}^{m} \alpha_{i_1}\alpha_{i_2}|W_{i_1,i_2}|\right\}^{k_2-k_3} \cdots (\alpha_1\alpha_2\cdots\alpha_m)^{k_m}|W|^{k_m}, \tag{4.3.8}$$

and the highest monomial term with respect to $\alpha_1, \alpha_2, \ldots, \alpha_m$ is

$$\begin{aligned} &\alpha_1^{k_1-k_2}(\alpha_1\alpha_2)^{k_2-k_3}\cdots(\alpha_1\alpha_2\cdots\alpha_m)^{k_m} \\ &\quad\times |W_1|^{k_1-k_2}|W_{1,2}|^{k_2-k_3}\cdots|W|^{k_m} \\ &= \alpha_1^{k_1}\alpha_2^{k_2}\cdots\alpha_m^{k_m}|W_1|^{k_1-k_2}|W_{1,2}|^{k_2-k_3}\cdots|W|^{k_m}. \end{aligned} \tag{4.3.9}$$

In view of Lemma 4.2.1, $E[a_\kappa(AW)]$ may be expressed as a linear combination of $a_\tau(A)$ for $\tau \leq \kappa$. The coefficient of the highest term is

$$d_{\kappa\kappa}^{(n)} = E\left[|W_1|^{k_1-k_2}|W_{1,2}|^{k_2-k_3}\cdots|W|^{k_m}\right]. \tag{4.3.10}$$

Let T be an $m \times m$ lower triangular matrix with positive diagonal elements and put $W = TT'$, then

$$\begin{aligned} d_{\kappa\kappa}^{(n)} &= E\left[(t_{11}^2)^{k_1-k_2}(t_{11}^2t_{22}^2)^{k_2-k_3}\cdots(t_{11}^2t_{22}^2\cdots t_{mm}^2)^{k_m}\right] \\ &= E\left[(t_{11}^2)^{k_1}(t_{22}^2)^{k_2}\cdots(t_{mm}^2)^{k_m}\right]. \end{aligned} \tag{4.3.11}$$

Since t_{ii}^2's are independently distributed as χ^2_{n-i+1} by Lemma 4.1.2,

$$\begin{aligned} d_{\kappa\kappa}^{(n)} &= \prod_{i=1}^{m} 2^{k_i} \frac{\Gamma\left(\frac{1}{2}(n-i+1)+k_i\right)}{\Gamma\left(\frac{1}{2}(n-i+1)\right)} \\ &= 2^k \frac{\Gamma_m\left(\frac{n}{2};\kappa\right)}{\Gamma_m\left(\frac{n}{2}\right)} \\ &= 2^k \left(\frac{n}{2}\right)_\kappa . \end{aligned} \tag{4.3.12}$$

Remark 4.3.1 It seems to be difficult to determine the coefficient $d_{\kappa\tau}^{(n)}$ $(\kappa > \tau)$ in an explicit form. Equation (4.3.3) implies that there exists an upper triangular matrix $D_n^{(k)}$ with diagonal element $d_{\kappa\kappa}^{(n)}$ such that

$$E\left[a^{(k)}(AW)\right] = D_n^{(k)} a^{(k)}(A). \tag{4.3.13}$$

Lemma 4.3.2

$$D_{n_1}^{(k)} D_{n_2}^{(k)} = D_{n_2}^{(k)} D_{n_1}^{(k)} . \tag{4.3.14}$$

PROOF Let S_1 and S_2 be independently distributed as $W_m(n_1, I_m)$ and $W_m(n_2, I_m)$, respectively. Consider

$$\begin{aligned} E_{s_1,s_2}\left[a^{(k)}\left(A^{1/2}S_1A^{1/2}S_2\right)\right] &= E_{s_1}\left[D_{n_2}^{(k)} a^{(k)}\left(A^{1/2}S_1A^{1/2}\right)\right] \\ &= D_{n_2}^{(k)} E_{s_1}\left[a^{(k)}(AS_1)\right] \\ &= D_{n_2}^{(k)} D_{n_1}^{(k)} a^{(k)}(A) . \end{aligned}$$

Similarly we have

$$E_{s_1,s_2}\left[a^{(k)}\left(A^{1/2}S_2A^{1/2}S_1\right)\right] = D_{n_1}^{(k)} D_{n_2}^{(k)} a^{(k)}(A)$$

for any A. This yields (4.3.14).

Lemma 4.3.3 For the upper triangular matrix $D_n^{(k)}$, there exists an upper triangular matrix $\Xi^{(k)}$ such that

$$\Xi^{(k)} D_n^{(k)} = \Lambda_n^{(k)} \Xi^{(k)} \tag{4.3.15}$$

where $\Lambda_n^{(k)} = \mathrm{diag}(d_{(k)(k)}^{(n)}, d_{(k-1,1)(k-1,1)}^{(n)}, \ldots, d_{(1^k)(1^k)}^{(n)})$ and $\Xi^{(k)}$ is uniquely determined up to a multiplication constant for each row.

PROOF The existence and uniqueness of $\Xi^{(k)}$ are proved by induction. The remaining problem is to show that $\Xi^{(k)}$ is independent of n. For fixed N such that $d_{\kappa\kappa}^{(N)}$ are all different, let $\Lambda_N^{(k)} = \mathrm{diag}(d_{\kappa\kappa}^{(N)})$. Then for any r, $\Lambda_N^{(k)}(\Xi^{(k)} D_r^{(k)}) = (\Lambda_N^{(k)} \Xi^{(k)}) D_r^{(k)} =$

$(\Xi^{(k)}D_N^{(k)})D_r^{(k)} = \Xi^{(k)}D_r^{(k)}D_N^{(k)} = (\Xi^{(k)}D_r^{(k)})D_N^{(k)}$. By setting $\Xi_0^{(k)} = \Xi^{(k)}D_r^{(k)}$, we have

$$\Lambda_N^{(k)}\Xi_0^{(k)} = \Xi_0^{(k)}D_N^{(k)}.$$

By the uniqueness of the decomposition (4.3.15), there exists a diagonal matrix $\Lambda^{(k)}$ such that $\Xi_0^{(k)} = \Lambda^{(k)}\Xi^{(k)}$, which implies that

$$\Lambda^{(k)}\Xi^{(k)} = \Xi^{(k)}D_r^{(k)}. \tag{4.3.16}$$

Comparing (4.3.15) and (4.3.16), we have $\Lambda^{(k)} = \Lambda_r^{(k)}$. Hence for any r

$$\Xi^{(k)}D_r^{(k)} = \Lambda_r^{(k)}\Xi^{(k)}.$$

Example 4.3.1 $D_n^{(k)}$ for $k = 3$ is given as

$$D_n^{(3)} = \begin{bmatrix} n(n+2)(n+4) & -12n(n+2) & 24n \\ 0 & n(n+2)(n-1) & -6n(n-1) \\ 0 & 0 & n(n-1)(n-2) \end{bmatrix}.$$

$\Xi^{(3)}$ is obtained by (4.3.15).

$$\Xi^{(3)} = \begin{bmatrix} 5a & -12a & 8a \\ 0 & 2b & -3b \\ 0 & 0 & c \end{bmatrix},$$

where a, b and c are arbitrary constants. These are explicitly determined by introducing a new normalization of zonal polynomials.

$D_n^{(4)}$ and $\Xi^{(4)}$ are obtained as

$$D_n^{(4)} = \left[\underset{\sim}{d}_1^{(4)}, \underset{\sim}{d}_2^{(4)}, \underset{\sim}{d}_3^{(4)}, \underset{\sim}{d}_4^{(4)}, \underset{\sim}{d}_5^{(4)}\right]$$

where

$$\underset{\sim}{d}_1^{(4)} = \begin{bmatrix} n(n+2)(n+4)(n+6) \\ 0 \\ 0 \\ 0 \\ 0 \end{bmatrix}, \quad \underset{\sim}{d}_2^{(4)} = \begin{bmatrix} -24n(n+2)(n+4) \\ n(n+2)(n+4)(n-1) \\ 0 \\ 0 \\ 0 \end{bmatrix}$$

$$\underset{\sim}{d}_3^{(4)} = \begin{bmatrix} 48n(n+2) \\ -4n(n+2)(n-1) \\ n(n+2)(n-1)(n+1) \\ 0 \\ 0 \end{bmatrix}, \quad \underset{\sim}{d}_4^{(4)} = \begin{bmatrix} 96n(n+2) \\ -12n(n+2)(n-1) \\ -4n(n+2)(n-1) \\ n(n+2)(n-1)(n-2) \\ 0 \end{bmatrix}$$

$$\underset{\sim}{d}_5^{(4)} = \begin{bmatrix} -192n \\ 48n(n-1) \\ -4n(n-1)(2n-5) \\ -8n(n-1)(n-2) \\ n(n-1)(n-2)(n-3) \end{bmatrix}$$

and

$$\Xi^{(4)} = \begin{bmatrix} 35a & -120a & 48a & 96a & -64a \\ 0 & 9b & -12b & -10b & 16b \\ 0 & 0 & 3c & -4c & c \\ 0 & 0 & 0 & 5d & -8d \\ 0 & 0 & 0 & 0 & e \end{bmatrix}.$$

Definition 4.3.1 *Zonal polynomial (I)* For $\Xi^{(k)}$ defined by (4.3.15), the zonal polynomials $Y_\kappa(A)$ for a partition $\kappa = (k_1, k_2, \ldots, k_m)$ of k into not more than m parts and a symmetric matrix A of order m are defined as

$$\mathbf{Y}^{(k)} = \begin{bmatrix} Y_{(k)}(A) \\ Y_{(k-1,1)}(A) \\ \vdots \\ Y_{(1^k)}(A) \end{bmatrix} = \Xi^{(k)} \boldsymbol{a}^{(k)}(A). \tag{4.3.17}$$

It should be noted that if rank $A = r < m$, then $Y_\kappa(A) = 0$ for $\kappa = (k_1, \ldots, k_r, k_{r+1}, \ldots, k_m)$ and $k_{r+1} > 0$, in view of (4.2.11) and the upper triangularity of $\Xi^{(k)}$.

It is worth noting that the following holds since $a_\kappa(AB)$'s are symmetric functions of the latent roots of AB where A and B are positive semidefinite matrices,

$$Y_\kappa(AB) = Y_\kappa(A^{1/2}BA^{1/2}) = Y_\kappa(B^{1/2}AB^{1/2}) = Y_\kappa(BA). \tag{4.3.18}$$

Theorem 4.3.1 Let W be distributed as $W_m(n, \Sigma)$, then

$$E_W\left[\mathbf{Y}^{(k)}(AW)\right] = \Lambda_n^{(k)} \mathbf{Y}^{(k)}(A\Sigma) \tag{4.3.19}$$

that is,

$$E_W\left[Y_\kappa(AW)\right] = d_{\kappa\kappa}^{(n)} Y_\kappa(A\Sigma).$$

PROOF Let $W = \Sigma^{1/2} Z \Sigma^{1/2}$, then Z is distributed as $W_m(n, I_m)$. With the help of

Lemma 4.3.1 we have

$$\begin{aligned} E_W\left[\boldsymbol{Y}^{(k)}(AW)\right] &= \Xi^{(k)} E_W\left[\boldsymbol{a}^{(k)}(AW)\right] \\ &= \Xi^{(k)} E_Z\left[\boldsymbol{a}^{(k)}\left(A\Sigma^{1/2}Z\Sigma^{1/2}\right)\right] \\ &= \Xi^{(k)} D_n^{(k)} \boldsymbol{a}^{(k)}\left(\Sigma^{1/2}A\Sigma^{1/2}\right) \\ &= \Lambda_n^{(k)} \boldsymbol{Y}^{(k)}(A\Sigma). \end{aligned}$$

Remark 4.3.2 Let R be an $m \times m$ complex symmetric matrix whose real part $\mathcal{R}e(R)$ is positive definite and let T be an arbitrary symmetric matrix. Then

$$\int_{W>0} \operatorname{etr}(-RW)|W|^{t-\frac{1}{2}(m+1)} Y_\kappa(TW)\mathrm{d}W$$

$$= \Gamma_m(t;\kappa)|R|^{-t}\, Y_\kappa(TR^{-1}), \tag{4.3.20}$$

where $t > \dfrac{1}{2}(m-1)$. This is obtained by replacing R with $(2\Sigma)^{-1}$, T with A, and t with $\dfrac{n}{2}$ in (4.3.19). Note that (4.3.20) is the Laplace transform of $|W|^{t-\frac{1}{2}(m+1)}\, Y_\kappa(TW)$.

One of the fundamental properties is the integration of zonal polynomials over the orthogonal group.

Theorem 4.3.2 Let A and B be $m \times m$ positive definite symmetric matrices and let H be an orthogonal matrix. Then

$$\int_{O(m)} Y_\kappa(AHBH')\mathrm{d}(H) = Y_\kappa(A)Y_\kappa(B)/Y_\kappa(I_m). \tag{4.3.21}$$

PROOF Let $H_1'AH_1 = \Lambda_1 = \operatorname{diag}(\alpha_1,\ldots,\alpha_m)$ and $H_2'BH_2 = \Lambda_2 = \operatorname{diag}(\beta_1,\ldots,\beta_m)$, where H_1 and H_2 are orthogonal matrices, respectively. As $\mathrm{d}(H)$ is invariant under the translation $H \to H_1'HH_2$,

$$\begin{aligned} &\int_{O(m)} Y_\kappa(H_1\Lambda_1H_1'HH_2\Lambda_2H_2'H')\mathrm{d}(H) \\ &\quad = \int_{O(m)} Y_\kappa\left(\Lambda_1(H_1'HH_2)\Lambda_2(H_1'HH_2)'\right)\mathrm{d}(H) \\ &\quad = \int_{O(m)} Y_\kappa(\Lambda_1H\Lambda_2H')\mathrm{d}(H)\,, \end{aligned}$$

which implies that it is a function of $\alpha_1,\ldots,\alpha_m$ and $\beta_1,\ldots,\beta_m$. For any permutation matrix P which is also an orthogonal matrix, it is invariant and it is symmetric function of $\alpha_1,\ldots,\alpha_m$ and $\beta_1,\ldots,\beta_m$, simultaneously. This yields that (4.3.21) is expressed by a bilinear form of $\boldsymbol{Y}^{(k)}(A)$ and $\boldsymbol{Y}^{(k)}(B)$ as

$$\int_{O(m)} Y_\kappa(AHBH')\mathrm{d}(H) = \boldsymbol{Y}^{(k)\prime}(A)C\boldsymbol{Y}^{(k)}(B) \tag{4.3.22}$$

where $C = (c_{\lambda\nu})$ is a symmetric matrix in view of (4.3.18). Let A be distributed as $W_m(r, \Sigma)$ where r is such that the $d_{\kappa\kappa}^{(r)}$'s for all partition of k are different, that is, the diagonal matrix $\Lambda_r^{(k)}$ has different diagonal elements. Then

$$\begin{aligned} E_A\left[\int_{O(m)} Y_\kappa(AHBH')\mathrm{d}(H)\right] &= E\left[\boldsymbol{Y}^{(k)}(A)\right]' C\boldsymbol{Y}^{(k)}(B) \\ &= \boldsymbol{Y}^{(k)\prime}(\Sigma)\Lambda_r^{(k)}C\boldsymbol{Y}^{(k)}(B). \end{aligned} \tag{4.3.23}$$

On the other hand (4.3.23) can also be expressed as

$$\begin{aligned} \int_{O(m)} E_A\left[Y_\kappa(AHBH')\right]\mathrm{d}(H) &= d_{\kappa\kappa}^{(r)} \int_{O(m)} Y_\kappa(\Sigma HBH')\mathrm{d}(H) \\ &= d_{\kappa\kappa}^{(r)}\boldsymbol{Y}^{(k)}(\Sigma)'C\boldsymbol{Y}^{(k)}(B). \end{aligned}$$

This yields

$$\boldsymbol{Y}^{(k)}(\Sigma)'\left[d_{\kappa\kappa}^{(r)}C - \Lambda_r^{(k)}C\right]\boldsymbol{Y}^{(k)}(B) = 0$$

for any Σ and B. Hence

$$(d_{\kappa\kappa}^{(r)} - d_{\lambda\lambda}^{(r)})c_{\lambda\nu} = 0 \qquad \text{for all } \lambda, \nu \quad .$$

This gives

$$c_{\lambda\nu} = 0 \qquad \text{for all } \kappa \neq \lambda \text{ and } \nu\,,$$

since $d_{\kappa\kappa}^{(r)} \neq d_{\lambda\lambda}^{(r)}$ for $\kappa \neq \lambda$. The symmetry of C yields $c_{\lambda\nu} = c_{\nu\lambda}$. This means that $c_{\lambda\nu} = 0$ unless $\kappa = \lambda = \nu$. Finally we have

$$\int_{O(m)} Y_\kappa(AHBH')\mathrm{d}(H) = c_{\kappa\kappa}Y_\kappa(A)Y_\kappa(B)\,.$$

By setting $B = I_m$, we have

$$Y_\kappa(A) = c_{\kappa\kappa}Y_\kappa(A)Y_\kappa(I_m),$$

and

$$c_{\kappa\kappa} = 1/Y_\kappa(I_m).$$

This proves the theorem.

Corollary 4.3.1 Let $U = (u_{ij})$ be an $m \times m$ matrix whose elements u_{ij}'s are independent standard normal variables. Then for symmetric matrices A and B,

$$E_U[Y_\kappa(AUBU')] = \frac{2^k\left(\frac{m}{2}\right)_\kappa}{Y_\kappa(I_m)}\,Y_\kappa(A)\,Y_\kappa(B). \tag{4.3.24}$$

PROOF We decompose $U = TH$, where T is a lower triangular matrix with positive diogonal elements and H is an $m\times m$ orthogonal matrix. Then TT' is a Wishart matrix distributed as $W_m(m, I_m)$.

$$\begin{aligned}
E_U[Y_\kappa(AUBU')] &= E_T\left[\int_{O(m)} Y_\kappa(ATHBH'T')\,\mathrm{d}(H)\right] \\
&= E_T\left[\int_{O(m)} Y_\kappa(T'ATHBH')\mathrm{d}(H)\right] \\
&= \frac{Y_\kappa(B)}{Y_\kappa(I_m)}\,E_T[Y_\kappa(T'AT)] \\
&= \frac{Y_\kappa(B)}{Y_\kappa(I_m)}\,E_T[Y_\kappa(ATT')] \\
&= 2^k\left(\frac{m}{2}\right)_\kappa \frac{Y_\kappa(A)Y_\kappa(B)}{Y_\kappa(I_m)}.
\end{aligned}$$

As we have noted, that it is hard to obtain an explicit expression for $\Xi^{(k)}$ in (4.3.15) for large k; here we give a numerical method to obtain $\Xi^{(k)}$ by using a generating function of zonal polynomials $\boldsymbol{Y}^{(k)}(A) = \Xi^{(k)}\boldsymbol{a}^{(k)}(A) = \Xi^{(k)}T_s^{(k)^{-1}}\boldsymbol{s}^{(k)}(A)$, where $T_s^{(k)}$ is given in (4.2.15).

Definition 4.3.2 *Zonal polynomials (II)* Let $Z^{(k)} = (Z_{\kappa\lambda}^{(k)}) = \Xi^{(k)}T_s^{(k)^{-1}}$ be a normalized matrix whose first column elements are all unities. Define

$$\boldsymbol{Z}^{(k)}(A) = Z^{(k)}\boldsymbol{s}^{(k)}(A) \tag{4.3.25}$$

where

$$\boldsymbol{Z}^{(k)}(A) = (Z_{(k)}(A), Z_{(k-1,1)}(A), \ldots, Z_{(1^k)}(A))'\,,$$
$$\boldsymbol{s}^{(k)}(A) = (s_1^k, s_1^{k-2}s_2, \ldots, s_1^k)', \quad s_r = \mathrm{tr}(A^r), \quad r = 1, 2, \ldots, k.$$

$Z_\kappa(A)$'s enjoy all properties of Theorem 4.3.1 and Theorem 4.3.2.

As $(\mathrm{tr}\,(G))^k$ is a homogeneous symmetric polynomial of k degrees, $(\mathrm{tr}\,(G))^k$ can be expressed with suitable constants c_κ's as

$$(\mathrm{tr}(G))^k = \sum_\kappa \frac{Z_\kappa(I_m)}{2^k\left(\frac{m}{2}\right)_\kappa}c_\kappa Z_\kappa(G)\,. \tag{4.3.26}$$

Let $G = AUBU'$ where $A = \text{diag}(\alpha_1, \ldots, \alpha_m)$, $B = \text{diag}(\beta_1, \ldots, \beta_m)$ and the elements of $U = (u_{ij})$ are independent standard normal random variables. Then by using Corollary 4.3.1,

$$E_U[\{\text{tr}(AUBU')\}^k] = \sum_\kappa c_\kappa Z_\kappa(A) Z_\kappa(B)$$
$$= \boldsymbol{Z}^{(k)}(A)' C_d^{(k)} \boldsymbol{Z}^{(k)}(B) \qquad (4.3.27)$$

where

$$C_d^{(k)} = \text{diag}(c_{(k)}, c_{(k-1,1)}, \ldots, c_{(1^k)}). \qquad (4.3.28)$$

This implies that for sufficiently small t

$$E_U[\text{etr}(tAUBU')] = \sum_{k=0}^{\infty} \frac{t^k}{k!} \boldsymbol{Z}^{(k)}(A)' C_d^{(k)} \boldsymbol{Z}^{(k)}(B). \qquad (4.3.29)$$

On the other hand

$$\text{tr}(AUBU') = \sum_{i=1}^{m} \sum_{j=1}^{m} \alpha_i \beta_j u_{ij}^2 .$$

Hence for small t

$$E_U[\text{etr}(tAUBU')] = \prod_{i=1}^{m} \prod_{j=1}^{m} E_{u_{ij}} [\exp(t\alpha_i \beta_j u_{ij}^2)]$$
$$= \prod_{i=1}^{m} \prod_{j=1}^{m} (1 - 2t\alpha_i \beta_j)^{-1/2}$$
$$= \exp\left[-\frac{1}{2} \sum_{i=1}^{m} \sum_{j=1}^{m} \ln(1 - 2t\alpha_i \beta_j)\right]$$
$$= \exp\left[\frac{1}{2} \sum_{i=1}^{m} \sum_{j=1}^{m} \sum_{\ell=1}^{\infty} \frac{(2t)^\ell}{\ell} \alpha_i^\ell \beta_j^\ell\right]$$
$$= \exp\left[\frac{1}{2} \sum_{\ell=1}^{\infty} \frac{(2t)^\ell}{\ell} s_\ell(A) s_\ell(B)\right]$$
$$= \sum_{k=0}^{\infty} \frac{t^k}{k!} \sum_\kappa \lambda_\kappa^{(k)} s_\kappa(A) s_\kappa(B)$$
$$= \sum_{k=0}^{\infty} \frac{t^k}{k!} \boldsymbol{s}^{(k)}(A)' \Lambda_d^{(k)} \boldsymbol{s}^{(k)}(B) \qquad (4.3.30)$$

where

$$\Lambda_d^{(k)} = \text{diag}\left(\lambda_{(k)}^{(k)}, \lambda_{(k-1,1)}^{(k)}, \ldots, \lambda_{(1^k)}^{(k)}\right) \qquad (4.3.31)$$

and

$$\lambda_{\kappa}^{(k)} = k!2^{k-k_1} \left\{ \prod_{\ell=1}^{m} \ell^{k_\ell - k_{\ell+1}} (k_\ell - k_{\ell+1})! \right\}^{-1} .$$

Comparing the coefficients of t^k in (4.3.27) and (4.3.30), we have

$$\boldsymbol{z}^{(k)\prime}(A)\, C_d^{(k)} \boldsymbol{z}^{(k)}(B) = \boldsymbol{s}^{(k)}(A)' \Lambda_d^{(k)} \boldsymbol{s}^{(k)}(B) , \tag{4.3.32}$$

and by using (4.3.25), this yields

$$\begin{aligned} &\boldsymbol{s}^{(k)}(A)' Z^{(k)\prime} C_d^{(k)} Z^{(k)} \boldsymbol{s}^{(k)}(B) \\ &\qquad = \boldsymbol{s}^{(k)}(A)' (T_s^{(k)^{-1}})' \Xi^{(k)\prime} C_d^{(k)} \Xi^{(k)} T_s^{(k)^{-1}} \boldsymbol{s}^{(k)}(B) \\ &\qquad = \boldsymbol{s}^{(k)}(A)' \Lambda_d^{(k)} \boldsymbol{s}^{(k)}(B) . \end{aligned} \tag{4.3.33}$$

Thus we have

$$\Xi^{(k)\prime} C_d^{(k)} \Xi^{(k)} = T_s^{(k)\prime} \Lambda_d^{(k)} T_s^{(k)} . \tag{4.3.34}$$

The right-hand side of (4.3.34) can be obtained by using Tables 5 in David, Kendall and Barton (1966). $\Xi^{(k)}$ is determined up to a multiplicative constant for each row.

Example 4.3.2 Here we consider the case $k = 3$. (4.3.31) and Tables in David, Kendall and Barton (1966) give the followings.

$$\Lambda_d^{(3)} = \begin{bmatrix} 1 & & \\ & 6 & \\ & & 8 \end{bmatrix} ,$$

$$T_s^{(3)} = \begin{bmatrix} 1 & 0 & 0 \\ 1 & -2 & 0 \\ 1 & -3 & 3 \end{bmatrix} , \qquad T_s^{(3)^{-1}} = \frac{1}{6} \begin{bmatrix} 6 & 0 & 0 \\ 3 & -3 & 0 \\ 1 & -3 & 2 \end{bmatrix} .$$

This yields

$$T_s^{(k)\prime} \Lambda_d^{(3)} T_s^{(k)} = \begin{bmatrix} 15 & -36 & 24 \\ -36 & 96 & -72 \\ 24 & -72 & 72 \end{bmatrix} .$$

Here we set and partition each matrix as follows:

$$\Xi^{(3)} = \begin{bmatrix} \xi_{11} & \xi_{12} & \xi_{13} \\ 0 & \xi_{22} & \xi_{23} \\ 0 & 0 & \xi_{33} \end{bmatrix} = \begin{bmatrix} \xi_{11} & \boldsymbol{\xi}_{12}' \\ \boldsymbol{0} & \Xi_{22} \end{bmatrix} ,$$

$$\Xi_{22} = \begin{bmatrix} \xi_{22} & \xi_{23} \\ 0 & \xi_{33} \end{bmatrix} , \quad \Xi_{33} = \xi_{33} , \quad \boldsymbol{\xi}_{12}' = (\xi_{12}, \xi_{13}) ,$$

$$C_d^{(3)} = \begin{bmatrix} c_1 & & \\ & c_2 & \\ & & c_3 \end{bmatrix} = \begin{bmatrix} c_1 & \\ & C_2 \end{bmatrix}, \qquad C_2 = \begin{bmatrix} c_2 & \\ & c_3 \end{bmatrix},$$

$$R = \begin{bmatrix} 15 & -36 & 24 \\ -36 & 96 & -72 \\ 24 & -72 & 72 \end{bmatrix} = \begin{bmatrix} r_{11} & \boldsymbol{r}'_{12} \\ \boldsymbol{r}_{21} & R_{22} \end{bmatrix}, \qquad \begin{array}{l} R_{22} = \begin{bmatrix} 96 & -72 \\ -72 & 72 \end{bmatrix} \\ \boldsymbol{r}_{12} = (-36, 24). \end{array}$$

This yields

$$\Xi^{(3)\prime} C_d^{(3)} \Xi^{(3)} = \begin{bmatrix} c_1 \xi_{11}^2 & c_1 \xi_{11} \boldsymbol{\xi}'_{12} \\ c_1 \xi_{11} \boldsymbol{\xi}_{12} & c_1 \boldsymbol{\xi}_{12} \boldsymbol{\xi}'_{12} + \Xi'_{22} C_2 \Xi_{22} \end{bmatrix}.$$

By fixing ξ_{11} we have

$$c_1 = r_{11}/\xi_{11}^2 = 15/\xi_{11}^2,$$

$$\boldsymbol{\xi}'_{12} = \frac{1}{r_{11}} \boldsymbol{r}'_{12} \xi_{11} = \frac{1}{15}(-36, 24)\xi_{11},$$

$$\Xi'_{22} C_2 \Xi_{22} = R_{22} - \frac{1}{r_{11}} \boldsymbol{r}_{21} \boldsymbol{r}'_{12} \equiv R_{22\cdot 1}$$

$$= \frac{1}{5} \begin{bmatrix} 48 & -72 \\ -72 & 168 \end{bmatrix}.$$

Similarly we obtain

$$c_2 = \frac{48}{5} \frac{1}{\xi_{22}^2}, \qquad \xi_{23} = -\frac{3}{2} \xi_{22}.$$

Here we will fix $\Xi^{(3)}$ by using a normalization such that the elements of the first column vector of $\Xi^{(3)} T_s^{(k)^{-1}}$ are unity. This yields

$$\frac{1}{6}\left(6\xi_{11} + (3)\left(\frac{-36}{15}\xi_{11}\right) + \frac{24}{15}\xi_{11}\right) = 1$$

and $\xi_{11} = 15$. Thus $(\xi_{11}, \xi_{12}, \xi_{13}) = (15, -36, 24)$ and $c_1 = \dfrac{1}{15}$. Similarly

$$\frac{1}{6}\left\{3\xi_{22} + (1)\left(-\frac{3}{2}\xi_{22}\right)\right\} = 1$$

and $\xi_{22} = 4$. Thus

$$(\xi_{22}, \xi_{23}) = (4, -6) \text{ and } c_2 = \frac{9}{15}.$$

Finally

$$\frac{1}{6}(1)\xi_{33} = 1 \qquad \text{which implies} \qquad \xi_{33} = 6.$$

and

$$c_3\, \xi_{33}^2 = \frac{168}{5} - c_2\, \xi_{23}^2 = \frac{168}{5} - \left(\frac{9}{15}\right)(-6)^2 = 12$$

$$c_3 = \frac{1}{3} = \frac{5}{15}.$$

Thus we have

$$\Xi^{(3)} = \begin{bmatrix} 15 & -36 & 24 \\ 0 & 4 & -6 \\ 0 & 0 & 6 \end{bmatrix}, \qquad C_d^{(3)} = \frac{1}{15}\begin{bmatrix} 1 & & \\ & 9 & \\ & & 5 \end{bmatrix}.$$

Lemma 4.3.4 Let $\boldsymbol{\epsilon}$ be an equiangular vector $(1, 1, \ldots, 1)'$, then

$$\boldsymbol{\epsilon}' C_d^{(k)} \Xi^{(k)} T_s^{(k)^{-1}} = (1, 0, \ldots, 0). \tag{4.3.35}$$

PROOF From (4.3.34) we have

$$\begin{aligned} C_d^{(k)} \Xi^{(k)} T_s^{(k)^{-1}} &= \left(\Xi^{(k)'}\right)^{-1} T_s^{(k)'} \Lambda_d^{(k)} \\ &= \left(\left(\Xi^{(k)} T_s^{(k)^{-1}}\right)^{-1}\right)' \Lambda_d^{(k)}. \end{aligned} \tag{4.3.36}$$

This yields

$$\begin{aligned} \boldsymbol{\epsilon}' C_d^{(k)} \Xi^{(k)} T_s^{(k)^{-1}} &= \boldsymbol{\epsilon}' \left((\Xi^{(k)} T_s^{(k)^{-1}})^{-1}\right)' \Lambda_d^{(k)} \\ &= \left((\Xi^{(k)} T_s^{(k)^{-1}})^{-1} \boldsymbol{\epsilon}\right)' \Lambda_d^{(k)}. \end{aligned}$$

Noting that the first column vector of $\Xi^{(k)} T_s^{(k)^{-1}}$ is $\boldsymbol{\epsilon}$ and the top order element of $\Lambda_d^{(k)}$ is $\lambda_{(k)}^{(k)} = 1$, we have

$$\boldsymbol{\epsilon}' C_d^{(k)} \Xi^{(k)} T_s^{(k)^{-1}} = (1, 0, \ldots, 0)\, \Lambda_d^{(k)} = (1, 0, \ldots, 0).$$

Here we define a type of normalized zonal polynomials as follows.

Definition 4.3.3 *Zonal polynomial (III)* Let $\Xi^{(k)}$ and $C_d^{(k)}$ be defined by (4.3.34) under the condition that the elements of the first column vector of $\Xi^{(k)} T_s^{(k)^{-1}}$ are unities. The zonal polynomials $C_\kappa(A)$ are defined by

$$C^{(k)}(A) = C_d^{(k)} \Xi^{(k)} T_s^{(k)^{-1}} \boldsymbol{s}^{(k)}(A) \tag{4.3.37}$$

where

$$C^{(k)}(A) = (C_{(k)}(A), C_{(k-1,1)}(A), \ldots, C_{(1^k)}(A))'.$$

Remark 4.3.3 Lemma 4.3.4 suggests that

$$\sum_\kappa C_\kappa(A) = s_1^k = (\mathrm{tr}(A))^k. \tag{4.3.38}$$

Thus $C^{(k)}(A)$ corresponds to the zonal polynomials due to Constantine (1963), and $Z^{(k)}(A)$ corresponds to those due to James (1961a).

Remark 4.3.4 Noting that $C_\kappa(G) = c_\kappa Z_\kappa(G)$, from (4.3.26) and (4.3.38) we have

$$\begin{aligned}(\operatorname{tr}(G))^k &= \sum_\kappa \frac{Z_\kappa(I_m)}{2^k\left(\frac{m}{2}\right)_\kappa} c_\kappa Z_\kappa(G) \\ &= \sum_\kappa C_\kappa(G). \end{aligned} \tag{4.3.39}$$

As the $C_\kappa(G)$'s are linear independent homogeneous polynomials, this yields

$$Z_\kappa(I_m) = 2^k \left(\frac{m}{2}\right)_\kappa. \tag{4.3.40}$$

Lemma 4.3.5 (Orthogonality)

$$\sum_\lambda c_\lambda z_{\lambda\nu}^{(k)} z_{\lambda\mu}^{(k)} = \delta_{\nu\mu} z_{(k)\nu}^{(k)}, \tag{4.3.41}$$

$$\sum_\nu z_{\lambda\nu}^{(k)} z_{\mu\nu}^{(k)} / z_{(k)\nu}^{(k)} = \delta_{\lambda\mu}/c_\lambda, \tag{4.3.42}$$

where $Z^{(k)} = (z_{\lambda\nu}^{(k)})$, $C_d^{(k)} = \operatorname{diag}(c_{(k)}, \ldots, c_{(1^k)})$ and $\Lambda_d^{(k)} = \operatorname{diag}(\lambda_{(k)}^{(k)}, \ldots, \lambda_{(1^k)}^{(k)})$ are defined in (4.3.34).

PROOF Put $B = \operatorname{diag}(1, 0, \ldots, 0)$, then $s^{(k)}(B) = (1, 1, \ldots, 1)'$ and $Z_\kappa(B) = 0$ for $(k) > \kappa$. (4.3.32) yields

$$Z_{(k)}(A) c_{(k)} Z_{(k)}(B) = \sum_\nu \lambda_\nu^{(k)} s_\nu(A). \tag{4.3.43}$$

(4.3.39) yields

$$1 = (\operatorname{tr}(B))^k = c_{(k)} Z_{(k)}(B).$$

Thus

$$\begin{aligned} Z_{(k)}(A) &= \sum_\nu \lambda_\nu^{(k)} s_\nu(A) \\ &= \sum_\nu z_{(k)\nu}^{(k)} s_\nu(A). \end{aligned} \tag{4.3.44}$$

Comparing the coefficients of $s_\nu(A)$, one obtains

$$\lambda_\nu^{(k)} = z_{(k)\nu}^{(k)}. \tag{4.3.45}$$

(4.3.33) yields

$$Z^{(k)'}C_d^{(k)}Z^{(k)} = \Lambda_d^{(k)}, \tag{4.3.46}$$

and

$$Z^{(k)}\Lambda_d^{(k)^{-1}}Z^{(k)'} = C_d^{(k)^{-1}}. \tag{4.3.47}$$

With the help of (4.3.45), (4.3.46) gives (4.3.41) and (4.3.47) gives (4.3.42), respectively. Equation (4.3.47) yields

$$Z^{(k)-1} = \Lambda_d^{(k)-1}Z^{(k)'}C_d^{(k)},$$

$$\Lambda_d^{(k)} = \mathrm{diag}\left(1, z^{(k)}_{(k)(k-1,1)}, \ldots, z^{(k)}_{(k)(1^k)}\right),$$

and

$$s^{(k)}(A) = \Lambda_d^{(k)-1}Z^{(k)'}C^{(k)}(A). \tag{4.3.48}$$

Theorem 4.3.3

$$\int_{O(m)} (\mathrm{tr}(AH))^{2k+1}\mathrm{d}(H) = 0 \tag{4.3.49}$$

and

$$\int_{O(m)} (\mathrm{tr}(AH))^{2k}\mathrm{d}(H) = \sum_\kappa \frac{(2k)!}{2^k k!\left(\frac{m}{2}\right)_\kappa} C_\kappa\left(\frac{1}{2}AA'\right). \tag{4.3.50}$$

PROOF Substitute $(-I)H$ for H. Since $\mathrm{d}(H)$ is the invariant measure on $O(m)$, $\mathrm{d}((-I)H) = \mathrm{d}(H)$. This yields

$$\int_{O(m)} (\mathrm{tr}(AH))^{2k+1}\mathrm{d}(H) = \int_{O(m)} \{\mathrm{tr}(A(-I)H)\}^{2k+1}\mathrm{d}(-I)H$$
$$= -\int_{O(m)} (\mathrm{tr}(AH))^{2k+1}\mathrm{d}(H).$$

By the singular value decomposition theorem, A can be expressed as $A = H_1\Gamma H_2$ where H_1 and H_2 are orthogonal matrices, $\Gamma = \mathrm{diag}(\gamma_1, \ldots, \gamma_m)$ and $\lambda_i = \gamma_i^2$, $i = 1, 2, \ldots, m$ are the latent roots of AA'. $(\mathrm{tr}(AH))^{2k} = (\mathrm{tr}(\Gamma H_2 H H_1))^{2k}$ and $\mathrm{d}(H)$ is invariant under the translation H_2HH_1. This implies that the integration of $(\mathrm{tr}(AH))^{2k}$ over $O(m)$ yields a $2k$-th degree homogeneous symmetric polynomial of $\gamma_1, \ldots, \gamma_m$, that is,

$$\int_{O(m)} (\mathrm{tr}(AH))^{2k}\mathrm{d}(H) = \sum_\kappa b_\kappa C_\kappa(AA').$$

Let $A = \mathrm{diag}(\alpha_1, \ldots, \alpha_m)$ and let the u_{ij}'s in $U = (u_{ij})$ be independent standard normal random variables. Then $\mathrm{tr}(AU) = \sum \alpha_i u_{ii}$ is distributed as $\mathrm{N}\left(0, \sum_{i=1}^m \alpha_i^2\right)$. Hence

$$E_U\left[\{\mathrm{tr}(AU)\}^{2k}\right] = \left(\sum_{i=1}^m \alpha_i^2\right)^k \frac{(2k)!}{2^k k!}$$

$$= \frac{(2k)!}{2^k k!}(\mathrm{tr}(AA'))^k$$
$$= \frac{(2k)!}{2^k k!}\sum_{\kappa} C_\kappa(AA'). \tag{4.3.51}$$

By the decomposition $U = TH$ where T is a lower triangular matrix with positive diagonal elements and H is an orthogonal matrix

$$\begin{aligned} E_U[\{\mathrm{tr}(AU)\}^{2k}] &= E_T\left[\int_{O(m)} \{\mathrm{tr}(ATH)\}^{2k} \mathrm{d}(H)\right] \\ &= E_T[\sum_{\kappa} b_\kappa C_\kappa(ATT'A')] \\ &= \sum_{\kappa} b_\kappa 2^k \left(\frac{m}{2}\right)_\kappa C_\kappa(AA'), \end{aligned} \tag{4.3.52}$$

since TT' is a Wishart matrix distributed as $W_m(m, I)$. Comparing (4.3.51) and (4.3.52) gives

$$b_\kappa = \frac{(2k)!}{2^{2k}k!\left(\frac{m}{2}\right)_\kappa}.$$

Remark 4.3.5 $C_d^{(k)} = \mathrm{diag}(c_{(k)}, c_{(k-1,1)}, \ldots, c_{(1^k)})$ is suitably defined by (4.3.26) and it plays a fundamental role for the definition of a new normalized zonal polynomials $C^{(k)}(A)$ for a symmetric matrix A. However it is difficult to determine the exact form with this definition. James (1961a) gave these c_κ's as

$$c_\kappa = \frac{2^k k! \prod_{i<j}(2k_i - 2k_j - i + j)}{\prod_{i=1}^{\ell_\kappa}(2k_i + \ell_\kappa - i)!}, \tag{4.3.53}$$

where ℓ_κ is the length of a partition $\kappa = (k_1, \ldots, k_{\ell_\kappa})$ of k, which was obtained with the help of the theory of group representation of a symmetric group.

Zonal Polynomials

Table 4.3.1

$$\boldsymbol{Z}^{(k)} = Z^{(k)}\boldsymbol{s}^{(k)}, \qquad \boldsymbol{C}^{(k)} = C_d^{(k)}\boldsymbol{Z}^{(k)},$$

$$Z_{(1)} = s_1,$$

$$\begin{bmatrix} Z_{(2)} \\ Z_{(1^2)} \end{bmatrix} = \begin{bmatrix} 1 & 2 \\ 1 & -1 \end{bmatrix}\begin{bmatrix} s_1^2 \\ s_2 \end{bmatrix}, \qquad C_d^{(2)} = \frac{1}{3}\mathrm{diag}(1, 2),$$

$$\begin{bmatrix} Z_{(3)} \\ Z_{(21)} \\ Z_{(1^3)} \end{bmatrix} = \begin{bmatrix} 1 & 6 & 8 \\ 1 & 1 & -2 \\ 1 & -3 & 2 \end{bmatrix} \begin{bmatrix} s_1^3 \\ s_1 s_2 \\ s_3 \end{bmatrix}, \qquad C_d^{(3)} = \frac{1}{15}\mathrm{diag}(1,9,5),$$

$$\begin{bmatrix} Z_{(4)} \\ Z_{(31)} \\ Z_{(2^2)} \\ Z_{(21^2)} \\ Z_{(1^4)} \end{bmatrix} = \begin{bmatrix} 1 & 12 & 12 & 32 & 48 \\ 1 & 5 & -2 & 4 & -8 \\ 1 & 2 & 7 & -8 & -2 \\ 1 & -1 & -2 & -2 & 4 \\ 1 & -6 & 3 & 8 & -6 \end{bmatrix} \begin{bmatrix} s_1^4 \\ s_1^2 s_2 \\ s_2^2 \\ s_1 s_3 \\ s_4 \end{bmatrix},$$

$$C_d^{(4)} = \frac{1}{105}\mathrm{diag}(1,20,14,56,14),$$

$$\begin{bmatrix} Z_{(5)} \\ Z_{(41)} \\ Z_{(32)} \\ Z_{(31^2)} \\ Z_{(2^21)} \\ Z_{(21^3)} \\ Z_{(1^4)} \end{bmatrix} = \begin{bmatrix} 1 & 20 & 60 & 80 & 160 & 240 & 384 \\ 1 & 11 & 6 & 26 & -20 & 24 & -48 \\ 1 & 6 & 11 & -4 & 20 & -26 & -8 \\ 1 & 3 & -10 & 2 & -4 & -8 & 16 \\ 1 & 0 & 5 & -10 & -10 & 10 & 4 \\ 1 & -4 & -3 & 2 & 10 & 6 & -12 \\ 1 & -10 & 15 & 20 & -20 & -30 & 24 \end{bmatrix} \begin{bmatrix} s_1^5 \\ s_1^3 s_2 \\ s_1 s_2^2 \\ s_1^2 s_3 \\ s_2 s_3 \\ s_1 s_4 \\ s_5 \end{bmatrix},$$

$$C_d^{(5)} = \frac{1}{945}\mathrm{diag}(1,35,90,225,252,300,42),$$

$$\begin{bmatrix} Z_{(6)} \\ Z_{(51)} \\ Z_{(42)} \\ Z_{(41^2)} \\ Z_{(3^2)} \\ Z_{(321)} \\ Z_{(31^3)} \\ Z_{(2^3)} \\ Z_{(2^21^2)} \\ Z_{(21^4)} \\ Z_{(1^6)} \end{bmatrix} = \begin{bmatrix} 1 & 30 & 180 & 160 & 120 & 960 & 720 & 640 & 1440 & 2304 & 3840 \\ 1 & 19 & 48 & 72 & -12 & 80 & 192 & -64 & -144 & 192 & -384 \\ 1 & 12 & 27 & 16 & 30 & 24 & -18 & -8 & 108 & -144 & -48 \\ 1 & 9 & -12 & 22 & -12 & -60 & 12 & 16 & -24 & -48 & 96 \\ 1 & 9 & 33 & -8 & -27 & 120 & -78 & 136 & -114 & -48 & -24 \\ 1 & 4 & 3 & -8 & -2 & 0 & -18 & -24 & -4 & 32 & 16 \\ 1 & 0 & -21 & 4 & 6 & 12 & -6 & 16 & 12 & 24 & -48 \\ 1 & 0 & 15 & -20 & 30 & -60 & 30 & 40 & -60 & 24 & 0 \\ 1 & -3 & 3 & -8 & -9 & 0 & 24 & 4 & 24 & -24 & -12 \\ 1 & -8 & 3 & 12 & 6 & 20 & -6 & -16 & -36 & -24 & 48 \\ 1 & -15 & 45 & 40 & -15 & -120 & -90 & 40 & 90 & 144 & -120 \end{bmatrix} \begin{bmatrix} s_1^6 \\ s_1^4 s_2 \\ s_1^2 s_2^2 \\ s_1^3 s_3 \\ s_2^3 \\ s_1 s_2 s_3 \\ s_1^2 s_4 \\ s_3^2 \\ s_2 s_4 \\ s_1 s_5 \\ s_6 \end{bmatrix},$$

$$C_d^{(6)} = \frac{1}{10395}\text{diag}(1, 54, 275, 616, 132, 2673, 1925, 462, 2640, 1485, 132),$$

$$\boldsymbol{s}^{(k)} = \Lambda_d^{(k)^{-1}} Z^{(k)'} C^{(k)},$$

$$s_1 = C_{(1)},$$

$$\begin{bmatrix} s_1^2 \\ s_2 \end{bmatrix} = \frac{1}{2}\begin{bmatrix} 2 & 2 \\ 2 & -1 \end{bmatrix}\begin{bmatrix} C_{(2)} \\ C_{(1^2)} \end{bmatrix},$$

$$\begin{bmatrix} s_1^3 \\ s_1 s_2 \\ s_3 \end{bmatrix} = \frac{1}{12}\begin{bmatrix} 12 & 12 & 12 \\ 12 & 2 & -6 \\ 12 & -3 & 3 \end{bmatrix}\begin{bmatrix} C_{(3)} \\ C_{(21)} \\ C_{(1^3)} \end{bmatrix},$$

$$\begin{bmatrix} s_1^4 \\ s_1^2 s_2 \\ s_2^2 \\ s_1 s_3 \\ s_4 \end{bmatrix} = \frac{1}{48}\begin{bmatrix} 48 & 48 & 48 & 48 & 48 \\ 48 & 20 & 8 & -4 & -24 \\ 48 & -8 & 28 & -8 & 12 \\ 48 & 6 & -12 & -3 & 12 \\ 48 & -8 & -2 & 4 & -6 \end{bmatrix}\begin{bmatrix} C_{(4)} \\ C_{(31)} \\ C_{(2^2)} \\ C_{(21^2)} \\ C_{(1^4)} \end{bmatrix},$$

$$\begin{bmatrix} s_1^5 \\ s_1^3 s_2 \\ s_1 s_2^2 \\ s_1^2 s_3 \\ s_2 s_3 \\ s_1 s_4 \\ s_5 \end{bmatrix} = \frac{1}{480}\begin{bmatrix} 480 & 480 & 480 & 480 & 480 & 480 & 480 \\ 480 & 264 & 144 & 72 & 0 & -96 & -240 \\ 480 & 48 & 88 & -80 & 40 & -24 & 120 \\ 480 & 156 & -24 & 12 & -60 & 12 & 120 \\ 480 & -60 & 60 & -12 & -30 & 30 & -60 \\ 480 & 48 & -52 & -16 & 20 & 12 & -60 \\ 480 & -60 & -10 & 20 & 5 & -15 & 30 \end{bmatrix}\begin{bmatrix} C_{(5)} \\ C_{(41)} \\ C_{(32)} \\ C_{(31^2)} \\ C_{(2^2 1)} \\ C_{(21^3)} \\ C_{(1^5)} \end{bmatrix},$$

$$\begin{bmatrix} s_1^6 \\ s_1^4 s_2 \\ s_1^2 s_2^2 \\ s_1^3 s_3 \\ s_2^3 \\ s_1 s_2 s_3 \\ s_1^2 s_4 \\ s_3^2 \\ s_2 s_4 \\ s_1 s_5 \\ s_6 \end{bmatrix} = \frac{1}{2880}\left[\begin{matrix} 2880 & 2880 & 2880 & 2880 & 2880 & 2880 & 2880 \\ 2880 & 1824 & 1152 & 864 & 864 & 384 & 0 \\ 2880 & 768 & 432 & -192 & 528 & 48 & -336 \\ 2880 & 1296 & 288 & 396 & -144 & -144 & 72 \\ 2880 & -288 & 720 & -288 & -648 & -48 & 144 \\ 2880 & 240 & 72 & -180 & 360 & 0 & 36 \\ 2880 & 768 & -72 & 48 & -312 & -72 & -24 \\ 2880 & -288 & -36 & 72 & 612 & -108 & 72 \\ 2880 & -288 & 216 & -48 & -228 & -8 & 24 \\ 2880 & 240 & -180 & -60 & -60 & 40 & 30 \\ 2880 & -288 & -36 & 72 & -18 & 12 & -36 \end{matrix}\right.$$

$$\begin{bmatrix} 2880 & 2880 & 2880 & 2880 \\ 0 & -288 & -764 & -1440 \\ 240 & 48 & 48 & 720 \\ -360 & -144 & 216 & 720 \\ 720 & -216 & 144 & -360 \\ -180 & 0 & 60 & -360 \\ 120 & 96 & -24 & -360 \\ 180 & 18 & -72 & 180 \\ -120 & 48 & -72 & 180 \\ 30 & -30 & -30 & 180 \\ 0 & -9 & 36 & -90 \end{bmatrix} \begin{bmatrix} C_{(6)} \\ C_{(51)} \\ C_{(42)} \\ C_{(41^2)} \\ C_{(3^2)} \\ C_{(321)} \\ C_{(31^3)} \\ C_{(2^3)} \\ C_{(2^21^2)} \\ C_{(21^4)} \\ C_{(1^6)} \end{bmatrix}.$$

Parkhurst and James (1974) tabulated zonal polynomials up to the order 12.

Lemma 4.3.6 Let S be distributed as $W_m(n, \Sigma)$. Suppose that $f\left(\frac{S}{n}\right)$ is an analytic function around $\frac{S}{n} = \Lambda = \mathrm{diag}(\lambda_1, \ldots, \lambda_m)$ and that it is invariant for the transformation $S \to HSH'$, $H \in O(m)$. Then the expectation of $f\left(\frac{S}{n}\right)$ with respect to S is expressed as

$$E\left[f\left(\frac{S}{n}\right)\right] = \left\{1 + \frac{1}{n}\mathrm{tr}((\Lambda\partial)^2) + o\left(\frac{1}{n}\right)\right\} f(\Sigma)\Big|_{\Sigma=\Lambda}, \tag{4.3.54}$$

where ∂ is an $m \times m$ symmetric differential operator matrix defined by

$$\partial = \left(\frac{1}{2}(1 + \delta_{ij})\frac{\partial}{\partial \sigma_{ij}}\right)$$

for a symmetric matrix Σ.

PROOF The Taylor expansion of $f\left(\frac{S}{n}\right)$ at $\frac{S}{n} = \Lambda$ is expressed as

$$f\left(\frac{S}{n}\right) = \mathrm{etr}\left\{\left(\frac{S}{n} - \Lambda\right)\partial\right\} f(\Sigma)\Big|_{\Sigma=\Lambda}. \tag{4.3.55}$$

Thus

$$\begin{aligned} E\left[f\left(\frac{S}{n}\right)\right] &= E\left[\mathrm{etr}\left\{\left(\frac{S}{n} - \Lambda\right)\partial\right\}\right] f(\Sigma)\Big|_{\Sigma=\Lambda} \\ &= \mathrm{etr}(-\Lambda\partial)\left|I - \frac{2}{n}\Lambda\partial\right|^{-\frac{n}{2}} f(\Sigma)\Big|_{\Sigma=\Lambda} \\ &= \left\{1 + \frac{1}{n}\,\mathrm{tr}((\Lambda\partial)^2) + o\left(\frac{1}{n}\right)\right\} f(\Sigma)\Big|_{\Sigma=\Lambda}. \end{aligned}$$

Corollary 4.3.2 Zonal polynomial $C_\kappa(\Lambda)$ of $\Lambda = \mathrm{diag}(\lambda_1, \ldots, \lambda_m)$ corresponding to the partition $\kappa = (k_1, \ldots, k_m)$ of k satisfies the differential equation

$$\mathrm{tr}\left((\Lambda\partial)^2\right)C_\kappa(\Sigma)|_{\Sigma=\Lambda} = a_1(\kappa)C_\kappa(\Lambda), \tag{4.3.56}$$

where

$$a_1(\kappa) = \sum_{\alpha=1}^{m} k_\alpha(k_\alpha - \alpha). \tag{4.3.57}$$

PROOF Simple calculation of the left-hand side of (4.3.54) for $f(S) = C_\kappa(S)$ gives

$$\left(\frac{2}{n}\right)^k \left(\frac{n}{2}\right)_\kappa C_\kappa(\Sigma) = \left\{1 + \frac{1}{n}a_1(\kappa) + O\left(\frac{1}{n}\right)\right\} C_\kappa(\Sigma).$$

Comparing the order of $1/n$ yields (4.3.56).

Theorem 4.3.4 For the differential operators

$$E = \sum_{i=1}^{m} \lambda_i \frac{\partial}{\partial \lambda_i} \tag{4.3.58}$$

and

$$D_\Lambda = \sum_{i=1}^{m} \lambda_i^2 \frac{\partial^2}{\partial \lambda_i^2} + \sum_{i\neq j}^{m} \frac{\lambda_i^2}{\lambda_i - \lambda_j} \frac{\partial}{\partial \lambda_i}, \tag{4.3.59}$$

the zonal polynomial $C_\kappa(\Lambda)$ of $\Lambda =$ diag $(\lambda_1, \ldots, \lambda_m)$ corresponding to the partition $\kappa = (k_1, \ldots, k_m)$ of k satisfies

$$EC_\kappa(\Lambda) = kC_\kappa(\Lambda) \tag{4.3.60}$$

$$D_\Lambda C_\kappa(\Lambda) = \{a_1(\kappa) + k(m-1)\}C_\kappa(\Lambda) \tag{4.3.61}$$

D_Λ is known as the Laplace-Beltrami operator (James, 1968).

PROOF Since $C_\kappa(\Lambda)$ is a homogeneous symmetric polynomial, (4.3.60) is obvious. The differential operator satisfies the following form.

$$\frac{\partial^2}{\partial \sigma_{ij}^2} = \sum_{\alpha=1}^{m} \frac{\partial^2 \lambda_\alpha}{\partial \sigma_{ij}^2} \frac{\partial}{\partial \lambda_\alpha} + \sum_{\alpha=1}^{m}\sum_{\beta=1}^{m} \frac{\partial \lambda_\alpha}{\partial \sigma_{ij}} \frac{\partial \lambda_\beta}{\partial \sigma_{ij}} \frac{\partial^2}{\partial \lambda_\alpha \partial \lambda_\beta}.$$

By using of the derivatives in Exercise 4.16 the differential operators at $\Sigma = \Lambda$ are expressed as

$$\frac{\partial^2}{\partial \sigma_{ii}^2} = \frac{\partial^2}{\partial \lambda_i^2},$$

$$\frac{\partial^2}{\partial \sigma_{ij}^2} = \frac{2}{\lambda_i - \lambda_j}\frac{\partial}{\partial \lambda_i} + \frac{2}{\lambda_j - \lambda_i}\frac{\partial}{\partial \lambda_j}.$$

Thus we have

$$\operatorname{tr}((\Lambda\partial)^2) = \sum_{i=1}^{m} \lambda_i^2 \frac{\partial^2}{\partial \sigma_{ii}^2} + \frac{1}{4}\sum_{i\neq j} \lambda_i\lambda_j \frac{\partial^2}{\partial \sigma_{ij}^2}$$

$$= \sum_{i=1}^{m} \lambda_i^2 \frac{\partial^2}{\partial \lambda_i^2} + \sum_{i \neq j} \frac{\lambda_i \lambda_j}{\lambda_i - \lambda_j} \frac{\partial}{\partial \lambda_i} . \qquad (4.3.62)$$

With the help of (4.3.60), (4.3.62) is reduced to (4.3.61).

Theorem 4.3.5 The zonal polynomial $C_\kappa(\Lambda)$ corresponding to the partition $\kappa = (k_1, \ldots, k_m)$ of k is expressed as

$$C_\kappa(\Lambda) = \sum_{\kappa \geq \mu} c_{\kappa\mu} m_\mu(\Lambda) ,$$

where $m_\mu(\Lambda)$ is a monomial symmetric function corresponding to the partition μ of k and $c_{(k)(k)} = 1$. The coefficients $c_{\kappa\mu}$'s satisfy a recurrence relation

$$c_{\kappa\lambda} = \sum_{\kappa \geq \mu > \lambda} \frac{(\ell_i + r) - (\ell_j - r)}{a_1(\kappa) - a_1(\lambda)} c_{\kappa\mu} , \qquad (4.3.63)$$

where $\lambda = (\ell_1, \ldots, \ell_m)$ and $\mu = (\ell_1, \ldots, \ell_i + r, \ldots, \ell_j - r, \ldots, \ell_m)$ for $r = 1, 2, \ldots, \ell_j$ such that, when the partition μ is arranged in descending order, μ is above λ and below or equal to κ. The summation is over all such μ, including possibly, nondescending ones. If there does not exist such μ satisfying the condition $\kappa \geq \mu > \lambda$, we will set $c_{\kappa\lambda} = 0$.

PROOF The effect of the operator $\sum_{i=1}^{m} \lambda_i^2 \frac{\partial^2}{\partial \lambda_i^2}$ on the monomial $\lambda_1^{\ell_1} \cdots \lambda_m^{\ell_m}$ is to multiply it by

$$\sum_{i=1}^{m} \ell_i(\ell_i - 1) .$$

The effect of the operator

$$\sum_{i \neq j} \frac{\lambda_i^2}{\lambda_i - \lambda_j} \frac{\partial}{\partial \lambda_i}$$

can be seen by considering the expression

$$\begin{aligned} \frac{1}{\lambda_i - \lambda_j} & \left\{ \lambda_i^2 \frac{\partial}{\partial \lambda_i} - \lambda_j^2 \frac{\partial}{\partial \lambda_j} \right\} \left(\lambda_i^{\ell_i} \lambda_j^{\ell_j} + \lambda_i^{\ell_j} \lambda_j^{\ell_i} \right) \\ &= \ell_i \left(\lambda_i^{\ell_i} \lambda_j^{\ell_j} + \text{symmetric terms} \right) \\ &\quad + (\ell_i - \ell_j) \left(\lambda_i^{\ell_i - 1} \lambda_j^{\ell_j + 1} + \text{symmetric terms} \right) \\ &\quad + (\ell_i - \ell_j) \left(\lambda_i^{\ell_i - 2} \lambda_j^{\ell_j + 2} + \text{symmetric terms} \right) + \cdots \end{aligned}$$

This implies that the operator Δ multiplies the monomial $\lambda_1^{\ell_1} \cdots \lambda_m^{\ell_m}$ by

$$\sum_{i=1}^{m} \ell_i(m - i)$$

and adds on $(\ell_i - \ell_j)$ times the monomial

$$\lambda_1^{\ell_1} \cdots \lambda_i^{\ell_i - r} \cdots \lambda_j^{\ell_j + r} \cdots \lambda_m^{\ell_m}$$

for any such admissible monomials. Thus with the help of (4.3.61) the recurrence relation between the coefficients $c_{\kappa\mu}$'s of the monomial symmetric functions m_λ, m_μ which corresponds to the differential equation is given by (4.3.63).

Remark 4.3.6 James (1968) states that all coefficients $c_{\kappa\lambda}$'s are positive for the partitions κ and λ satisfying the condition $\kappa \geq \mu > \lambda$. Takemura (1984) notes in Section 4.5, Remark 1 that $a_1(\kappa) > a_1(\lambda)$ for $\kappa \succ \lambda$, where $\kappa \succ \lambda$ is defined if and only if $k_1 \geq \ell_1$, $k_1 + k_2 \geq \ell_1 + \ell_2, \ldots, k_1 + \cdots + k_m = \ell_1 + \cdots + \ell_m$, and he also mentions this fact in Section 4.2, Remark 2 from a different point of view. Hence the zonal polynomials are positive on the domain of positive definite symmetric matrices.

Definition 4.3.4 *Differential operator* ∂_κ Let $\kappa = (k_1, \ldots, k_m)$ be a partition of k and let $\nu = (1^{\nu_1} 2^{\nu_2} \cdots k^{\nu_k})$, $\nu_1 + 2\nu_2 + \cdots + k\nu_k = k$ be another expression of κ. For $|\nu| = \nu_1 + \nu_2 + \cdots + \nu_k$ define a differential operator ∂_κ for κ as

$$\partial_\kappa = \sum_\nu \frac{z_{\kappa\nu}^{(k)}}{z_{(k)\nu}^{(k)}} \frac{1}{\nu_1! \cdots \nu_k!} \frac{\partial^{|\nu|}}{\partial s_1^{\nu_1} \cdots \partial s_k^{\nu_k}}, \tag{4.3.64}$$

where $z_{\kappa\nu}^{(k)}$'s are the elements of $Z^{(k)}$ in (4.3.25).

Definition 4.3.5 *Coefficients* $a_{\kappa,\lambda}^\phi$ *and* $b_{\kappa,\lambda}^\phi$ Let κ and λ be partitions of k and ℓ, respectively. Define coefficients $a_{\kappa,\lambda}^\phi$ and $b_{\kappa,\lambda}^\phi$ corresponding to the partition ϕ of $f = k + \ell$ as

$$C_\kappa(S) C_\lambda(S) = \sum_\phi a_{\kappa,\lambda}^\phi C_\phi(S) \tag{4.3.65}$$

$$C_\phi(S \oplus T) = \sum_{k=0}^{f} \sum_\kappa \sum_\lambda b_{\kappa,\lambda}^\phi C_\kappa(S) C_\lambda(T), \tag{4.3.66}$$

where $\oplus$ denotes the direct sum of S and T.

Theorem 4.3.6 Denote $C_\kappa(S)$ as

$$C_\kappa(S) = \sum_\nu c_{\kappa\nu}^{(k)} s_1^{\nu_1} \cdots s_k^{\nu_k}.$$

Then the coefficients $a_{\kappa,\lambda}^\phi$ and $b_{\kappa,\lambda}^\phi$ are given as follows.

$$a_{\kappa,\lambda}^\phi = \sum_\mu \frac{z_{\phi\mu}^{(f)}}{z_{(f)\mu}^{(f)}} \sum_\nu \sum_\rho c_{\kappa\nu}^{(k)} c_{\lambda\rho}^{(\ell)}, \tag{4.3.67}$$

where ν and ρ satisfy the relation $\nu + \rho = (1^{\nu_1+\rho_1} 2^{\nu_2+\rho_2} \ldots)$.

$$b^{\phi}_{\kappa,\lambda} = \sum_{\mu}\sum_{\nu}\sum_{\rho} \frac{\mu_1!\mu_2!\cdots}{\nu_1!\nu_2!\cdots\rho_1!\rho_2!\cdots} \frac{z^{(k)}_{\kappa\nu}\, z^{(f-k)}_{\lambda\rho}}{z^{(k)}_{(k)\nu}\, z^{(f-k)}_{(f-k)\rho}} c^{(f)}_{\phi\mu}, \tag{4.3.68}$$

where $\mu = \nu + \rho$.

PROOF Applying the differential operator ∂_ϕ corresponding to the partition ϕ of $f = k + \ell$ to both sides of (4.3.65) and noting the orthogonality relations (Lemma 4.3.5), we have

$$\begin{aligned} a^{\phi}_{\kappa,\lambda} &= \sum_{\mu} \frac{z^{(f)}_{\phi\mu}}{z^{(f)}_{(f)\mu}} \frac{1}{\mu_1!\mu_2!\cdots\mu_f!} \frac{\partial^{|\mu|}}{\partial s_1^{\mu_1}\cdots\partial s_f^{\mu_f}} \\ &\quad \times \left(\sum_{\nu} c^{(k)}_{\kappa\nu} s_1^{\nu_1}\cdots s_k^{\nu_k}\right)\left(\sum_{\rho} c^{(\ell)}_{\lambda\rho} s_1^{\rho_1}\cdots s_\ell^{\rho_\ell}\right) \\ &= \sum_{\mu} \frac{z^{(f)}_{\phi\mu}}{z^{(f)}_{(f)\mu}} \sum_{\nu}\sum_{\rho} c^{(k)}_{\kappa\nu} c^{(\ell)}_{\lambda_\rho}, \end{aligned}$$

where the last sum with respect to μ is the one for $\mu = \nu + \rho = (1^{\nu_1+\rho_1}2^{\nu_2+\rho_2}\ldots)$.

The proof of (4.3.68) is left as an exercise (Exercise 4.15). The coefficients $a^{\phi}_{\kappa,\lambda}$ and $b^{\phi}_{\kappa,\lambda}$ are respectively given in Tables 4.3.2(a) and 4.3.2(b) at the end of this chapter.

Constantine (1963) discovers the remarkable reproductivity property (4.3.20) of the zonal polynomial under expectation taken with respect to the Wishart distribution. There are several ways to approach zonal polynomials. Kushner and Meisner (1980), Kushner, Lebow and Meisner (1981) and Kushner and Meisner (1984) consider those by using the idea of eigenfunctions of expected value operator. Farrel (1985) gives detailed discussion of these polynomials from the view point of tensor representation of the general linear group. Muirhead (1982) uses the partial differential equation technique by using the Laplace-Beltrami operator due to James (1968). Takeuchi and Takemura (1985) give their coefficients by using the relationship algebra of a partially balanced incomplete block design. Richards (1979) studies zonal polynomial for the special case of a 3×3 matrix.

Theorem 4.3.4 and Theorem 4.3.5 are given by James (1968). The differential operator (4.3.56) is given by Sugiura and Fujikoshi (1969) and some related differential operators are found in Sugiura (1973). The differential operator ∂_κ is introduced by Richards (1982a, 1982b), and the coefficients $a^{\phi}_{\kappa,\lambda}$ and $b^{\phi}_{\kappa,\lambda}$ are introduced by Hayakawa (1967) and Khatri and Pillai (1968). Other representations are given by Davis (1979).

Zonal polynomials can be extended to the case of Hermitian matrices. For references in this area, see James (1964), Khatri (1966), Fujikoshi (1971), Farrel (1980) and Conradie and Troskie (1982).

4.4 Laplace Transform and Hypergeometric Function

The Laplace transform plays the fundamental role in the derivation of certain formulas involving positive definite symmetric matrices.

Definition 4.4.1 *Laplace transform* Let $f(S)$ be a function of the positive definite symmetric $m \times m$ matrix S, the Laplace transform of $f(S)$ is defined to be

$$g(Z) = \int_{S>0} \operatorname{etr}(-ZS) f(S) \mathrm{d}S, \tag{4.4.1}$$

where $Z = X + iY$ is a complex symmetric matrix, $\mathcal{R}e(Z) = X$, Y is real, and it is assumed that the integral converges in the half-plane $\mathcal{R}e(Z) = X > X_0$ for some positive definite matrix X_0. If this is so, $g(Z)$ is an analytic function of Z in the half-plane.

If $g(Z)$ satisfies the conditions

$$\int |g(X+iY)|\, \mathrm{d}Y < \infty \tag{4.4.2}$$

and

$$\lim_{X\to\infty} \int |g(X+iY)|\, \mathrm{d}Y = 0, \tag{4.4.3}$$

then the inverse Laplace transform is given by

$$f(S) = \frac{2^{\frac{1}{2}m(m-1)}}{(2\pi i)^{\frac{1}{2}m(m+1)}} \int_{\mathcal{R}e(Z)=X_0>0} \operatorname{etr}(XZ) g(Z) \mathrm{d}Z. \tag{4.4.4}$$

Lemma 4.4.1 *(Convolution theorem)* If $g_1(Z)$ and $g_2(Z)$ are the Laplace transforms of $f_1(S)$ and $f_2(S)$, then $g_1(Z)\, g_2(Z)$ is the Laplace transform of

$$f(R) = \int_{0<S<R} f_1(S) f_2(R-S) \mathrm{d}S. \tag{4.4.5}$$

PROOF The Laplace transform of $f(R)$ is given by

$$\int_{R>0} \operatorname{etr}(-ZR) f(R) \mathrm{d}R = \int_{R>0} \operatorname{etr}(-ZR) \int_{0<S<R} f_1(S) f_2(R-S) \mathrm{d}S \mathrm{d}R.$$

By setting $R - S = U$ the above integral becomes

$$\int_{S>0} \int_{U>0} \operatorname{etr}(-Z(S+U)) f_1(S) f_2(U) \mathrm{d}S \mathrm{d}U$$

$$= \int_{S>0} \operatorname{etr}(-ZS) f_1(S) \mathrm{d}S \int_{U>0} \operatorname{etr}(-ZU) f_2(U) \mathrm{d}U$$
$$= g_1(Z)\, g_2(Z)\,.$$

Theorem 4.4.1 Let R be a complex symmetric matrix whose real part is positive, and let T be an arbitrary complex symmetric matrix. Then

$$\int_{S>0} \operatorname{etr}(-RS)\, |S|^{t-\frac{1}{2}(m+1)}\, C_\kappa(ST) \mathrm{d}S$$
$$= \Gamma_m(t;\kappa) |R|^{-t}\, C_\kappa(TR^{-1})\,. \tag{4.4.6}$$

The integration is valid for all complex number t which satisfies $\mathcal{R}e(t) > \frac{1}{2}(m-1)$. The corresponding inverse transform is given by

$$\frac{2^{\frac{1}{2}m(m-1)}}{(2\pi i)^{\frac{1}{2}m(m+1)}} \int_{\mathcal{R}e(Z)=X_0>0} \operatorname{etr}(SZ)\, |Z|^{-t} C_\kappa(Z) \mathrm{d}Z$$
$$= \frac{1}{\Gamma_m(t;\kappa)} |S|^{t-\frac{1}{2}(m+1)} C_\kappa(S)\,. \tag{4.4.7}$$

PROOF Equation (4.4.6) is essentially the same as (4.3.20). To prove (4.4.7), it has to be shown that $|Z|^{-t} C_\kappa(Z^{-1})$ satisfies (4.4.2) and (4.4.3). To do this note that since $\mathcal{R}e(Z) > 0$, $C_\kappa(Z^{-1})$ is bounded so that it is sufficient to verify the results for $|Z|^{-t}$. This was done by Herz (1955).

Theorem 4.4.2 If R is a positive definite $m \times m$ matrix, then

$$\int_{I>S>0} |S|^{t-\frac{1}{2}(m+1)}\, |I-S|^{u-\frac{1}{2}(m+1)} C_\kappa(RS) \mathrm{d}S$$
$$= \frac{\Gamma_m(t;\kappa)\Gamma_m(u)}{\Gamma_m(t+u;\kappa)} C_\kappa(R) \tag{4.4.8}$$

PROOF The left-hand side of (4.4.8) is a symmetric function $F(R)$ of R, so that the integrand of (4.4.8) is invariant under the translation $S \to HSH'$, where H is an $m \times m$ orthogonal matrix. The integration over $O(m)$ gives by using Theorem 4.3.2

$$F(R) = (F(I)/C_\kappa(I_m)) C_\kappa(R)\,.$$

Making the transformation $S = R^{-1/2} T R^{-1/2}$ yields

$$F(R)\, |R|^{t+u-\frac{1}{2}(m+1)}$$
$$= \int_{0<T<R} |T|^{t-\frac{1}{2}(m+1)}\, |R-T|^{u-\frac{1}{2}(m+1)} C_\kappa(T) \mathrm{d}T\,. \tag{4.4.9}$$

Taking the Laplace transform of both sides of (4.4.9) and using the convolution theorem

(Lemma 4.4.1) one gets

$$\int_{R>0} \operatorname{etr}(-RZ)\frac{F(I)}{C_\kappa(I_m)}C_\kappa(R)\,|R|^{t+u-\frac{1}{2}(m+1)}dR$$

$$= \frac{F(I)}{C_\kappa(I)}\Gamma_m(t+u;\kappa)\,|Z|^{-t-u}C_\kappa(Z^{-1})$$

$$= \int_{T>0} \operatorname{etr}(-ZT)\,|T|^{t-\frac{1}{2}(m+1)}C_\kappa(T)dT$$

$$\times \int_{T>0} \operatorname{etr}(-ZT)|T|^{u-\frac{1}{2}(m+1)}dT$$

$$= \Gamma_m(t;\kappa)\,|Z|^{-t}C_\kappa(Z^{-1})\Gamma_m(u)\,|Z|^{-u}.$$

This gives

$$\frac{F(I)}{C_\kappa(I)} = \frac{\Gamma_m(t;\kappa)\Gamma_m(u)}{\Gamma_m(t+u;\kappa)},$$

which completes the proof.

Corollary 4.4.1

$$\int_{0<S<\Omega} |S|^{t-\frac{1}{2}(m+1)}C_\kappa(WS)dS$$

$$= \frac{\Gamma_m(t;\kappa)\Gamma_m\left(\frac{m+1}{2}\right)}{\Gamma_m\left(t+\frac{m+1}{2};\kappa\right)}\,|\Omega|^t C_\kappa(W\Omega). \tag{4.4.10}$$

PROOF Setting $u = \frac{1}{2}(m+1)$ in (4.4.8) yields

$$\int_{0<T<I} |T|^{t-\frac{1}{2}(m+1)}C_\kappa(RT)dT = \frac{\Gamma_m(t;\kappa)\Gamma_m\left(\frac{m+1}{2}\right)}{\Gamma_m\left(t+\frac{m+1}{2};\kappa\right)}C_\kappa(R).$$

Putting $T = \Omega^{-1/2}S\Omega^{-1/2}$ and $R = \Omega^{1/2}W\Omega^{1/2}$, we have (4.4.10).

The classical hypergeometric function is defined as

$${}_pF_q(a_1,\ldots,a_p;b_1,\ldots,b_q;z) = \sum_{k=0}^{\infty}\frac{(a_1)_k\cdots(a_p)_k}{(b_1)_k\cdots(b_q)_k}\frac{z^k}{k!} \tag{4.4.11}$$

where $(a)_k = a(a+1)\cdots(a+k-1)$ and z is a complex argument. It is well known that the series converges for all finite z if $p \le q$, it converges for $|z| < 1$ and diverges for $|z| > 1$ if $p = q+1$, and it diverges for all $z \neq 0$ if $p > q+1$.

The hypergeometric function of matrix argument is considered below.

Definition 4.4.2 *Hypergeometric function* The hypergeometric function of matrix argument is defined as

$$ {}_pF_q(a_1,\ldots,a_p;b_1,\ldots,b_q;Z)=\sum_{k=0}^{\infty}\sum_{\kappa}\frac{(a_1)_\kappa\cdots(a_p)_\kappa}{(b_1)_\kappa\cdots(b_q)_\kappa}\frac{C_\kappa(Z)}{k!} \tag{4.4.12} $$

where

$$ (a)_\kappa=\prod_{j=1}^{m}\left(a-\frac{1}{2}(j-1)\right)_{k_j}=\Gamma_m(a;\kappa)/\Gamma_m(a) \tag{4.4.13} $$

and Z is a complex symmetric $m\times m$ matrix.

The parameters a_i and b_j are arbitrary complex numbers so long as none of the b_j's is an integer or half-integer less than or equal to $\frac{1}{2}(m-1)$. If one of the a_i is a negative integer, say $a_1=-n$, then for $k\geq mn+1$ $(a_1)_\kappa=(-n)_\kappa=0$ and the function is reduced to a polynomial of degree mn.

Example 4.4.1

$$ {}_0F_0(Z)=\sum_{k=0}^{\infty}\sum_{\kappa}\frac{C_\kappa(Z)}{k!}=\sum_{k=0}^{\infty}\frac{1}{k!}(\mathrm{tr}(Z))^k=\mathrm{etr}\,(Z) \tag{4.4.14} $$

This series converges for all Z.

First suppose that Z is a real positive definite matrix. By using (4.4.6) and (4.4.14),

$$ \begin{aligned} {}_1F_0(a;Z)&=\sum_{k=0}^{\infty}\sum_{\kappa}\frac{(a)_\kappa}{k!}C_\kappa(Z)\\ &=\frac{1}{\Gamma_m(a)}|Z|^{-a}\int_{S>0}\mathrm{etr}\,(-SZ^{-1})\,|S|^{a-\frac{1}{2}(m+1)}\,\mathrm{etr}\,(S)\mathrm{d}S\\ &=\frac{|Z|^{-a}}{\Gamma_m(a)}\int_{S>0}\mathrm{etr}\,\left\{-(Z^{-1}-I)S\right\}\,|S|^{a-\frac{1}{2}(m+1)}\mathrm{d}S\,. \end{aligned} $$

For the integrability of the right-hand side it is required that $Z^{-1}>I$, that is, $I>Z$, which implies $1>\ell_1$ where ℓ_1 is the largest of the latent roots of Z. Thus

$$ {}_1F_0(a;Z)=|Z|^{-a}\,|Z^{-1}-I|^{-a}=|I-Z|^{-a}\,. \tag{4.4.15} $$

Secondly, without loss of generality we assume $Z=\mathrm{diag}(\ell_1,\ldots,\ell_r,\ \ell_{r+1},\ldots,\ell_m)$ where ℓ_i's are latent roots of Z satisfying $\ell_1>\cdots>\ell_r>0>\ell_{r+1}>\cdots>\ell_m$. Let $Z=D_ID_L$, where $D_I=\mathrm{diag}(I_r,-I_{m-r})$ and $D_L=\mathrm{diag}(\ell_1,\ldots,\ell_r,|\ell_{r+1}|,\ldots,|\ell_m|)$. By noting

$$ |C_\kappa(Z)|\leq C_\kappa(D_L)\,, $$

we have $I>D_L$ by a similar way, which implies $1>\parallel Z\parallel$ where $\parallel Z\parallel$ is the maximum of the absolute value of the latent roots of Z.

The analytic continuation yields the extension to the case of a complex symmetric matrix with $\mathcal{R}e(Z)>0$.

Let X be an $m \times m$ matrix and let H be an $m \times m$ orthogonal matrix; then by using Theorem 4.3.3 we have

$$\begin{aligned}\int_{O(m)} \operatorname{etr}(XH)d(H) &= \sum_{k=0}^{\infty}\frac{1}{k!}\int_{O(m)}(\operatorname{tr}(XH))^k d(H)\\ &= \sum_{k=0}^{\infty}\frac{1}{(2k)!}\sum_{\kappa}\frac{(2k)!}{2^k k!\left(\frac{m}{2}\right)_{\kappa}}C_{\kappa}\left(\frac{1}{2}XX'\right)\\ &= \sum_{k=0}^{\infty}\sum_{\kappa}\frac{1}{k!\left(\frac{m}{2}\right)_{\kappa}}C_{\kappa}\left(\frac{1}{4}XX'\right)\\ &= {}_0F_1\left(\frac{m}{2};\frac{1}{4}XX'\right).\end{aligned} \tag{4.4.16}$$

${}_0F_1$ is called the Bessel function and converges for all X.

Theorem 4.4.3

(1) If $p \leq q$, then (4.4.12) converges for all Z.

(2) If $p = q + 1$, then (4.4.12) converges for $\| Z \| < 1$.

(3) If $p > q + 1$, then (4.4.12) diverges unless it terminates.

PROOF Since $C_{\kappa}(S)$ is expressed by a linear combination of monomial symmetric functions $m_{\kappa}(S)$ of latent roots of S with positive coefficients (Remark 4.3.6), we have

$$C_{\kappa}(S) \leq C_{\kappa}(T) \qquad \text{for} \quad S < T\,.$$

This yields

$$|C_{\kappa}(S)| \leq x^k C_{\kappa}(I_m) \leq x^k \sum_{\kappa} C_{\kappa}(I_m) = x^k m^k\,,$$

when $\| S \| < x$.

Let

$$\begin{aligned} a_{i\alpha} &= a_i - \frac{1}{2}(\alpha - 1)\,, \quad i = 1,2,\ldots,p\,,\\ & \qquad\qquad\qquad\qquad j = 1,2,\ldots,q\\ b_{j\alpha} &= b_j - \frac{1}{2}(\alpha - 1)\,, \quad \alpha = 1,2,\ldots,m\end{aligned}$$

and set

$$\begin{aligned}&{}_pF_q^{(\alpha)}(a_{1\alpha},\ldots,a_{p\alpha};\, b_{1\alpha},\ldots,b_{q\alpha};\, x)\\ &\quad = \sum_{k_\alpha=0}^{\infty}\frac{(a_{1\alpha})_{k_\alpha}(a_{2\alpha})_{k_\alpha}\cdots(a_{p\alpha})_{k_\alpha}}{(b_{1\alpha})_{k_\alpha}(b_{2\alpha})_{k_\alpha}\cdots(b_{q\alpha})_{k_\alpha}}\frac{x^{k_\alpha}}{k_\alpha!}\,.\end{aligned}$$

It should be noted that

$$k! \geq k_1!\,k_2!\cdots k_m!$$

for $k = k_1 + k_2 + \cdots + k_m$, $k_1 \geq k_2 \geq \cdots \geq k_m \geq 0$. Thus

$$\begin{aligned}
&\sum_{k=0}^{\infty}\sum_{\kappa}\left|\frac{(a_1)_\kappa \cdots (a_p)_\kappa}{(b_1)_\kappa \cdots (b_q)_\kappa}\frac{C_\kappa(S)}{k!}\right| \\
&\qquad \leq \sum_{k=0}^{\infty}\sum_{\kappa}\left|\frac{(a_1)_\kappa \cdots (a_p)_\kappa}{(b_1)_\kappa \cdots (b_q)_\kappa}\right| \frac{(mx)^k}{k_1!\,k_2!\cdots k_m!} \\
&\qquad \leq \prod_{\alpha=1}^{m}\left\{\sum_{k_\alpha=0}^{\infty}\left|\frac{(a_{1\alpha})_{k_\alpha}\cdots(a_{p\alpha})_{k_\alpha}}{(b_{1\alpha})_{k_\alpha}\cdots(b_{q\alpha})_{k_\alpha}}\right|\frac{(mx)^k}{k_\alpha!}\right\}.
\end{aligned}$$

This implies that the series converges for all S if $p \leq q$. On the other hand

$$\begin{aligned}
&\sum_{k=0}^{\infty}\sum_{\kappa}\left|\frac{(a_1)_\kappa(a_2)_\kappa \cdots (a_p)_\kappa}{(b_1)_\kappa(b_2)_\kappa \cdots (b_q)_\kappa}\right|\frac{|C_\kappa(xI_m)|}{k!} \\
&\qquad \geq \sum_{k=0}^{\infty}\left|\frac{(a_1)_k(a_2)_k \cdots (a_p)_k}{(b_1)_k(b_2)_k \cdots (b_q)_k}\right|\frac{|x|^k|C_{(k)}(I_m)|}{k!} \\
&\qquad = \sum_{k=0}^{\infty}\left|\frac{(a_1)_k \cdots (a_p)_k}{(b_1)_k \cdots (b_q)_k}\right|\frac{|x|^k\left(\frac{m}{2}\right)_k}{\left(\frac{1}{2}\right)_k k!}.
\end{aligned}$$

This implies that the divergence of ${}_pF_q(a_1, \ldots, a_p;\ b_1, \ldots, b_q;\ xI_m)$ depends on the divergence of ${}_{p+1}F_{q+1}(a_1, \ldots, a_p,\ \frac{m}{2};\ b_1, \ldots, b_q, \frac{1}{2};\ x)$. Thus when $p > q$, the series diverges for all $Z \neq O$. The convergence of ${}_{p+1}F_p$ on $\| Z \| < 1$ is proved by induction.

Set

$$\begin{aligned}
&G_p\,(a_1, \ldots, a_p; b_1, \ldots, b_{p-1}; Z) \\
&\qquad = \left(\prod_{j=1}^{p-1}\Gamma_m(b_j)\right)^{-1} {}_pF_{p-1}(a_1, \ldots, a_p; b_1, \ldots, b_{p-1}; Z).
\end{aligned}$$

Let $p = 1$, then

$$G_1(a_1; Z) = {}_1F_0(a_1; Z) = |I - Z|^{-a_1},$$

which converges for $\| Z \| < 1$. By the induction hypothesis we assume that $G_p(a_1, \ldots, a_p; b_1, \ldots, b_{p-1}; Z)$ converges for $\| Z \| < 1$. By using (4.4.8) for $b_p > a_{p+1} + \frac{1}{2}(m-1) > m-1$,

$$\begin{aligned}
&G_{p+1}\,(a_1, \ldots, a_p, a_{p+1}; b_1, \ldots, b_p; Z) \\
&\qquad = \frac{1}{\Gamma_m(a_{p+1})\Gamma_m(b_p - a_{p+1})}\int_{0<X<I}|X|^{a_{p+1}-\frac{1}{2}(m+1)} \\
&\qquad \times |I - X|^{b_p - a_{p+1} - \frac{1}{2}(m+1)} G_p(a_1, \ldots, a_p; b_1, \ldots, b_{p-1}; XZ)\mathrm{d}X.
\end{aligned}$$

The existence of the integral on the right-hand side is clear. This implies that the hypergeometric function ${}_{p+1}F_p$ converges for $\| Z \| < 1$.

Using (4.4.6) and (4.4.7) and integrating the series term by term, it follows that

$$\begin{aligned} {}_{p+1}F_q\,(a_1,\dots,a_p,c\,;b_1,\dots,b_q;Z^{-1})\,|Z|^{-c} \\ = \frac{1}{\Gamma_m(c)}\int_{S>0}\operatorname{etr}(-SZ)\,|S|^{c-\frac{1}{2}(m+1)} \\ \times\ {}_pF_q(a_1,\dots,a_p;b_1,\dots,b_q;S)\mathrm{d}S\,; \end{aligned} \tag{4.4.17}$$

$$\begin{aligned} {}_pF_{q+1}\,(a_1,\dots,a_p;b_1,\dots,b_q,c\,;S)\,|S|^{c-\frac{1}{2}(m+1)} \\ = \Gamma_m(c)\frac{2^{\frac{1}{2}m(m-1)}}{(2\pi i)^{\frac{1}{2}m(m+1)}}\int_{\mathcal{R}e(Z)=X_0>0}\operatorname{etr}(SZ) \\ \times\ |Z|^{-c}\,{}_pF_q(a_1,\dots,a_p;b_1,\dots,b_q;Z^{-1})\mathrm{d}Z\,. \end{aligned} \tag{4.4.18}$$

Some integral representations of the hypergeometric functions for low p and q are given below.

Lemma 4.4.2

$$\begin{aligned} {}_1F_1(a;b;S) = \frac{\Gamma_m(b)}{\Gamma_m(a)\Gamma_m(b-a)}\int_{0<T<I_m}\operatorname{etr}(ST) \\ \times\ |T|^{a-\frac{1}{2}(m+1)}\,|I-T|^{c-a-\frac{1}{2}(m+1)}\mathrm{d}T\,, \end{aligned} \tag{4.4.19}$$

for $\mathcal{R}e(a) > \frac{1}{2}(m-1)$, $\mathcal{R}e(c) > \mathcal{R}e(a) + \frac{1}{2}(m-1) > m-1$;

$$\begin{aligned} {}_2F_1(a,b;c;S) = \frac{\Gamma_m(c)}{\Gamma_m(a)\Gamma_m(c-a)}\int_{0<T<I}|I-ST|^{-b} \\ \times\ |T|^{a-\frac{1}{2}(m+1)}\,|I-T|^{c-a-\frac{1}{2}(m+1)}\mathrm{d}T\,, \end{aligned} \tag{4.4.20}$$

for $\| S \| < 1$, $\mathcal{R}e(a) > \frac{1}{2}(m-1)$ and $\mathcal{R}e(c-a) > \frac{1}{2}(m-1)$.

PROOF (4.4.19) is proved by expanding $\operatorname{etr}(ST)$ and using (4.4.8) while (4.4.20) is proved by expanding $|I-ST|^{-b}$ and using (4.4.8).

Lemma 4.4.3 (Kummer formula)

$${}_1F_1\,(a;b;S) = \operatorname{etr}(S)\,{}_1F_1(b-a;b\,;-S)\,; \tag{4.4.21}$$

$$\begin{aligned} {}_2F_1\,(a_1,a_2;b;S) \\ = |I-S|^{-a_2}\,{}_2F_1(b-a_1,a_2;b\,;-S(I-S)^{-1}) \\ = |I-S|^{b-a_1-a_2}\,{}_2F_1(b-a_1,b-a_2;b\,;S)\,. \end{aligned} \tag{4.4.22}$$

PROOF The integral representation of ${}_1F_1$ gives

$$\operatorname{etr}(S)\,{}_1F_1(b-a;b\,;-S) = \operatorname{etr}(S)\frac{\Gamma_m(b)}{\Gamma_m(b-a)\Gamma_m(a)}$$

$$\times \int_{0<T<I} \operatorname{etr}(-ST)\,|T|^{b-a-\frac{1}{2}(m+1)}\,|I-T|^{a-\frac{1}{2}(m+1)}\mathrm{d}T\,.$$

Letting $I-T=U$ yields

$$\frac{\Gamma_m(b)}{\Gamma_m(b-a)\Gamma_m(a)}\int_{0<U<I} \operatorname{etr}(SU)\,|U|^{a-\frac{1}{2}(m+1)}\,|I-U|^{b-a-\frac{1}{2}(m+1)}\mathrm{d}U$$
$$= {}_1F_1(a;b;S)\,.$$

Similarly by using (4.4.20),

$$|I-S|^{-a_2}\,{}_2F_1(b-a_1,a_2;b;-S(I-S)^{-1})$$
$$= |I-S|^{-a_2}\frac{\Gamma_m(b)}{\Gamma_m(b-a_1)\Gamma_m(a_1)}\int_{0<T<I}|I+S(I-S)^{-1}T|^{-a_2}$$
$$\times\,|T|^{\,b-a_1-\frac{1}{2}(m+1)}\,|I-T|^{\,a_1-\frac{1}{2}(m+1)}\mathrm{d}T\,.$$

Noting that

$$|I+S(I-S)^{-1}T|^{-a_2} = |I-S|^{a_2}\,|I-S+ST|^{-a_2}$$
$$= |I-S|^{a_2}\,|I-S(I-T)|^{-a_2}$$

and letting $I-T=U$, we have

$$\frac{\Gamma_m(b)}{\Gamma_m(b-a_1)\Gamma_m(a_1)}\int_{0<U<I}|I-SU|^{-a_2}\,|U|^{\,a_1-\frac{1}{2}(m+1)}$$
$$\times\,|I-U|^{\,b-a_1-\frac{1}{2}(m+1)}\mathrm{d}U = {}_2F_1(a_1,a_2;b;S)\,.$$

The second equality in (4.4.22) is obtained by replacing the step from the end of the proof.

Remark 4.4.1 Multiplying both side of (4.4.21) by $|S|^{\,b-\frac{1}{2}(m+1)}$ and taking the Laplace transform we have $\Gamma_m(b)\,|Z|^{\,-b}\,{}_1F_0(a;Z^{-1})$ on both sides, which implies that (4.4.21) holds.

Lemma 4.4.4 Let S be a $m\times m$ Wishart matrix with $W_m(n,\Sigma)$. Put $v_j = \operatorname{tr}((SA)^j)$ and $\sigma_j = \operatorname{tr}((A\Sigma)^j)$. Let

$$\boldsymbol{v}^{(1)} = v_1\,,\quad \boldsymbol{v}^{(2)\prime} = (v_1^2, v_2),\quad \boldsymbol{v}^{(3)\prime} = (v_1^3, v_1v_2, v_3),$$
$$\boldsymbol{v}^{(4)\prime} = (v_1^4, v_1^2v_2, v_2^2, v_1v_3, v_4),$$
$$\boldsymbol{v}^{(5)\prime} = (v_1^5, v_1^3v_2, v_1v_2^2, v_1^2v_3, v_2v_3, v_1v_4, v_5),$$
$$\boldsymbol{v}^{(6)\prime} = (v_1^6, v_1^4v_2, v_1^2v_2^2, v_2^3, v_1^3v_3, v_1v_2v_3, v_3^2, v_1^2v_4, v_2v_4, v_1v_5, v_6)$$

and

$$\boldsymbol{\sigma}^{(1)} = \sigma_1\,,\quad \boldsymbol{\sigma}^{(2)\prime} = (\sigma_1^2,\sigma_2),\quad \boldsymbol{\sigma}^{(3)\prime} = (\sigma_1^3,\sigma_1\sigma_2,\sigma_3),$$
$$\boldsymbol{\sigma}^{(4)\prime} = (\sigma_1^4,\sigma_1^2\sigma_2,\sigma_2^2,\sigma_1\sigma_3,\sigma_4),$$

$$\boldsymbol{\sigma}^{(5)\prime} = (\sigma_1^5, \sigma_1^3\sigma_2, \sigma_1\sigma_2^2, \sigma_1^2\sigma_3, \sigma_2\sigma_3, \sigma_1\sigma_4, \sigma_5),$$
$$\boldsymbol{\sigma}^{(6)\prime} = (\sigma_1^6, \sigma_1^4\sigma_2, \sigma_1^2\sigma_2^2, \sigma_2^3, \sigma_1^3\sigma_3, \sigma_1\sigma_2\sigma_3, \sigma_3^2, \sigma_1^2\sigma_4, \sigma_2\sigma_4, \sigma_1\sigma_5, \sigma_6).$$

Then the following equalities hold.

$$E[\boldsymbol{v}^{(1)}] = n\boldsymbol{\sigma}^{(1)}, \tag{4.4.23}$$
$$E[\boldsymbol{v}^{(2)}] = (n^2 I_2 + nA_1^{(2)})\boldsymbol{\sigma}^{(2)}, \tag{4.4.24}$$
$$E[\boldsymbol{v}^{(3)}] = (n^3 I_3 + n^2 A_2^{(3)} + nA_1^{(3)})\boldsymbol{\sigma}^{(3)}, \tag{4.4.25}$$
$$E[\boldsymbol{v}^{(4)}] = (n^4 I_5 + n^3 A_3^{(4)} + n^2 A_2^{(4)} + nA_1^{(4)})\boldsymbol{\sigma}^{(4)}, \tag{4.4.26}$$
$$E[\boldsymbol{v}^{(5)}] = (n^5 I_7 + n^4 A_4^{(5)} + n^3 A_3^{(5)} + n^2 A_2^{(5)} + nA_1^{(5)})\boldsymbol{\sigma}^{(5)}, \tag{4.4.27}$$
$$E[\boldsymbol{v}^{(6)}] = (n^6 I_{11} + n^5 A_5^{(6)} + n^4 A_4^{(6)} + n^3 A_3^{(6)} + n^2 A_2^{(6)} + nA_1^{(6)})\boldsymbol{\sigma}^{(6)} \tag{4.4.28}$$

where

$$A_1^{(2)} = \begin{bmatrix} 0 & 2 \\ 1 & 1 \end{bmatrix}, \quad A_2^{(3)} = \begin{bmatrix} 0 & 6 & 0 \\ 1 & 1 & 4 \\ 0 & 3 & 3 \end{bmatrix}, \quad A_1^{(3)} = \begin{bmatrix} 0 & 0 & 8 \\ 0 & 4 & 4 \\ 1 & 3 & 4 \end{bmatrix},$$

$$A_3^{(4)} = \begin{bmatrix} 0 & 12 & 0 & 0 & 0 \\ 1 & 1 & 2 & 8 & 0 \\ 0 & 2 & 2 & 0 & 8 \\ 0 & 3 & 0 & 3 & 6 \\ 0 & 0 & 2 & 4 & 6 \end{bmatrix}, \quad A_2^{(4)} = \begin{bmatrix} 0 & 0 & 12 & 32 & 0 \\ 0 & 10 & 2 & 8 & 24 \\ 1 & 2 & 5 & 16 & 20 \\ 1 & 3 & 6 & 16 & 18 \\ 0 & 6 & 5 & 12 & 21 \end{bmatrix},$$

$$A_1^{(4)} = \begin{bmatrix} 0 & 0 & 0 & 0 & 48 \\ 0 & 0 & 8 & 16 & 24 \\ 0 & 8 & 4 & 16 & 20 \\ 0 & 6 & 6 & 12 & 24 \\ 1 & 6 & 5 & 16 & 20 \end{bmatrix},$$

$$A_4^{(5)} = \begin{bmatrix} 0 & 20 & 0 & 0 & 0 & 0 & 0 \\ 1 & 1 & 6 & 12 & 0 & 0 & 0 \\ 0 & 2 & 2 & 0 & 8 & 8 & 0 \\ 0 & 3 & 0 & 3 & 2 & 12 & 0 \\ 0 & 0 & 3 & 1 & 4 & 0 & 12 \\ 0 & 0 & 2 & 4 & 0 & 6 & 8 \\ 0 & 0 & 0 & 0 & 5 & 5 & 10 \end{bmatrix},$$

$$A_3^{(5)} = \begin{bmatrix} 0 & 0 & 60 & 80 & 0 & 0 & 0 \\ 0 & 18 & 6 & 12 & 32 & 72 & 0 \\ 1 & 2 & 13 & 24 & 16 & 20 & 64 \\ 1 & 3 & 18 & 28 & 6 & 36 & 48 \\ 0 & 4 & 6 & 3 & 31 & 36 & 60 \\ 0 & 6 & 5 & 12 & 24 & 45 & 48 \\ 0 & 0 & 10 & 10 & 25 & 30 & 65 \end{bmatrix},$$

$$A_2^{(5)} = \begin{bmatrix} 0 & 0 & 0 & 0 & 160 & 240 & 0 \\ 0 & 0 & 48 & 56 & 32 & 72 & 192 \\ 0 & 16 & 12 & 24 & 72 & 116 & 160 \\ 0 & 14 & 18 & 24 & 80 & 120 & 144 \\ 1 & 4 & 27 & 40 & 64 & 96 & 168 \\ 1 & 6 & 29 & 40 & 64 & 92 & 168 \\ 0 & 10 & 25 & 30 & 70 & 105 & 160 \end{bmatrix},$$

$$A_1^{(5)} = \begin{bmatrix} 0 & 0 & 0 & 0 & 0 & 0 & 384 \\ 0 & 0 & 0 & 0 & 96 & 96 & 192 \\ 0 & 0 & 32 & 32 & 64 & 96 & 160 \\ 0 & 0 & 24 & 24 & 72 & 72 & 192 \\ 0 & 12 & 24 & 36 & 60 & 108 & 144 \\ 0 & 8 & 24 & 24 & 72 & 96 & 160 \\ 1 & 10 & 25 & 40 & 60 & 100 & 148 \end{bmatrix},$$

$$A_5^{(6)} = \begin{bmatrix} 0 & 30 & 0 & 0 & 0 & 0 & 0 & 0 & 0 & 0 & 0 \\ 1 & 1 & 12 & 0 & 16 & 0 & 0 & 0 & 0 & 0 & 0 \\ 0 & 2 & 2 & 2 & 0 & 16 & 0 & 8 & 0 & 0 & 0 \\ 0 & 0 & 3 & 3 & 0 & 0 & 0 & 0 & 24 & 0 & 0 \\ 0 & 3 & 0 & 0 & 3 & 6 & 0 & 18 & 0 & 0 & 0 \\ 0 & 0 & 3 & 0 & 1 & 4 & 4 & 0 & 6 & 12 & 0 \\ 0 & 0 & 0 & 0 & 0 & 6 & 6 & 0 & 0 & 0 & 18 \\ 0 & 0 & 2 & 0 & 4 & 0 & 0 & 6 & 2 & 16 & 0 \\ 0 & 0 & 0 & 2 & 0 & 4 & 0 & 1 & 7 & 0 & 16 \\ 0 & 0 & 0 & 0 & 0 & 5 & 0 & 5 & 0 & 10 & 10 \\ 0 & 0 & 0 & 0 & 0 & 0 & 3 & 0 & 6 & 6 & 15 \end{bmatrix},$$

$$A_4^{(6)} = \begin{bmatrix} 0 & 0 & 180 & 0 & 160 & 0 & 0 & 0 & 0 & 0 & 0 \\ 0 & 28 & 12 & 12 & 16 & 128 & 0 & 144 & 0 & 0 & 0 \\ 1 & 2 & 25 & 4 & 32 & 32 & 32 & 20 & 64 & 128 & 0 \\ 0 & 3 & 6 & 15 & 0 & 48 & 0 & 24 & 84 & 0 & 160 \\ 1 & 3 & 36 & 0 & 40 & 18 & 8 & 54 & 36 & 144 & 0 \\ 0 & 4 & 6 & 6 & 3 & 59 & 16 & 42 & 24 & 60 & 120 \\ 0 & 0 & 9 & 0 & 2 & 24 & 35 & 0 & 54 & 72 & 144 \\ 0 & 6 & 5 & 4 & 12 & 56 & 0 & 69 & 12 & 96 & 80 \\ 0 & 0 & 8 & 7 & 4 & 16 & 24 & 6 & 75 & 64 & 136 \\ 0 & 0 & 10 & 0 & 10 & 25 & 20 & 30 & 40 & 105 & 100 \\ 0 & 0 & 0 & 5 & 0 & 30 & 24 & 15 & 51 & 60 & 155 \end{bmatrix},$$

$$A_3^{(6)} = \begin{bmatrix} 0 & 0 & 0 & 120 & 0 & 960 & 0 & 720 & 0 & 0 & 0 \\ 0 & 0 & 156 & 12 & 128 & 128 & 128 & 144 & 336 & 768 & 0 \\ 0 & 26 & 24 & 26 & 32 & 272 & 64 & 260 & 136 & 320 & 640 \\ 1 & 3 & 39 & 25 & 48 & 96 & 128 & 60 & 312 & 384 & 704 \\ 0 & 24 & 36 & 36 & 36 & 336 & 24 & 288 & 108 & 432 & 480 \\ 1 & 4 & 51 & 12 & 56 & 124 & 124 & 114 & 258 & 456 & 600 \\ 0 & 6 & 18 & 24 & 6 & 186 & 114 & 108 & 252 & 360 & 726 \\ 1 & 6 & 65 & 10 & 64 & 152 & 96 & 164 & 234 & 528 & 480 \\ 0 & 7 & 17 & 26 & 12 & 172 & 112 & 117 & 265 & 336 & 736 \\ 0 & 10 & 25 & 20 & 30 & 190 & 100 & 165 & 210 & 400 & 650 \\ 0 & 0 & 30 & 22 & 20 & 150 & 121 & 90 & 276 & 390 & 701 \end{bmatrix},$$

$$A_2^{(6)} = \begin{bmatrix} 0 & 0 & 0 & 0 & 0 & 0 & 640 & 0 & 1440 & 2304 & 0 \\ 0 & 0 & 0 & 96 & 0 & 704 & 128 & 432 & 336 & 768 & 1920 \\ 0 & 0 & 128 & 24 & 96 & 256 & 288 & 240 & 664 & 1088 & 1600 \\ 0 & 24 & 36 & 44 & 48 & 432 & 288 & 348 & 540 & 960 & 1664 \\ 0 & 0 & 108 & 36 & 80 & 312 & 320 & 216 & 720 & 1152 & 1440 \\ 0 & 22 & 48 & 54 & 52 & 436 & 256 & 348 & 576 & 912 & 1680 \\ 1 & 6 & 81 & 54 & 80 & 384 & 268 & 288 & 630 & 1008 & 1584 \\ 0 & 18 & 60 & 58 & 48 & 464 & 256 & 336 & 568 & 896 & 1680 \\ 1 & 7 & 83 & 45 & 80 & 384 & 280 & 284 & 604 & 1008 & 1608 \\ 1 & 10 & 85 & 50 & 80 & 380 & 280 & 280 & 630 & 988 & 1600 \\ 0 & 15 & 75 & 52 & 60 & 420 & 264 & 315 & 603 & 960 & 1620 \end{bmatrix},$$

$$A_1^{(6)} = \begin{bmatrix} 0 & 0 & 0 & 0 & 0 & 0 & 0 & 0 & 0 & 0 & 3840 \\ 0 & 0 & 0 & 0 & 0 & 0 & 384 & 0 & 768 & 768 & 1920 \\ 0 & 0 & 0 & 64 & 0 & 384 & 256 & 192 & 576 & 768 & 1600 \\ 0 & 0 & 96 & 32 & 64 & 384 & 224 & 288 & 480 & 960 & 1312 \\ 0 & 0 & 0 & 48 & 0 & 288 & 288 & 144 & 576 & 576 & 1920 \\ 0 & 0 & 72 & 48 & 48 & 336 & 240 & 216 & 576 & 864 & 1440 \\ 0 & 18 & 72 & 42 & 72 & 360 & 216 & 324 & 504 & 864 & 1368 \\ 0 & 0 & 48 & 48 & 32 & 288 & 288 & 144 & 624 & 768 & 1600 \\ 0 & 16 & 72 & 40 & 64 & 384 & 224 & 312 & 488 & 896 & 1344 \\ 0 & 10 & 60 & 50 & 40 & 360 & 240 & 240 & 560 & 800 & 1480 \\ 1 & 15 & 75 & 41 & 80 & 360 & 228 & 300 & 504 & 888 & 1348 \end{bmatrix},$$

These are given by Fujikoshi (1973).

Remark 4.4.2 The sum of the elements of any row of $A_\ell^{(k)}$ is equal to the coefficient of n^ℓ of $2^k\left(\frac{n}{2}\right)_k = n(n+2)\cdots(n+2k-2)$.

PROOF By using (4.3.37) we have

$$\begin{aligned} E_s[v^{(k)}] &= Z^{(k)^{-1}} C_d^{(k)^{-1}} E_s[C^{(k)}(SA)] \\ &= Z^{(k)^{-1}} C_d^{(k)^{-1}} \Lambda_n^{(k)} C^{(k)}(A\Sigma^{-1}) \\ &= Z^{(k)^{-1}} C_d^{(k)^{-1}} \Lambda_n^{(k)} C_d^{(k)} Z^{(k)} \sigma^{(k)} \\ &= Z^{(k)^{-1}} \Lambda_n^{(k)} Z^{(k)} \sigma^{(k)}, \end{aligned}$$

where

$$\begin{aligned} \Lambda_n^{(k)} &= \operatorname{diag}\left(d_{(k)(k)}^{(n)}, d_{(k-1,1)(k-1,1)}^{(n)}, \ldots, d_{(1^k)(1^k)}^{(n)}\right) \\ &= \operatorname{diag}\left(2^k\left(\frac{n}{2}\right)_k, 2^k\left(\frac{n}{2}\right)_{(k-1,1)}, \ldots, 2^k\left(\frac{n}{2}\right)_{(1^k)}\right). \end{aligned}$$

The sum of the elements of the κ-th row of $Z^{(k)^{-1}}\Lambda_n^{(k)} Z^{(k)}$ is

$$\sum_\nu \left(\sum_\mu z_{(k)}^{\kappa\mu} d_{\mu\mu}^{(n)} z_{\mu\nu}^{(k)}\right) = \sum_\mu z_{(k)}^{\kappa\mu} d_{\mu\mu}^{(n)} \left(\sum_\nu z_{\mu\nu}^{(k)}\right), \tag{4.4.29}$$

where $Z^{(k)^{-1}} = (z_{(k)}^{\kappa\mu})$.

The orthogonality relation (4.3.41) of $Z^{(k)}$ yields the following equality by putting $z_{\lambda\nu}^{(k)} = z_{(k)\nu}^{(k)}$ (=1) that

$$\sum_\nu z_{\mu\nu}^{(k)} = \delta_{(k)\mu}/c_{(k)}.$$

Hence we have

$$\sum_\mu z_{(k)}^{\kappa\mu} d_{\mu\mu}^{(n)} \delta_{(k)\mu}/c_{(k)} = z_{(k)}^{\kappa(k)} d_{(k)(k)}^{(n)}/c_{(k)}.$$

As

$$Z^{(k)^{-1}} = \Lambda_d^{(k)^{-1}} Z^{(k)'} C_d^{(k)}$$

and

$$\Lambda_d^{(k)^{-1}} = \text{diag}\left(1, 1/z^{(k)}_{(k)(k-1,1)}, \ldots, 1/z^{(k)}_{(k)(1^k)}\right),$$

the element of the first column of $Z^{(k)^{-1}}$ is

$$z^{\kappa(k)}_{(k)} = c_{(k)}.$$

This yields

$$z^{\kappa(k)}_{(k)} d^{(n)}_{(k)(k)} / c_{(k)} = d^{(n)}_{(k)(k)} = 2^k \left(\frac{n}{2}\right)_k,$$

which completes the proof.

Here we shall consider some weighted sums of zonal polynomials which are useful for handling the asymptotic distribution theory of multivariate statistical analysis. The asymptotic formula for the gamma function due to Barnes (1899) is as follows.

$$\begin{aligned}\ln\Gamma(z+h) = {} & \left(z+h-\frac{1}{2}\right)\ln z - z + \frac{1}{2}\ln(2\pi) \\ & - \sum_{r=1}^{m} \frac{(-1)^r B_{r+1}(h)}{r(r+1)z^r} + O\left(1/z^{m+1}\right)\end{aligned} \tag{4.4.30}$$

where $|\arg z| < \pi$ and $B_r(h)$ is the Bernoulli polynomial of degree r; for example

$$\begin{aligned} & B_0(h) = 1, \quad B_1(h) = h - \frac{1}{2}, \quad B_2(h) = h^2 - h + \frac{1}{6}, \\ & B_3(h) = h^3 - \frac{3}{2}h^2 + \frac{1}{2}h, \quad B_4(h) = h^4 - 2h^3 + h^2 - \frac{1}{30}, \\ & B_5(h) = h^5 - \frac{5}{2}h^4 + \frac{5}{3}h^3 - \frac{1}{6}h. \end{aligned}$$

By using (4.4.30),

$$\begin{aligned}(n)_\kappa = n^k \Bigg[& 1 + \frac{a_1(\kappa)}{2n} + \frac{1}{24n^2}\{3a_1^2(\kappa) - a_2(\kappa) + k\} \\ & + \frac{1}{48n^3}\{a_1^3(\kappa) - a_1(\kappa)(a_2(\kappa) - k + 2) + 2a_3(\kappa)\} \\ & + \frac{1}{5760n^4}\{120a_4(\kappa) + 120a_3(\kappa)a_1(\kappa) \\ & \quad + 5a_2^2(\kappa) - 30a_1^2(\kappa)a_2(\kappa) \\ & \quad + 15a_1^4(\kappa) - 36a_3(\kappa) - 120a_1^2(\kappa) \\ & - 10ka_2(\kappa) + 30ka_1^2(\kappa) + 5k^2 - 84k\} + o\left(\frac{1}{n^4}\right)\Bigg]\end{aligned} \tag{4.4.31}$$

where

$$a_1(\kappa) = \sum_{\alpha=1}^{m} k_\alpha(k_\alpha - \alpha), \quad a_2(\kappa) = \sum_{\alpha=1}^{m} k_\alpha(4k_\alpha^2 - 6k_\alpha\alpha + 3\alpha^2),$$

$$a_3(\kappa) = \sum_{\alpha=1}^{m} k_\alpha(2k_\alpha^3 - 4k_\alpha^2\alpha + 3k_\alpha\alpha^2 - \alpha^3),$$

$$a_4(\kappa) = \sum_{\alpha=1}^{m} k_\alpha(16k_\alpha^4 - 40k_\alpha^3\alpha + 40k_\alpha^2\alpha^2 - 20k_\alpha\alpha^3 + 5\alpha^4).$$

Lemma 4.4.5 Let Z be a symmetric matrix and put $s_j = \operatorname{tr}(Z^j)$. Then the following identities hold:

$$\sum_{k=\ell}^{\infty}\sum_{\kappa} C_\kappa(Z)/(k-\ell)! = s_1^\ell \operatorname{etr}(Z), \tag{4.4.32}$$

$$\sum_{k=0}^{\infty}\sum_{\kappa} a_1(\kappa)C_\kappa(Z)/k! = s_2 \operatorname{etr}(Z), \tag{4.4.33}$$

$$\sum_{k=0}^{\infty}\sum_{\kappa} a_1^2(\kappa)C_\kappa(Z)/k! = \{s_1^2 + s_2 + 4s_3 + s_2^2\} \operatorname{etr}(Z), \tag{4.4.34}$$

$$\sum_{k=0}^{\infty}\sum_{\kappa} a_2(\kappa)C_\kappa(Z)/k! = \{s_1 + 3s_1^2 + 3s_2 + 4s_3\} \operatorname{etr}(Z), \tag{4.4.35}$$

$$\sum_{k=0}^{\infty}\sum_{\kappa} a_3(\kappa)C_\kappa(Z)/k! = \{s_1^2 + 4s_2 + 4s_1s_2 + 4s_3 + 2s_4\} \operatorname{etr}(Z), \tag{4.4.36}$$

$$\sum_{k=0}^{\infty}\sum_{\kappa} a_1(\kappa)a_2(\kappa)C_\kappa(Z)/k! = \{3s_1^2 + 11s_2 + 25s_1s_2 + 24s_3 + 3s_1^2s_2 + 3s_2^2 + 24s_4 + 4s_2s_3\} \operatorname{etr}(Z), \tag{4.4.37}$$

$$\sum_{k=0}^{\infty}\sum_{\kappa} a_1^3(\kappa)C_\kappa(Z)/k! = \{s_1^2 + 3s_2 + 16s_1s_2 + 16s_3 + 3s_1^2s_2 + 3s_2^2 + 32s_4 + 12s_2s_3 + s_2^3\} \operatorname{etr}(Z), \tag{4.4.38}$$

$$\sum_{k=0}^{\infty}\sum_{\kappa} a_4(\kappa)C_\kappa(Z)/k! = \{s_1 + 25s_1^2 + 35s_2 + \frac{40}{3}s_1^3 + 60s_1s_2 + \frac{380}{3}s_3 + 20s_2^2 + 40s_1s_3 + 60s_4 + 16s_5\} \operatorname{etr}(Z), \tag{4.4.39}$$

$$\sum_{k=0}^{\infty}\sum_{\kappa} a_3(\kappa)a_1(\kappa)C_\kappa(Z)/k! = \{4s_1^2 + 6s_2 + 4s_1^3 + 20s_1s_2 + 44s_3 + s_1^2s_2 + 16s_2^2 + 24s_1s_3 + 36s_4 + 4s_1s_2^2$$

$$+ 4s_2s_3 + 16s_5 + 2s_2s_4\} \operatorname{etr}(Z), \tag{4.4.40}$$

$$\sum_{k=0}^{\infty}\sum_{\kappa} a_2^2(\kappa)C_\kappa(Z)/k! = \{s_1 + 40s_1^2 + 57s_2 + 54s_1^3 + 186s_1s_2 + 324s_3$$
$$+ 9s_1^4 + 18s_1^2s_2 + 81s_2^2 + 224s_1s_3 + 288s_4 + 24s_1^2s_3$$
$$+ 24s_2s_3 + 144s_5 + 16s_3^2\} \operatorname{etr}(Z), \tag{4.4.41}$$

$$\sum_{k=0}^{\infty}\sum_{\kappa} a_1^2(\kappa)a_2(\kappa)C_\kappa(Z)/k! = \{11s_1^2 + 17s_2 + 25s_1^3$$
$$+ 109s_1s_2 + 216s_3 + 3s_1^4 + 12s_1^2s_2 + 121s_2^2 + 220s_1s_3$$
$$+ 312s_4 + 16s_1^2s_3 + 64s_2s_3 + 49s_1s_2^2 + 240s_5 + 3s_1^2s_2^2$$
$$+ 48s_2s_4 + 3s_2^3 + 16s_3^2 + 4s_2^2s_3\} \operatorname{etr}(Z), \tag{4.4.42}$$

$$\sum_{k=0}^{\infty}\sum_{\kappa} a_1^4(\kappa)C_\kappa(Z)/k! = \{3s_1^2 + 5s_2 + 16s_1^3 + 68s_1s_2$$
$$+ 124s_3 + 3s_1^4 + 10s_1^2s_2 + 111s_2^2 + 216s_1s_3 + 312s_4$$
$$+ 24s_1^2s_3 + 88s_2s_3 + 64s_1s_2^2 + 400s_5 + 6s_1^2s_2^2$$
$$+ 6s_2^3 + 128s_2s_4 + 48s_3^2 + 24s_2^2s_3 + s_2^4\} \operatorname{etr}(Z). \tag{4.4.43}$$

PROOF (4.4.32) follows by differentiating

$$\sum_{k=0}^{\infty}\sum_{\kappa} x^k C_\kappa(Z)/k! = \operatorname{etr}(xZ)$$

ℓ times with respect to x and then setting $x = 1$. By using (4.4.31) we have

$$\begin{aligned}\left|I - \frac{Z}{n}\right|^{-n} &= \sum_{k=0}^{\infty}\sum_{\kappa} \frac{(n)_\kappa}{k!}\frac{C_\kappa(Z)}{n^k} \\ &= \sum_{k=0}^{\infty}\sum_{\kappa} \frac{C_\kappa(Z)}{k!}\left[1 + \frac{a_1(\kappa)}{2n} + \frac{1}{24n^2}\{k - a_2(\kappa) + 3a_1^2(\kappa)\}\right. \\ &\quad \left. + \frac{1}{48n^3}\{a_1^3(\kappa) - a_1(\kappa)(a_2(\kappa) - k + 2) + 2a_3(\kappa)\}\right] \\ &\quad + o\left(\frac{1}{n^3}\right).\end{aligned} \tag{4.4.44}$$

The left-hand side has also the following asymptotic expansion:

$$\begin{aligned}\left|I - \frac{Z}{n}\right|^{-n} &= \operatorname{etr}(Z)\left[1 + \frac{1}{2n}\operatorname{tr}(Z^2) + \frac{1}{24n^2}\{3(\operatorname{tr}(Z^2))^2 + 8\operatorname{tr}(Z^3)\}\right. \\ &\quad \left. + \frac{1}{48n^3}\{12\operatorname{tr}(Z^4) + 8\operatorname{tr}(Z^2)\operatorname{tr}(Z^3) + (\operatorname{tr}(Z^2))^3\}\right]\end{aligned}$$

$$+o\left(\frac{1}{n^3}\right). \tag{4.4.45}$$

Comparing the coefficients of the term of order $1/n$ in (4.4.44) and (4.4.45) we have (4.4.33). From the terms of order $1/n^2$ in (4.4.44) and (4.4.45) and with the help of (4.4.32) we have

$$\sum_{k=0}^{\infty}\sum_{\kappa}\{3a_1^2(\kappa)-a_2(\kappa)\}C_\kappa(Z)/k!$$
$$= \operatorname{etr}(Z)\{8\operatorname{tr}(Z^3)+3(\operatorname{tr}(Z^2))^2-\operatorname{tr}(Z)\}.$$

By using (4.4.6) we have

$$\frac{1}{\Gamma_m(n)|Z|^n}\int_{S>0}\operatorname{etr}(-Z^{-1}S)|S|^{n-\frac{1}{2}(m+1)}C_\kappa\left(\frac{S}{n}\right)\mathrm{d}S$$
$$=\left(\frac{1}{n}\right)^k(n)_\kappa C_\kappa(Z)$$
$$=\left[1+\frac{a_1(\kappa)}{2n}+O\left(\frac{1}{n^2}\right)\right]C_\kappa(Z). \tag{4.4.46}$$

Multiplying both sides of (4.4.46) by $a_1(\kappa)/k!$ and using (4.4.32) we have

$$\frac{1}{\Gamma_m(n)|Z|^n}\int_{S>0}\operatorname{etr}\left\{-\left(I-\frac{Z}{n}\right)Z^{-1}S\right\}$$
$$\times|S|^{n-\frac{1}{2}(m+1)}\operatorname{tr}\left(\frac{S}{n}\right)^2\mathrm{d}S$$
$$=\sum_{k=0}^{\infty}\sum_{\kappa}\frac{C_\kappa(Z)}{k!}\left\{a_1(\kappa)+\frac{a_1^2(\kappa)}{2n}+O\left(\frac{1}{n^2}\right)\right\}$$
$$=\operatorname{etr}(Z)s_2+\frac{1}{2n}\sum_{k=0}\sum_{\kappa}\frac{a_1^2(\kappa)}{k!}C_\kappa(Z)+O\left(\frac{1}{n^2}\right). \tag{4.4.47}$$

The left-hand side is expressed as

$$\frac{1}{\Gamma_m(n)|Z|^n}\int_{S>0}\operatorname{etr}\left\{-\left(I-\frac{Z}{n}\right)Z^{-1}S\right\}$$
$$\times|S|^{n-\frac{1}{2}(m+1)}\left(\frac{1}{n}\right)^2\left\{C_{(2)}(S)-\frac{1}{2}C_{(1^2)}(S)\right\}\mathrm{d}S$$
$$=\left|I-\frac{Z}{n}\right|^{-n}\left(\frac{1}{n}\right)^2\left[(n)_{(2)}C_{(2)}\left\{\left(I-\frac{Z}{n}\right)^{-1}Z\right\}\right.$$
$$\left.-\frac{1}{2}(n)_{(1^2)}C_{(1^2)}\left\{\left(I-\frac{Z}{n}\right)^{-1}Z\right\}\right]$$

$$= \operatorname{etr}(Z)\left[1+\frac{s_2}{2n}+O\left(\frac{1}{n^2}\right)\right]$$
$$\times\left[s_2+\frac{1}{2n}\{4s_3+s_1^2+s_2\}+O\left(\frac{1}{n^2}\right)\right]$$
$$= \operatorname{etr}(Z)\left[s_2+\frac{1}{2n}\{s_2^2+4s_3+s_1^2+s_2\}+O\left(\frac{1}{n^2}\right)\right]. \tag{4.4.48}$$

Comparing the terms of order $1/n$ in (4.4.47) and (4.4.48), we have (4. 4.34). The other formulae are obtained similarly (Exercise 4.13).

Lemma 4.4.6 Let Z be a symmetric matrix with $\| Z \| < 1$ and put $V = Z(I-Z)^{-1}$. Then the following identities hold.

$$\sum_{k=1}^{\infty}\sum_{\kappa}(b)_\kappa C_\kappa(Z)/(k-1)! = b\operatorname{tr}(V)\,|I-Z|^{-b}, \tag{4.4.49}$$

$$\sum_{k=2}^{\infty}\sum_{\kappa}(b)_\kappa C_\kappa(Z)/(k-2)!$$
$$= \{b^2(\operatorname{tr}(V))^2+b\operatorname{tr}(V^2)\}\,|I-Z|^{-b}, \tag{4.4.50}$$

$$\sum_{k=3}^{\infty}\sum_{\kappa}(b)_\kappa C_\kappa(Z)/(k-3)!$$
$$= \{b^3(\operatorname{tr}(V))^3+3b^2\operatorname{tr}(V)\operatorname{tr}(V^2)+2b\operatorname{tr}(V^3)\}\,|I-Z|^{-b}, \tag{4.4.51}$$

$$\sum_{k=0}^{\infty}\sum_{\kappa}a_1(\kappa)(b)_\kappa C_\kappa(Z)/k!$$
$$= \frac{1}{2}\{b(\operatorname{tr}(V))^2+b(2b+1)\operatorname{tr}(V^2)\}\,|I-Z|^{-b}, \tag{4.4.52}$$

$$\sum_{k=1}^{\infty}\sum_{\kappa}a_1(\kappa)(b)_\kappa C_\kappa(Z)/(k-1)!$$
$$= \frac{1}{2}\{2b(\operatorname{tr}(V))^2+2b(2b+1)\operatorname{tr}(V^2)+b^2(\operatorname{tr}(V))^3$$
$$+b(2b^2+b+2)\operatorname{tr}(V)\operatorname{tr}(V^2)+2b(2b+1)\operatorname{tr}(V^3)\}$$
$$\times|I-Z|^{-b}, \tag{4.4.53}$$

$$\sum_{k=0}^{\infty}\sum_{\kappa}a_2(\kappa)(b)_\kappa C_\kappa(Z)/k!$$
$$= \frac{1}{2}\{2b\operatorname{tr}(V)+3b(2b+1)(\operatorname{tr}(V))^2+3b(2b+3)\operatorname{tr}(V^2)$$
$$+2b(\operatorname{tr}(V))^3+6b(2b+1)\operatorname{tr}(V)\operatorname{tr}(V^2)$$
$$+4b(2b^2+3b+2)\operatorname{tr}(V^3)\}\,|I-Z|^{-b}, \tag{4.4.54}$$

$$\sum_{k=0}^{\infty}\sum_{\kappa}a_1^2(\kappa)(b)_\kappa C_\kappa(Z)/k!$$

$$= \frac{1}{4}\{2b(2b+1)(\operatorname{tr}(V))^2 + 2b(2b+3)\operatorname{tr}(V^2) + 4b(\operatorname{tr}(V))^3$$
$$+ 12b(2b+1)\operatorname{tr}(V)\operatorname{tr}(V^2) + 8b(2b^2+3b+2)\operatorname{tr}(V^3)$$
$$+ b^2(\operatorname{tr}(V)^4) + 2b(2b^2+b+2)(\operatorname{tr}(V))^2\operatorname{tr}(V^2)$$
$$+ b(2b+1)(2b^2+b+2)(\operatorname{tr}(V^2))^2 + 8b(2b+1)\operatorname{tr}(V)\operatorname{tr}(V^3)$$
$$+ 2b(8b^2+10b+5)\operatorname{tr}(V^4)\}\, |I-Z|^{-b}. \tag{4.4.55}$$

PROOF (4.4.32) gives

$$\sum_{k=1}^{\infty}\sum_{\kappa}\frac{C_\kappa(S)}{(k-1)!} = \operatorname{tr}(S)\operatorname{etr}(S).$$

Multiplying both sides by $\operatorname{etr}(-Z^{-1}S)\,|S|^{b-\frac{1}{2}(m+1)}/(\Gamma_m(b)\,|Z|^b)$ and integrating over $S>0$, we have

$$\sum_{k=1}^{\infty}\sum_{\kappa}\frac{(b)_\kappa C_\kappa(Z)}{(k-1)!} = (b)_1 C_{(1)}(V)\,|I-Z|^{-b}$$
$$= b\operatorname{tr}(V)\,|I-Z|^{-b},$$

which yields (4.4.49). The remaining formulae are obtained similarly (Exercise 4.15).

For topics related to the Laplace transform and other transforms of a matrix argument, see Bochner (1944), Bochner (1952), Bochner and Martin (1948), Herz (1955) and Constantine (1963, 1966). Muirhead (1970a) gives partial differential operators for hypergeometric function. In addition relevant materials can be found in Muirhead (1970b, 1972), Constantine and Muirhead (1972), Fujikoshi (1975), Sugiura (1972) and Sugiura (1974). Lemma 4.4.5 appears in Sugiura and Fujikoshi (1969) and Fujikoshi (1970, 1973). Muirhead (1970a) also derives some results using partial differential operators. Sugiura (1971) discusses weighted sums of zonal polynomials with respect to the partitions κ of k. Lemma 4.4.6 is obtained with the help of Exercise 4.14 by Fujikoshi (1970). Additional materials for the hypergeometric functions of Hermitian matrices can be found in James (1964), Khatri (1966, 1970), Hayakawa (1972c, 1972d) and Chikuse (1976).

4.5 Binomial Coefficients

Constantine (1966) introduced generalized binomial type coefficients as follows:

Definition 4.5.1 *Binomial coefficient* The binomial coefficient $\binom{\kappa}{\sigma}$ is defined by the expansion

$$C_\kappa(I+S)/C_\kappa(I_m) = \sum_{\ell=0}^{k}\sum_{\sigma}\binom{\kappa}{\sigma}C_\sigma(S)/C_\sigma(I_m), \tag{4.5.1}$$

where κ is a partition of k and σ is a partition of ℓ, respectively.

This is a generalization of the usual binomial expansion. Let $S = xI$ in (4.5.1), then

$$(1+x)^k = \sum_{\ell=0}^{k} x^\ell \sum_{\sigma} \binom{\kappa}{\sigma}$$

which yields

$$\binom{k}{\ell} = \sum_{\sigma} \binom{\kappa}{\sigma}. \tag{4.5.2}$$

Constantine (1966) tabulated the binomial coefficients to $k = 4$ and Pillai and Jouris (1969) tabulated them up to $k = 8$, see Table 4.4.1 at the end of this chapter.

Muirhead (1982) derives many formulae involving weighted sums of binomial coefficients in connection with the partial differential equation for zonal polynomials.

For $m \times m$ symmetric matrices Z and T, define ${}_pF_q^{(m)}\,(a_1,\ldots,a_p;\, b_1,\ldots,b_q;\, Z,T)$ as

$$\begin{aligned}
&{}_pF_q^{(m)}\,(a_1,\ldots,a_p;b_1,\ldots,b_q;Z,T)\\
&\quad = \int_{O(m)} {}_pF_q(a_1,\ldots,a_p;b_1,\ldots,b_q;ZHTH')\mathrm{d}(H)\\
&\quad = \sum_{k=0}^{\infty}\sum_{\kappa} \frac{(a_1)_\kappa\cdots(a_p)_\kappa}{(b_1)_\kappa\cdots(b_q)_\kappa}\,\frac{C_\kappa(Z)C_\kappa(T)}{k!\,C_\kappa(I_m)}.
\end{aligned} \tag{4.5.3}$$

Lemma 4.5.1 Let u and v be arbitrary scalars, then

$$\begin{aligned}
&\mathrm{etr}\,(-ZT)\,{}_0F_0^{(m)}(Z,T)\\
&\quad = \mathrm{etr}\,(-(Z+uI)(T+vI))\ {}_0F_0^{(m)}(Z+uI,T+vI).
\end{aligned} \tag{4.5.4}$$

PROOF Since

$$\mathrm{tr}\,(ZHTH') = \mathrm{tr}\,((Z+uI)H(T+vI)H') + \mathrm{tr}\,(ZT) - \mathrm{tr}\,((Z+uI)(T+vI)),$$

the result follows.

Theorem 4.5.1

$$\mathrm{etr}\,(Z)C_\kappa(Z)/k! = \sum_{\ell=k}^{\infty}\sum_{\sigma}\binom{\sigma}{\kappa} C_\sigma(Z)/\ell!, \tag{4.5.5}$$

where σ is a partition of ℓ.

PROOF We have

$$\mathrm{etr}\,(Z)\ {}_0F_0^{(m)}(Z,T)$$

$$= \sum_{k=0}^{\infty} \frac{1}{k!} \sum_{\kappa} \{C_\kappa(Z)\,\mathrm{etr}\,(Z)\} C_\kappa(T)/C_\kappa(I_m). \tag{4.5.6}$$

Setting $u = 0$ and $v = 1$ in (4.5.4) we have

$$\begin{aligned}
\mathrm{etr}\,(Z)\, {}_0F_0^{(m)}(Z,T) &= {}_0F_0^{(m)}(Z, I+T) \\
&= \sum_{\ell=0}^{\infty} \frac{1}{\ell!} \sum_{\sigma} C_\sigma(Z) C_\sigma(I+T)/C_\sigma(I) \\
&= \sum_{\ell=0}^{\infty} \frac{1}{\ell!} \sum_{\sigma} C_\sigma(Z) \left\{ \sum_{k=0}^{\ell} \sum_{\kappa} \binom{\sigma}{\kappa} C_\kappa(T)/C_\sigma(I) \right\} \\
&= \sum_{k=0}^{\infty} \sum_{\kappa} \left[\sum_{\ell=k}^{\infty} \sum_{\sigma} \binom{\sigma}{\kappa} \frac{C_\sigma(Z)}{\ell!} \right] \frac{C_\kappa(T)}{C_\kappa(I)}.
\end{aligned} \tag{4.5.7}$$

Comparing the coefficients of $C_\kappa(T)/C_\kappa(I)$ on both sides of (4.5.6) and (4.5.7), we have (4.5.5).

Theorem 4.5.2 Let Z be such that $\| Z \| < 1$ and let b be an arbitrary number. Then for a partition κ of k,

$$\sum_{\ell=k}^{\infty} \sum_{\sigma} \binom{\sigma}{\kappa} \frac{(b)_\sigma C_\sigma(Z)}{\ell!} = |I - Z|^{-b} (b)_\kappa \frac{C_\kappa((I-Z)^{-1}Z)}{k!}. \tag{4.5.8}$$

PROOF Define

$${}_1F_0^{(m)}(b; Z, T) = \int_{O(m)} |I - ZHTH'|^{-b} \mathrm{d}(H) \tag{4.5.9}$$

for $\| Z \| < 1$ and $\| T \| < 1$. Then by noting

$${}_1F_0^{(m)}(b; Z, I+T) = |I - Z|^{-b}\, {}_1F_0^{(m)}(b; (I-Z)^{-1}Z, T), \tag{4.5.10}$$

we obtain (4.5.8) by steps similar to those used in the proof of Theorem 4.5.1.

Bingham (1974) introduced coefficients of a new type which constitute an extension of the binomial coefficients.

Definition 4.5.2 *Generalized binomial coefficients* Let $P(Z)$ be a homogeneous symmetric polynomial of degree k in Z. Then the coefficients $\binom{\sigma}{P}$ are defined as

$$P(Z)\,\mathrm{etr}\,(Z)/k! = \sum_{\ell=k}^{\infty} \sum_{\sigma} \binom{\sigma}{P} \frac{C_\sigma(Z)}{\ell!}. \tag{4.5.11}$$

Lemma 4.5.2 Let σ be a partition of ℓ, then the following identities hold:

$$\binom{\sigma}{s_1^k} = \binom{\ell}{\ell-k},$$

$$\binom{\sigma}{s_2} = \frac{1}{2}a_1(\sigma), \quad \binom{\sigma}{s_1s_2} = \frac{1}{6}(\ell-2)a_1(\sigma), \tag{4.5.12}$$

$$\binom{\sigma}{s_3} = \frac{1}{24}(a_2(\sigma)-\ell) - \frac{1}{8}a_1(\sigma) - \frac{1}{4}\binom{\ell}{2},$$

$$\binom{\sigma}{s_1^2s_2} = \frac{1}{12}\left[\binom{\ell}{2}a_1(\sigma) - 2\binom{\ell}{1}a_1(\sigma) + 3a_1(\sigma)\right],$$

$$\binom{\sigma}{s_2^2} = \frac{1}{24}\left[4\binom{\ell}{2} + \binom{\ell}{1} + a_1^2(\sigma) + 2a_1(\sigma) - a_2(\sigma)\right],$$

$$\binom{\sigma}{s_1s_3} = \frac{1}{96}\left[-18\binom{\ell}{3} + 4\binom{\ell}{2} + 2\binom{\ell}{1} - 3\binom{\ell}{1}a_1(\sigma) + \binom{\ell}{1}a_2(\sigma) + 9a_1(\sigma) - 3a_2(\sigma)\right],$$

$$\binom{\sigma}{s_4} = \frac{1}{48}\left[4\binom{\ell}{2} + \binom{\ell}{1} - 4\binom{\ell}{1}a_1(\sigma) + 7a_1(\sigma) - a_2(\sigma) + a_3(\sigma)\right].$$

PROOF Comparing (4.5.11) and the weighted sums in Lemma 4.4.5, we have (4.5.12).

Lemma 4.5.3 The explicit expressions of the binomial coefficients $\binom{\sigma}{\kappa}$ up to $k = 4$, where σ is a partition of ℓ, and κ is a partition of k, are given as follows.

$$\binom{\sigma}{(0)} = 1, \quad \binom{\sigma}{(1)} = \ell,$$

$$\binom{\sigma}{(2)} = \frac{1}{3}\left[\binom{\ell}{2} + a_1(\sigma)\right], \quad \binom{\sigma}{(1^2)} = \frac{1}{3}\left[2\binom{\ell}{2} - a_1(\sigma)\right],$$

$$\binom{\sigma}{(3)} = \frac{1}{15}\left[\binom{\ell}{3} - 2\binom{\ell}{2} - \frac{1}{3}\binom{\ell}{1} + \binom{\ell}{1}a_1(\sigma) - 3a_1(\sigma) + \frac{1}{3}a_2(\sigma)\right],$$

$$\binom{\sigma}{(21)} = \frac{1}{15}\left[9\binom{\ell}{3} + \frac{9}{2}\binom{\ell}{2} + \frac{3}{4}\binom{\ell}{1} + \frac{3}{2}\binom{\ell}{1}a_1(\sigma)\right.$$

$$-\frac{3}{4}a_1(\sigma) - \frac{3}{4}a_2(\sigma)\Big] ,$$

$$\binom{\sigma}{(1^3)} = \frac{1}{15}\Big[5\binom{\ell}{3} - \frac{5}{2}\binom{\ell}{2} - \frac{5}{12}\binom{\ell}{1} - \frac{5}{2}\binom{\ell}{1}a_1(\sigma) + \frac{15}{4}a_1(\sigma) + \frac{5}{12}a_2(\sigma)\Big] ,$$

$$\binom{\sigma}{(4)} = \frac{1}{105}\Big[\binom{\ell}{4} - 6\binom{\ell}{3} + \frac{22}{3}\binom{\ell}{2} + \frac{13}{6}\binom{\ell}{1} + \binom{\ell}{2}a_1(\sigma) - 7\binom{\ell}{1}a_1(\sigma) + \frac{1}{3}\binom{\ell}{1}a_2(\sigma) + \frac{1}{2}a_1(\sigma)^2 + 14a_1(\sigma) - \frac{5}{2}a_2(\sigma) + a_3(\sigma)\Big] ,$$

$$\binom{\sigma}{(31)} = \frac{1}{105}\Big[20\binom{\ell}{4} - 15\binom{\ell}{3} - \frac{50}{3}\binom{\ell}{2} - \frac{10}{3}\binom{\ell}{1} + \frac{25}{3}\binom{\ell}{2}a_1(\sigma) - \frac{35}{6}\binom{\ell}{1}a_1(\sigma) + \frac{5}{6}\binom{\ell}{1}a_2(\sigma) - \frac{5}{3}a_1^2(\sigma) + \frac{35}{6}a_1(\sigma) + \frac{5}{2}a_2(\sigma) - \frac{10}{3}a_3(\sigma)\Big] ,$$

$$\binom{\sigma}{(2^2)} = \frac{1}{105}\Big[14\binom{\ell}{4} + 21\binom{\ell}{3} + \frac{28}{3}\binom{\ell}{2} + \frac{7}{6}\binom{\ell}{1} + \frac{7}{3}\binom{\ell}{2}a_1(\sigma) + \frac{7}{6}\binom{\ell}{1}a_1(\sigma) - \frac{7}{6}\binom{\ell}{1}a_2(\sigma) + \frac{49}{12}a_1^2(\sigma) + \frac{7}{12}a_1(\sigma) - \frac{7}{12}a_3(\sigma)\Big] ,$$

$$\binom{\sigma}{(21^2)} = \frac{1}{105}\Big[56\binom{\ell}{4} + 21\binom{\ell}{3} - \frac{14}{3}\binom{\ell}{2} - \frac{7}{3}\binom{\ell}{1} - \frac{14}{3}\binom{\ell}{2}a_1(\sigma) - \frac{35}{6}\binom{\ell}{1}a_1(\sigma) - \frac{7}{6}\binom{\ell}{1}a_2(\sigma) - \frac{14}{3}a_1^2(\sigma) - \frac{7}{6}a_1(\sigma) + \frac{7}{2}a_2(\sigma) + \frac{14}{3}a_3(\sigma)\Big] ,$$

$$\binom{\sigma}{(1^4)} = \frac{1}{105}\left[14\binom{\ell}{4} - 21\binom{\ell}{3} + \frac{14}{3}\binom{\ell}{2} + \frac{7}{3}\binom{\ell}{1}\right.$$
$$- 7\binom{\ell}{2}a_1(\sigma) + \frac{35}{2}\binom{\ell}{1}a_1(\sigma) + \frac{7}{6}\binom{\ell}{1}a_2(\sigma)$$
$$\left. + \frac{7}{4}a_1^2(\sigma) - \frac{77}{4}a_1(\sigma) - \frac{7}{2}a_2(\sigma) - \frac{7}{4}a_3(\sigma)\right] .$$

It should be noted that

$$\sum_{\kappa}\binom{\sigma}{\kappa} = \binom{\ell}{k}, \quad k = 1, 2, 3, 4.$$

PROOF The proof is left as an exercise (Exercise 4.19).

Some partial differential forms with respect to zonal polynomial are expressed by the weighted sums of them with binomial coefficients, Muirhead (1970a, 1972). Muirhead (1974) gives the explicit forms of binomial coefficients for particular partitions. Kushner (1985) also gives the expansion of a zonal polynomial.

4.6 Some Special Functions

In this section we discuss some orthogonal polynomials with respect to certain weighted functions used in certain distributional results.

Definition 4.6.1 *Bessel function* Let R be an $m \times m$ symmetric matrix, then the Bessel function of a matrix argument is defined as

$$A_\gamma(R) = \frac{1}{\Gamma_m\left(\gamma + \frac{1}{2}(m+1)\right)}\, {}_0F_1\left(\gamma + \frac{1}{2}(m+1); -R\right) . \tag{4.6.1}$$

By using the inversion formula (4.4.7), $A_\gamma(R)$ has the integral representation

$$A_\gamma(R) = \frac{2^{\frac{1}{2}m(m-1)}}{(2\pi i)^{\frac{1}{2}m(m+1)}} \int_{\mathcal{R}e(Z)=X_0>0} \operatorname{etr}(Z)$$
$$\times \operatorname{etr}(-RZ^{-1})|Z|^{-\gamma-\frac{1}{2}(m+1)}dZ . \tag{4.6.2}$$

Definition 4.6.2 *Laguerre polynomial (I)* Let S be an $m \times m$ symmetric matrix, then the Laguerre polynomial $L_\kappa^\gamma(S)$, $\gamma > -1$, is defined as

$$\operatorname{etr}(-S)L_\kappa^\gamma(S) = \int_{R>0} A_\gamma(RS)\operatorname{etr}(-R)|R|^\gamma C_\kappa(R)dR . \tag{4.6.3}$$

Lemma 4.6.1 The Laguerre polynomial $L_\kappa^\gamma(S)$ is expressed as

$$L_\kappa^\gamma(S) = \Gamma_m\left(\gamma + \frac{1}{2}(m+1); \kappa\right) \frac{2^{\frac{1}{2}m(m-1)}}{(2\pi i)^{\frac{1}{2}m(m+1)}}$$

$$\times \int_{\mathcal{R}e(Z)=X_0>0} \operatorname{etr}(Z)\,|Z|^{-\gamma-\frac{1}{2}(m+1)} C_\kappa(I - SZ^{-1})\mathrm{d}Z \tag{4.6.4}$$

$$= (\gamma + \frac{1}{2}(m+1))_\kappa C_\kappa(I_m)$$

$$\times \sum_{\ell=0}^{k}\sum_{\sigma} \binom{\kappa}{\sigma} \frac{C_\sigma(-S)}{(\gamma+\frac{1}{2}(m+1))_\sigma C_\sigma(I_m)}. \tag{4.6.5}$$

The Laplace transform of $L_\kappa^\gamma(S)$ is given by

$$\int_{S>0} \operatorname{etr}(-SZ)\,|S|^{\gamma} L_\kappa^\gamma(S)\mathrm{d}S$$

$$= \Gamma_m(\gamma + \frac{1}{2}(m+1); \kappa)\,|Z|^{-\gamma-\frac{1}{2}(m+1)} C_\kappa(I - Z^{-1}). \tag{4.6.6}$$

PROOF Substituting (4.6.2) into (4.6.3) and reversing the order of integration, we have

$$L_\kappa^\gamma(S) = \operatorname{etr}(S)\frac{2^{\frac{1}{2}m(m-1)}}{(2\pi i)^{\frac{1}{2}m(m+1)}} \int_{\mathcal{R}e(Z)=X_0>0} \operatorname{etr}(Z)\,|Z|^{-\gamma-\frac{1}{2}(m+1)}$$

$$\times \int_{R>0} \operatorname{etr}(-(I+SZ^{-1})R)|R|^{\gamma} C_\kappa(R)\mathrm{d}R\mathrm{d}Z$$

$$= \frac{2^{\frac{1}{2}m(m-1)}}{(2\pi i)^{\frac{1}{2}m(m+1)}} \int_{\mathcal{R}e(Z)=X_0>0} \operatorname{etr}(S+Z)\,|Z|^{-\gamma-\frac{1}{2}(m+1)}$$

$$\times \Gamma_m(\gamma + \frac{1}{2}(m+1); \kappa)|I + SZ^{-1}|^{-\gamma-\frac{1}{2}(m+1)}$$

$$\times C_\kappa((I+SZ^{-1})^{-1})\mathrm{d}Z.$$

By setting $S + Z = U$,

$$L_\kappa^\gamma(S) = \Gamma_m(\gamma + \frac{1}{2}(m+1); \kappa)\frac{2^{\frac{1}{2}m(m-1)}}{(2\pi i)^{\frac{1}{2}m(m+1)}}$$

$$\times \int_{\mathcal{R}e(U)=X_0>0} \operatorname{etr}(U)|U|^{-\gamma-\frac{1}{2}(m+1)} C_\kappa(I - SU^{-1})\mathrm{d}U,$$

which is (4.6.5). Putting $S^{1/2}U^{-1}S^{1/2} = Z^{-1}$, we have

$$L_\kappa^\gamma(S)\,|S|^\gamma = \Gamma_m\left(\gamma + \frac{1}{2}(m+1);\kappa\right)\frac{2^{\frac{1}{2}m(m-1)}}{(2\pi i)^{\frac{1}{2}m(m+1)}}$$

$$\times \int_{\mathcal{R}e(Z)=X_0>0} \operatorname{etr}(SZ)\,|Z|^{-\gamma-\frac{1}{2}(m+1)} C_\kappa(I - Z^{-1})\mathrm{d}Z\,;$$

the inverse Laplace transform gives (4.6.6).

Expanding $C_\kappa(I - SZ^{-1})$ in (4.6.4) by using (4.5.1) and performing the integration in (4.6.4) using (4.4.7) yield (4.6.5).

The value of L_κ^γ at the origin is

$$L_\kappa^\gamma(0) = (\gamma + \frac{1}{2}(m+1))_\kappa\, C_\kappa(I_m)\,. \tag{4.6.7}$$

The Laguerre polynomials for lower orders are expressed as follows when $\gamma = \frac{1}{2}(n-m-1)$ and $s_j = \operatorname{tr}(S^j)$:

$$L_{(1)}^\gamma(S) = \frac{mn}{2} - s_1\,,$$

$$L_{(2)}^\gamma(S) = \frac{1}{12}mn(m+2)(n+2) - \frac{1}{3}(m+2)(n+2)s_1 + \frac{1}{3}\{s_1^2 + 2s_2\}\,,$$

$$L_{(1^2)}^\gamma(S) = \frac{1}{6}mn(m-1)(n-1) - \frac{2}{3}(m-1)(n-1)s_1 + \frac{2}{3}\{s_1^2 - s_2\}\,,$$

$$\begin{aligned} L_{(3)}^\gamma(S) = {} & \frac{1}{120}mn(m+2)(m+4)(n+2)(n+4) \\ & - \frac{1}{20}(m+2)(m+4)(n+2)(n+4)s_1 \\ & + \frac{1}{10}(m+4)(n+4)\{s_1^2 + 2s_2\} \\ & + \frac{1}{15}\{s_1^3 + 6s_1s_2 + 8s_3\}\,, \end{aligned} \tag{4.6.8}$$

$$\begin{aligned} L_{(21)}^\gamma(S) = {} & \frac{3}{40}mn(m+2)(m-1)(n+2)(n-1) \\ & - \frac{9}{20}(m+2)(m-1)(n+2)(n-1)s_1 \\ & + \frac{2}{5}(m-1)(n-1)\{s_1^2 + 2s_2\} + \frac{1}{2}(m+2)(n+2)\{s_1^2 - s_2\} \\ & - \frac{3}{5}\{s_1^3 + s_1s_2 - 2s_3\}\,, \end{aligned}$$

$$\begin{aligned}L^{\gamma}_{(1^3)}(S) = &\frac{1}{24}mn(m-1)(m-2)(n-1)(n-2)\\ &-\frac{1}{4}(m-1)(m-2)(n-1)(n-2)s_1\\ &+\frac{1}{2}(m-2)(n-2)\{s_1^2 - s_2\} + \frac{1}{3}\{s_1^3 - 3s_1s_2 + 2s_3\}.\end{aligned}$$

Theorem 4.6.1 The generating function of the generalized Laguerre polynomials is

$$\begin{aligned}&|I-Z|^{-\gamma-\frac{1}{2}(m+1)}\int_{O(m)} \operatorname{etr}\{-SH'Z(I-Z)^{-1}H\}\,\mathrm{d}(H)\\ &= \sum_{k=0}^{\infty}\sum_{\kappa}\frac{L^{\gamma}_{\kappa}(S)C_{\kappa}(Z)}{k!C_{\kappa}(I)}, \quad \| Z \| < 1\,. \end{aligned} \tag{4.6.9}$$

PROOF Substituting (4.6.3) into the right-hand side of (4.6.9), we have

$$\begin{aligned}&\operatorname{etr}(S)\int_{R>0} A_{\gamma}(RS)\operatorname{etr}(-R)|R|^{\gamma}\,{}_0F_0^{(m)}(Z,R)\mathrm{d}R\\ &= \operatorname{etr}(S)\int_{R>0} A_{\gamma}(RS)\operatorname{etr}(-R)|R|^{\gamma}\\ &\qquad\times\int_{O(m)}\operatorname{etr}(ZHRH')\mathrm{d}(H)\mathrm{d}R\\ &= \operatorname{etr}(S)\int_{O(m)}\int_{R>0}\operatorname{etr}\{-(I-H'ZH)R\}|R|^{\gamma}A_{\gamma}(RS)\mathrm{d}R\mathrm{d}(H)\\ &= \int_{O(m)}|I-H'ZH|^{-\gamma-\frac{1}{2}(m+1)}\\ &\qquad\times\operatorname{etr}(S)\,{}_0F_0(-S(I-H'ZH)^{-1})\mathrm{d}(H)\\ &= |I-Z|^{-\gamma-p}\int_{O(m)}\operatorname{etr}\{-SH'Z(I-Z)^{-1}H\}\mathrm{d}(H).\end{aligned}$$

Corollary 4.6.1

$$\sum_{\kappa} L^{\gamma}_{\kappa}(S) = L_k^{m\left(\gamma+\frac{1}{2}(m+1)\right)-1}(\operatorname{tr}(S)) \tag{4.6.10}$$

where $L_k^a(x)$ is a univariate Laguerre polynomial.

PROOF Putting $Z = xI_m$, $|x| < 1$, in (4.6.9), we have

$$(1-x)^{-m\left(\gamma+\frac{1}{2}(m+1)\right)}\exp\left\{-\frac{x}{1-x}\operatorname{tr}(S)\right\}$$

$$= \sum_{k=0}^{\infty} \frac{x^k}{k!} \sum_{\kappa} L_{\kappa}^{\gamma}(S). \tag{4.6.11}$$

The left-hand side is nothing but the generating function of univariate Laguerre polynomials $L_k^{m(\gamma+\frac{1}{2}(m+1))-1}(\operatorname{tr}(S))$. Comparing the coefficients of x^k one gets (4.6.10).

Theorem 4.6.2 Let x be any number such that $|x| < 1$ and $\gamma = \frac{1}{2}(n-m-1)$ (> -1), then the following identities hold :

$$\sum_{k=1}^{\infty} \sum_{\kappa} \frac{x^k L_{\kappa}^{\gamma}(S)}{(k-1)!} = (1-x)^{-\frac{1}{2}mn} \frac{x}{1-x} \times \left\{ \frac{mn}{2} - \frac{s_1}{1-x} \right\} \operatorname{etr}\left(-\frac{x}{1-x} S\right), \tag{4.6.12}$$

$$\sum_{k=2}^{\infty} \sum_{\kappa} \frac{x^k L_{\kappa}^{\gamma}(S)}{(k-2)!} = (1-x)^{-\frac{1}{2}mn} \left(\frac{x}{1-x}\right)^2 \times \left[\frac{mn}{2}\left(\frac{mn}{2}+1\right) - (mn+2)\frac{s_1}{1-x} + \left(\frac{s_1}{1-x}\right)^2 \right] \times \operatorname{etr}\left(-\frac{x}{1-x} S\right), \tag{4.6.13}$$

$$\sum_{k=0}^{\infty} \sum_{\kappa} \frac{x^k a_1(\kappa) L_{\kappa}^{\gamma}(S)}{k!} = (1-x)^{-\frac{1}{2}mn} \left(\frac{x}{1-x}\right)^2 \times \left[\frac{1}{4} mn(m+n+1) - (m+n+1)\frac{s_1}{1-x} + \frac{s_2}{(1-x)^2} \right] \times \operatorname{etr}\left(-\frac{x}{1-x} S\right), \tag{4.6.14}$$

$$\sum_{k=0}^{\infty} \sum_{\kappa} \frac{x^k a_1^2(\kappa) L_{\kappa}^{\gamma}(S)}{k!} = (1-x)^{-\frac{1}{2}mn} \left(\frac{x}{1-x}\right)^2$$

$$\times \left[\frac{mn}{16} \left\{ 4((n+1)m+n+3) + 8(m^2+3(n+1)m \right.\right.$$

$$+ n^2 + 3n + 4)\frac{x}{1-x} + (nm^3 + 2(n^2+n+4)m^2$$

$$\left. + (n+1)(n^2+n+20)m + 4(2n^2+5n+5)\left(\frac{x}{1-x}\right)^2 \right\}$$

$$- \left\{ (n+1)m + n + 3 + 3(m^2 + 3(n+1)m + n^2 + 3n + 4) \right.$$

$$\times \frac{x}{1-x} + \frac{1}{2}(nm^3 + 2(n^2+n+4)m^2 + (n+1)$$
$$\times (n^2+n+20)m + 4(2n^2+5n+5))\left(\frac{x}{1-x}\right)^2\Bigg\} \frac{s_1}{1-x}$$
$$+\Bigg\{1 + \frac{6x}{1-x} + (m^2 + 2(n+1)m + n^2 + 2n + 7)$$
$$\times \left(\frac{x}{1-x}\right)^2\Bigg\} \left(\frac{s_1}{1-x}\right)^2 + \Bigg\{1 + 6(m+n+2)\frac{x}{1-x}$$
$$+ \frac{1}{2}(nm^2 + (n^2+n+20)m + 4(5n+8))\left(\frac{x}{1-x}\right)^2\Bigg\}$$
$$\times \frac{s_2}{(1-x)^2} - 2(m+n+1)\left(\frac{x}{1-x}\right)^2 \frac{s_1 s_2}{(1-x)^3} - \frac{4x(1+x)}{(1-x)^2}$$
$$\times \frac{s_3}{(1-x)^3} + \left(\frac{x}{1-x}\right)^2 \frac{s_2^2}{(1-x)^4}\Bigg]$$
$$\times \operatorname{etr}\left(-\frac{x}{1-x}S\right), \tag{4.6.15}$$

$$\sum_{k=0}^{\infty}\sum_{\kappa} \frac{x^k a_2(\kappa) L_\kappa^\gamma(S)}{k!} = (1-x)^{-\frac{1}{2}mn} \frac{x}{1-x}$$
$$\times \Bigg[\frac{mn}{4}\Bigg\{2 + 3((n+1)m + n + 3)\frac{x}{1-x}$$
$$+2(m^2 + 3(n+1)m + n^2 + 3n + 4)\left(\frac{x}{1-x}\right)^2\Bigg\}$$
$$-\Bigg\{1 + 3((n+1)m + n + 3)\frac{x}{1-x} + 3(m^2 + 3(n+1)m$$
$$+ n^2 + 3n + 4)\left(\frac{x}{1-x}\right)^2\Bigg\} \frac{s_1}{1-x}$$
$$+ \frac{3x}{1-x}\Bigg\{\frac{x}{1-x}\cdot\left(\frac{s_1}{1-x}\right)^2$$
$$+ \left(1 + 2(m+n+2)\frac{x}{1-x}\right)\frac{s_2}{(1-x)^2}$$
$$-4\left(\frac{x}{1-x}\right)^2 \frac{s_3}{(1-x)^3}\Bigg] \operatorname{etr}\left(-\frac{x}{1-x}S\right). \tag{4.6.16}$$

PROOF We obtain (4.6.12) by differentiating (4.6.11) with respect to x. The definition (4.6.3) and (4.4.33) yield

$$\sum_{k=0}^{\infty}\sum_{\kappa}\frac{x^k a_1(\kappa)L_\kappa^\gamma(S)}{k!}$$

$$= \operatorname{etr}(S)\int_{R>0} A_\gamma(RS)|R|^\gamma \operatorname{etr}(-R)x^2 \operatorname{tr}(R^2)\operatorname{etr}(xR)\mathrm{d}R$$

$$= \operatorname{etr}(S)(1-x)^{-m(\gamma+\frac{1}{2}(m+1))}\left(\frac{x}{1-x}\right)^2$$

$$\times \int_{R>0} A_\gamma\left(\frac{S}{1-x}R\right)|R|^\gamma \operatorname{etr}(-R)\left[C_{(2)}(R)-\frac{1}{2}C_{(1^2)}(R)\right]\mathrm{d}R$$

$$= \operatorname{etr}\left(-\frac{x}{1-x}S\right)(1-x)^{-\frac{mn}{2}}\left(\frac{x}{1-x}\right)^2$$

$$\times\left[L_{(2)}^\gamma\left(\frac{S}{1-x}\right)-\frac{1}{2}L_{(1^2)}^\gamma\left(\frac{S}{1-x}\right)\right].$$

From (4.6.8) we have

$$\operatorname{etr}\left(-\frac{x}{1-x}S\right)(1-x)^{-\frac{mn}{2}}\left(\frac{x}{1-x}\right)^2\left[\frac{1}{4}mn(m+n+1)\right.$$

$$\left.-(m+n+1)\frac{s_1}{1-x}+\frac{s_2}{(1-x)^2}\right].$$

The remaining results are obtained in a similar way.

Corollary 4.6.2 The following equalities hold for $\gamma=\frac{1}{2}(n-m-1)$.

$$\sum_{\kappa} a_1(\kappa)\,L_\kappa^\gamma(S)=k(k-1)\left[\frac{1}{4}mn(n+m+1)L_{k-2}^{\frac{1}{2}mn+1}(s_1)\right.$$

$$\left.-(n+m+1)s_1\,L_{k-2}^{\frac{1}{2}mn+2}(s_1)+s_2\,L_{k-2}^{\frac{1}{2}mn+3}(s_1)\right], \tag{4.6.17}$$

$$\sum_{\kappa} a_2(\kappa)\,L_\kappa^\gamma(S)=k\left[\frac{mn}{2}L_{k-1}^{\frac{1}{2}mn}(s_1)-s_1L_{k-1}^{\frac{1}{2}mn+1}(s_1)\right]$$

$$+3k(k-1)\left[\frac{1}{4}mn\{(m+1)n+m+3\}\,L_{k-2}^{\frac{1}{2}mn+1}(s_1)\right.$$

$$\left.-\{(m+1)n+m+3\}\,s_1L_{k-2}^{\frac{1}{2}mn+2}(s_1)+(s_1^2+s_2)L_{k-2}^{\frac{1}{2}mn+3}(s_1)\right]$$

$$+4k(k-1)(k-2)\left[\frac{1}{8}mn\{n^2+3(m+1)n\right.$$

$$+m^2+3m+4\}L_{k-3}^{\frac{1}{2}mn+2}(s_1)$$

$$
\begin{aligned}
&-\frac{3}{4}\{n^2+3(m+1)n+m^2+3m+4\}s_1L_{k-3}^{\frac{1}{2}mn+3}(s_1)\\
&+\frac{3}{2}\{s_1^2+(n+m+2)s_2\}L_{k-3}^{\frac{1}{2}mn+4}(s_1)\\
&\qquad\qquad -s_3L_{k-3}^{\frac{1}{2}mn+5}(s_1)\Big], \qquad (4.6.18)
\end{aligned}
$$

$$
\begin{aligned}
\sum_\kappa a_1^2(\kappa)L_\kappa^\gamma(S) &= k(k-1)\Big[\frac{1}{4}mn\{(m+1)n+m+3\}L_{k-2}^{\frac{1}{2}mn+1}(s_1)\\
&-\{(m+1)n+m+3\}s_1L_{k-2}^{\frac{1}{2}mn+2}(s_1)+(s_1^2+s_2)L_{k-2}^{\frac{1}{2}mn+3}(s_1)\Big]\\
&+4k(k-1)(k-2)\Big[\frac{mn}{8}\{n^2+3(m+1)n\\
&\qquad +m^2+3m+4\}L_{k-3}^{\frac{1}{2}mn+2}(s_1)\\
&-\frac{3}{4}\{n^2+3(m+1)n+m^2+3m+4\}s_1L_{k-3}^{\frac{1}{2}mn+3}(s_1)\\
&+\frac{3}{2}\{s_1^2+(n+m+2)s_2\}L_{k-3}^{\frac{1}{2}mn+4}(s_1)-s_3L_{k-3}^{\frac{1}{2}mn+5}(s_1)\Big]\\
&+k(k-1)(k-2)(k-3)\Big[\frac{mn}{16}\{mn^3+2(m^2+m+4)n^2\\
&+(m^3+2m^2+21m+20)n+4(2m^2+5m+5)\}L_{k-4}^{\frac{1}{2}mn+3}(s_1)\\
&-\frac{1}{2}\{mn^3+2(m^2+m+4)n^2+(m^3+2m^2+21m+20)n\\
&\qquad +4(2m^2+5m+5)\}s_1L_{k-4}^{\frac{1}{2}mn+4}(s_1)\\
&+\frac{1}{2}\{2\left\{(n+m+1)^2+6\right\}s_1^2+\{(mn+20)\\
&\qquad \times(n+m+1)+12\}s_2\}L_{k-4}^{\frac{1}{2}mn+5}(s_1)\\
&-\{2(n+m+1)s_1s_2+8s_3\}L_{k-4}^{\frac{1}{2}mn+6}(s_1)\\
&\qquad +s_2^2L_{k-4}^{\frac{1}{2}mn+7}(s_1)\Big]. \qquad (4.6.19)
\end{aligned}
$$

PROOF With the help of the generating function of univariate Laguerre polynomials, $L_k^{\alpha-1}(s_1)$, the right-hand side of (4.6.14) is expressed as

$$
\sum_{k=0}^{\infty}\frac{x^{k+2}}{k!}\Big\{\frac{mn}{4}(n+m+1)L_k^{\frac{1}{2}mn+1}(s_1)
-(n+m+1)s_1L_k^{\frac{1}{2}mn+2}(s_1)+s_2L_k^{\frac{1}{2}mn+3}(s_1)\Big\}.
$$

Comparing the coefficients of x^k on both sides, we have (4.6.17). The identities (4.6.18) and (4.6.19) are obtained in a similar way (Exercise 4.23).

Theorem 4.6.3 Let

$$\eta(\alpha, x) = x^{\alpha-1}/2^{\alpha}\Gamma(\alpha)$$
$$h_{2\alpha+2j}(x,z) = h_{2\alpha+2j} = \eta(\alpha+j,x)\exp(-x/2)\,{}_0F_1\left(\alpha+j, \frac{xz}{2}\right)$$

and

$$g_{2\alpha+2j}(x,z) = g_{2\alpha+2j} = \exp(-z)h_{2\alpha+2j}(x,z)$$

where $g_{2\alpha+2j}(x,z)$ denotes the probability density function of a chi-square random variable with $2\alpha + 2j$ degrees of freedom and noncentrality parameter z, and ${}_0F_1\left(\alpha+j, \frac{1}{2}xz\right)$ is a classical univariate Bessel function. The following equalities hold for the univariate Laguerre polynomials $L_k^{\alpha-1}(z)$.

$$\eta(\alpha,x)\sum_{k=0}^{\infty}\frac{(-x/2)^k L_k^{\alpha-1}(z)}{k!(\alpha)_k} = h_{2\alpha}, \tag{4.6.20}$$

$$\eta(\alpha,x)\sum_{k=0}^{\infty}\frac{(-x/2)^k L_k^{\alpha}(z)}{k!(\alpha)_k} = -h_{2\alpha+2} + h_{2\alpha}, \tag{4.6.21}$$

$$\eta(\alpha,x)\sum_{k=0}^{\infty}\frac{(-x/2)^k L_k^{\alpha+1}(z)}{k!(\alpha)_k} = h_{2\alpha+4} - 2h_{2\alpha+2} + h_{2\alpha}, \tag{4.6.22}$$

$$\eta(\alpha,x)\sum_{k=0}^{\infty}\frac{(-x/2)^k L_k^{\alpha+2}(z)}{k!(\alpha)_k} = -h_{2\alpha+6} + 3h_{2\alpha+4} - 3h_{2\alpha+2} + h_{2\alpha}, \tag{4.6.23}$$

$$\eta(\alpha,x)\sum_{k=0}^{\infty}\frac{(-x/2)^k L_k^{\alpha+3}(z)}{k!(\alpha)_k} = h_{2\alpha+8} - 4h_{2\alpha+6} + 6h_{2\alpha+4} - 4h_{2\alpha+2} + h_{2\alpha}, \tag{4.6.24}$$

$$\eta(\alpha,x)\sum_{k=1}^{\infty}\frac{(-x/2)^k L_k^{\alpha-1}(z)}{(k-1)!(\alpha)_k} = -zh_{2\alpha+4} + (z-\alpha)h_{2\alpha+2}, \tag{4.6.25}$$

$$\eta(\alpha,x)\sum_{k=2}^{\infty}\frac{(-x/2)^k L_k^{\alpha-1}(z)}{(k-2)!(\alpha)_k} = z^2h_{2\alpha+8} + 2\{(\alpha+1)z - z^2\}h_{2\alpha+6} + \{z^2 - 2(\alpha+1)z + \alpha(\alpha+1)\}h_{2\alpha+4}, \tag{4.6.26}$$

$$\eta(\alpha,x)\sum_{k=1}^{\infty}\frac{(-x/2)^k L_k^{\alpha}(z)}{(k-1)!(\alpha)_k} = zh_{2\alpha+6} - (2z-\alpha-1)h_{2\alpha+4} + (z-\alpha-1)h_{2\alpha+2}, \tag{4.6.27}$$

$$\eta(\alpha, x) \sum_{k=1}^{\infty} \frac{(-x/2)^k L_k^{\alpha+1}(z)}{(k-1)!(\alpha)_k}$$
$$= -z h_{2\alpha+8} + (3z - \alpha - 2) h_{2\alpha+6}$$
$$- (3z - 2\alpha - 4) h_{2\alpha+4} + (z - \alpha - 2) h_{2\alpha+2} . \tag{4.6.28}$$

PROOF Note that (4.6.20) is another type of generating function for $L_k^{\alpha-1}(z)$ (see Erdélyi et al (1953b)). As (4.6.21), (4.6.22), (4.6.23) and (4.6.24) are obtained similarly, we shall only derive (4.6.22) as an example.

$$\sum_{k=0}^{\infty} \frac{(-x/2)^k L_k^{\alpha+1}(z)}{k!(\alpha)_k}$$
$$= \frac{1}{(\alpha)_2} \sum_{k=0}^{\infty} \frac{(-x/2)^k L_k^{\alpha+1}(z)}{k!(\alpha+2)_k} \{k(k-1) + 2(\alpha+1)k + \alpha(\alpha+1)\}$$
$$= \frac{1}{(\alpha)_2} \sum_{k=2}^{\infty} \frac{(-x/2)^k L_k^{\alpha+1}(z)}{(k-2)!(\alpha+2)_k} + \frac{2(\alpha+1)}{(\alpha)_2} \sum_{k=1}^{\infty} \frac{(-x/2)^k L_k^{\alpha+1}(z)}{(k-1)!\,(\alpha+2)_k}$$
$$+ \sum_{k=0}^{\infty} \frac{(-x/2)^k L_k^{\alpha+1}(z)}{k!(\alpha+2)_k} .$$

The third term is obtained from (4.6.20) by replacing $\alpha - 1$ by $\alpha + 1$.

Consider

$$\sum_{k=0}^{\infty} \frac{(-x/2)^k L_k^{\alpha+1}(z)}{k!\,(\alpha+2)_k} = \exp(-x/2) \sum_{k=0}^{\infty} \frac{(xz/2)^k}{k!\,(\alpha+2)_k} . \tag{4.6.29}$$

Differentiating both sides of (4.6.29) with respect to x and multiplying both sides by x yield

$$\sum_{k=1}^{\infty} \frac{(-x/2)^k L_k^{\alpha+1}(z)}{(k-1)!\,(\alpha+2)_k}$$
$$= \exp\left(-\frac{x}{2}\right) \left[-\frac{x}{2} \sum_{k=0}^{\infty} \frac{(xz/2)^k}{k!(\alpha+2)_k} + \sum_{k=1}^{\infty} \frac{(xz/2)^k}{(k-1)!(\alpha+2)_k} \right] .$$

Differentiating both sides of (4.6.29) twice and multiplying by x^2 yield

$$\sum_{k=2}^{\infty} \frac{(-x/2)^k L_k^{\alpha+1}(z)}{(k-2)!\,(\alpha+2)_k} = \exp\left(-\frac{x}{2}\right) \left[\left(\frac{x}{2}\right)^2 \sum_{k=0}^{\infty} \frac{(xz/2)^k}{k!\,(\alpha+2)_k} \right.$$
$$\left. -2\left(\frac{x}{2}\right) \sum_{k=1}^{\infty} \frac{(xz/2)^k}{(k-1)!\,(\alpha+2)_k} + \sum_{k=2}^{\infty} \frac{(xz/2)^k}{(k-2)!\,(\alpha+2)_k} \right] .$$

Adding these, we have

$$\exp\left(-\frac{x}{2}\right)\left[\frac{x^2}{2^2(\alpha)_2}\sum_{k=0}^{\infty}\frac{(xz/2)^k}{k!(\alpha+2)_k}\right.$$
$$-2\left(\frac{x}{2\alpha}\right)\sum_{k=1}^{\infty}\frac{(xz/2)^k}{(k-1)!(\alpha+2)_k}\left\{\frac{1}{\alpha+1}+\frac{1}{k}\right\}$$
$$\left.+\sum_{k=2}^{\infty}\frac{(xz/2)^k}{(k-2)!(\alpha+2)_k}\left\{\frac{1}{(\alpha)_2}+\frac{2}{(k-1)\alpha}+\frac{1}{k(k-1)}\right\}\right]$$
$$=\exp\left(-\frac{x}{2}\right)\left[\frac{x^2}{2^2(\alpha)_2}\sum_{k=0}^{\infty}\frac{(xz/2)^k}{k!(\alpha+2)_k}-2\left(\frac{x}{2\alpha}\right)\sum_{k=0}^{\infty}\frac{(xz/2)^k}{k!(\alpha+1)_k}\right.$$
$$\left.+\sum_{k=0}^{\infty}\frac{(xz/2)^k}{k!\,(\alpha)_k}\right].$$

Hence we have (4.6.22) by multiplying both sides by $\eta(\alpha, x)$. The derivation of the remaining formulae is left as an exercise (Exercise 4.24).

Corollary 4.6.3

$$\frac{x^{\alpha+3}}{2^{\alpha+4}\,\Gamma(\alpha+4)}\sum_{k=4}^{\infty}\frac{(-x/2)^{k-4}\,L_{k-4}^{\alpha+5}(z)}{(k-4)!(\alpha+4)_{k-4}}$$
$$=h_{2\alpha+12}-2h_{2\alpha+10}+h_{2\alpha+8}\,, \tag{4.6.30}$$

$$\frac{x^{\alpha+1}}{2^{\alpha+2}\Gamma(\alpha+2)}\sum_{k=3}^{\infty}\frac{(-x/2)^k L_{k-2}^{\alpha+3}(z)}{(k-3)!(\alpha+2)_{k-2}}$$
$$=-zh_{2\alpha+12}+(3z-\alpha-4)h_{2\alpha+10}$$
$$-(3z-2\alpha-8)h_{2\alpha+8}+(z-\alpha-4)h_{2\alpha+6}\,. \tag{4.6.31}$$

PROOF The proof is left as an exercise (Exercise 4.25).

Theorem 4.6.4 Let κ be a partition of k, and ν a partition of n $(k \neq n)$, then $L_\kappa^\gamma(S)$ and $L_\nu^\gamma(S)$ are orthogonal on $S > 0$ with respect to the weight function

$$W(S) = \operatorname{etr}(-S)\,|S|^{\gamma}$$

unless $\kappa = \nu$. The L^2-norm of $L_\kappa^\gamma(S)$ is given as

$$\| L_\kappa^\gamma \|^2 = k!\, C_\kappa(I_m)\Gamma_m(\gamma+\frac{1}{2}(m+1);\kappa)\,. \tag{4.6.32}$$

PROOF Multiply both sides of (4.6.9) by $\operatorname{etr}(-S)\,|S|^{\gamma}C_\nu(S)$ and integrate over $S >$

0. The left-hand side becomes

$$\Gamma_m(\gamma + \frac{1}{2}(m+1); \nu) C_\nu(I - Z)$$
$$= \Gamma_m(\gamma + \frac{1}{2}(m+1); \nu) \left[(-1)^n C_\nu(Z) + \text{ terms of lower degree}\right] .$$

The right-hand side becomes

$$\sum_{k=0}^{\infty} \sum_{\kappa} \frac{C_\kappa(Z)}{k!\, C_\kappa(I_m)} \int_{S>0} \operatorname{etr}(-S)\, |S|^\gamma\, C_\nu(S) L_\kappa^\gamma(S) \mathrm{d}S .$$

Comparing the coefficients of $C_\kappa(S)$ on both sides, we have

$$\int_{S>0} \operatorname{etr}(-S)\, |S|^\gamma\, C_\nu(S) L_\kappa^\gamma(S) \mathrm{d}S = 0 \quad \text{for} \quad k \geq n$$

unless $\kappa = \nu$. Hence $L_\kappa^\gamma(S)$ is orthogonal to any Laguerre polynomial of lower degree, and from (4.6.5)

$$L_\kappa^\gamma(S) = (-1)^k C_\kappa(S) + \text{ terms of lower degrees};$$

it is also orthogonal to all Laguerre polynomials $L_\nu^\gamma(S)$ of the same degree unless $\kappa = \nu$. Comparing the coefficients of $C_\nu(Z)$ gives the L^2-norm of L_ν^γ;

$$\int_{S>0} \operatorname{etr}(-S)\, |S|^{\gamma} L_\kappa^\gamma(S) L_\nu^\gamma(S) \mathrm{d}S$$
$$= \delta_{\kappa\nu} k!\, C_\kappa(I_m) \Gamma_m(\gamma + \frac{1}{2}(m+1); \kappa) . \tag{4.6.33}$$

Hayakawa (1969) introduced a Hermite polynomial of a rectangular matrix as follows.

Definition 4.6.3 *Hermite polynomial* Let T and U be $m \times n$ $(m \leq n)$ matrices. Define an Hermite polynomial $H_\kappa(T)$ for the partition κ of k as

$$\operatorname{etr}(-TT') H_\kappa(T)$$
$$= \frac{(-1)^k}{\pi^{mn/2}} \int_U \operatorname{etr}(-2iTU') \operatorname{etr}(-UU') C_\kappa(UU') \mathrm{d}U . \tag{4.6.34}$$

Theorem 4.6.5

$$H_\kappa(T) = (-1)^k L_\kappa^{\frac{n}{2} - \frac{1}{2}(m+1)}(TT') \tag{4.6.35}$$

and

$$H_\kappa(T) = H_\kappa(H_1 T) = H_\kappa(T H_2) \tag{4.6.36}$$

where $H_1 \in O(m)$ and $H_2 \in O(n)$, respectively.

The $H_\kappa(T)$'s are orthogonal functions with respect to the weight function $\operatorname{etr}(-TT')$

and

$$\int_T \operatorname{etr}(-TT')H_\kappa(T)H_\tau(T)\mathrm{d}T = \delta_{\kappa\tau}\pi^{\frac{1}{2}mn}k!\left(\frac{n}{2}\right)_\kappa C_\kappa(I_m), \tag{4.6.37}$$

$$|H_\kappa(T)| \le \left(\frac{n}{2}\right)_\kappa C_\kappa(I_m)\operatorname{etr}(TT'). \tag{4.6.38}$$

PROOF The region of integration is invariant under the right transformation $U \to UH$, $H \in O(n)$. In view of (4.4.16), the average over $O(n)$ yields

$$\begin{aligned}
&\frac{(-1)^k}{\pi^{\frac{1}{2}mn}}\int_U \operatorname{etr}(-UU')C_\kappa(UU')\left[\int_{O(n)} \operatorname{etr}(-2iU'TH)\mathrm{d}(H)\right]\mathrm{d}U \\
&= \frac{(-1)^k}{\pi^{\frac{1}{2}mn}}\int_U \operatorname{etr}(-UU')C_\kappa(UU')\,{}_0F_1\left(\frac{n}{2}; -TT'UU'\right)\mathrm{d}U \\
&= \frac{(-1)^k}{\Gamma_m\left(\frac{n}{2}\right)}\int_{S>0} \operatorname{etr}(-S)\,|S|^{\frac{1}{2}(n-m-1)} \\
&\qquad\qquad \times C_\kappa(S)\,{}_0F_1\left(\frac{n}{2}; -TT'S\right)\mathrm{d}S \\
&= (-1)^k\int_{S>0} \operatorname{etr}(-S)\,|S|^{\frac{1}{2}(n-m-1)}C_\kappa(S)A_{\frac{1}{2}(n-m-1)}(TT'S)\mathrm{d}S \\
&= (-1)^k L_\kappa^{\frac{1}{2}(n-m-1)}(TT')\operatorname{etr}(-TT').
\end{aligned}$$

The proof of (4.6.36) is simple and (4.6.37) is the same as (4.6.33). (4.6.38) is estimated as

$$\begin{aligned}
|H_\kappa(T)| &\le \operatorname{etr}(TT)\frac{1}{\pi^{\frac{1}{2}mn}}\int_U \operatorname{etr}(-UU')\,C_\kappa(UU)\mathrm{d}U \\
&= \operatorname{etr}(TT')\left(\frac{n}{2}\right)_\kappa C_\kappa(I_m).
\end{aligned}$$

Theorem 4.6.6 Let S and T be $m \times n$ $(m \le n)$ matrices. Then the generating function of $H_\kappa(T)$'s is given by

$$\begin{aligned}
&\int_{O(m)}\int_{O(n)} \operatorname{etr}(-SS' + 2H_1TH_2S')\mathrm{d}(H_1)\mathrm{d}(H_2) \\
&\qquad = \sum_{k=0}^{\infty}\sum_\kappa \frac{H_\kappa(T)C_\kappa(SS')}{k!\left(\frac{n}{2}\right)_\kappa C_\kappa(I_m)};
\end{aligned} \tag{4.6.39}$$

Mehler formula is given by

$$\begin{aligned}
\frac{1}{(1-\rho^2)^{\frac{mn}{2}}}\int_{O(m)}\int_{O(n)} \operatorname{etr}\Bigg\{&-\frac{\rho^2}{1-\rho^2}(SS' + TT') \\
&+ \frac{2\rho}{1-\rho^2}H_1SH_2T'\Bigg\}\mathrm{d}(H_1)\mathrm{d}(H_2)
\end{aligned}$$

$$= \sum_{k=0}^{\infty} \sum_{\kappa} \frac{H_\kappa(S) H_\kappa(T)}{k! \left(\frac{n}{2}\right)_\kappa C_\kappa(I_m)} \rho^{2k}, \quad |\rho| < 1. \tag{4.6.40}$$

PROOF Substituting (4.6.34) into the right-hand side of (4.6.39) yields by using (4.4.16)

$$\begin{aligned}
&\frac{1}{\pi^{\frac{1}{2}mn}} \int_U \operatorname{etr}(-UU' - 2iUT' + TT') \sum_{k=0}^{\infty} \sum_{\kappa} \frac{C_\kappa(-UU') C_\kappa(SS')}{k! \left(\frac{n}{2}\right)_\kappa C_\kappa(I_m)} \mathrm{d}U \\
&= \frac{1}{\pi^{\frac{1}{2}mn}} \int_U \operatorname{etr}(-UU' - 2iUT' + TT') \\
&\quad \times \sum_{k=0}^{\infty} \sum_{\kappa} \frac{1}{k! \left(\frac{n}{2}\right)_\kappa} \int_{O(m)} C_\kappa(-UU' H_1 SS' H_1') \mathrm{d}(H_1) \mathrm{d}U \\
&= \frac{1}{\pi^{\frac{1}{2}mn}} \int_U \operatorname{etr}(-UU' - 2iUT' + TT') \\
&\quad \times \int_{O(m)} \int_{O(n)} \operatorname{etr}(2iU' H_1 S H_2) \mathrm{d}(H_1) \mathrm{d}(H_2) \mathrm{d}U,
\end{aligned}$$

where $H_1 \in O(m)$ and $H_2 \in O(n)$. Thus we have

$$\begin{aligned}
&\operatorname{etr}(TT') \int_{O(m)} \int_{O(n)} \frac{1}{\pi^{\frac{1}{2}mn}} \int_U \operatorname{etr}\{-UU' - 2iU(T - H_1 S H_2)'\} \mathrm{d}U \\
&= \int_{O(m)} \int_{O(n)} \operatorname{etr}(-SS' + 2H_1 S H_2 T') \mathrm{d}(H_1) \mathrm{d}(H_2).
\end{aligned}$$

(4.6.40) is obtained similarly. The proof is left as an exercise (Exercise 4.26).

Corollary 4.6.4 Let X and Y be $m \times m$ positive definite symmetric matrices and $\gamma > -1$. The following equalities hold.

$$\begin{aligned}
&\operatorname{etr}(X)\, {}_0F_1^{(m)}\left(\gamma + \frac{1}{2}(m+1); X, -Y\right) \\
&= \sum_{k=0}^{\infty} \sum_{\kappa} \frac{L_\kappa^\gamma(Y) C_\kappa(X)}{\left(\gamma + \frac{1}{2}(m+1)\right)_\kappa k! C_\kappa(I_m)}
\end{aligned} \tag{4.6.41}$$

and

$$\begin{aligned}
&(1-\rho^2)^{-m(\gamma + \frac{1}{2}(m+1))} \operatorname{etr}\left\{-\frac{\rho^2}{1-\rho^2}(X+Y)\right\} \\
&\quad \times {}_0F_1^{(m)}\left(\gamma + \frac{1}{2}(m+1); \frac{\rho^2}{(1-\rho^2)^2} X, Y\right) \\
&= \sum_{k=0}^{\infty} \sum_{\kappa} \frac{L_\kappa^\gamma(X) L_\kappa^\gamma(Y)}{\left(\gamma + \frac{1}{2}(m+1)\right)_\kappa k!\, C_\kappa(I_m)} \rho^{2k}.
\end{aligned} \tag{4.6.42}$$

PROOF The proof is left to the reader as an exercise (Exercise 4.27).

Definition 4.6.4 *Polynomial* $P_\kappa(T, A, B)$ Let T and U be $m \times n$ ($m \le n$) matrices and A be an $n \times n$ symmetric matrix and B be an $m \times m$ symmetric matrix. Define $P_\kappa(T, A, B)$ for a partition κ of k as

$$\begin{aligned} &\operatorname{etr}(-TT')P_\kappa(T, A, B) \\ &\quad = \frac{1}{\pi^{mn/2}} \int_U \operatorname{etr}(-UU' - 2iTU')C_\kappa(-BUAU')dU \end{aligned} \tag{4.6.43}$$

where $P_\kappa(T, A, B)$ was introduced by Crowther (1975). Hayakawa (1969) introduced $P_\kappa(T, A)$, which is identical to $P_\kappa(T, A, I_m)$.

Lemma 4.6.2

$$P_\kappa(T, I_n) = P_\kappa(T, I_n, I_m) = H_\kappa(T), \tag{4.6.44}$$

$$\int_{O(m)} P_\kappa(T, A, HBH')d(H) = \frac{C_\kappa(B)}{C_\kappa(I_m)} P_\kappa(T, A), \tag{4.6.45}$$

$$P_\kappa(T, A, B) = E_V[C_\kappa(-B(V - iT)A(V - iT)')], \tag{4.6.46}$$

where $V = (v_{ij})_{m \times n} \sim N_{m,n}(O, I_m/2, I_n)$.

PROOF The proof is left as an exercise (Exercise 4.28).

The explicit expressions of $P_\kappa(T, A, B)$ for lower degrees are given as follows:

$$\begin{aligned} P_{(1)}(T, A, B) &= -\frac{1}{2}\operatorname{tr}(A)\operatorname{tr}(B) + \operatorname{tr}(BTAT'), \\ 3P_{(2)}(T, A, B) &= \frac{1}{4}\{(\operatorname{tr}(A))^2 + 2\operatorname{tr}(A^2)\}\{(\operatorname{tr}(B))^2 + 2\operatorname{tr}(B^2)\} \\ &\quad - \operatorname{tr}(A)\operatorname{tr}(B)\operatorname{tr}(BTAT') - 2\operatorname{tr}(A)\operatorname{tr}(B^2TAT') \\ &\quad - 2\operatorname{tr}(B)\operatorname{tr}(BTA^2T') - 4\operatorname{tr}(B^2)\operatorname{tr}(TA^2T') \\ &\quad + (\operatorname{tr}(BTAT'))^2 + 2\operatorname{tr}((BTAT')^2), \\ 3P_{(1^2)}(T, A, B) &= \frac{1}{2}\{(\operatorname{tr}(A))^2 - \operatorname{tr}(A^2)\}\{(\operatorname{tr}(B))^2 - \operatorname{tr}(B^2)\} \\ &\quad - 2\operatorname{tr}(A)\operatorname{tr}(B)\operatorname{tr}(BTA'T) + 2\operatorname{tr}(A)\operatorname{tr}(B^2TAT') \\ &\quad + 2\operatorname{tr}(B)\operatorname{tr}(BTA^2T') - 2\operatorname{tr}(B^2TA^2T') \\ &\quad + 2(\operatorname{tr}(BTAT'))^2 - 2\operatorname{tr}((BTAT')^2). \end{aligned} \tag{4.6.47}$$

Crowther (1975) gave $P_\kappa(T, A, B)$ for $k = 3$. The explicit formula of $P_\kappa(T, A)$ is given in Appendix, Davis (1979).

Theorem 4.6.7 The generating function of $P_\kappa(T,A,B)$ is expressed as

$$\int_{O(m)}\int_{O(n)} \operatorname{etr}(-BH_1SH_2AH_2'S'H_1' + 2B^{1/2}H_1SH_2A^{1/2}T')\mathrm{d}(H_1)\mathrm{d}(H_2) = \sum_{k=0}^{\infty}\sum_{\kappa}\frac{P_\kappa(T,A,B)C_\kappa(SS')}{k!\left(\frac{n}{2}\right)_\kappa C_\kappa(I_m)}. \tag{4.6.48}$$

The generating function of $P_\kappa(T,A)$ is obtained by setting $B = I_m$ in (4.6.48). The generating function of $\sum_\kappa P_\kappa(T,A)$ for k is expressed as

$$|I+xA|^{-m/2}\operatorname{etr}\{xTA(I+xA)^{-1}T'\} = \sum_{k=0}^{\infty}\frac{x^k}{k!}\sum_{\kappa}P_\kappa(T,A), \quad \| xA \| < 1, \tag{4.6.49}$$

and

$$\sum_\kappa P_\kappa(T,A) = k!\left\{A_k + \frac{1}{2}\sum_{i_1+i_2=k}A_{i_1}A_{i_2} + \frac{1}{3!}\sum_{i_1+i_2+i_3=k}A_{i_1}A_{i_2}A_{i_3} + \cdots + \frac{1}{k!}A_1^k\right\}, \tag{4.6.50}$$

where the i_α's are positive integers greater than or equal to 1, and

$$A_\ell = (-1)^\ell\left\{\frac{m}{2\ell}\operatorname{tr}(A)^\ell - \operatorname{tr}(TA)^\ell T'\right\}, \quad \ell = 1,2,\ldots,k$$
$$A_0 \equiv 0, \qquad \text{for convenience}. \tag{4.6.51}$$

PROOF (4.6.48) is obtained by steps similar to those used in the proof of Theorem 4.6.6. Noting $\| xA \| < 1$ for the convergence of the right-hand side one has (4.6.49). Expanding the left-hand side for $\| xA \| < 1$ and comparing the coefficients of x^k one gets (4.6.50).

Khatri (1977) introduced the following two types of Laguerre polynomials.

Definition 4.6.5 *Laguerre polynomials (II)* Let S and B be $m \times m$ symmetric matrices and let T be an $m \times n$ $(m \le n)$ matrix. Define $L_\kappa^\gamma(S,B)$ and $L_\kappa^\gamma(T,S,B)$ as follows:

$$\operatorname{etr}(-S)L_\kappa^\gamma(S,B) = \int_{R>0}A_\gamma(SR)\operatorname{etr}(-R)|R|^\gamma C_\kappa(BR)\mathrm{d}R \tag{4.6.52}$$

$$\operatorname{etr}(-S)L_\kappa^\gamma(S,A,T)$$

$$= \int_{R>0} A_\gamma(SR)\,\mathrm{etr}(-R)|R|^\gamma P_\kappa(T,-A,R)\mathrm{d}R. \tag{4.6.53}$$

Lemma 4.6.3

$$L_\kappa^\gamma(S,A,T) = E_V[L_\kappa^\gamma(S,(V-iT)A(V-iT)')] \tag{4.6.54}$$

where $V=(v_{ij})_{m\times n} \sim N_{m,n}(O, I_m/2, I_n)$.

$$L_\kappa^\gamma(H_1SH_1', A, T) = L_\kappa^\gamma(S, A, H_1T),$$
$$L_\kappa^\gamma(S, H_2AH_2', T) = L_\kappa^\gamma(S, A, TH_2),$$
$$\int_{O(m)} L_\kappa^\gamma(HSH', A, T)\mathrm{d}(H) = \frac{L_\kappa^\gamma(S)}{C_\kappa(I_m)} P_\kappa(T,-A) \tag{4.6.55}$$

where $H_1 \in O(m)$ and $H_2 \in O(n)$, respectively.

PROOF These are obtained from Definition 4.6.5.

The explicit forms of these for $\gamma = \frac{1}{2}(n-m-1)$ are listed below.

$$\begin{aligned}
L_{(1)}^\gamma(S,B) &= \frac{n}{2}\,\mathrm{tr}(B) - \mathrm{tr}(BS),\\
L_{(2)}^\gamma(S,B) &= \frac{1}{12}n(n+2)\{(\mathrm{tr}(B))^2 + 2\,\mathrm{tr}(B^2)\}\\
&\quad - \frac{1}{3}(n+2)\{\mathrm{tr}(B)\,\mathrm{tr}(BS) + 2\,\mathrm{tr}(B^2S)\}\\
&\quad + \frac{1}{3}\{(\mathrm{tr}(BS))^2 + 2\,\mathrm{tr}((BS)^2)\},
\end{aligned} \tag{4.6.56}$$

$$\begin{aligned}
L_{(1^2)}(S,B) &= \frac{1}{6}n(n-1)\{(\mathrm{tr}(B))^2 - \mathrm{tr}(B^2)\}\\
&\quad - \frac{2}{3}(n-1)\{\mathrm{tr}(B)\,\mathrm{tr}(BS) - \mathrm{tr}(B^2S)\}\\
&\quad + \frac{2}{3}\{(\mathrm{tr}(BS))^2 - \mathrm{tr}((BS)^2)\},
\end{aligned}$$

$$\begin{aligned}
L_{(1)}^\gamma(S,A,T) &= \frac{mn}{4}\,\mathrm{tr}(A) - \frac{n}{2}\,\mathrm{tr}(TAT')\\
&\quad - \frac{1}{2}\,\mathrm{tr}(A)\,\mathrm{tr}(S) + \mathrm{tr}(STAT'),\\
L_{(2)}^\gamma(S,A,T) &= \frac{1}{12}n(n+2)\left\{\frac{m(m+2)}{4}(\mathrm{tr}(A))^2 + \frac{1}{2}m(m+1)\,\mathrm{tr}(A^2)\right.\\
&\quad - (m+2)\,\mathrm{tr}(A)\,\mathrm{tr}(TAT') - 2(m+1)\,\mathrm{tr}(TA^2T')\\
&\quad \left. + (\mathrm{tr}(TAT'))^2 + 2\,\mathrm{tr}((TAT')^2)\right\}\\
&\quad - \frac{1}{3}(n+2)\left\{\frac{1}{4}(m+2)\,\mathrm{tr}(S)(\mathrm{tr}(A))^2 + \frac{1}{2}(m+1)\,\mathrm{tr}(S)\,\mathrm{tr}(A^2)\right.
\end{aligned}$$

$$
\begin{aligned}
&-\frac{1}{2}(m+4)\operatorname{tr}(A)\operatorname{tr}(STAT')-\frac{1}{2}\operatorname{tr}(A)\operatorname{tr}(S)\operatorname{tr}(TAT')\\
&-(m+2)\operatorname{tr}(STA^2T')-\operatorname{tr}(S)\operatorname{tr}(TA^2T')\\
&+\operatorname{tr}(TAT')\operatorname{tr}(STAT')+2\operatorname{tr}((TAT')^2S)\Big\}\\
&+\frac{1}{3}\Big\{\frac{1}{4}(\operatorname{tr}(A))^2(\operatorname{tr}(S))^2+\frac{1}{2}\operatorname{tr}(A^2)\operatorname{tr}(S^2)+\frac{1}{2}(\operatorname{tr}(A))^2\operatorname{tr}(S^2)\\
&+\frac{1}{2}\operatorname{tr}(A^2)(\operatorname{tr}(S))^2-\operatorname{tr}(A)\operatorname{tr}(S)\operatorname{tr}(STAT')-2\operatorname{tr}(S^2TA^2T')\\
&-2\operatorname{tr}(A)\operatorname{tr}(S^2TAT')-2\operatorname{tr}(S)\operatorname{tr}(STA^2T')\\
&+(\operatorname{tr}(STAT'))^2+2\operatorname{tr}((STAT')^2)\Big\}, \qquad (4.6.57)
\end{aligned}
$$

$$
\begin{aligned}
L^{\gamma}_{(1^2)}(S,A,T)=&\frac{1}{6}n(n-1)\Big\{\frac{1}{4}m(m-1)(\operatorname{tr}(A))^2-\frac{1}{4}m(m+1)\operatorname{tr}(A^2)\\
&-(m-2)\operatorname{tr}(A)\operatorname{tr}(TAT')+(m+1)\operatorname{tr}(TA^2T')\\
&+(\operatorname{tr}(TAT')^2)-\operatorname{tr}((TAT')^2)\Big\}\\
&-\frac{2}{3}(n-1)\Big\{\frac{1}{4}(m-1)\operatorname{tr}(S)(\operatorname{tr}(A))^2-\frac{1}{4}(m+1)\operatorname{tr}(S)\operatorname{tr}(A^2)\\
&-\frac{1}{2}(m-2)\operatorname{tr}(A)\operatorname{tr}(STAT')-\frac{1}{2}\operatorname{tr}(A)\operatorname{tr}(S)\operatorname{tr}(TAT')\\
&+\frac{1}{2}(m+2)\operatorname{tr}(STA^2T')+\frac{1}{2}\operatorname{tr}(S)\operatorname{tr}(TA^2T')\\
&+\operatorname{tr}(TAT')\operatorname{tr}(STAT')-\operatorname{tr}((TAT')^2S)\Big\}\\
&+\frac{2}{3}\Big\{\frac{1}{4}(\operatorname{tr}(A))^2(\operatorname{tr}(S))^2-\frac{1}{4}\operatorname{tr}(A^2)\operatorname{tr}(S^2)-\frac{1}{4}(\operatorname{tr}(A))^2\operatorname{tr}(S^2)\\
&-\frac{1}{4}\operatorname{tr}(A^2)(\operatorname{tr}(S))^2\\
&-\operatorname{tr}(A)\operatorname{tr}(S)\operatorname{tr}(STAT')+\operatorname{tr}(S^2TA^2T')+\operatorname{tr}(A)\operatorname{tr}(S^2TAT')\\
&+\operatorname{tr}(S)\operatorname{tr}(STA^2T')+(\operatorname{tr}(STAT'))^2-\operatorname{tr}((STAT')^2)\Big\}.
\end{aligned}
$$

Theorem 4.6.8 The generating function of $L^{\gamma}_{\kappa}(S,B)$ is given by

$$
\begin{aligned}
\int_{O(m)}&|I-B^{1/2}HZH'B^{1/2}|^{-\gamma-\frac{1}{2}(m+1)}\\
&\times\operatorname{etr}\{-SB^{1/2}HZH'B^{1/2}(I-B^{1/2}HZH'B^{1/2})^{-1}\}\mathrm{d}(H)\\
&=\sum_{k=0}^{\infty}\sum_{\kappa}\frac{L^{\gamma}_{\kappa}(S,B)\,C_{\kappa}(Z)}{k!\,C_{\kappa}(I_m)}. \qquad (4.6.58)
\end{aligned}
$$

The generating function of $\sum_\kappa L^\gamma_\kappa(S,B)$ is obtained by setting $Z = xI$, $|x| < 1$ in (4.6.58).

$L^\gamma_\kappa(S,B)$'s are orthogonal polynomials with respect to the weight function $\operatorname{etr}(-S)\,|S|^\gamma$ and

$$\int_{S>0} \operatorname{etr}(-S)\,|S|^\gamma\, L^\gamma_\kappa(S,A) L^\gamma_\nu(S,B)\mathrm{d}S$$
$$= \delta_{\kappa\nu} k!\,\Gamma_m(\gamma+\frac{1}{2}(m+1);\kappa) C_\kappa(AB). \qquad (4.6.59)$$

The Laplace transforms of $|S|^\gamma L^\gamma_\kappa(S,B)$ and $|S|^\gamma L^\gamma_\kappa(S,A,T)$ are respectively

$$\int_{S>0} \operatorname{etr}(-ZS)\,|S|^\gamma\, L^\gamma_\kappa(S,B)\mathrm{d}S$$
$$= \Gamma_m(\gamma+\frac{1}{2}(m+1);\kappa)\,|Z|^{-\gamma-\frac{1}{2}(m+1)}$$
$$\times\, C_\kappa((I-Z^{-1})B) \qquad (4.6.60)$$

and

$$\int_{S>0} \operatorname{etr}(-ZS)\,|S|^\gamma\, L^\gamma_\kappa(S,A,T)\mathrm{d}S$$
$$= \Gamma_m(\gamma+\frac{1}{2}(m+1);\kappa)\,|Z|^{-\gamma-\frac{1}{2}(m+1)}$$
$$\times\, P(T,-A,I-Z^{-1}) \qquad (4.6.61)$$

for $\mathcal{R}e(Z) > I$.

PROOF The proofs of (4.6.58), (4.6.59), (4.6.60) and (4.6.61) are obtained by steps similar to those used in the proofs of Theorem 4.6.1, Theorem 4.6.4 and Exercise 4.31.

Corollary 4.6.5 $\sum_\kappa L^\gamma_\kappa(S,B)$ is given below for lower degrees and $\gamma = \frac{1}{2}(n-m-1)$.

$$k=1:\quad \frac{n}{2}\operatorname{tr}(B) - \operatorname{tr}(SB),$$

$$k=2:\quad \frac{n^2}{4}(\operatorname{tr}(B))^2 + \frac{n}{2}\operatorname{tr}(B^2) - 2\operatorname{tr}(SB^2)$$
$$- n\operatorname{tr}(B)\operatorname{tr}(SB) + (\operatorname{tr}(SB))^2, \qquad (4.6.62)$$

$$k=3:\quad n\operatorname{tr}(B^3) + \frac{3}{4}n^2\operatorname{tr}(B)\operatorname{tr}(B^2) + \frac{n^3}{8}(\operatorname{tr}(B))^3$$
$$- 6\operatorname{tr}(SB^3) + 6\operatorname{tr}(SB)\operatorname{tr}(SB^2) - (\operatorname{tr}(SB))^3$$
$$- n\left\{\frac{3}{2}\operatorname{tr}(B^2)\operatorname{tr}(SB) + 3\operatorname{tr}(B)\operatorname{tr}(SB^2) - \frac{3}{2}\operatorname{tr}(B)(\operatorname{tr}(SB))^2\right\}$$
$$- \frac{3}{4}n^2(\operatorname{tr}(B))^2\operatorname{tr}(SB).$$

PROOF Setting $Z = xI$ and $\gamma = \frac{1}{2}(n-m-1)$ in (4.6.61) yields

$$|I - xB|^{-\frac{n}{2}} \operatorname{etr}(-xSB(I-xB)^{-1}) = \sum_{k=0}^{\infty} \frac{x^k}{k!} \sum_{\kappa} L_{\kappa}^{\gamma}(S, B).$$

The left-hand side can be expressed as

$$\exp\left\{-\frac{n}{2}\ln|I - xB| - \sum_{k=1}^{\infty} \frac{x^k}{k!} \operatorname{tr}(SB^{k+1})\right\} = \exp\left\{\sum_{k=1}^{\infty} \frac{x^k}{k!}\left(\frac{n}{2}\operatorname{tr}(B^k) - \operatorname{tr}(SB^{k+1})\right)\right\};$$

then comparing the coefficient of x^k for lower degrees gives (4.6.62).

Theorem 4.6.2 is given by Fujikoshi (1970), and Corollary 4.6.2 and Theorem 4.6.3 are given by Hayakawa (1972b). The orthogonality of Laguerre polynomials is based on Constantine (1966). The Jacobi polynomials are studied in James and Constantine (1974). The relevant summary for special functions of a matrix argument is given in James (1976) and Mathai (1993). Additional materials for Hermitian matrix arguments can be found in Khatri (1970), Fujikoshi (1971) and Hayakawa (1972c).

EXERCISES

4.1 Show that $m_{(k)}(\alpha), m_{(k-1,1)}(\alpha), \ldots, m_{(1^k)}(\alpha)$ are linearly independent. Show the linear independence of $\{a_\kappa(\alpha)\}$, $\{s_\kappa(\alpha)\}$ and $\{C_\kappa(A)\}$, respectively.

4.2 Find $T_t^{(k)}$ of (4.2.21) for $k = 3, 4$ and 5 with the help of Table 4.2.1.

4.3 Find $\Xi^{(4)}$ and $\Xi^{(5)}$.

4.4 Find $\Xi^{(k)}T_A^{(k)}$ for $k = 3, 4$ and 5, and note that all elements are positive.

4.5 Show that sum of the elements of each row except the first row of $Z^{(k)}$ is zero and that sum of the elements of the first row is $\xi_{11}^{(k)}$ of $\Xi^{(k)}$.

4.6 Let W be distributed as $W_m(n, I_m)$ and let A be an $m \times m$ symmetric non-singular matrix. Let $Q_n^{(k)}$ be a matrix defined in Lemma 4.4.4 as

$$E_W[\boldsymbol{v}^{(\kappa)}(AW)] = Q_n^{(k)} \boldsymbol{v}^{(k)}(A).$$

Show that $Q_n^{(k)}$ is given by

$$Q_n^{(k)} = T_A^{(k)} D_n^{(k)} T_A^{(k)^{-1}},$$

where $D_n^{(k)}$ is defined by (4.3.13).

4.7 Let W be distributed as $W_m(n, I_m)$ and let A be an $m \times m$ symmetric matrix. Prove for a partition $\kappa = (k_1, \ldots, k_m)$ of k

$$E_W[a_\kappa(AW^{-1})] = \sum_{\kappa \geq \tau} \bar{d}^{(n)}_{\kappa\tau} a_\tau(A)$$

and

$$\begin{aligned}\bar{d}^{(n)}_{\kappa\kappa} &= \Gamma_m\left(\frac{n}{2}; -\kappa\right) \\ &= \pi^{\frac{1}{4}m(m-1)} \prod_{\alpha=1}^{m} \Gamma\left(\frac{n}{2} - k_i - \frac{1}{2}(m-i)\right) \\ &= \frac{(-1)^k \Gamma_m\left(\frac{n}{2}\right)}{\left(-\frac{n}{2} + \frac{1}{2}(m+1)\right)_\kappa}\end{aligned}$$

for $\frac{n}{2} > k_1 + \frac{1}{2}(m-1)$.

4.8 Let Z be an $m \times m$ complex symmetric matrix with $\mathcal{R}e(Z) > 0$, and let T be an $m \times m$ arbitrary complex symmetric matrix. Show that

$$\begin{aligned}&\int_{S>0} \operatorname{etr}(-ZS)\,|S|^{a-\frac{1}{2}(m+1)} C_\kappa(TS^{-1}) dS \\ &\quad = \Gamma_m(a; -\kappa)\,|Z|^{-a} C_\kappa(TZ)\end{aligned}$$

for $\mathcal{R}e(a) > k_1 + \frac{1}{2}(m-1)$ (Khatri (1966), Constantine (1966)).

4.9 Let T be an $m \times m$ arbitrary symmetric matrix. Then

(1)
$$\begin{aligned}&\int_{S>0} |S|^{t-\frac{1}{2}(m+1)}\,|I+S|^{-(t+u)} C_\kappa(TS) dS \\ &\quad = \frac{\Gamma_m(t;\kappa)\,\Gamma_m(u;-\kappa)}{\Gamma_m(t+u)} C_\kappa(T)\end{aligned}$$

(2)
$$\begin{aligned}&\int_{S>0} |S|^{t-\frac{1}{2}(m+1)}\,|I+S|^{-(t+u)} C_\kappa(TS^{-1}) dS \\ &\quad = \frac{\Gamma_m(t;-\kappa)\,\Gamma_m(u;\kappa)}{\Gamma_m(t+u)} C_\kappa(T)\,.\end{aligned}$$

Let R be an $m \times m$ positive definite matrix. Then for $t \geq \frac{1}{2}(m + k_1)$ show that

(3)
$$\begin{aligned}&\int_{0<S<I} |S|^{t-\frac{1}{2}(m+1)}\,|I-S|^{u-\frac{1}{2}(m+1)} C_\kappa(RS^{-1}) dS \\ &\quad = \frac{\Gamma_m(t;-\kappa)\,\Gamma_m(u)}{\Gamma_m(t+u;-\kappa)} C_\kappa(R).\end{aligned}$$

Hint: Multiplying both sides of (4.4.6) by $\operatorname{etr}(-R)$ $|R|^{t+u-\frac{1}{2}(m+1)}$ and integrating over $R > 0$ give (1). Transforming S to S^{-1} and interchanging t and u in (1) yield (2). Multiply both sides of the formula in Exercise 4.8 for $a = t$ and $T = Z^{-1}$ by

$\operatorname{etr}(-Z)\,|Z|^{t+u-\frac{1}{2}(m+1)}$ and integrate it over $Z > 0$, and transform $(I + S^{-1})^{-1}$ to S (Khatri (1966)).

4.10 Show that the following hold. Let T be any $m \times m$ complex symmetric matrix, then

(1)
$$\int_{S>0} \operatorname{etr}(-S)\ |S|^{t-\frac{1}{2}(m+1)}(\operatorname{tr}(S))^j C_\kappa(TS)\mathrm{d}S = \frac{\Gamma_m(t;\kappa)\Gamma(mt+j+k)}{\Gamma(mt+k)} C_\kappa(T)$$

(2)
$$\int_{S>0} \operatorname{etr}(-S)\ |S|^{t-\frac{1}{2}(m+1)}(\operatorname{tr}(S))^j C_\kappa(TS^{-1})\mathrm{d}S = \frac{\Gamma_m(t;-\kappa)\Gamma(mt+j-k)}{\Gamma(mt-k)} C_\kappa(T)$$

(Khatri (1966)).

4.11 Prove (4.3.68).

4.12 Let $f\left(\frac{S}{n}\right)$ be a function defined in Lemma 4.3.6. Show that (4.3.54) is extended as follows:

$$E\left[f\left(\frac{1}{n}S\right)\right] = \Theta f(\Sigma)|_{\Sigma=\Lambda}\,,$$

where

$$\Theta = 1 + \frac{1}{n}\operatorname{tr}((\Lambda\partial)^2) + \frac{1}{6n^2}\left[3\left\{\operatorname{tr}((\Lambda\partial)^2\right\}^2) + 8\operatorname{tr}((\Lambda\partial)^3)\right] + o\left(\frac{1}{n^2}\right).$$

4.13 Derive all formulae from (4.4.35) to (4.4.43).

4.14 Show that the following hold for $V = \Lambda(I-\Lambda)^{-1}$.

$$\operatorname{tr}((\Lambda\partial)^2)\ |I-\Sigma|^{-b}\big|_{\Sigma=\Lambda} = \frac{b}{2}\{(\operatorname{tr}(V))^2 + (2b+1)\operatorname{tr}(V^2)\}\,|I-\Lambda|^{-b}$$

$$\operatorname{tr}((\Lambda\partial)^3)\ |I-\Sigma|^{-b}\big|_{\Sigma=\Lambda} = \frac{b}{4}\{(\operatorname{tr}(V))^3 + 3(2b+1)\operatorname{tr}(V)\operatorname{tr}(V^2)$$
$$+2(2b^2+3b+2)\operatorname{tr}(V^3)\}\,|I-\Lambda|^{-b}$$

$$\{\operatorname{tr}((\Lambda\partial)^2)\}^2\ |I-\Sigma|^{-b}\big|_{\Sigma=\Lambda}$$
$$= \frac{b}{4}\{b(\operatorname{tr}(V))^4 + 2(2b^2+b+2)(\operatorname{tr}(V))^2\operatorname{tr}(V^2)$$
$$+(2b+1)(2b^2+b+2)(\operatorname{tr}(V^2))^2 + 8(2b+1)\operatorname{tr}(V)\operatorname{tr}(V^3)$$
$$+2(8b^2+10b+5)\operatorname{tr}(V^4)\}\,|I-\Lambda|^{-b}$$

(Fujikoshi (1970)).

4.15 Prove Lemma 4.4.6 except (4.4.49).

4.16 Let ∂ be an $m \times m$ differential operator matrix defined in Lemma 4.3.6 and let A be an $m \times m$ symmetric matrix. For the zonal polynomial $C_\kappa(A)$ corresponding to the partition κ prove that

$$C_\kappa(\partial)\,\mathrm{etr}\,(A\Sigma) = C_\kappa(A)\,\mathrm{etr}\,(A\Sigma).$$

4.17 Let Σ be an $m \times m$ symmetric matrix and let $\ell_\alpha(\Sigma)$ be the α-th latent root of the determinantal equation

$$f(\ell, \Sigma) = |\Sigma - \ell I_m| = 0.$$

Let $\Lambda = \mathrm{diag}\,(\lambda_1, \ldots, \lambda_m)$, $\lambda_1 > \lambda_2 > \cdots > \lambda_m$. With the help of the implicit function theorem under the condition

$$\frac{\partial}{\partial \ell} f(\ell, \Sigma) \neq 0,$$

show that the derivatives of $\ell_\alpha \equiv \ell_\alpha(\Sigma)$ with respect to $\sigma_{ij}\,(i \leq j)$ at the point $\Sigma = \Lambda$ are given as

$$\frac{\partial \ell_\alpha}{\partial \sigma_{\alpha\alpha}} = 1, \qquad \frac{\partial \ell_\alpha}{\partial \sigma_{ij}} = 0 \quad \text{for } (i,j) \neq (\alpha, \alpha)$$

$$\frac{\partial^2 \ell_\alpha}{\partial \sigma_{\alpha j}^2} = \frac{2}{\lambda_\alpha - \lambda_j} \quad \text{for } j \neq \alpha$$

$$\frac{\partial^2 \ell_\alpha}{\partial \sigma_{ij}\partial \sigma_{k\ell}} = 0 \quad \text{except for the above case,}$$

$$\frac{\partial^3 \ell_\alpha}{\partial \sigma_{\alpha\alpha}\partial \sigma_{\alpha j}^2} = \frac{-2}{(\lambda_\alpha - \lambda_j)^2} \quad \text{for } j \neq \alpha$$

$$\frac{\partial^3 \ell_\alpha}{\partial \sigma_{jj}\partial \sigma_{\alpha j}^2} = \frac{2}{(\lambda_\alpha - \lambda_j)^2} \quad \text{for } j \neq \alpha$$

$$\frac{\partial^3 \ell_\alpha}{\partial \sigma_{\alpha j}\partial \sigma_{\alpha k}\partial \sigma_{jk}} = \frac{2}{(\lambda_\alpha - \lambda_j)(\lambda_\alpha - \lambda_k)} \quad \text{for } j \neq k \neq \alpha$$

$$\frac{\partial^3 \ell_\alpha}{\partial \sigma_{jk}\partial \sigma_{pq}\partial \sigma_{rs}} = 0 \quad \text{except for the above three cases.}$$

The Taylor expansion of $\ell_\alpha(\Sigma)$ at Λ is

$$\ell_\alpha(\Sigma) = \lambda_\alpha + (\sigma_{\alpha\alpha} - \lambda_\alpha) + \frac{1}{2} \sum_{j \neq \alpha} \frac{2}{\lambda_\alpha - \lambda_j} \sigma_{\alpha j}^2$$

$$+ \frac{1}{3!} \left[\sum_{j \neq \alpha} \frac{-2 \times 3}{(\lambda_\alpha - \lambda_j)^2} (\sigma_{\alpha\alpha} - \lambda_\alpha)\sigma_{\alpha j}^2 \right.$$

$$+\sum_{j\neq\alpha}\frac{2\times 3}{(\lambda_\alpha-\lambda_j)^2}\sigma_{\alpha j}^2(\sigma_{jj}-\lambda_j)$$

$$\left.+\sum_{\substack{j<k\\ j,k\neq\alpha}}\frac{2\times 6}{(\lambda_\alpha-\lambda_j)(\lambda_\alpha-\lambda_k)}\sigma_{\alpha j}\sigma_{\alpha k}\sigma_{jk}\right]$$

+ terms of higher order (Sugiura (1973)).

4.18 Show (4.5.8) with the help of (4.5.5).

4.19 Show Lemma 4.5.3 with the help of Theorem 4.5.1.

4.20 Prove that for the partition σ of ℓ

$$\sum_{k=0}^{\ell}(-1)^k\sum_{\kappa}\binom{\sigma}{\kappa}\frac{(a)_\kappa}{(b)_\kappa}=\frac{(b-a)_\sigma}{(b)_\sigma}\quad\text{(Bingham (1974))}.$$

4.21 Let $P(Z)$ be a homogeneous symmetric polynomial of degree k in Z and $s_1=\mathrm{tr}\,(Z)$.

(i) Show for the partition σ of $k+r$

$$s_1^rP(Z)=\binom{k+r}{r}^{-1}\sum_{\sigma}\binom{\sigma}{P}C_\sigma(Z).$$

(ii) If σ is a partition of s and $s\geq k+r$, then show that

$$\binom{\sigma}{s_1^rP}=\binom{k+r}{r}^{-1}\binom{s-k}{r}\binom{\sigma}{P}.$$

(iii) For a partition σ of s and $k\leq\ell\leq s$, show that

$$\sum_{\lambda}\binom{\sigma}{\lambda}\binom{\lambda}{P}=\binom{s-k}{\ell-k}\binom{\sigma}{P}$$

and

$$\sum_{\lambda}\binom{\sigma}{\lambda}\binom{\lambda}{\kappa}=\binom{s-k}{\ell-k}\binom{\sigma}{\kappa},$$

where λ is a partition of S (Bingham (1974)).

4.22 Let S be distributed as $W(I_m,n)$ and define

$$Z_n=(S-nI_m)/\sqrt{2n}.$$

(i) Find the exact probability density function $g_n(Z_n)$ and show that

$$\lim_{n\to\infty}g_n(Z_n)=\frac{1}{2^{\frac{m}{2}}\pi^{\frac{1}{4}m(m+1)}}\,\mathrm{etr}\left(-\frac{1}{2}Z^2\right),$$

where $Z=Z'=(z_{ij})$.

(ii) Find the exact moment $E[\operatorname{tr}(Z_n^2)(\operatorname{tr}(Z_n))^2]$ with the help of Lemma 4.4.4 and give

$$\lim_{n\to\infty} E[\operatorname{tr}(Z_n^2)(\operatorname{tr}(Z_n))^2]\,.$$

(iii) If $Z = (z_{ij})$, $z_{ii} \sim N(0,1)$ and $z_{ij} \sim N(0,1/2)$ are independently distributed, $i,j = 1,2,\ldots,m$, show that

$$E[\operatorname{tr}(Z^2)(\operatorname{tr}(Z))^2] = \lim_{n\to\infty} E[\operatorname{tr}(Z_n^2)(\operatorname{tr}(Z_n))^2]\,.$$

Let $\boldsymbol{s}^{(k)}(Z) = [(\operatorname{tr}(Z))^k, (\operatorname{tr}(Z))^{k-2}\operatorname{tr}(Z^2), \ldots, \operatorname{tr}(Z^k)]'$, $\boldsymbol{m}^{(k)} = (m^k, m^{k-2}, \ldots, m)$ and

$$E[\boldsymbol{s}^{(k)}(Z)] = L^{(k)}\boldsymbol{m}^{(k)}\,.$$

The $L^{(k)}$'s for $k = 2, 4, 6$, and 8 are given below.

$$L^{(8)} = \frac{1}{16}\begin{bmatrix}
0 & 0 & 0 & 0 & 1680 & 0 & 0 & 0 \\
0 & 0 & 0 & 120 & 120 & 1440 & 0 & 0 \\
0 & 0 & 12 & 24 & 252 & 240 & 1152 & 0 \\
0 & 2 & 6 & 54 & 98 & 400 & 352 & 768 \\
1 & 4 & 30 & 76 & 249 & 376 & 560 & 384 \\
0 & 0 & 0 & 0 & 360 & 360 & 960 & 0 \\
0 & 0 & 0 & 36 & 72 & 516 & 480 & 576 \\
0 & 0 & 6 & 18 & 138 & 246 & 696 & 576 \\
0 & 0 & 0 & 0 & 120 & 252 & 732 & 576 \\
0 & 0 & 0 & 24 & 78 & 384 & 690 & 504 \\
0 & 0 & 0 & 24 & 60 & 636 & 576 & 384 \\
0 & 0 & 4 & 14 & 116 & 226 & 744 & 576 \\
0 & 2 & 9 & 57 & 155 & 397 & 580 & 480 \\
0 & 0 & 0 & 12 & 42 & 348 & 702 & 576 \\
0 & 0 & 4 & 20 & 117 & 354 & 689 & 496 \\
0 & 0 & 0 & 0 & 120 & 300 & 780 & 480 \\
0 & 0 & 0 & 20 & 70 & 340 & 650 & 600 \\
0 & 0 & 0 & 0 & 90 & 360 & 720 & 510 \\
0 & 0 & 0 & 10 & 44 & 344 & 682 & 600 \\
0 & 0 & 5 & 27 & 134 & 357 & 665 & 492 \\
0 & 0 & 0 & 0 & 70 & 308 & 728 & 574 \\
0 & 0 & 0 & 14 & 93 & 374 & 690 & 509
\end{bmatrix}$$

$$L^{(2)} = \frac{1}{2}\begin{bmatrix} 0 & 2 \\ 1 & 1 \end{bmatrix}, \qquad L^{(4)} = \frac{1}{4}\begin{bmatrix} 0 & 0 & 12 & 0 \\ 0 & 2 & 2 & 8 \\ 1 & 2 & 5 & 4 \\ 0 & 0 & 6 & 6 \\ 0 & 2 & 5 & 5 \end{bmatrix}$$

$$L^{(6)} = \frac{1}{8}\begin{bmatrix} 0 & 0 & 0 & 120 & 0 & 0 \\ 0 & 0 & 12 & 12 & 96 & 0 \\ 0 & 2 & 4 & 26 & 24 & 64 \\ 1 & 3 & 15 & 25 & 44 & 32 \\ 0 & 0 & 0 & 36 & 36 & 48 \\ 0 & 0 & 6 & 12 & 54 & 48 \\ 0 & 0 & 0 & 24 & 54 & 42 \\ 0 & 0 & 4 & 10 & 58 & 48 \\ 0 & 2 & 7 & 26 & 45 & 40 \\ 0 & 0 & 0 & 20 & 50 & 50 \\ 0 & 0 & 5 & 22 & 52 & 41 \end{bmatrix}$$

The $L^{(k)}$'s for $k = 10, 12$ are given in Hayakawa and Kikuchi (1978).

4.23 Show (4.6.18) and (4.6.19).

4.24 Show (4.6.24), (4.6.26) and (4.6.28).

4.25 Show (4.6.30) and (4.6.31).

4.26 Show Mehler formula for $H_\kappa(T)$.

4.27 Show (4.6.41) and (4.6.42).

4.28 Show Lemma 4.6.2.

4.29 Let κ be a partition of k and $\gamma = \frac{1}{2}(n-m-1)$. Let Z be an $m \times m$ symmetric matrix. Show

$$\sum_{\ell=k}^{\infty}\sum_{\lambda}\binom{\lambda}{\kappa}\frac{x^\ell L_\lambda^\gamma(Z)}{\ell!}$$
$$= (1-x)^{-\frac{1}{2}mn}\operatorname{etr}\left(-\frac{x}{1-x}Z\right)\left(\frac{x}{1-x}\right)^k\frac{1}{k!}L_\kappa^\gamma\left(\frac{1}{1-x}Z\right).$$

4.30 Let κ be a partition of k and let A be an $n \times n$ symmetric matrix satisfying $\| A \| < 1$. Let T be an $m \times m$ rectangular matrix. Show

$$\sum_{\ell=k}^{\infty}\sum_{\lambda}(-1)^\ell\binom{\lambda}{\kappa}\frac{P_\lambda(T,A)}{\ell!}$$
$$= (-1)^k \operatorname{etr}\{-(I-A)^{-1}ATT'\}|I_n - A|^{-\frac{1}{2}m}$$
$$\times P_\kappa\left(T(I-A)^{-1/2}, (I-A)^{-1/2}A(I-A)^{-1/2}\right)/k!$$

(Bingham (1974)).

4.31 Prove the orthogonality of $\{L_\kappa^\gamma(S,A)\}$ and $\{L_\kappa^\gamma(S,A,T)\}$ with respect to the weight function $\operatorname{etr}(-S)\,|S|^{\gamma}$, respectively.

4.32 Let X be an $m \times m$ symmetric matrix with the probability density function

$$\varphi(X) = \frac{1}{2^{\frac{1}{2}m}\pi^{\frac{1}{4}m(m+1)}}\operatorname{etr}\left(-\frac{1}{2}X^2\right)$$

and let ∂X be defined as

$$\partial X = \left(\frac{1}{2}(1+\delta_{ij})\frac{\partial}{\partial x_{ij}}\right).$$

Define the coefficients $\eta_{\kappa,\phi}$ as

$$C_\kappa(X^2) = \sum_{\phi\in\kappa\cdot\kappa} \eta_{\kappa,\phi} C_\phi(X),$$

where κ is a partition of k, and ϕ is a partition of $2k$, respectively.
Define the Hermite polynomial $H_\kappa^{(m)}(X)$ of a symmetric matrix X for a partition κ of k as

$$H_\kappa^{(m)}(X)\varphi(X) = C_\kappa(-\partial X)\varphi(X).$$

(i) Show that the generating function of $H_\kappa^{(m)}(X)$ is given by

$$\sum_{k=0}^{\infty}\sum_{\kappa} \frac{H_\kappa^{(m)}(X)C_\kappa(T)}{k!C_\kappa(I)} = \operatorname{etr}\left(-\frac{1}{2}T^2\right)\int_{O(m)} \operatorname{etr}(XHTH')d(H).$$

(ii) Show that

$$\frac{C_\lambda(\partial X)H_\kappa^{(m)}(X)}{k!C_\kappa(I)} = \begin{cases} \sum_{\sigma} \dfrac{a_{\lambda,\sigma}^{\kappa} H_\sigma^{(m)}(X)}{(k-\ell)!C_\sigma(I)}; & k \ge \ell \\ 0 \quad ; & k < \ell, \end{cases}$$

where λ is a partition of ℓ and σ is a partition of $(k-\ell)$, respectively. $a_{\lambda,\sigma}^{\kappa}$ is defined by (4.3.65).

(iii) Show that

$$\frac{H_\kappa^{(m)}(X)}{k!C_\kappa(I)} = \sum_{s=0}^{k}\sum_{\sigma}\left[\sum_{\tau}\sum_{\rho} \frac{a_{\sigma,\rho}^{\kappa}\eta_{\tau,\rho}}{(-2)^t t!}\right]\frac{C_\sigma(X)}{s!C_\sigma(I)},$$

where τ is a partition of t and ρ is a partition of $2t = k - s$.

(iv) Show that $H_\kappa(X)$'s for low degree are given as follows:

$$H_{(1)}^{(m)}(X) = C_{(1)},$$
$$H_{(2)}^{(m)}(X) = C_{(2)}(X) - \frac{1}{3}m(m+2),$$
$$H_{(1^2)}^{(m)}(X) = C_{(1^2)}(X) + \frac{1}{3}m(m-1),$$
$$H_{(3)}^{(m)}(X) = C_{(3)}(X) - \frac{1}{5}(m+2)(m+4)C_{(1)}(X),$$
$$H_{(21)}^{(m)}(X) = C_{(21)}(X) - \frac{3}{10}(m-1)(m-2)C_{(1)}(X),$$
$$H_{(1^3)}^{(m)}(X) = C_{(1^3)}(X) + \frac{1}{2}(m-1)(m-2)C_{(1)}(X)$$

(Chikuse (1992a)).

4.33 Let X be an $m \times n$ matrix with the probability density function

$$\varphi_{mn}(X) = \frac{1}{(2\pi)^{\frac{1}{2}mn}} \operatorname{etr}\left(-\frac{1}{2}XX'\right)$$

and let ∂X be an $m \times n$ differential operator matrix defined as

$$\partial X = \left(\frac{\partial}{\partial x_{ij}}\right), \quad \text{respectively.}$$

Define the Hermite polynomial $H_\kappa^{(mn)}(X)$ of X for a partition κ of k as

$$H_\kappa^{(mn)}(X)\varphi_{mn}(X) = \frac{1}{4^k \left(\frac{n}{2}\right)_\kappa} C_\kappa(\partial X \partial X')\varphi_{mn}(X).$$

(i) Show that the generating function of $H_\kappa^{(mn)}(X)$ is given by

$$\sum_{k=0}^{\infty}\sum_{\kappa} \frac{H_\kappa^{(mn)}(X)C_\kappa(TT')}{k!C_\kappa(I_m)}$$
$$= \operatorname{etr}\left(-\frac{1}{2}TT'\right)\int_{O(m)}\int_{O(n)} \operatorname{etr}(H_1XH_2T')d(H_1)d(H_2),$$

where T is an $m \times m$ matrix and $H_1 \in O(m)$, $H_2 \in O(n)$, respectively.

(ii) Show that

$$\frac{H_\kappa^{(mn)}(X)}{k!C_\kappa(I)} = \sum_{s=0}^{k}\sum_{\sigma}\sum_{\tau} \frac{a_{\tau,\sigma}^{\kappa}}{(-2)^{k+s}(k-s)!} \frac{C_\sigma(XX')}{s!\left(\frac{n}{2}\right)_\sigma C_\sigma(I)},$$

where τ is a partition of $(k-s)$, and σ is a partition of s, respectively.

(iii) Show that the correspondence between $H_\kappa(T)$ defined by (4.6.34) and $H_\kappa^{(mn)}(T)$ is given by

$$H_\kappa(T) = 2^k\left(\frac{n}{2}\right)_\kappa H_\kappa^{(mn)}(\sqrt{2}T)$$

(Chikuse (1992a)).

Table 4.3.2(a) Table of $a^{\phi}_{\kappa,\lambda}$, $a^{\phi}_{\kappa,\lambda} = a^{\phi}_{\lambda,\kappa}$

κ	$\lambda \setminus \phi$	(2)	(1^2)	(3)	(21)	(1^3)	(4)	(31)	(2^2)
(1)	(1)	1	1						
(1)	(2)			1	4/9				
(1)	(1^2)				5/9	1			
(1)	(3)						1	3/10	
(1)	(21)							7/10	1
(1)	(1^3)								
(2)	(2)						1	2/9	4/9
(2)	(1^2)							7/18	
(1^2)	(1^2)								5/9
(1)	(4)								
(1)	(31)								
(1)	(2^2)								
(1)	(21^2)								
(1)	(1^4)								
(2)	(3)								
(2)	(21)								
(2)	(1^3)								
(1^2)	(3)								
(1^2)	(21)								
(1^2)	(1^3)								

κ	$\lambda \setminus \phi$	(21^2)	(1^4)	(5)	(41)	(32)	(31^2)	(2^21)	(21^3)	(1^5)
(1)	(1)									
(1)	(2)									
(1)	(1^2)									
(1)	(3)									
(1)	(21)	5/8								
(1)	(1^3)	3/8	1							
(2)	(2)									
(2)	(1^2)	5/18								
(1^2)	(1^2)	4/9	1							
(1)	(4)			1	8/35					
(1)	(31)				27/35	8/15	7/15			
(1)	(2^2)					7/15	1/3			
(1)	(21^2)						8/15	2/3	18/25	
(1)	(1^4)								7/25	1
(2)	(3)			1	4/25	4/25				
(2)	(21)				27/50	28/75	7/30	1/3		
(2)	(1^3)						1/5		1/5	
(1^2)	(3)				3/10		7/50			
(1^2)	(21)					7/15	32/75	5/12	9/20	
(1^2)	(1^3)							1/4	7/20	1

Table 4.3.2(b) Table of $b^{\phi}_{\kappa,\lambda}$, $b^{\phi}_{\kappa,\lambda} = b^{\phi}_{\lambda,\kappa}$

$\phi \backslash \kappa\,\lambda$	(2)(0)	$(1^2)(0)$	(1)(1)
(2)	1		2/3
(1^2)		1	4/3

$\phi \backslash \kappa\,\lambda$	(3)(0)	(21)(0)	$(1^3)(0)$	(2)(1)	$(1^2)(1)$
(3)	1			3/5	
(21)		1		12/5	3/2
(1^3)			1		3/2

$\phi \backslash \kappa\,\lambda$	(4)(0)	(31)(0)	$(2^2)(0)$	$(21^2)(0)$	$(1^4)(0)$	(3)(1)	(21)(1)	$(1^3)(1)$	(2)(2)	$(2)(1^2)$	$(1^2)(1^2)$
(4)	1					4/7			18/35		
(31)		1				24/7	8/9		16/7	2	
(2^2)			1				8/9		16/5		1
(21^2)				1			20/9	12/5		4	16/5
(1^4)					1			8/5			9/5

φ \ κλ	(5)(0)	(41)(0)	(32)(0)	(31²)(0)	(2²1)(0)	(21³)(0)	(1⁵)(0)
(5)	1						
(41)		1					
(32)			1				
(31²)				1			
(2²1)					1		
(21³)						1	
(1⁵)							1

(4)(1)	(31)(1)	(2²)(1)	(21²)(1)	(1⁴)(1)
5/9				
40/9	3/4			
	4/3	5/3		
	35/12		25/21	
		10/3	5/3	
			15/7	10/3
				5/3

(3)(2)	(21)(2)	(1³)(2)	(3)(1²)	(21)(1²)	(1³)(1)
10/21					
8/3	1		5/2		
48/7	16/9			10/9	
	25/9	30/7	15/2	160/63	
	40/9			25/9	3
		40/7		25/7	5
					2

Table 4.4.1 The values of $\binom{\kappa}{\sigma}$ up to order 5

$\kappa \setminus \sigma$	(0)	(1)	(2)	(1^2)	(3)	(21)	(1^3)	(4)	(31)	(2^2)	(21^2)	(1^4)	(5)	(41)	(32)	(31^2)	(2^21)	(21^3)	(1^5)
(1)	1	1																	
(2)	1	2	1																
(1^2)	1	2		1															
(3)	1	3	3		1														
(21)	1	3	4/3	5/3		1													
(1^3)	1	3		3			1												
(4)	1	4	6		4			1											
(31)	1	4	11/3	7/3	6/5	14/5			1										
(2^2)	1	4	8/3	10/3		4				1									
(21^2)	1	4	5/3	13/3		5/2	3/2				1								
(1^4)	1	4		6			4					1							
(5)	1	5	10		10			5					1						
(41)	1	5	7	3	23/5	27/5		8/7	27/7					1					
(32)	1	5	16/3	14/3	8/5	42/5			8/3	7/3					1				
(31^2)	1	5	13/3	17/3	7/5	33/5	2		7/3		8/3					1			
(2^21)	1	5	10/3	20/3		15/2	5/2			5/3	10/3						1		
(21^3)	1	5	2	8		9/2	11/2				18/5	7/5						1	
(1^5)	1	5		10			10					5							1

Constantine (1966) and Pillai and Jouris (1969).

CHAPTER 5

Generalized Quadratic Forms

5.0 Introduction

In this chapter we consider some distributional problems associated with generalized quadratic forms in normal matrix variates and some of their applications in multivariate statistical analysis. The exact probability density function of a generalized quadratic form is given in Section 5.1. An alternate representation of the probability density function is obtained in terms of orthogonal polynomials of matrix arguments in Section 5.2. Section 5.3 gives some representations of the joint probability density function of the latent roots and of the largest latent root of a generalized quadratic form. Section 5.4 deals with the distributions of certain functions of a generalized quadratic form, including a matrix t-variate distribution. Hotelling's generalized T_0^2-statistic is a fundamental statistic in multivariate analysis of variance. Two representations of its exact probability density function are given in Section 5.5. The first one is obtained in terms of P-polynomials and Laguerre polynomials of matrix arguments and converges for a certain range; the second one is expressed in terms of invariant polynomials and converges everywhere. The asymptotic expansion of the distribution function of T_0^2 in the non-null case is also obtained and its range of convergence is determined. Anderson's linear discriminant function plays a fundamental role in discriminant analysis. Section 5.6 gives its exact moments in terms of P-polynomials; the distribution of the normalized Anderson's linear discriminant function is obtained as an Edgeworth expansion. The multivariate calibration problem is considered in Section 5.7 and the distributions of certain statistics are approximated by the central F distribution. As the exact distribution of a generalized quadratic form is complicated and difficult to evaluate, two types of asymptotic expansions of the distribution function are proposed in Section 5.8.

5.1 A Representation of the Distribution of a Generalized Quadratic Form

Definition 5.1.1 *Generalized quadratic form* Let $X = [\boldsymbol{x}_1, \boldsymbol{x}_2, \ldots, \boldsymbol{x}_n]$ be an $m \times n$ random matrix whose probability density function is

$$\frac{1}{\pi^{\frac{1}{2}mn} |2\Sigma_1|^{\frac{n}{2}} |\Sigma_2|^{\frac{m}{2}}} \operatorname{etr}\left\{-\frac{1}{2}\Sigma_1^{-1}(X-M)\Sigma_2^{-1}(X-M)'\right\}, \tag{5.1.1}$$

where M is an $m \times n$ matrix, Σ_1 is an $m \times m$ symmetric nonsingular matrix and Σ_2 is an $n \times n$ symmetric nonsingular matrix. Let $A = (a_{\alpha\beta})$ be an $n \times n$ positive definite symmetric matrix. A generalized quadratic form in normal matrices is defined as

$$S = XAX' = \sum_{\alpha=1}^{n}\sum_{\beta=1}^{n} a_{\alpha\beta}\boldsymbol{x}_\alpha \boldsymbol{x}'_\beta . \tag{5.1.2}$$

The distribution will be denoted by $Q_m(A, \Sigma_1, \Sigma_2, M, n)$.

Remark 5.1.1 It is well known that if $A = A^2$, rank $A = p$, and $\Sigma_2 = I_n$, then S is a noncentral Wishart matrix with p degrees of freedom.

Theorem 5.1.1 The probability density function of S as defined in (5.1.2) is given by

$$\begin{aligned}
&\frac{1}{\Gamma_m\left(\frac{n}{2}\right) |2\Sigma_1|^{\frac{n}{2}} |\Sigma_2|^{\frac{m}{2}} |A|^{\frac{m}{2}}} \operatorname{etr}\left(-\frac{1}{2}\Sigma_1^{-1}M\Sigma_2^{-1}M'\right) \\
&\quad \times \operatorname{etr}\left(-\frac{q}{2}\Sigma_1^{-1}S\right) |S|^{\frac{1}{2}(n-m-1)} \\
&\quad \times \sum_{k=0}^{\infty}\sum_{\kappa} \frac{1}{k!\left(\frac{n}{2}\right)_\kappa} P_\kappa\left(\frac{1}{\sqrt{2}}M^*,\ A^*,\ \frac{1}{2}\Sigma_1^{-1/2}S\Sigma_1^{-1/2}\right),
\end{aligned} \tag{5.1.3}$$

where $M^* = \Sigma_1^{-1/2}M\Sigma_2^{-1}\left(\Sigma_2^{-1} - qA\right)^{-\frac{1}{2}}$, $A^* = \left(\Sigma_2^{-1} - qA\right)^{\frac{1}{2}} A^{-1}(\Sigma_2^{-1} - qA)^{\frac{1}{2}}$ for $q \geq 0$, $\Sigma_2^{-1} - qA > 0$, and $P_\kappa(\cdot,\cdot,\cdot)$ is defined by (4.6.43).

The probability density function for $q > 0$ is called the Wishart type representation and the probability density function for $q = 0$ is called the power series type representation.

PROOF The probability density function of S is obtained by integrating over $S = XAX'$.

$$\begin{aligned}
f(S) &= (2\pi)^{-\frac{1}{2}mn} |\Sigma_1|^{-n/2} |\Sigma_2|^{-m/2} \\
&\quad \times \int_{S=XAX'} \operatorname{etr}\left[-\frac{1}{2}\Sigma_1^{-1}(X-M)\Sigma_2^{-1}(X-M)'\right] dX
\end{aligned}$$

$$= (2\pi)^{-\frac{1}{2}mn} |\Sigma_1|^{-n/2} |\Sigma_2|^{-m/2}$$
$$\times \int_{S=XAX'} \operatorname{etr}\left[-\frac{1}{2}\Sigma_1^{-1}\left\{qXAX' + X\left(\Sigma_2^{-1} - qA\right)X'\right.\right.$$
$$\left.\left. - 2M\Sigma_2^{-1}X' + M\Sigma_2^{-1}M'\right\}\right] \pi^{-\frac{1}{2}mn} \int_U \operatorname{etr}\left[-\left\{U - \frac{i}{\sqrt{2}}\Sigma_1^{-1/2}\right.\right.$$
$$\left.\times \left(X\left(\Sigma_2^{-1} - qA\right)^{\frac{1}{2}} - M\Sigma_2^{-1}\left(\Sigma_2^{-1} - qA\right)^{-\frac{1}{2}}\right)\right\}$$
$$\times \left\{U - \frac{i}{\sqrt{2}}\Sigma_1^{-1/2}\left(X\left(\Sigma_2^{-1} - qA\right)^{\frac{1}{2}}\right.\right.$$
$$\left.\left.\left. - M\Sigma_2^{-1}\left(\Sigma_2^{-1} - qA\right)^{-\frac{1}{2}}\right)\right\}'\right] dU dX,$$

where $i = \sqrt{-1}$. Note that the integral with respect to the $m \times n$ matrix U is unity. Then

$$f(S) = (2\pi^2)^{-\frac{1}{2}mn} |\Sigma_1|^{-\frac{n}{2}} |\Sigma_2|^{-\frac{m}{2}}$$
$$\times \operatorname{etr}\left\{-\frac{1}{2}\Sigma_1^{-1}M\Sigma_2^{-1}M' + \frac{1}{2}\Sigma_1^{-1}M\Sigma_2^{-1}\left(\Sigma_2^{-1} - qA\right)^{-1}\Sigma_2^{-1}M'\right\}$$
$$\times \int_U \operatorname{etr}\left\{-UU' - \sqrt{2}iU\left(\Sigma_2^{-1} - qA\right)^{-\frac{1}{2}}\Sigma_2^{-1}M'\Sigma_1^{-1/2}\right\}$$
$$\times \int_{S=XAX'} \operatorname{etr}\left\{-\frac{q}{2}\Sigma_1^{-1}XAX'\right.$$
$$\left. + \sqrt{2}iU\left(\Sigma_2^{-1} - qA\right)^{\frac{1}{2}}X'\Sigma_1^{-1/2}\right\} dX dU. \tag{5.1.4}$$

By using (4.4.16) one has

$$\int_{S=XAX'} \operatorname{etr}\left\{-\frac{q}{2}\Sigma_1^{-1}XAX' + \sqrt{2}i\Sigma_1^{-1/2}X\left(\Sigma_2^{-1} - qA\right)^{\frac{1}{2}}U'\right\} dX$$
$$= |A|^{-m/2} \int_{S=YY'} \operatorname{etr}\left\{-\frac{q}{2}\Sigma_1^{-1}YY'\right.$$
$$\left. + \sqrt{2}i\Sigma_1^{-1/2}YA^{-\frac{1}{2}}\left(\Sigma_2^{-1} - qA\right)^{\frac{1}{2}}U'\right\} dY$$

$$= |A|^{-m/2} \int_{S=YY'} \operatorname{etr}\left\{-\frac{q}{2}\Sigma_1^{-1}YY'\right\}$$
$$\times \int_{O(n)} \operatorname{etr}\left\{\sqrt{2}iA^{-\frac{1}{2}}\left(\Sigma_2^{-1} - qA\right)^{\frac{1}{2}}U'\Sigma_1^{-1/2}YH\right\} d(H) dY$$

$$= |A|^{-m/2} \int_{S=YY'} \operatorname{etr}\left\{-\frac{q}{2}\Sigma_1^{-1}YY'\right\}$$

$$\times {}_0F_1\left(\frac{n}{2}; -\frac{1}{2}U\left(\Sigma_2^{-1} - qA\right)^{\frac{1}{2}} A^{-1} \left(\Sigma_2^{-1} - qA\right)^{\frac{1}{2}} U'\right.$$

$$\left. \times \Sigma_1^{-1/2} YY' \Sigma_1^{-1/2}\right) \mathrm{d}Y$$

$$= \frac{\pi^{\frac{1}{2}mn}}{\Gamma_m\left(\frac{n}{2}\right)} |A|^{-m/2} \operatorname{etr}\left(-\frac{q}{2}\Sigma_1^{-1}S\right) |S|^{\frac{1}{2}(n-m-1)}$$

$$\times {}_0F_1\left(\frac{n}{2}; -\frac{1}{2}U\left(\Sigma_2^{-1} - qA\right)^{\frac{1}{2}} A^{-1} \left(\Sigma_2^{-1} - qA\right)^{\frac{1}{2}} U'\right.$$

$$\left. \times \Sigma_1^{-1/2} S \Sigma_1^{-1/2}\right).$$

Substituting this in (5.1.4) and using the Definition 4.6.4, the theorem follows from the integration with respect to U.

Lemma 5.1.1

$$\int_{S>0} \operatorname{etr}(-GS)|S|^{\frac{1}{2}(n-m-1)} P_\kappa(T, A, B^{-\frac{1}{2}}SB^{-\frac{1}{2}}) \mathrm{d}S$$

$$= \Gamma_m\left(\frac{n}{2}; \kappa\right) |G|^{-\frac{n}{2}} P_\kappa(T, A, B^{-\frac{1}{2}}G^{-1}B^{-\frac{1}{2}}), \tag{5.1.5}$$

$$\int_{\Omega>S>0} |S|^{\frac{1}{2}(n-m-1)} P_\kappa(T, A, B^{-\frac{1}{2}}SB^{-\frac{1}{2}}) \mathrm{d}S$$

$$= \frac{\Gamma_m\left(\frac{n}{2}; \kappa\right)\Gamma_m\left(\frac{m+1}{2}\right)}{\Gamma_m\left(\frac{n+m+1}{2}; \kappa\right)} |\Omega|^{\frac{n}{2}} P_\kappa(T, A, B^{-\frac{1}{2}}\Omega B^{-\frac{1}{2}}), \tag{5.1.6}$$

$$\int_{O(m)} P_\kappa(T, A, B^{-\frac{1}{2}}H\Lambda H'B^{-\frac{1}{2}}) \mathrm{d}(H)$$

$$= \frac{C_\kappa(\Lambda)}{C_\kappa(I_m)} P_\kappa(T, A, B^{-1}), \tag{5.1.7}$$

$$\lim_{q\to 1} P_\kappa\left(\frac{1}{\sqrt{2}}\frac{T}{\sqrt{1-q}}, (1-q)A, B\right) = C_\kappa(BTAT'). \tag{5.1.8}$$

PROOF Substituting (4.6.43) in (5.1.5) and using (4.4.6) yield (5.1.5). Equations (5.1.6) and (5.1.7) are obtained by using (4.4.10) and (4.6.45), respectively.

$$\lim_{q\to 1} P_\kappa\left(\frac{1}{\sqrt{2}}\frac{T}{\sqrt{1-q}}, (1-q)A, B\right)$$

$$= \lim_{q\to 1} \pi^{-mn/2} \int_U \operatorname{etr}(-UU')$$

$$\times C_\kappa\left(-B\left(U - \frac{i}{\sqrt{1-q}}T\right)A\left(U - \frac{i}{\sqrt{1-q}}T\right)'(1-q)\right) dU$$

$$= C_\kappa(BTAT'),$$

which proves (5.1.8).

Remark 5.1.2 The derivation of (5.1.3) is due to Crowther (1975). When $A = \Sigma_2 = I_n$ and $q = 1$, (5.1.3) can be reduced by using (5.1.8) to the probability density function of the noncentral Wishart distribution with n degrees of freedom and the noncentrality parameter matrix $\frac{1}{2}\Sigma_1^{-1}MM'$, and can be expressed as

$$\frac{1}{\Gamma_m\left(\frac{n}{2}\right)|2\Sigma_1|^{\frac{n}{2}}} \operatorname{etr}\left(-\frac{1}{2}\Sigma_1^{-1}MM'\right) \operatorname{etr}\left(-\frac{1}{2}\Sigma_1^{-1}S\right)$$

$$\times |S|^{\frac{1}{2}(n-m-1)} {}_0F_1\left(\frac{n}{2}; \frac{1}{4}\Sigma_1^{-1}MM'\Sigma_1^{-1}S\right). \tag{5.1.9}$$

The noncentral distribution will be denoted as $W_m\left(n, \Sigma_1, \frac{1}{2}\Sigma_1^{-1}MM'\right)$.

Theorem 5.1.2 Let S be distributed as $Q_m(A, \Sigma_1, \Sigma_2, M, n)$, then the distribution function of S is given, for any positive definite Ω, as

$$\Pr\{S < \Omega\} = \frac{\Gamma_m\left(\frac{m+1}{2}\right)}{\Gamma_m\left(\frac{n+m+1}{2}\right)|2\Sigma_1|^{\frac{n}{2}}|\Sigma_2|^{\frac{m}{2}}|A|^{\frac{m}{2}}}$$

$$\times \operatorname{etr}\left(-\frac{1}{2}\Sigma_1^{-1}M\Sigma_2^{-1}M'\right)|\Omega|^{\frac{n}{2}} \sum_{k=0}^{\infty}\sum_{\kappa} \frac{1}{k!\left(\frac{n+m+1}{2}\right)_\kappa}$$

$$\times P_\kappa\left(\frac{1}{\sqrt{2}}\Sigma_1^{-1/2}M\Sigma_2^{-1}, \Sigma_2^{-1/2}A^{-1}\Sigma_2^{-1/2}, \frac{1}{2}\Sigma_1^{-1/2}\Omega\Sigma_1^{-1/2}\right). \tag{5.1.10}$$

The distribution function of the largest latent root λ_1 of S is given by

$$\Pr\{\lambda_1 < x\} = \frac{\Gamma_m\left(\frac{m+1}{2}\right)}{\Gamma_m\left(\frac{n+m+1}{2}\right)|2\Sigma_1|^{\frac{n}{2}}|\Sigma_2|^{\frac{m}{2}}|A|^{\frac{m}{2}}}$$

$$\times \operatorname{etr}\left(-\frac{1}{2}\Sigma_1^{-1}M\Sigma_2^{-1}M'\right)x^{\frac{1}{2}mn} \sum_{k=0}^{\infty}\sum_{\kappa} \frac{x^k}{k!\left(\frac{n+m+1}{2}\right)_\kappa}$$

$$\times P_\kappa\left(\frac{1}{\sqrt{2}}\Sigma_1^{-1/2}M\Sigma_2^{-1/2}, \Sigma_2^{-1/2}A^{-1}\Sigma_2^{-1/2}, \frac{1}{2}\Sigma_1^{-1}\right). \tag{5.1.11}$$

PROOF (5.1.10) is obtained with the help of (5.1.6). When $\Omega = xI_m$, (5.1.10) is reduced to the distribution function of the largest latent root λ_1 of S.

Remark 5.1.3 When $M = 0$, $\Sigma_2 = A = I_n$, (5.1.10) reduces to

$$\frac{\Gamma_m\left(\frac{m+1}{2}\right)|\Omega|^{\frac{n}{2}}}{\Gamma_m\left(\frac{n+m+1}{2}\right)|2\Sigma_1|^{\frac{n}{2}}}\,{}_1F_1\left(\frac{n}{2};\frac{n+m+1}{2};-\frac{1}{2}\Sigma_1^{-1}\Omega\right), \tag{5.1.12}$$

which is obtained by Constantine (1963, (62)). Note that (5.1.12) reduces to the following by applying Kummer's formula (4.4.21).

$$\frac{\Gamma_m\left(\frac{m+1}{2}\right)}{\Gamma_m\left(\frac{n+m+1}{2}\right)|2\Sigma_1|^{\frac{n}{2}}}\operatorname{etr}\left(-\frac{x}{2}\Sigma_1^{-1}\right)x^{\frac{1}{2}mn}$$
$$\times\,{}_1F_1\left(\frac{m+1}{2};\frac{n+m+1}{2};\frac{x}{2}\Sigma_1^{-1}\right). \tag{5.1.13}$$

This result was obtained independently by Muirhead (1970b, (5.2)) and Sugiyama (1967a, (3.5)). The asymptotic expansion of the distribution of λ_1 for large n was studied by Sugiura (1973) and Muirhead and Chikuse (1975).

Remark 5.1.4 When $\Sigma_2 = A = I_n$, (5.1.10) is the distribution of the noncentral Wishart matrix, and noting that

$$P_\kappa(T, I_n, B) = (-1)^k L_\kappa^{\frac{1}{2}(n-m-1)}(TT', B), \tag{5.1.14}$$

(5.1.10) reduces to the form

$$\Pr\{S < \Omega\} = \frac{\Gamma_m\left(\frac{m+1}{2}\right)}{2^{\frac{mn}{2}}\Gamma_m\left(\frac{n+m+1}{2}\right)|\Sigma_1|^{\frac{n}{2}}}\operatorname{etr}\left(-\frac{1}{2}\Sigma_1^{-1}MM'\right)$$
$$\times|\Omega|^{\frac{n}{2}}\sum_{k=0}^{\infty}\sum_{\kappa}\frac{(-1)^k}{k!\left(\frac{n+m+1}{2}\right)_\kappa}$$
$$\times L_\kappa^{\frac{1}{2}(n-m-1)}\left(\frac{1}{2}\Sigma_1^{-1}MM'\Sigma_1^{-1/2}, \frac{1}{2}\Sigma_1^{-1/2}\Omega\Sigma_1^{-1/2}\right), \tag{5.1.15}$$

which was obtained by Davis (1979, (4.9)).

The distribution function of λ_1 is obtained by taking $\Omega = xI$.

Lemma 5.1.2 For $S > 0$, let $x = \operatorname{tr}(S)$ and $S = xS_1$, then the Jacobian of this transformation is $x^{\frac{1}{2}m(m+1)-1}$ by Theorem 1.1a.8. Let $\mathcal{D} = \{S_1;\ \operatorname{tr}(S_1) = 1\}$, then

$$\int_{\mathcal{D}}|S_1|^{a-\frac{1}{2}(m+1)}C_\kappa(AS_1)\mathrm{d}S_1 = \frac{\Gamma_m(a;\kappa)C_\kappa(A)}{\Gamma(ma+k)} \tag{5.1.16}$$

and

$$\int_{\mathcal{D}}|S_1|^{a-\frac{1}{2}(m+1)}P_\kappa\left(T, A, \frac{1}{2}B^{-\frac{1}{2}}S_1B^{-\frac{1}{2}}\right)\mathrm{d}S_1$$

$$= \frac{\Gamma_m(a;\kappa)}{\Gamma(ma+k)} P_\kappa\left(T, A, \frac{1}{2}B^{-1}\right). \tag{5.1.17}$$

For $\Lambda = \text{diag}(\lambda_1, \ldots, \lambda_m)$, $\lambda_1 > \cdots > \lambda_m > 0$, let $x = \text{tr}(\Lambda)$, $\Lambda = x\Lambda_L$, $\Lambda_L = \text{diag}(\ell_1, \ldots, \ell_m)$. Let $\mathcal{D}_L = \{\Lambda_L; \text{tr}(\Lambda_L) = 1\}$. Then

$$\int_{\mathcal{D}_L} |\Lambda_L|^{a-\frac{1}{2}(m+1)} \prod_{i<j}(\ell_i - \ell_j) C_\kappa(\Lambda_L) d\Lambda_L$$
$$= \frac{\Gamma_m(a;\kappa)\Gamma_m\left(\frac{m}{2}\right)}{\pi^{\frac{1}{2}m^2}\Gamma(ma+k)} C_\kappa(I_m). \tag{5.1.18}$$

PROOF Since

$$\int_{S>0} \text{etr}(-S)\, |S|^{a-\frac{1}{2}(m+1)} C_\kappa(AS) dS = \Gamma_m(a;\kappa) C_\kappa(A), \tag{5.1.19}$$

by the transformation $S = xS_1$ with $x = \text{tr}(S)$, (5.1.18) is expressed as

$$\int_0^\infty \exp(-x) x^{ma+k-1} dx \int_{\mathcal{D}} |S_1|^{a-\frac{1}{2}(m+1)} C_\kappa(AS_1) dS_1$$
$$= \Gamma(ma+k) \int_{\mathcal{D}} |S_1|^{a-\frac{1}{2}(m+1)} C_\kappa(AS_1) dS_1,$$

which yields (5.1.16). (5.1.17) is obtained by using (5.1.16). (5.1.18) is obtained in a similar way as (5.1.16).

Theorem 5.1.3 Let S be distributed as $Q_m(A, \Sigma_1, \Sigma_2, M, n)$, then the probability density function of $z = \text{tr}(S)$ is given by

$$\frac{1}{\Gamma\left(\frac{mn}{2}\right) |2\Sigma_1|^{\frac{n}{2}} |\Sigma_2|^{\frac{m}{2}} |A|^{\frac{m}{2}}} \text{etr}\left(-\frac{1}{2}\Sigma_1^{-1} M \Sigma_2^{-1} M'\right)$$
$$\times z^{\frac{1}{2}mn-1} \sum_{k=0}^{\infty} \frac{z^k}{k!\left(\frac{mn}{2}\right)_k}$$
$$\times \sum_\kappa P_\kappa\left(\frac{1}{\sqrt{2}}\Sigma_1^{-1/2} M \Sigma_2^{-1}, \Sigma_2^{-1/2} A^{-1} \Sigma_2^{-1/2}, \frac{1}{2}\Sigma_1^{-1}\right) \tag{5.1.20}$$

PROOF (5.1.17) yields (5.1.20).

Remark 5.1.5 Let $X' = [\boldsymbol{x}_{(1)}, \ldots, \boldsymbol{x}_{(m)}]$ and $M' = [\boldsymbol{\mu}_{(1)}, \ldots, \boldsymbol{\mu}_{(m)}]$ and set $\boldsymbol{x}' = (\boldsymbol{x}'_{(1)}, \ldots, \boldsymbol{x}'_{(m)})$ and $\boldsymbol{\mu}' = (\boldsymbol{\mu}'_{(1)}, \ldots, \boldsymbol{\mu}'_{(m)})$, then

$$z = \text{tr}(XAX') = \boldsymbol{x}'(I_m \otimes A)\boldsymbol{x} \tag{5.1.21}$$

and

$$\boldsymbol{x} \sim N(\boldsymbol{\mu}, \Sigma_1 \otimes \Sigma_2).$$

This implies

$$z = \boldsymbol{y}' \left(\Sigma_1 \otimes \Sigma_2^{\frac{1}{2}} A \Sigma_2^{\frac{1}{2}}\right) \boldsymbol{y}, \quad \boldsymbol{y} \sim N\left(\left(\Sigma_1^{-\frac{1}{2}} \otimes \Sigma_2^{-\frac{1}{2}}\right) \boldsymbol{\mu}, I_{mn}\right). \tag{5.1.22}$$

This is a noncentral univariate quadratic form in normal random vectors whose distribution theory was studied extensively by Johnson and Kotz (1970, chapter 29) and Mathai and Provost (1992).

Corollary 5.1.1 When $\Sigma_1 = I_m$, the probability density function of $z = \operatorname{tr}(S)$ is given by

$$\frac{1}{\Gamma\left(\frac{mn}{2}\right) |2A\Sigma_2|^{\frac{m}{2}}} \operatorname{etr}\left(-\frac{1}{2} M \Sigma_2^{-1} M'\right) z^{\frac{1}{2}mn-1}$$
$$\times \sum_{k=0}^{\infty} \frac{z^k}{k! \left(\frac{mn}{2}\right)_k} \sum_{\kappa} P_\kappa\left(\frac{1}{\sqrt{2}} M \Sigma_2^{-1}, \left(2\Sigma_2^{\frac{1}{2}} A \Sigma_2^{\frac{1}{2}}\right)^{-1}\right). \tag{5.1.23}$$

When $A = \Sigma_2 = I_n$, then the probability density function of $z = \operatorname{tr}(S)$ is given by

$$\frac{1}{\Gamma\left(\frac{mn}{2}\right) |2\Sigma_1|^{\frac{n}{2}}} \operatorname{etr}\left(-\frac{1}{2} \Sigma_1^{-1} M M'\right) z^{\frac{1}{2}mn-1}$$
$$\times \sum_{k=0}^{\infty} \frac{(-z)^k}{k! \left(\frac{mn}{2}\right)_k} \sum_{\kappa} L_\kappa^\gamma\left(\frac{1}{2} \Sigma_1^{-1/2} M M' \Sigma_1^{-1/2}, \frac{1}{2} \Sigma_1^{-1}\right) \tag{5.1.24}$$

and $\gamma = \frac{1}{2}(n - m - 1)$.

Remark 5.1.6 When $A = \Sigma_2 = I_n$, the probability density function of $z = \operatorname{tr}(\Sigma_1^{-1} X A X')$ is given as

$$\frac{1}{2^{\frac{mn}{2}} \Gamma\left(\frac{mn}{2}\right)} \operatorname{etr}\left(-\frac{1}{2} \Sigma_1^{-1} M M'\right) z^{\frac{1}{2}mn-1}$$
$$\times \sum_{k=0}^{\infty} \frac{\left(-\frac{1}{2} z\right)^k}{k! \left(\frac{mn}{2}\right)_k} \sum_{\kappa} L_\kappa^\gamma\left(\frac{1}{2} \Sigma_1^{-1} M M'\right)$$
$$= \frac{1}{2^{\frac{mn}{2}} \Gamma\left(\frac{mn}{2}\right)} \operatorname{etr}\left(-\frac{1}{2} \Sigma_1^{-1} M M'\right) \exp\left(-\frac{1}{2} z\right) z^{\frac{1}{2}mn-1}$$
$$\times \sum_{k=0}^{\infty} \frac{1}{k! \left(\frac{mn}{2}\right)_k} \left(\frac{z}{2} \operatorname{tr}(\Sigma_1^{-1} M M')\right)^k, \tag{5.1.25}$$

which is the probability density function of the noncentral chi-square random variable with mn degrees of freedom and noncentrality parameter $\frac{1}{2} \operatorname{tr}(\Sigma_1^{-1} M M')$. The second formula in (5.1.25) is obtained with the help of (4.6.10) and (4.6.20).

5.2 An Alternate Representation

An alternate representation of the probability density function of a generalized quadratic form is given in this section.

Lemma 5.2.1 Let S be distributed as $Q_m(A, \Sigma_1, I_n, M, n)$, then the Laplace transform of S is given by

$$E[\operatorname{etr}(-ZS)] = |\Theta|^{n/2} \sum_{k=0}^{\infty} \sum_{\kappa} \frac{1}{k!} P_\kappa(W_1, B_1, \Theta - I)\,, \tag{5.2.1}$$

$$E[\operatorname{etr}(-ZS)] = |\Theta|^{n/2} |qA|^{-m/2} \operatorname{etr}\left(-\frac{1}{2}\Sigma_1 M M'\right) \times \sum_{k=0}^{\infty} \sum_{\kappa} \frac{1}{k!} P_\kappa(W_2, B_2, \Theta), \tag{5.2.2}$$

where

$$\Sigma_1^{1/2} Z \Sigma_1^{1/2} = \frac{q}{2}(\Theta^{-1} - I), \quad q \neq 0,$$

$$B_1 = I - qA, \qquad B_2 = (qA)^{-1} - I,$$

$$W_1 = \frac{1}{\sqrt{2}} \Sigma_1^{-1/2} M((qA)^{-1} - I)^{-1/2},$$

$$W_2 = \frac{1}{\sqrt{2}} \Sigma_1^{-1/2} M(I - qA)^{-1/2}.$$

PROOF Let $A = HD_\alpha H'$, H being an orthogonal matrix and $D_\alpha = \operatorname{diag}(\alpha_1, \alpha_2, \ldots, \alpha_n)$. Let $MH = N = [\boldsymbol{\nu}_1, \boldsymbol{\nu}_2, \ldots, \boldsymbol{\nu}_n]$. Then the Laplace transform of the probability density function of $S = XAX'$ is given by

$$E[\operatorname{etr}(-ZS)] = \prod_{j=1}^{n} \left[\left| I + 2\alpha_j \Sigma_1^{1/2} Z \Sigma_1^{1/2} \right|^{-1/2} \times \exp\left\{ \frac{1}{2} \boldsymbol{\nu}_j' \Sigma_1^{-1} \left(2\alpha_j Z + \Sigma_1^{-1} \right)^{-1} \Sigma_1^{-1} \boldsymbol{\nu}_j - \frac{1}{2} \boldsymbol{\nu}_j' \Sigma_1^{-1} \boldsymbol{\nu}_j \right\} \right]. \tag{5.2.3}$$

Letting

$$\Sigma_1^{1/2} Z \Sigma_1^{1/2} = \frac{q}{2}(\Theta^{-1} - I), \quad q \neq 0,$$

$$b_{j(1)} = 1 - \alpha_j q, \qquad b_{j(2)} = (\alpha_j q)^{-1} - 1,$$

$$\omega_{j(1)} = \Sigma_1^{-1/2} \boldsymbol{\nu}_j / \sqrt{2b_{j(2)}}, \qquad \omega_{j(2)} = \Sigma_1^{-1/2} \boldsymbol{\nu}_j / \sqrt{2b_{j(1)}}$$

for $j = 1, 2, \ldots, n$, then (5.2.3) can be written in two ways:

$$E[\operatorname{etr}(-ZS)] = |\Theta|^{n/2} \prod_{j=1}^{n} \left[|I + b_{j(1)}(\Theta - I)|^{-1/2} \right.$$

$$\times \exp\left\{b_{j(1)}\omega'_{j(1)}(\Theta - I)(I + b_{j(1)}(\Theta - I))^{-1}\omega_{j(1)}\right\}\Big]$$
$$= |\Theta|^{n/2} E_V\left[\operatorname{etr}\{-(\Theta - I)(V - iW_1)B_1(V - iW_1)'\}\right] \tag{5.2.4}$$

and

$$E[\operatorname{etr}(-ZS)]$$
$$= |\Theta|^{n/2}|qA|^{-m/2}\operatorname{etr}\left(-\frac{1}{2}\Sigma_1^{-1}MM'\right)\prod_{j=1}^{n}\Big[|I + b_{j(2)}\Theta|^{-1/2}$$
$$\times \exp\left\{b_{j(2)}\omega'_{j(2)}\Theta(I + b_{j(2)}\Theta)^{-1}\omega_{j(2)}\right\}\Big]$$
$$= |\Theta|^{n/2}|qA|^{-m/2}\operatorname{etr}\left(-\frac{1}{2}\Sigma_1^{-1}MM'\right)$$
$$\times E_V\left[\operatorname{etr}\{-\Theta(V - iW_2)B_2(V - iW_2)'\}\right], \tag{5.2.5}$$

with

$$V = (v_{ij})_{m\times n} \sim N_{m,n}\left(O, \frac{1}{2}I_m, I_n\right),$$
$$B_1 = I - qA, \qquad B_2 = (qA)^{-1} - I,$$
$$W_1 = \frac{1}{\sqrt{2}}\Sigma_1^{-1/2}M((qA)^{-1} - I)^{-1/2},$$
$$W_2 = \frac{1}{\sqrt{2}}\Sigma_1^{-1/2}M(I - qA)^{-1/2}.$$

Then (4.6.46) yields (5.2.1) and (5.2.2).

Theorem 5.2.1 Let S be distributed as $Q_m(A, \Sigma_1, I_n, M, n)$, then the probability density function of S is

$$\frac{1}{\Gamma_m\left(\frac{n}{2}\right)|2\Sigma_1/q|^{\frac{n}{2}}}\operatorname{etr}\left(-\frac{q}{2}\Sigma_1^{-1}S\right)|S|^{\frac{1}{2}(n-m-1)}$$
$$\times \sum_{k=0}^{\infty}\sum_{\kappa}\frac{1}{k!\left(\frac{n}{2}\right)_\kappa}L_\kappa^{\frac{1}{2}(n-m-1)}\left(\frac{q}{2}\Sigma_1^{-1/2}S\Sigma_1^{-1/2},\right.$$
$$\left. I - qA, \frac{1}{\sqrt{2}}\Sigma_1^{-1/2}M((qA)^{-1} - I)^{-1/2}\right). \tag{5.2.6}$$

PROOF The Laplace transforms of

$$\frac{1}{\Gamma_m\left(\frac{n}{2};\kappa\right)|2\Sigma_1/q|^{\frac{n}{2}}}\operatorname{etr}\left(-\frac{q}{2}\Sigma_1^{-1}S\right)|S|^{\frac{1}{2}(n-m-1)}$$
$$\times L_\kappa^{\frac{1}{2}(n-m-1)}\left(\frac{q}{2}\Sigma_1^{-1/2}S\Sigma_1^{-1/2}, B_1, W_1\right)$$

and

$$\frac{1}{\Gamma_m\left(\frac{n}{2};\kappa\right)|2\Sigma_1/q|^{\frac{n}{2}}}\operatorname{etr}\left(-\frac{q}{2}\Sigma_1^{-1}S\right)|S|^{\frac{1}{2}(n-m-1)} \times P_\kappa\left(W_2,B_2,\frac{q}{2}\Sigma_1^{-1/2}S\Sigma_1^{-1/2}\right)$$

are respectively given by

$$|\Theta|^{n/2}P_\kappa(W_1,B_1,\Theta-I)$$

and

$$|\Theta|^{n/2}P_\kappa(W_2,B_2,\Theta).$$

The uniqueness property of the Laplace transform gives the probability density function of S as (5.2.6) and (5.1.3), respectively.

Corollary 5.2.1 Let $M=O$ in (5.2.6), then the probability density function of S reduces to

$$\frac{1}{\Gamma_m\left(\frac{n}{2}\right)|2\Sigma_1/q|^{\frac{n}{2}}}\operatorname{etr}\left(-\frac{q}{2}\Sigma_1^{-1}S\right)|S|^{\frac{1}{2}(n-m-1)} \times\sum_{k=0}^{\infty}\sum_{\kappa}\frac{1}{k!}\frac{C_\kappa(I_n-qA)}{C_\kappa(I_n)}L_\kappa^{\frac{1}{2}(n-m-1)}\left(\frac{q}{2}\Sigma_1^{-1}S\right). \tag{5.2.7}$$

Let $A=I_n$ and $q=1$ in (5.2.6), then the probability density function of the noncentral Wishart matrix is expressed as

$$\frac{1}{\Gamma_m\left(\frac{n}{2}\right)|2\Sigma_1|^{n/2}}\operatorname{etr}\left(-\frac{1}{2}\Sigma_1^{-1}S\right)|S|^{\frac{1}{2}(n-m-1)} \times\sum_{k=0}^{\infty}\sum_{\kappa}\frac{1}{k!\left(\frac{n}{2}\right)_\kappa}L_\kappa^{\frac{1}{2}(n-m-1)}\left(\frac{1}{2}\Sigma_1^{-1/2}S\Sigma_1^{-1/2}, -\frac{1}{2}\Sigma_1^{-1/2}MM'\Sigma^{-1/2}\right). \tag{5.2.8}$$

PROOF Noting that

$$L_\kappa^{\frac{1}{2}(n-m-1)}(S,A,O)=\left(\frac{n}{2}\right)_\kappa\frac{C_\kappa(A)}{C_\kappa(I_n)}L_\kappa^{\frac{1}{2}(n-m-1)}(S) \tag{5.2.9}$$

and

$$\lim_{q\to 1}L_\kappa^{\frac{1}{2}(n-m-1)}\left(S,(1-q)I_n,\sqrt{\frac{q}{1-q}}T\right)=L_\kappa^{\frac{1}{2}(n-m-1)}(S,-TT'), \tag{5.2.10}$$

(5.2.7) and (5.2.8) are obtained.

Remark 5.2.1 The probability density functions given in (5.2.6), (5.2.7) and (5.2.8) were obtained by Khatri (1977) and Shah (1970), respectively. Davis (1979) obtained

(5.2.8) by using invariant polynomials. For discussions on the convergence of the series representation of the probability density function, see Khatri (1971, 1975). The distribution of Hermitian quadratic forms including the noncentral complex Wishart matrix is available from James (1964) and Khatri (1966, 1970).

5.3 The Distribution of the Latent Roots of a Quadratic Form

In this section we give several representations of the joint probability density function of the latent roots of a quadratic form. Lemma 5.3.1 will be used for deriving the probability density function of the largest latent root.

Lemma 5.3.1

$$\int_{I_{m-1}>W>0} |W|^{\alpha-\frac{1}{2}(m+1)}|I-W|C_\kappa\left(\begin{bmatrix}1 & 0' \\ 0 & W\end{bmatrix}\right) \mathrm{d}W$$
$$= \frac{\Gamma_m(\alpha;\kappa)\Gamma_m\left(\frac{m+1}{2}\right)}{\Gamma_m\left(\alpha+\frac{m+1}{2};\kappa\right)} \frac{\Gamma\left(\frac{m}{2}\right)}{\pi^{m/2}}(\alpha m+k)C_\kappa(I_m) \tag{5.3.1}$$

and

$$\int_{1>\omega_2>\cdots>\omega_m>0}\left(\prod_{i=2}^m \omega_i\right)^{\alpha-\frac{1}{2}(m+1)}\left(\prod_{i=2}^m(1-\omega_i)\right)$$
$$\times\left(\prod_{2\le i<j\le m}(\omega_i-\omega_j)\right) C_\kappa\left(\begin{bmatrix}1 & 0 & \cdots & 0 \\ 0 & \omega_2 & & \vdots \\ \vdots & & \ddots & \vdots \\ 0 & \cdots & \cdots & \omega_m\end{bmatrix}\right)\prod_{i=2}^m \mathrm{d}\omega_i$$
$$= \frac{\Gamma_m(\alpha;\kappa)\Gamma_m\left(\frac{m+1}{2}\right)}{\Gamma_m\left(\alpha+\frac{m+1}{2};\kappa\right)} \frac{\Gamma_m\left(\frac{m}{2}\right)}{\pi^{m^2/2}}(\alpha m+k)C_\kappa(I_m). \tag{5.3.2}$$

PROOF We know that

$$\int_{I>S>0}|S|^{\alpha-\frac{1}{2}(m+1)}C_\kappa(S)dS$$
$$= \frac{\Gamma_m(\alpha;\kappa)\Gamma_m\left(\frac{m+1}{2}\right)}{\Gamma_m\left(\alpha+\frac{m+1}{2};\kappa\right)}C_\kappa(I_m). \tag{5.3.3}$$

Decompose S as

$$S = H\begin{bmatrix}\lambda_1 & 0' \\ 0 & \lambda_1 W\end{bmatrix}H'$$

where λ_1 is the largest latent root of S whose range is $1>\lambda_1>0$, W is an $(m-1)\times(m-1)$ positive definite matrix in the range $I_{m-1}>W>0$, and the elements of the first column of H are all positive to ensure the uniqueness of this decomposition, then the Jacobian is given by

$$\lambda_1^{\frac{1}{2}m(m+1)-1}|I_{m-1}-W|\left(1-\sum_{i=2}^{m}h_{i1}^2\right)^{-1/2}.$$

Hence the left-hand side of (5.3.3) becomes

$$\int_0^1 \lambda_1^{\alpha m+k-1}\mathrm{d}\lambda_1\int_{I_{m-1}>W>0}|W|^{\alpha-\frac{1}{2}(m+1)}|I-W|C_\kappa\left(\begin{bmatrix}1 & 0'\\ 0 & W\end{bmatrix}\right)\mathrm{d}W$$

$$\times\int_{\substack{\sum_{i=2}^m h_{i1}^2\le 1\\ h_{i1}>0,\, i=2,\dots,m}}\left(1-\sum_{i=2}^m h_{i1}^2\right)^{-\frac{1}{2}}\prod_{i=2}^m \mathrm{d}h_{i1}$$

$$=\frac{1}{\alpha m+k}\frac{\pi^{\frac{1}{2}m}}{\Gamma\left(\frac{m}{2}\right)}\int_{I>W>0}|W|^{\alpha-\frac{1}{2}(m+1)}|I-W|C_\kappa\left(\begin{bmatrix}1 & 0'\\ 0 & W\end{bmatrix}\right)\mathrm{d}W,$$

which completes the proof. We decompose W further as $W=H_1\Lambda_\omega H_1'$, where $\Lambda_\omega=\mathrm{diag}(\omega_2,\dots,\omega_m)$, $\omega_2>\cdots>\omega_m$ and the elements of the first column of H_1 satisfying $H_1'H_1=I_{m-1}$ are positive. The Jacobian is

$$\mathrm{d}W=\left\{\prod_{2\le i<j\le m}(\omega_i-\omega_j)\right\}\prod_{i=2}^m \mathrm{d}\omega_i\ \mathrm{d}(\tilde{H}_1)$$

and

$$\int_{\substack{H_1'H_1=I_{m-1}\\ h_{i2}>0,\, i=3,\dots,m}}\mathrm{d}(\tilde{H}_1)=\frac{\pi^{\frac{1}{2}(m-1)^2}}{\Gamma_{m-1}\left(\frac{m-1}{2}\right)}.$$

We obtain (5.3.2) with the help of

$$\frac{\pi^{\frac{m}{2}}}{\Gamma\left(\frac{m}{2}\right)}\frac{\pi^{\frac{1}{2}(m-1)^2}}{\Gamma_{m-1}\left(\frac{m-1}{2}\right)}=\frac{\pi^{\frac{m^2}{2}}}{\Gamma_m\left(\frac{m}{2}\right)}. \tag{5.3.4}$$

Theorem 5.3.1 Let S be distributed as $Q_m(A,\Sigma_1,\Sigma_2,M,n)$, then the joint probability density function of its latent roots $\lambda_1>\lambda_2>\cdots>\lambda_m>0$, with $\Lambda=\mathrm{diag}(\lambda_1,\lambda_2,\dots,\lambda_m)$, is given by

$$\frac{\pi^{\frac{1}{2}m^2}}{\Gamma_m\left(\frac{n}{2}\right)\Gamma_m\left(\frac{m}{2}\right)|2\Sigma_1|^{n/2}|A\Sigma_2|^{m/2}}$$

$$\times\operatorname{etr}\left(-\frac{1}{2}\Sigma_1^{-1}M\Sigma_2^{-1}M'\right)|\Lambda|^{\frac{1}{2}(n-m-1)}\left\{\prod_{1\le i<j\le m}(\lambda_i-\lambda_j)\right\}$$

$$\times\sum_{k=0}^{\infty}\sum_{\kappa}\frac{1}{k!\left(\frac{n}{2}\right)_\kappa}\frac{C_\kappa(\Lambda)}{C_\kappa(I_m)}P_\kappa\left(\frac{1}{\sqrt{2}}\Sigma_1^{-1/2}M\Sigma_2^{-1/2},\right.$$

$$\left(\Sigma_2^{1/2} A \Sigma_2^{1/2}\right)^{-1}, \frac{1}{2}\Sigma_1^{-1}\right). \tag{5.3.5}$$

The distribution function of the largest latent root λ_1 of S is obtained with the help of (5.3.2) and is given by (5.1.11).

PROOF By the spectral decomposition $S = H\Lambda H'$, where $\Lambda = \operatorname{diag}(\lambda_1, \ldots, \lambda_m)$ and H is an orthogonal matrix with positive elements in the first column, we have

$$\mathrm{d}S = \frac{2^m \pi^{\frac{m^2}{2}}}{\Gamma_m\left(\frac{m}{2}\right)} \left\{ \prod_{1 \le i < j \le m} (\lambda_i - \lambda_j) \right\} \mathrm{d}\Lambda \mathrm{d}(H)$$

and

$$\int_{\substack{H'H = I_m \\ h_{i1} > 0,\, i=1,\ldots,m}} \mathrm{d}(H) = \frac{1}{2^m} \int_{O(m)} \mathrm{d}(H),$$

where $\mathrm{d}(H)$ is the normalized orthogonal invariant measure over $O(m)$. Substituting for S in (5.1.3) and integrating over $O(m)$ using (5.1.7), the probability density function of Λ is obtained as

$$\begin{aligned}
&\frac{\pi^{\frac{1}{2}m^2}}{\Gamma_m\left(\frac{n}{2}\right) |2\Sigma_1|^{n/2} |A\Sigma_2|^{m/2}} \operatorname{etr}\left(-\frac{1}{2}\Sigma_1^{-1} M \Sigma_2^{-1} M'\right) \\
&\quad \times |\Lambda|^{\frac{1}{2}(n-m-1)} \left\{ \prod_{i<j} (\lambda_i - \lambda_j) \right\} \\
&\quad \times \sum_{k=0}^{\infty} \sum_{\kappa} \frac{1}{k! \left(\frac{n}{2}\right)_\kappa} \frac{1}{2^m} \int_{O(m)} P_\kappa \left(\frac{1}{\sqrt{2}} \Sigma_1^{-1/2} M \Sigma_2^{-1/2}, \right. \\
&\qquad \left. \left(\Sigma_2^{1/2} A \Sigma_2^{1/2}\right)^{-1}, \frac{1}{2} \Sigma_1^{-1/2} H \Lambda H' \Sigma_1^{-1/2} \right) \mathrm{d}(H) \\
&= \frac{\pi^{\frac{1}{2}m^2}}{\Gamma_m\left(\frac{n}{2}\right) \Gamma_m\left(\frac{m}{2}\right) |2\Sigma_1|^{n/2} |A\Sigma_2|^{n/2}} \operatorname{etr}\left(-\frac{1}{2}\Sigma_1^{-1} M \Sigma_2^{-1} M'\right) \\
&\quad \times |\Lambda|^{\frac{1}{2}(n-m-1)} \left\{ \prod_{i<j} (\lambda_i - \lambda_j) \right\} \\
&\quad \times \sum_{k=0}^{\infty} \sum_{\kappa} \frac{1}{k! \left(\frac{n}{2}\right)_\kappa} \frac{C_\kappa(\Lambda)}{C_\kappa(I_m)} P_\kappa \left(\frac{1}{\sqrt{2}} \Sigma_1^{-1/2} M \Sigma_2^{-1/2}, \right. \\
&\qquad \left. \left(\Sigma_2^{1/2} A \Sigma_2^{1/2}\right)^{-1}, \frac{1}{2} \Sigma_1^{-1} \right).
\end{aligned}$$

On setting $\Lambda = \lambda_1 \operatorname{diag}(1, \omega_2, \ldots, \omega_m)$ and using (5.3.2), we obtain the probability density function of λ_1. The distribution function of λ_1 obtained by integrating over $0 < \lambda_1 < x$, is as given in (5.1.11).

Definition 5.3.1 *Noncentral mean distribution with known covariance* Let X be an $m \times n$ matrix distributed as $N_{m,n}(M, \Sigma_1, \Sigma_2)$, then the joint probability density function of the latent roots of $\Sigma_1^{-1/2} XAX'\Sigma_1^{-1/2}$, is called the noncentral mean distribution with known covariance.

Theorem 5.3.2 The joint probability density function of the latent roots of $\Sigma_1^{-1/2} XAX'\Sigma_1^{-1/2}$ has the following representations.

(i) Power series representation

$$\frac{\pi^{\frac{1}{2}m^2}}{\Gamma_m\left(\frac{n}{2}\right)\Gamma_m\left(\frac{m}{2}\right)|2A\Sigma_2|^{m/2}} \operatorname{etr}\left(-\frac{1}{2}\Sigma_1^{-1}M\Sigma_2^{-1}M'\right) \times |\Lambda|^{\frac{1}{2}(n-m-1)}\left\{\prod_{i<j}(\lambda_i-\lambda_j)\right\} \times \sum_{k=0}^{\infty}\sum_{\kappa}\frac{1}{k!\left(\frac{n}{2}\right)_\kappa}\frac{C_\kappa\left(\frac{1}{2}\Lambda\right)}{C_\kappa(I_m)} \times P_\kappa\left(\frac{1}{\sqrt{2}}\Sigma_1^{-1/2}M\Sigma_2^{-1/2}, \left(\Sigma_2^{1/2}A\Sigma_2^{1/2}\right)^{-1}\right); \quad (5.3.6)$$

(ii) Γ-type representation

$$\frac{\pi^{\frac{1}{2}m^2}}{\Gamma_m\left(\frac{n}{2}\right)\Gamma_m\left(\frac{m}{2}\right)|2A\Sigma_2/q|^{m/2}} \operatorname{etr}\left(-\frac{1}{2}\Sigma_1^{-1}M\Sigma_2^{-1}M'\right) \times \operatorname{etr}\left(-\frac{q}{2}\Lambda\right)|\Lambda|^{\frac{1}{2}(n-m-1)}\left\{\prod_{i<j}(\lambda_i-\lambda_j)\right\} \times \sum_{k=0}^{\infty}\sum_{\kappa}\frac{1}{k!\left(\frac{n}{2}\right)_\kappa}\frac{C_\kappa\left(\frac{1}{2}\Lambda\right)}{C_\kappa(I_m)} P_\kappa\left(\frac{1}{\sqrt{2}}M^*, A^*\right), \quad (5.3.7)$$

where $M^* = \Sigma_1^{-1/2}M\Sigma_2^{-1}(\Sigma_2^{-1} - qA)^{-1/2}$ and $A^* = (\Sigma_2^{-1} - qA)^{1/2} A^{-1}(\Sigma_2^{-1} - qA)^{1/2}$;

(iii) Laguerre type representation

$$\frac{\pi^{\frac{1}{2}m^2}}{\Gamma_m\left(\frac{n}{2}\right)\Gamma_m\left(\frac{m}{2}\right)(2/q)^{mn/2}} \operatorname{etr}\left(-\frac{q}{2}\Lambda\right)|\Lambda|^{\frac{1}{2}(n-m-1)} \times \left\{\prod_{i<j}(\lambda_i-\lambda_j)\right\}\sum_{k=0}^{\infty}\sum_{\kappa}\frac{1}{k!\left(\frac{n}{2}\right)_\kappa}\frac{L_\kappa^{\frac{1}{2}(n-m-1)}\left(\frac{q}{2}\Lambda\right)}{C_\kappa(I_m)} \times P_\kappa\left(\frac{1}{\sqrt{2}}\Sigma_1^{-1/2}M\left((qA)^{-1}-I\right)^{-1/2}, qA-I\right); \quad (5.3.8)$$

(iv) mixture type representation

$$\sum_{k=0}^{\infty}\sum_{\kappa}R_\kappa f_\kappa(\Lambda), \quad (5.3.9)$$

where

$$f_\kappa(\Lambda) = \frac{\pi^{\frac{1}{2}m^2}}{\Gamma_m\left(\frac{m}{2}\right)\Gamma_m\left(\frac{n}{2};\kappa\right)(2/q)^{mn/2}} \operatorname{etr}\left(-\frac{q}{2}\Lambda\right) \times |\Lambda|^{\frac{1}{2}(n-m-1)}\left\{\prod_{i<j}(\lambda_i - \lambda_j)\right\}\frac{C_\kappa\left(\frac{q}{2}\Lambda\right)}{C_\kappa(I_m)},$$

$$k!R_\kappa = \frac{1}{(2\pi)^{mn/2}}\int \operatorname{etr}\left(-\frac{1}{2}ZZ'\right) \times \operatorname{etr}\left(-\frac{1}{2}\Omega(Z)\right) C_\kappa\left(\frac{1}{2}\Omega(Z)\right) dZ \tag{5.3.10}$$

and

$$\Omega(Z) = \left\{Z(Aq - I)^{1/2} - \Sigma_1^{-1/2}M(Aq)^{1/2}\right\} \times \left\{Z(Aq - I)^{1/2} - \Sigma_1^{-1/2}M(Aq)^{1/2}\right\}'.$$

The R_κ's satisfy the following relations: $R_\kappa > 0$ for all partition κ of k and

$$\sum_{k=0}^{\infty}\sum_{\kappa} R_\kappa = 1. \tag{5.3.11}$$

PROOF On transforming $\Sigma^{-1/2}S\Sigma^{-1/2} = H\Lambda H'$, H having positive elements in its first column, in (5.1.3) and integrating over $O(m)$ with the help of (5.1.7), we obtain (5.3.6) and (5.3.7), respectively. The representation (5.3.8) is obtained similarly by using (4.6.55).

Let Z and Y be $m \times n$ independent matrices whose elements are independently distributed standard normal random variables. On setting

$$X = \Sigma^{1/2}Y(Aq)^{-1/2} - \Sigma^{1/2}Z(I - (Aq)^{-1})^{1/2} + M,$$

we have

$$\begin{aligned} qS &= q\Sigma^{-1/2}XAX'\Sigma^{-1/2} \\ &= \left\{Y - Z(Aq - I)^{1/2} + \Sigma^{-1/2}M(Aq)^{1/2}\right\} \\ &\quad \times \left\{Y - Z(Aq - I)^{1/2} + \Sigma^{-1/2}M(Aq)^{1/2}\right\}'. \end{aligned}$$

For fixed Z the right-hand side has the noncentral Wishart distribution $W_m(n, I_m, \Omega(Z))$ with

$$\Omega(Z) = \left\{Z(Aq - I)^{1/2} - \Sigma^{-1/2}M(Aq)^{1/2}\right\} \times \left\{Z(Aq - I)^{1/2} - \Sigma^{-1/2}M(Aq)^{1/2}\right\}'.$$

Hence the conditional probability density function of the latent roots Λ of $\Sigma^{-1/2}XAX'$ $\Sigma^{-1/2}$ for Z fixed is given by

$$\frac{\pi^{m^2/2}}{\Gamma_m\left(\frac{n}{2}\right)\Gamma_m\left(\frac{m}{2}\right)(2/q)^{mn/2}} \operatorname{etr}\left(-\frac{1}{2}\Omega(Z)\right)$$
$$\times \operatorname{etr}\left(-\frac{q}{2}\Lambda\right)|\Lambda|^{\frac{1}{2}(n-m-1)}\Big\{\prod_{i<j}(\lambda_i-\lambda_j)\Big\}$$
$$\times\ {}_0F_1^{(m)}\left(\frac{n}{2};\frac{1}{2}\Omega(Z),\frac{q}{2}\Lambda\right)$$
$$=\sum_{k=0}^{\infty}\sum_{\kappa}D_\kappa(Z)f_\kappa(\Lambda),$$

where

$$f_\kappa(\Lambda)=\frac{\pi^{\frac{1}{2}m^2}}{\Gamma_m\left(\frac{n}{2};\kappa\right)\Gamma_m\left(\frac{m}{2}\right)(2/q)^{mn/2}}\operatorname{etr}\left(-\frac{q}{2}\Lambda\right)|\Lambda|^{\frac{1}{2}(n-m-1)}$$
$$\times\Big\{\prod_{i<j}(\lambda_i-\lambda_j)\Big\}\frac{C_\kappa\left(\frac{q}{2}\Lambda\right)}{C_\kappa(I_m)},$$

and

$$D_\kappa(Z)=\operatorname{etr}\left(-\frac{1}{2}\Omega(Z)\right)C_\kappa\left(\frac{1}{2}\Omega(Z)\right)/k!\,.$$

Therefore the probability density function of Λ is given by

$$\sum_{k=0}^{\infty}\sum_{\kappa}R_\kappa f_\kappa(\Lambda),$$

where

$$k!R_\kappa=\frac{1}{(2\pi)^{mn/2}}\int_Z\operatorname{etr}\left(-\frac{1}{2}ZZ'\right)\operatorname{etr}\left(-\frac{1}{2}\Omega(Z)\right)$$
$$\times C_\kappa\left(\frac{1}{2}\Omega(Z)\right)dZ\,.$$

Since $\Omega(Z)>0$, $C_\kappa(\Omega(Z))>0$, $R_\kappa>0$ for all partition κ of k, clearly

$$\sum_{k=0}^{\infty}\sum_{\kappa}R_\kappa=1$$

and

$$\int_{\lambda_1>\cdots>\lambda_m>0}f_\kappa(\Lambda)d\Lambda=1\,.$$

Corollary 5.3.1 R_κ is expressed as

$$k!R_\kappa = |Aq|^{-m/2}\operatorname{etr}\left(-\frac{1}{2}\Sigma^{-1}MM'\right)$$
$$\times(-1)^k P_\kappa\left(\frac{i}{\sqrt{2}}\Sigma^{-1/2}M(Aq-I)^{-1/2}, I-(Aq)^{-1}\right). \tag{5.3.12}$$

The proof is left to the reader (Exercise 5.8).

Remark 5.3.1 The integrations with respect to the latent roots in Lemma 5.3.1 are due to Hayakawa (1967) and Sugiyama (1967a, 1967b). (5.3.5) was obtained by Crowther (1975) and Theorem 5.3.2 was proved by Hayakawa (1969, 1972a). The non-central mean case for a Wishart matrix was studied by James (1961b). Further materials for the case of Hermitian matrices can be found in Hayakawa (1972c) and Khatri (1970).

5.4 Distributions of Some Functions of XAX'

Theorem 5.4.1 Let the column vectors of $X_{m\times n_1}$ and $Y_{m\times n_2}$ be independently distributed as $N(\mathbf{0},\Sigma)$. Then the probability density function of $F = Y'(XAX')^{-1}Y$, for $n_2 \le m \le n_1$, is given by

$$\frac{\Gamma_m\left(\frac{n_1+n_2}{2}\right)}{\Gamma_m\left(\frac{n_1}{2}\right)\Gamma_{n_2}\left(\frac{m}{2}\right)|A|^{\frac{m}{2}}q^{\frac{1}{2}(m-n_2)(n_1+n_2)}}\frac{|F|^{\frac{1}{2}(m-n_2-1)}}{|qI_{n_2}+F|^{\frac{1}{2}(n_1+n_2)}}$$
$$\times\ {}_1F_0^{(n_1)}\left(\frac{n_1+n_2}{2}; I_{n_1}-(qA)^{-1}, qR\right) \tag{5.4.1}$$

where

$$q>0, \qquad R=\begin{bmatrix}(qI_{n_2}+F)^{-1} & O\\ O & \frac{1}{q}I_{m-n_2}\end{bmatrix}.$$

PROOF Since F is invariant under the simultaneous transformations $X\to\Sigma^{1/2}X$ and $Y\to\Sigma^{1/2}Y$, we shall take $\Sigma=I_m$. By using (5.2.1), the joint probability density function of $Z=S^{-1/2}Y$ and $S=XAX'$ is given by

$$\frac{1}{2^{\frac{1}{2}m(n_1+n_2)}\Gamma_m\left(\frac{n_1}{2}\right)|A|^{\frac{m}{2}}\pi^{\frac{1}{2}mn_2}}|S|^{\frac{1}{2}(n_1+n_2-m-1)}$$

$$\times\operatorname{etr}\left\{-\frac{1}{2}(qI_m+ZZ')S\right\}\ {}_0F_0^{(n_1)}\left(I-(qA)^{-1}, \frac{q}{2}S\right).$$

The integration with respect to S yields

$$\frac{\Gamma_m\left(\frac{n_1+n_2}{2}\right)}{\Gamma_m\left(\frac{n_1}{2}\right)|A|^{\frac{m}{2}}\pi^{\frac{1}{2}mn_2}}|qI_m+ZZ'|^{-\frac{1}{2}(n_1+n_2)}$$
$$\times\ {}_1F_0^{(n_1)}\left(\frac{n_1+n_2}{2};I-(qA)^{-1},q(qI_m+ZZ')^{-1}\right).$$

Note that

$$|qI_m+ZZ'|=q^{m-n_2}|qI_{n_2}+Z'Z|,$$

$$C_\kappa((qI_m+ZZ')^{-1})=C_\kappa(R),$$

$$R=\begin{bmatrix}(qI_{n_2}+Z'Z)^{-1} & O\\ O & \frac{1}{q}I_{m-n_2}\end{bmatrix}.$$

Since $m\geq n_2$, by Lemma 4.1.1, we obtain the probability density function of $F=Z'Z$ given in (5.4.1).

Theorem 5.4.2 Let X be distributed as $N_{p,m}(O,\Sigma_1,\Sigma_2)$ and let Y be independently distributed as $N_{p,n}(O,\Sigma,I_n)$, then the density function of $F=X'(YY')^{-1}X$ for $m\leq p\leq n$ is given by

$$\frac{\Gamma_p\left(\frac{m+n}{2}\right)}{\Gamma_p\left(\frac{n}{2}\right)\Gamma_m\left(\frac{p}{2}\right)}|\Omega|^{-\frac{1}{2}m}|\Sigma_2|^{-\frac{1}{2}p}\frac{|F|^{\frac{1}{2}(p-m-1)}}{|I_m+(q\Sigma_2)^{-1}F|^{\frac{1}{2}(m+n)}}$$
$$\times\ {}_1F_0^{(p)}\left(\frac{m+n}{2};\Omega^*,F(q\Sigma_2+F)^{-1}\right), \tag{5.4.2}$$

where $\Omega^*=I_p-q\Omega^{-1}$, $\Omega=\Sigma_1^{1/2}\Sigma^{-1}\Sigma_1^{1/2}$, $q>0$.

PROOF Since F is invariant under $X\to\Sigma_1^{1/2}X$ and $Y\to\Sigma_1^{1/2}Y$, we may take $\Sigma_1\to I_p$ and $\Sigma\to\Omega=\Sigma_1^{1/2}\Sigma^{-1}\Sigma_1^{1/2}$, respectively. On transforming X to Z by letting $Z=(YY')^{-1/2}X$ in the joint probability density function of X and Y and integrating with respect to Y, one has the probability density function of Z as

$$\frac{\Gamma_p\left(\frac{m+n}{2}\right)}{\Gamma_p\left(\frac{n}{2}\right)\pi^{\frac{1}{2}pm}}|\Omega|^{\frac{1}{2}n}|\Sigma_2|^{-\frac{1}{2}p}|\Omega+Z\Sigma_2^{-1}Z'|^{-\frac{1}{2}(m+n)}.$$

Noting that

$$|\Omega+Z\Sigma_2^{-1}Z'|=|\Omega||\Sigma_2|^{-1}|\Sigma_2+Z'Z/q|$$
$$\times|I_p-Z(q\Sigma_2+Z'Z)^{-1}Z'\Omega^*|,$$

where $\Omega^*=I_p-q\Omega^{-1}$, $q>0$, and integrating out Z such that $F=Z'Z$ is fixed, we have

$$\frac{\Gamma_p\left(\frac{m+n}{2}\right)}{\Gamma_p\left(\frac{n}{2}\right)\pi^{\frac{1}{2}pm}}|\Omega|^{-\frac{1}{2}m}|\Sigma_2|^{-\frac{1}{2}p}\int_{F=Z'Z}\left\{|I_n+(q\Sigma_2)^{-1}Z'Z|^{-\frac{1}{2}(m+n)}\right.$$
$$\left.\times\,|I_p-Z(q\Sigma_2+Z'Z)^{-1}Z'\Omega^*|^{-\frac{1}{2}(m+n)}\right\}\mathrm{d}Z$$

$$=\frac{\Gamma_p\left(\frac{m+n}{2}\right)}{\Gamma_p\left(\frac{n}{2}\right)\pi^{\frac{1}{2}pm}}|\Omega|^{-\frac{1}{2}m}|\Sigma_2|^{-\frac{1}{2}p}\int_{F=Z'Z}|I_n+(q\Sigma_2)^{-1}Z'Z|^{-\frac{1}{2}(m+n)}$$
$$\times\left\{\int_{O(p)}|I_p-Z(q\Sigma_2+Z'Z)^{-1}Z'H\Omega^*H'|^{-\frac{1}{2}(m+n)}\mathrm{d}(H)\right\}\mathrm{d}Z$$

$$=\frac{\Gamma_p\left(\frac{m+n}{2}\right)}{\Gamma_p\left(\frac{n}{2}\right)\pi^{\frac{1}{2}pm}}|\Omega|^{-\frac{1}{2}m}|\Sigma_2|^{-\frac{1}{2}p}\int_{F=Z'Z}|I_n+(q\Sigma_2)^{-1}Z'Z|^{-\frac{1}{2}(m+n)}$$
$$\times\,{}_1F_0^{(p)}\left(\frac{m+n}{2};\Omega^*,(q\Sigma_2+Z'Z)^{-1}Z'Z\right)\mathrm{d}Z$$

$$=\frac{\Gamma_p\left(\frac{m+n}{2}\right)}{\Gamma_p\left(\frac{n}{2}\right)\Gamma_m\left(\frac{p}{2}\right)}|\Omega|^{-\frac{1}{2}m}|\Sigma_2|^{-\frac{1}{2}p}\frac{|F|^{\frac{1}{2}(p-m-1)}}{|I_m+(q\Sigma_2)^{-1}F|^{\frac{1}{2}(m+n)}}$$
$$\times\,{}_1F_0^{(p)}\left(\frac{m+n}{2};\Omega^*,(q\Sigma_2+F)^{-1}F\right),$$

which completes the proof.

Theorem 5.4.3 Let X be distributed as $N_{p,m}(M,\Sigma_1,\Sigma_2)$ and let A be independently distributed as $W_p(n,\Sigma)$. The matric t-variate is defined as $Y=A^{-1/2}X$. The probability density function of $R=Y'Y=X'A^{-1}X$ is given for $m\le p\le n$ as

$$\frac{\Gamma_p\left(\frac{m+n}{2}\right)}{\Gamma_m\left(\frac{p}{2}\right)\Gamma_p\left(\frac{n}{2}\right)}|\Sigma_1\Sigma^{-1}|^{-\frac{1}{2}m}|\Sigma_2|^{-\frac{1}{2}p}$$
$$\times\,|R|^{\frac{1}{2}(p-m-1)}\operatorname{etr}\left(-\frac{1}{2}\Sigma_1^{-1}M\Sigma_2^{-1}M'\right)$$
$$\times\sum_{k=0}^{\infty}\sum_{\kappa}\frac{\left(\frac{m+n}{2}\right)_\kappa}{k!\left(\frac{p}{2}\right)_\kappa}P_\kappa\left(\frac{1}{\sqrt{2}}\Sigma_2^{-1/2}M'\Sigma_1^{-1/2},\right.$$
$$\left.\Sigma_1^{-1/2}\Sigma\Sigma_1^{-1/2},\Sigma_2^{-1/2}R\Sigma_2^{-1/2}\right). \tag{5.4.3}$$

PROOF The conditional probability density function of $R=Y'Y$ given A is considered first. Noting that the distribution of X' is $N_{m,p}(M',\Sigma_2,\Sigma_1)$ and using Theorem 5.1.1, the conditional probability density function of $R=X'A^{-1}X$ given A can be expressed as

$$\frac{1}{\Gamma_m\left(\frac{p}{2}\right)|2\Sigma_1|^{m/2}|\Sigma_2|^{p/2}}|A|^{\frac{m}{2}}\operatorname{etr}\left(-\frac{1}{2}\Sigma_2^{-1}M'\Sigma_1^{-1}M\right)|R|^{\frac{1}{2}(p-m-1)}$$
$$\times\sum_{k=0}^{\infty}\sum_{\kappa}\frac{1}{k!\left(\frac{p}{2}\right)_\kappa}P_\kappa\left(\frac{1}{\sqrt{2}}\Sigma_2^{-1/2}M'\Sigma_1^{-1/2},\right.$$
$$\left.\Sigma_1^{-1/2}A\Sigma_1^{-1/2},\frac{1}{2}\Sigma_2^{-1/2}R\Sigma_2^{-1/2}\right).$$

The expectation with respect to $A \sim W_p(\Sigma, n)$ gives

$$E_A\left[|A|^{\frac{m}{2}}P_\kappa\left(\frac{1}{\sqrt{2}}\Sigma_2^{-1/2}M'\Sigma_1^{-1/2},\Sigma_1^{-1/2}A\Sigma_1^{-1/2},\frac{1}{2}\Sigma_2^{-1/2}R\Sigma_2^{-1/2}\right)\right]$$
$$=\frac{\Gamma_p\left(\frac{m+n}{2};\kappa\right)}{\Gamma_p\left(\frac{n}{2}\right)}|2\Sigma|^{\frac{m}{2}}$$
$$\times P_\kappa\left(\frac{1}{\sqrt{2}}\Sigma_2^{-1/2}M'\Sigma_1^{-1/2},\Sigma_1^{-1/2}\Sigma\Sigma_1^{-1/2},\Sigma_2^{-1/2}R\Sigma_2^{-1/2}\right)$$

and thus (5.4.3) follows.

Remark 5.4.1 If $\Sigma_1 = \Sigma$ and $\Sigma_2 = I_m$, then with the help of the identity

$$\Gamma_p\left(\frac{m+n}{2}\right)\Gamma_m\left(\frac{m+n-p}{2}\right)=\Gamma_p\left(\frac{n}{2}\right)\Gamma_m\left(\frac{m+n}{2}\right)$$

and Exercise 5.9, (5.4.3) can be expressed as

$$\frac{\Gamma_m\left(\frac{m+n}{2}\right)}{\Gamma_m\left(\frac{p}{2}\right)\Gamma_m\left(\frac{m+n-p}{2}\right)}\frac{|R|^{\frac{1}{2}(p-m-1)}}{|I+R|^{\frac{1}{2}(m+n)}}\operatorname{etr}\left(-\frac{1}{2}M'\Sigma^{-1}M\right)$$
$$\times{}_1F_1\left(\frac{m+n}{2};\frac{p}{2};\frac{1}{2}M'\Sigma^{-1}M(I+R^{-1})^{-1}\right). \quad (5.4.4)$$

This is the probability density function of the noncentral F matrix due to James (1964).

Corollary 5.4.1 Under the assumptions of Theorem 5.4.3, the joint probability density function of the latent roots $\lambda_1, \lambda_2, \ldots, \lambda_m$ of R, the probability density function of $x = \operatorname{tr}(R)$ and the distribution function of the largest latent root λ_1 of R are respectively given by

$$\frac{\pi^{\frac{1}{2}m^2}\Gamma_p\left(\frac{m+n}{2}\right)}{\Gamma_m\left(\frac{p}{2}\right)\Gamma_m\left(\frac{m}{2}\right)\Gamma_p\left(\frac{n}{2}\right)}|\Sigma_1\Sigma^{-1}|^{-m/2}|\Sigma_2|^{-p/2}$$
$$\times\operatorname{etr}\left(-\frac{1}{2}\Sigma_1^{-1}M\Sigma_2^{-1}M'\right)|\Lambda|^{\frac{1}{2}(p-m-1)}\left\{\prod_{i<j}(\lambda_i-\lambda_j)\right\}$$

$$\times \sum_{k=0}^{\infty} \sum_{\kappa} \frac{\left(\frac{m+n}{2}\right)_\kappa}{k!\left(\frac{p}{2}\right)_\kappa} \frac{C_\kappa(\Lambda)}{C_\kappa(I_m)}$$
$$\times P_\kappa\left(\frac{1}{\sqrt{2}}\Sigma_2^{-1/2}M'\Sigma_1^{-1/2}, \Sigma_1^{-1/2}\Sigma\Sigma_1^{-1/2}, \Sigma_2^{-1}\right); \tag{5.4.5}$$

$$\frac{\Gamma_p\left(\frac{m+n}{2}\right)}{\Gamma_p\left(\frac{n}{2}\right)\Gamma\left(\frac{pm}{2}\right)}|\Sigma_1\Sigma^{-1}|^{-m/2}|\Sigma_2|^{-p/2}$$
$$\times \operatorname{etr}\left(-\frac{1}{2}\Sigma_1^{-1}M\Sigma_2^{-1}M'\right) x^{\frac{1}{2}pm-1} \sum_{k=0}^{\infty} \frac{x^k}{k!\left(\frac{pm}{2}\right)_k} \sum_{\kappa}\left(\frac{m+n}{2}\right)_\kappa$$
$$\times P_\kappa\left(\frac{1}{\sqrt{2}}\Sigma_2^{-1/2}M'\Sigma_1^{-1/2}, \Sigma_1^{-1/2}\Sigma\Sigma_1^{-1/2}, \Sigma_2^{-1}\right); \tag{5.4.6}$$

$$\frac{\Gamma_p\left(\frac{m+n}{2}\right)\Gamma_m\left(\frac{m+1}{2}\right)}{\Gamma_p\left(\frac{n}{2}\right)\Gamma_m\left(\frac{p+m+1}{2}\right)}|\Sigma_1\Sigma^{-1}|^{-m/2}|\Sigma_2|^{-p/2}$$
$$\times \operatorname{etr}\left(-\frac{1}{2}\Sigma_1^{-1}M\Sigma_2^{-1}M'\right) \lambda_1^{\frac{1}{2}pm-1} \sum_{k=0}^{\infty} \frac{\lambda_1^k}{k!} \sum_{\kappa} \frac{\left(\frac{m+n}{2}\right)_\kappa}{\left(\frac{p+m+1}{2}\right)_\kappa}$$
$$\times P_\kappa\left(\frac{1}{\sqrt{2}}\Sigma_2^{-1/2}M'\Sigma_1^{-1/2}, \Sigma_1^{-1/2}\Sigma\Sigma_1^{-1/2}, \Sigma_2^{-1}\right). \tag{5.4.7}$$

PROOF These expressions are obtained with the help of Lemma 5.1.1 and Lemma 5.1.2.

5.5 Generalized Hotelling's T_0^2

Definition 5.5.1 *Generalized Hotelling's* T_0^2 Let S_1 be an $m \times m$ noncentral Wishart matrix with n_1 degrees of freedom, covariance matrix Σ and noncentrality parameter matrix Ω, and let S_2 be an $m \times m$ central Wishart matrix with n_2 degree of freedom and the same covariance matrix Σ and S_1 and S_2 be independently distributed. The generalized Hotelling's T_0^2 is defined as

$$T_0^2 = \operatorname{tr}(S_1 S_2^{-1}). \tag{5.5.1}$$

Note 5.5.1 Let $X = [\boldsymbol{x}_1, \boldsymbol{x}_2, \ldots, \boldsymbol{x}_{n_1}]$ be distributed as $N_{m,n_1}(M, \Sigma, I_{n_1})$ with $M = [\boldsymbol{\mu}_1, \boldsymbol{\mu}_2, \ldots, \boldsymbol{\mu}_{n_1}]$, then S_1 can be expressed as $S_1 = XX'$.
Let

$$\boldsymbol{x} = \begin{bmatrix} \boldsymbol{x}_1 \\ \boldsymbol{x}_2 \\ \vdots \\ \boldsymbol{x}_{n_1} \end{bmatrix}, \quad \boldsymbol{\mu} = \begin{bmatrix} \boldsymbol{\mu}_1 \\ \boldsymbol{\mu}_2 \\ \vdots \\ \boldsymbol{\mu}_{n_1} \end{bmatrix},$$

then $\boldsymbol{x}$ is an $mn_1 \times 1$ column vector distributed as $N(\boldsymbol{\mu}, I_{n_1} \otimes \Sigma)$.

Thus T_0^2 can be expressed as

$$T_0^2 = \operatorname{tr}(XX'S_2^{-1}) = \boldsymbol{x}'\,(I_{n_1} \otimes S_2^{-1})\,\boldsymbol{x}\,. \tag{5.5.2}$$

T_0^2 is invariant under the simultaneous transformations $\boldsymbol{x} \to (I_{n_1} \otimes \Sigma^{-1/2})\boldsymbol{x}$ and $S_2 \to \Sigma^{-1/2}S_2\Sigma^{-1/2}$. Hence, without loss of generality we may assume that $\boldsymbol{x}$ is distributed as $N(\boldsymbol{\nu}, I_{mn_1})$ with $\boldsymbol{\nu} = (I_{n_1} \otimes \Sigma^{-1/2})\boldsymbol{\mu}$.

Theorem 5.5.1 The probability density function of $T_0^2 = \operatorname{tr}(S_1S_2^{-1})$ is given by

$$\frac{\Gamma_m\left(\frac{n_1+n_2}{2}\right)}{\Gamma_m\left(\frac{n_2}{2}\right)\Gamma\left(\frac{mn_1}{2}\right)} \operatorname{etr}(-\Omega)T^{\frac{1}{2}mn_1-1} \times \sum_{k=0}^{\infty} \frac{(-T)^k}{k!\left(\frac{mn_1}{2}\right)_k} \sum_{\kappa} \left(\frac{n_1+n_2}{2}\right)_{\kappa} L_{\kappa}^{\gamma}(\Omega)\,, \tag{5.5.3}$$

where $\Omega = \frac{1}{2}\Sigma^{-1}MM'$ and $\gamma = \frac{1}{2}(n_1 - m - 1)$. This series converges for $||T|| < 1$.

PROOF The conditional probability density function of $T = \boldsymbol{x}'(I_{n_1} \otimes S_2^{-1})\boldsymbol{x}$ given S_2 is obtained from (5.1.3), by making the substitutions

$$m \to 1, \qquad n \to mn_1\,, \qquad S \to T\,,$$
$$\Sigma \to 1, \qquad \Sigma^{-1/2}M \to \boldsymbol{\nu}\,, \qquad A \to I_{n_1} \otimes S_2^{-1}\,,$$

as follows:

$$\frac{|S_2|^{\frac{1}{2}n_1}}{2^{\frac{1}{2}mn_1}\Gamma\left(\frac{mn_1}{2}\right)} \exp(-\frac{1}{2}\boldsymbol{\nu}'\boldsymbol{\nu})T^{\frac{1}{2}mn_1-1} \times \sum_{k=0}^{\infty} \frac{1}{k!\left(\frac{mn_1}{2}\right)_k} P_k\left(\frac{1}{\sqrt{2}}\boldsymbol{\nu}, I_{n_1} \otimes S_2, \frac{T}{2}\right). \tag{5.5.4}$$

Noting that

$$\begin{aligned}
P_k&\left(\frac{1}{\sqrt{2}}\boldsymbol{\nu}, I_{n_1} \otimes S_2, \frac{T}{2}\right) \\
&= \left(\frac{T}{2}\right)^k \frac{(-1)^k}{\pi^{\frac{1}{2}mn_1}} \exp\left(\frac{1}{2}\boldsymbol{\nu}'\boldsymbol{\nu}\right) \times \int_{\boldsymbol{u}} \exp(-\boldsymbol{u}'\boldsymbol{u} - \sqrt{2}i\boldsymbol{u}'\boldsymbol{\nu})C_{(k)}\left(\boldsymbol{u}'(I_{n_1} \otimes S_2)\boldsymbol{u}\right)d\boldsymbol{u} \\
&= \left(\frac{T}{2}\right)^k \frac{(-1)^k}{\pi^{\frac{1}{2}mn_1}} \operatorname{etr}\left(\frac{1}{2}\Sigma^{-1}MM'\right) \times \int_U \operatorname{etr}(-UU' - \sqrt{2}iUM'\Sigma^{-1/2})(\operatorname{tr}(S_2UU'))^k dU
\end{aligned}$$

$$= \left(\frac{T}{2}\right)^k \frac{(-1)^k}{\pi^{\frac{1}{2}mn_1}} \operatorname{etr}\left(\frac{1}{2}\Sigma^{-1}MM'\right)$$
$$\times \sum_\kappa \int_U \operatorname{etr}(-UU' - \sqrt{2}iUM'\Sigma^{-1/2})C_\kappa(S_2UU')\mathrm{d}U\,,$$

we have

$$\int_{S_2>0} \operatorname{etr}\left(-\frac{1}{2}S_2\right)|S_2|^{\frac{1}{2}(n_1+n_2-m-1)}$$
$$\times P_k\left(\frac{1}{\sqrt{2}}\boldsymbol{\nu}, I_{n_1}\otimes S_2, \frac{T}{2}\right)\mathrm{d}S_2$$
$$= 2^{\frac{1}{2}m(n_1+n_2)}T^k\Gamma_m\left(\frac{n_1+n_2}{2}\right)\sum_\kappa\left(\frac{n_1+n_2}{2}\right)_\kappa$$
$$\times H_\kappa\left(\frac{1}{\sqrt{2}}\Sigma^{-1/2}M\right). \tag{5.5.5}$$

Then from (4.6.35) the unconditional probability density function of $\operatorname{tr}(S_1S_2^{-1})$ can be expressed as (5.5.3). Since (4.6.38) and $\left(\frac{n_1}{2}\right)_\kappa / \left(\frac{mn_1}{2}\right)_k \leq 1$ hold, the series converges for $||T|| < 1$.

Remark 5.5.1 The derivation of (5.5.3) is based on Phillips (1984). The representation (5.5.3) was given by Constantine (1966) in a different way.

Corollary 5.5.1 The limiting distribution of $n_2 \operatorname{tr}(S_1S_2^{-1})$ as $n_2 \to \infty$ is a noncentral chi-square distribution with mn_1 degrees of freedom and noncentrality parameter $\frac{1}{2}\operatorname{tr}(\Sigma^{-1}MM')$.

PROOF Since S_2/n_2 converges in probability to I_m as $n_2 \to \infty$, $\lim_{n_2\to\infty} \boldsymbol{x}'(I_{n_1} \otimes (S_2/n_2)^{-1})\boldsymbol{x} = \boldsymbol{x}'\boldsymbol{x}$, which implies that the limiting distribution is that of a noncentral chi-square with mn_1 degrees of freedom and noncentrality parameter $\boldsymbol{\nu}'\boldsymbol{\nu}/2$.

Lemma 5.5.1 Let $\boldsymbol{h}$ be an $mn \times 1$ vector satisfying $\boldsymbol{h}'\boldsymbol{h} = 1$ and let $\mathrm{d}(\boldsymbol{h})$ be the normalized invariant measure. Let $\boldsymbol{h}_i$ $(i = 1, 2, \ldots, n)$ be the $m \times 1$ vectors taken from the partition of $\boldsymbol{h}' = (\boldsymbol{h}_1', \boldsymbol{h}_2', \ldots, \boldsymbol{h}_n')$ into n component vectors. Put $Q = \sum_{i=1}^n \boldsymbol{h}_i\boldsymbol{h}_i'$. Then the following holds.

$$\int_{\boldsymbol{h}'\boldsymbol{h}=1} C_\kappa(Q)\mathrm{d}(\boldsymbol{h}) = \frac{\left(\frac{n}{2}\right)_\kappa C_\kappa(I_m)}{\left(\frac{mn}{2}\right)_k}. \tag{5.5.6}$$

PROOF We note that

$$\int_{\boldsymbol{h}'\boldsymbol{h}=1} C_\kappa((I_n \otimes Z)\boldsymbol{h}\boldsymbol{h}')\mathrm{d}(\boldsymbol{h})$$

$$= \int_{h'h=1} \int_{O(mn)} C_\kappa((I_n \otimes Z)Hhh'H')\mathrm{d}(H)\mathrm{d}(h)$$
$$= \int_{h'h=1} \frac{C_\kappa(I_n \otimes Z)C_\kappa(hh')}{C_\kappa(I_{mn})}\mathrm{d}(h)$$
$$= \frac{C_{(k)}(I_n \otimes Z)}{C_{(k)}(I_{mn})}.$$

From the expansion of $|I_{mn} - (I_n \otimes Z)|^{-1/2}$, we have

$$= \sum_\kappa \frac{\left(\frac{n}{2}\right)_\kappa C_\kappa(Z)}{\left(\frac{1}{2}\right)_k C_{(k)}(I_{mn})}.$$

The left-hand side is also given by

$$\int_{h'h=1} \{\mathrm{tr}\,((I_n \otimes Z)hh')\}^k \mathrm{d}(h)$$
$$= \int_{h'h=1} (\mathrm{tr}\,(ZQ))^k \mathrm{d}(h)$$
$$= \sum_\kappa \frac{C_\kappa(Z)}{C_\kappa(I_m)} \int_{h'h=1} C_\kappa(Q)\mathrm{d}(h),$$

since the integrand is invariant under the transformation $(I_n \otimes H)h$, $H \in O(m)$. Equating the coefficients of $C_\kappa(Z)$ we have (5.5.6).

Lemma 5.5.2 Let $P = nI - Q$, then

$$\int_{h'h=1} C_\kappa(P)\mathrm{d}(h) = C_\kappa(I_m) \sum_{t=0}^{k} \frac{(-1)^t n^{k-t}}{\left(\frac{mn}{2}\right)_t} \sum_\tau \binom{\kappa}{\tau} \left(\frac{n}{2}\right)_\tau. \tag{5.5.7}$$

PROOF By using the binomial expansion the result follows by Lemma 5.5.1.

Theorem 5.5.2 The probability density function of $T_0^2 = \mathrm{tr}\,(S_1 S_2^{-1})$ is given by

$$\frac{\Gamma_m\left(\frac{n_1+n_2}{2}\right)}{\Gamma_m\left(\frac{n_2}{2}\right)\Gamma\left(\frac{mn_1}{2}\right)} \mathrm{etr}\,(-\Omega) \frac{T^{\frac{1}{2}mn_1 - 1}}{(1+n_1T)^{\frac{1}{2}m(n_1+n_2)}} \tag{5.5.8}$$
$$\times \sum_{k=0}^{\infty}\sum_{\ell=0}^{\infty} \frac{\{T/(1+n_1T)\}^{k+\ell}}{k!\ell!} \sum_{\kappa,\lambda} \frac{C_\lambda(\Omega)}{\left(\frac{n_1}{2}\right)_\lambda C_\lambda(I_m)}$$
$$\times \sum_{\phi \in \kappa\cdot\lambda} \theta_\phi^{\kappa,\lambda} \left(\frac{n_1+n_2}{2}\right)_\phi d_\phi^{\kappa,\lambda} C_\phi(I_m),$$

where

$$d_\phi^{\kappa,\lambda} = \sum_{r=0}^{k} \frac{(-1)^r n_1^{k-r}}{\left(\frac{mn_1}{2}\right)_{r+\ell}} \sum_{\rho,\tau \in \rho\cdot\lambda} \left(\begin{matrix}\kappa : \phi \\ \rho : \tau\end{matrix} \middle| \lambda\right) \theta_\tau^{\rho,\lambda} \left(\frac{n_1}{2}\right)_\tau, \tag{5.5.9}$$

$\theta_\phi^{\kappa,\lambda}$ is defined in Appendix A.3 and $\left(\begin{matrix}\kappa : \phi \\ \rho : \tau\end{matrix}\middle| \lambda\right)$ is a binomial coefficient defined by (A.3.17), respectively. This converges for all $T > 0$.

PROOF As the statistic $T = \operatorname{tr}(XX'S_2^{-1})$ is invariant under the simultaneous transformations $X \to H_1XH_2$, $S_2 \to H_1S_2H_1'$, $H_1 \in O(m)$, $H_2 \in O(n_1)$, we may assume that the joint probability density function of (X, S_2) is given by

$$\frac{1}{2^{\frac{1}{2}m(n_1+n_2)}\pi^{\frac{1}{2}mn_1}\Gamma_m\left(\frac{n_2}{2}\right)}\operatorname{etr}(-\Omega)\operatorname{etr}\left(-\frac{1}{2}XX'\right)$$
$$\times\, {}_0F_1^{(m)}\left(\frac{n_1}{2};\frac{1}{2}XX',\Omega\right)\operatorname{etr}\left(-\frac{1}{2}S_2\right)|S_2|^{\frac{1}{2}(n_2-m-1)}.$$

Letting $X = S_2^{1/2}Y$ yields

$$\frac{1}{2^{\frac{1}{2}m(n_1+n_2)}\pi^{\frac{1}{2}mn_1}\Gamma_m\left(\frac{n_2}{2}\right)}\operatorname{etr}(-\Omega)\operatorname{etr}\left(-\frac{1}{2}S_2YY'\right)$$
$$\times\, {}_0F_1^{(m)}\left(\frac{n_1}{2};\frac{1}{2}S_2YY',\Omega\right)\operatorname{etr}\left(-\frac{1}{2}S_2\right)|S_2|^{\frac{1}{2}(n_1+n_2-m-1)}.$$

Let the column vectors of $Y = [\boldsymbol{y}_1, \boldsymbol{y}_2, \ldots, \boldsymbol{y}_{n_1}]$ be arranged as $\boldsymbol{y}' = (\boldsymbol{y}_1', \boldsymbol{y}_2', \ldots, \boldsymbol{y}_{n_1}') \in R^{mn_1}$ and let $\boldsymbol{y} = T^{1/2}\boldsymbol{h}$, where $\boldsymbol{h}'\boldsymbol{h} = 1$ and $T = \boldsymbol{y}'\boldsymbol{y} = \operatorname{tr}(XX'S_2^{-1})$. The Jacobian is expressed as

$$\mathrm{d}\boldsymbol{y} = \frac{1}{2}T^{\frac{1}{2}mn_1-1}\frac{2\pi^{\frac{1}{2}mn_1}}{\Gamma\left(\frac{mn_1}{2}\right)}\mathrm{d}T\,\mathrm{d}(\boldsymbol{h}),$$

where $\mathrm{d}(\boldsymbol{h})$ is the normalized invariant measure. Thus the probability density function of T is obtained as

$$\frac{1}{2^{\frac{1}{2}m(n_1+n_2)}\Gamma\left(\frac{mn_1}{2}\right)\Gamma_m\left(\frac{n_2}{2}\right)}\operatorname{etr}(-\Omega)T^{\frac{1}{2}mn_1-1} \tag{5.5.10}$$
$$\times\int_{\boldsymbol{h}'\boldsymbol{h}=1}\int_{S_2>0}\operatorname{etr}\left\{-\frac{1}{2}(I+TQ)S_2\right\}|S_2|^{\frac{1}{2}(n_1+n_2-m-1)}$$
$$\times\, {}_0F_1^{(m)}\left(\frac{n_1}{2};\frac{T}{2}S_2Q,\Omega\right)\mathrm{d}S_2\mathrm{d}(\boldsymbol{h})$$
$$=\frac{\Gamma_m\left(\frac{n_1+n_2}{2}\right)}{\Gamma\left(\frac{mn_1}{2}\right)\Gamma_m\left(\frac{n_2}{2}\right)}\operatorname{etr}(-\Omega)T^{\frac{1}{2}mn_1-1}$$
$$\times\int_{\boldsymbol{h}'\boldsymbol{h}=1}{}_1F_1^{(m)}\left(\frac{n_1+n_2}{2};\frac{n_1}{2};T(I+TQ)^{-1}Q,\Omega\right)$$
$$\times\,|I+TQ|^{-\frac{1}{2}(n_1+n_2)}\mathrm{d}(\boldsymbol{h}),$$

where $Q = \sum_{i=1}^{n_1}\boldsymbol{h}_i\boldsymbol{h}_i'$ and $\boldsymbol{h}_i$'s are corresponding vectors of the partitioned vector of $\boldsymbol{y}$. Put $I + TQ = (1 + n_1T)I_m - TP$. The integral with respect to $\boldsymbol{h}$ and S_2 in (5.5.10) becomes, by using (A.3.1) in the Appendix,

$$\int_{\boldsymbol{h}'\boldsymbol{h}=1}\int_{S_2>0} \operatorname{etr}\left(-\frac{1}{2}(1+n_1T)S_2\right) \operatorname{etr}\left(\frac{T}{2}S_2P\right)$$

$$\times\ {}_0F_1^{(m)}\left(\frac{n_1}{2};\frac{T}{2}S_2Q,\Omega\right)|S_2|^{\frac{1}{2}(n_1+n_2-m-1)}\mathrm{d}(\boldsymbol{h})\mathrm{d}S_2$$

$$= \int_{\boldsymbol{h}'\boldsymbol{h}=1}\int_{S_2>0} \operatorname{etr}\left(-\frac{1}{2}(1+n_1T)S_2\right)|S_2|^{\frac{1}{2}(n_1+n_2-m-1)}$$

$$\times\left[\int_{O(m)} \operatorname{etr}\left(\frac{T}{2}HS_2H'P\right)\right.$$

$$\left.\times\ {}_0F_1^{(m)}\left(\frac{n_1}{2};\frac{T}{2}HS_2H'Q,\Omega\right)\mathrm{d}(H)\right]\mathrm{d}(\boldsymbol{h})\mathrm{d}S_2$$

$$= \int_{\boldsymbol{h}'\boldsymbol{h}=1}\int_{S_2>0} \operatorname{etr}\left(-\frac{1}{2}(1+n_1T)S_2\right)|S_2|^{\frac{1}{2}(n_1+n_2-m-1)}$$

$$\times\sum_{k=0}^{\infty}\sum_{\ell=0}^{\infty}\frac{\left(\frac{T}{2}\right)^{k+\ell}}{k!\ell!}\sum_{\kappa}\sum_{\lambda}\frac{C_\lambda(\Omega)}{\left(\frac{n_1}{2}\right)_\lambda C_\lambda(I_m)}$$

$$\times\sum_{\phi\in\kappa\cdot\lambda}\theta_\phi^{\kappa,\lambda}\frac{C_\phi(S_2)}{C_\phi(I)}C_\phi^{\kappa,\lambda}(P,Q)\mathrm{d}S_2\mathrm{d}(\boldsymbol{h})$$

$$= \Gamma_m\left(\frac{n_1+n_2}{2}\right)\left(\frac{2}{1+n_1T}\right)^{\frac{1}{2}m(n_1+n_2)}$$

$$\times\sum_{k,\ell=0}^{\infty}\frac{1}{k!\ell!}\left(\frac{T}{1+n_1T}\right)^{k+\ell}\sum_{\kappa}\sum_{\lambda}\frac{C_\lambda(\Omega)}{\left(\frac{n_1}{2}\right)_\lambda C_\lambda(I_m)}$$

$$\times\sum_{\phi\in\kappa\cdot\lambda}\theta_\phi^{\kappa,\lambda}\left(\frac{n_1+n_2}{2}\right)_\phi\int_{\boldsymbol{h}'\boldsymbol{h}=1}C_\phi^{\kappa,\lambda}(P,Q)\mathrm{d}(\boldsymbol{h})\,.$$

Now by the binomial expansion (A.3.17) and Lemma 5.5.1 we have

$$\int_{\boldsymbol{h}'\boldsymbol{h}=1}C_\phi^{\kappa,\lambda}(P,Q)\mathrm{d}(\boldsymbol{h}) = \int_{\boldsymbol{h}'\boldsymbol{h}=1}C_\phi^{\kappa,\lambda}(n_1I-Q,Q)\mathrm{d}(\boldsymbol{h})$$

$$= n_1^kC_\phi(I)\int_{\boldsymbol{h}'\boldsymbol{h}=1}\sum_{r=0}^{k}\sum_{\rho,\tau\in\rho\cdot\lambda}\left(\begin{matrix}\kappa:\phi\\ \rho:\tau\end{matrix}\middle|\lambda\right)\frac{C_\tau^{\rho,\lambda}(-Q/n_1,Q)}{C_\tau(I)}\mathrm{d}(\boldsymbol{h})$$

$$= C_\phi(I)\sum_{r=0}^{k}\sum_{\rho,\tau\in\rho\cdot\lambda}\left(\begin{matrix}\kappa:\phi\\ \rho:\tau\end{matrix}\middle|\lambda\right)\frac{(-1)^rn_1^{k-r}\theta_\tau^{\rho,\lambda}}{C_\tau(I)}\int_{\boldsymbol{h}'\boldsymbol{h}=1}C_\tau(Q)\mathrm{d}(\boldsymbol{h})$$

$$= C_\phi(I)\sum_{r=0}^{k}\frac{(-1)^rn_1^{k-r}}{\left(\frac{mn_1}{2}\right)_{k+\ell}}\sum_{\rho,\tau\in\rho\cdot\lambda}\left(\begin{matrix}\kappa:\phi\\ \rho:\tau\end{matrix}\middle|\lambda\right)\theta_\tau^{\rho,\lambda}\left(\frac{n_1}{2}\right)_\tau$$

$$= d_\phi^{\kappa,\lambda}C_\phi(I)\qquad\text{(say)}\,.$$

It can be proved that this probability density function converges everywhere for $T > 0$ (Exercise 5.11).

Theorem 5.5.3 Let S_1 be an $m \times m$ noncentral Wishart matrix with n_1 degrees of freedom, covariance matrix Σ_1 and noncentrality parameter matrix Ω and let S_2 be an $m \times m$ central Wishart matrix with n_2 degrees of freedom and covariance matrix Σ_2, $\Sigma_1 \neq \Sigma_2$, and S_1 and S_2 be independently distributed. The joint probability density function of the latent roots of $F = S_2^{-1/2} S_1 S_2^{-1/2}$ is given by

$$\frac{\pi^{\frac{m^2}{2}} \Gamma_m\left(\frac{n_1+n_2}{2}\right)}{\Gamma_m\left(\frac{m}{2}\right)\Gamma_m\left(\frac{n_1}{2}\right)\Gamma_m\left(\frac{n_2}{2}\right)} \operatorname{etr}(-\Omega)|\Sigma_1\Sigma_2^{-1}|^{-\frac{1}{2}n_2} \tag{5.5.11}$$
$$\times |\Lambda|^{\frac{1}{2}(n_1-m-1)}\Big\{\prod_{i<j}(\lambda_i-\lambda_j)\Big\} \sum_{\kappa,\lambda:\phi} \frac{\left(\frac{n_1+n_2}{2}\right)_\phi}{k!\ell!\left(\frac{n_1}{2}\right)_\lambda}\theta_\phi^{\kappa,\lambda}$$
$$\times \frac{C_\phi(\Lambda) C_\phi^{\kappa,\lambda}\left(-\Sigma_2^{1/2}\Sigma_1^{-1}\Sigma_2^{1/2}, \Sigma_2^{1/2}\Sigma_1^{-1/2}\Omega\Sigma_1^{-1/2}\Sigma_2^{1/2}\right)}{C_\phi(I)}$$
$$= \frac{\pi^{\frac{m^2}{2}} \Gamma_m\left(\frac{n_1+n_2}{2}\right)}{\Gamma_m\left(\frac{m}{2}\right)\Gamma_m\left(\frac{n_1}{2}\right)\Gamma_m\left(\frac{n_2}{2}\right)} \operatorname{etr}(-\Omega)|\Sigma_1\Sigma_2^{-1}|^{-\frac{1}{2}n_1}$$
$$\times |\Lambda|^{\frac{1}{2}(n_1-m-1)}\Big\{\prod_{i<j}(\lambda_i-\lambda_j)\Big\} \sum_{f=0}^{\infty}\sum_{\phi} \frac{(-1)^f\left(\frac{n_1+n_2}{2}\right)_\phi}{f!\left(\frac{n_1}{2}\right)_\phi}\frac{C_\phi(\Lambda)}{C_\phi(I_m)}$$
$$\times L_\phi^{\frac{1}{2}(n_1-m-1)}\left(\Omega, \Sigma_2^{-1/2}\Sigma_1\Sigma_2^{-1/2}\right). \tag{5.5.12}$$

The probability density function of $T = \operatorname{tr}(S_1 S_2^{-1})$ is given by

$$\frac{\Gamma_m\left(\frac{n_1+n_2}{2}\right)}{\Gamma\left(\frac{mn_1}{2}\right)\Gamma_m\left(\frac{n_2}{2}\right)} \operatorname{etr}(-\Omega)|\Sigma_1\Sigma_2^{-1}|^{-\frac{1}{2}n_1} T^{\frac{1}{2}mn_1-1} \tag{5.5.13}$$
$$\times \sum_{f=0}^{\infty} \frac{(-T)^f}{f!\left(\frac{mn_1}{2}\right)_f} \sum_{\phi}\left(\frac{n_1+n_2}{2}\right)_\phi L_\phi^{\frac{1}{2}(n_1-m-1)}\left(\Omega, \Sigma_2^{-1/2}\Sigma_1\Sigma_2^{-1/2}\right).$$

PROOF The joint probability density function of (S_2, F) is given as

$$\frac{\operatorname{etr}(-\Omega)}{\Gamma_m\left(\frac{n_1}{2}\right)\Gamma_m\left(\frac{n_2}{2}\right)} |2\Sigma_1|^{-\frac{1}{2}n_1}|2\Sigma_2|^{-\frac{1}{2}n_2}|F|^{\frac{1}{2}(n_1-m-1)}$$
$$\times \operatorname{etr}\left(-\frac{1}{2}\Sigma_2^{-1}S_2\right)|S_2|^{\frac{1}{2}(n_1+n_2-m-1)} \operatorname{etr}\left(-\frac{1}{2}\Sigma_1^{-1}S_2^{1/2}FS_2^{1/2}\right)$$
$$\times {}_0F_1\left(\frac{n_1}{2};\frac{1}{2}\Sigma_1^{-1/2}\Omega\Sigma_1^{-1/2}S_2^{1/2}FS_2^{1/2}\right).$$

Let $F = H\Lambda H'$, $\Lambda = \operatorname{diag}(\lambda_1, \ldots, \lambda_m)$, $H \in O(m)$; then by Theorem A.3 in the

Appendix we have

$$\int_{O(m)} \operatorname{etr}\left(-\frac{1}{2}\Lambda H' S_2^{1/2}\Sigma_1^{-1}S_2^{1/2}H\right)$$
$$\times {}_0F_1\left(\frac{n_1}{2}; \frac{1}{2}\Lambda H' S_2^{1/2}\Sigma_1^{-1/2}\Omega\Sigma_1^{-1/2}S_2^{1/2}H\right) \mathrm{d}(H)$$
$$= \sum_{\kappa,\lambda:\phi} \frac{(-1)^k \theta_\phi^{\kappa,\lambda}}{k!\ell!\left(\frac{n_1}{2}\right)_\lambda} \frac{C_\phi(\Lambda)}{C_\phi(I_m)} C_\phi^{\kappa,\lambda}\left(\frac{1}{2}\Sigma_1^{-1}S_2, \frac{1}{2}\Sigma_1^{-1/2}\Omega\Sigma_1^{-1/2}S_2\right).$$

By making use of (A.3.10) the integration with respect to S_2 gives (5.5.11). (5.5.12) is obtained with the help of Exercise 5.13 and (5.5.13) is obtained by making use of (5.1.18).

Corollary 5.5.2 (5.5.13) reduces to (5.5.3) for $\Sigma_1 = \Sigma_2$.

Remark 5.5.2 Theorem 5.5.2 and Theorem 5.5.3 were obtained by Phillips (1984) and Davis (1979), respectively.

Theorem 5.5.4 The asymptotic expansion of the probability density function $f(x)$ of $x = n_2 \operatorname{tr}(S_1 S_2^{-1})$ for large n_2 is given below where $g_{mn_1}(x, \sigma_1)$ is the probability density function of a noncentral chi-square random variable with mn_1 degrees of freedom and noncentrality parameter σ_1. This series converges for $0 \leq x < n_2$.

$$f(x) = g_{mn_1}(x,\sigma_1) + \frac{1}{4n_2}\sum_{\alpha=0}^{4} \ell_{1\alpha} g_{mn_1+2\alpha}(x,\sigma_1)$$
$$+ \frac{1}{96n_2^2}\sum_{\alpha=0}^{8} \ell_{2\alpha} g_{mn_1+2\alpha}(x,\sigma_1) + o\left(\frac{1}{n_2^2}\right), \tag{5.5.14}$$

where $\sigma_j = \operatorname{tr}(\Sigma^{-1}MM')^j$, $j = 1, 2, 3$,

$$\ell_{10} = mn_1(n_1 - m - 1), \quad \ell_{11} = -2n_1(mn_1 - \sigma_1),$$
$$\ell_{12} = mn_1(m + n_1 + 1) - 2(m + 2n_1 + 1)\sigma_1 + \sigma_2,$$
$$\ell_{13} = 2(m + n_1 + 1)\sigma_1 - 2\sigma_2, \quad \ell_{14} = \sigma_2,$$
$$\ell_{20} = mn_1\left\{3n_1m^3 - 2(3n_1^2 - 3n_1 + 4)m^2 + 3(n_1 - 1)(n_1^2 - n_1 + 4)m - 4(2n_1^2 - 3n_1 - 1)\right\},$$
$$\ell_{21} = -12mn_1^2(n_1 - m - 1)(mn_1 - \sigma_1),$$
$$\ell_{22} = 6mn_1^2\left\{3mn_1^2 + 8n_1 - (m+1)(m^2 + m - 4)\right\} - 12n_1\left\{4mn_1^2 - (m^2 + m - 8)n_1 - (m^3 + 2m^2 - 3m - 4)\right\}\sigma_1 + 12n_1^2\sigma_1^2 + 6n_1\left\{mn_1 - (m^2 + m - 4)\right\}\sigma_2,$$

$$\begin{aligned}
\ell_{23} = {} & -4mn_1\left\{3mn_1^3+(3m^2+3m+16)n_1^2\right.\\
& \qquad \left.+\,24(m+1)n_1+4(m^2+3m+4)\right\}\\
& +12\left\{6mn_1^3+3(m^2+m+8)n_1^2\right.\\
& \qquad \left.-(m^3+2m^2-27m-28)n_1+4(m^2+3m+4)\right\}\sigma_1\\
& -24(2n_1^2+(m+1)n_1+2)\sigma_1^2\\
& \qquad -12\left\{2mn_1^2-(m^2+m-16)n_1+4(m+2)\right\}\sigma_2\\
& +12n_1\sigma_1\sigma_2+16\sigma_3\,,\\
\ell_{24} = {} & 3m^2n_1^4+6m(m^2+m+4)n_1^3\\
& \quad +3m(m^3+2m^2+21m+20)n_1^2+12m(2m^2+5m+5)n_1\\
& -12\left\{4mn_1^3+(5m^2+5m+24)n_1^2\right.\\
& \qquad \left.+(m^3+2m^2+45m+44)n_1+4(3m^2+8m+9)\right\}\sigma_1\\
& +12\left\{6n_1^2+6(m+1)n_1+m^2+2m+15\right\}\sigma_1^2\\
& +12\left\{3mn_1^2+36n_1+18m+32\right\}\sigma_2\\
& \quad -12(4n_1+m+1)\sigma_1\sigma_2-96\sigma_3+6\sigma_2^2\,,\\
\ell_{25} = {} & 12\left\{mn_1^3+2(m^2+m+4)n_1^2+(m^3+2m^2+21m+20)n_1\right.\\
& \qquad \left.+\,4(2m^2+5m+5)\right\}\sigma_1\\
& -24\left\{2n_1^2+3(m+1)n_1+m^2+2m+9\right\}\sigma_1^2\\
& \quad -12\left\{2mn_1^2+(m^2+m+32)n_1+8(3m+5)\right\}\sigma_2\\
& \quad +36(2n_1+m+1)\sigma_1\sigma_2+192\sigma_3-12\sigma_2^2\,,\\
\ell_{26} = {} & 12(n_1^2+2(m+1)n_1+m^2+2m+7)\sigma_1^2\\
& +6(mn_1^2+(m^2+m+20)n_1+20m+32)\sigma_2\\
& -12(4n_1+3(m+1))\sigma_1\sigma_2-160\sigma_3+18\sigma_2^2\,,\\
\ell_{27} = {} & 12(n_1+m+1)\sigma_1\sigma_2+48\sigma_3-12\sigma_2^2\,,\\
\ell_{28} = {} & 3\sigma_2^2\,.
\end{aligned}$$

PROOF With the help of (5.5.3) the probability density function of $x = n_2 \operatorname{tr}(S_1S_2^{-1})$ can be expressed as follows:

$$\begin{aligned}
& \frac{\Gamma_m\left(\frac{n_1+n_2}{2}\right)}{\Gamma_m\left(\frac{n_2}{2}\right)\Gamma\left(\frac{mn_1}{2}\right)n_2^{\frac{1}{2}mn_1}} \operatorname{etr}\left(-\frac{1}{2}\Sigma^{-1}MM'\right)x^{\frac{1}{2}mn_1-1}\\
& \qquad \times \sum_{k=0}^{\infty}\frac{1}{k!\left(\frac{mn_1}{2}\right)_k}\left(-\frac{x}{n_2}\right)^k\sum_{\kappa}\left(\frac{n_1+n_2}{2}\right)_\kappa\\
& \qquad \times L_\kappa^{\frac{1}{2}(n_1-m-1)}\left(\frac{1}{2}\Sigma^{-1}MM'\right).
\end{aligned}$$

This series converges for $|x| < n_2$. Using (4.4.30) we have

$$\begin{aligned}
&\frac{n_2^{-\frac{1}{2}mn_1}\Gamma_m\left(\frac{n_1+n_2}{2}\right)}{\Gamma_m\left(\frac{n_2}{2}\right)\Gamma\left(\frac{mn_1}{2}\right)} \\
&= \frac{1}{2^{\frac{mn_1}{2}}\Gamma\left(\frac{mn_1}{2}\right)}\left[1+\frac{mn_1}{4n_2}(n_1-m-1)\right. \\
&\quad + \frac{mn_1}{96n_2^2}\left\{3m^3n_1-2m^2(3n_1^2-3n_1+4)\right. \\
&\quad \left.\left. + 3m(n_1^3-2n_1^2+5n_1-4)-8n_1^2+12n_1+4\right\} + o\left(\frac{1}{n_2^2}\right)\right],
\end{aligned}$$

and

$$\begin{aligned}
&\left(\frac{n_1+n_2}{2}\right)_\kappa \\
&= \left(\frac{n_2}{2}\right)^k\left[1+\frac{1}{n_2}(a_1(\kappa)+n_1k)+\frac{1}{6n_2^2}\left\{3a_1^2(\kappa)-a_2(\kappa)\right.\right. \\
&\qquad \left.\left.+6n_1(k-1)a_1(\kappa)+3n_1^2k(k-1)+k\right\}+o\left(\frac{1}{n_2^2}\right)\right].
\end{aligned}$$

Thus the probability density function of x, expanded up to order $1/n_2^2$ for $\gamma = \frac{1}{2}(n_1 - m - 1)$ is

$$\begin{aligned}
&\operatorname{etr}\left(-\frac{1}{2}\Sigma^{-1}MM'\right)\frac{x^{\frac{1}{2}mn_1-1}}{2^{\frac{1}{2}mn_1}\Gamma\left(\frac{mn_1}{2}\right)}\sum_{k=0}^{\infty}\frac{1}{k!\left(\frac{mn_1}{2}\right)_k}\left(-\frac{x}{2}\right)^k \\
&\times\left[\sum_\kappa L_\kappa^\gamma\left(\frac{1}{2}\Sigma^{-1}MM'\right)+\frac{1}{4n_2}\left\{\sum_\kappa\{mn_1(n_1-m-1)\right.\right. \\
&\quad \left. + 4(a_1(\kappa)+n_1k)\}\,L_\kappa^\gamma\left(\frac{1}{2}\Sigma^{-1}MM'\right)\right\} \\
&\quad +\frac{1}{96n_2^2}\left\{\sum_\kappa\{mn_1\{3m^3n_1-2m^2(3n_1^2-3n_1+4)\right. \\
&\qquad +3m(n_1^3-2n_1^2+5n_1-4)-8n_1^2+12n_1+4\} \\
&\quad +24mn_1(n_1-m-1)(a_1(\kappa)+n_1k) \\
&\quad +16\left\{3a_1^2(\kappa)-a_2(\kappa)+6n(k-1)a_1(\kappa)+3n_1^2k(k-1)+k\right\}\} \\
&\qquad \left.\left.\times L_\kappa^\gamma\left(\frac{1}{2}\Sigma^{-1}MM'\right)\right\}+o\left(\frac{1}{n_2^2}\right)\right].
\end{aligned}$$

The first term of (5.5.14) By using Corollary 4.6.1, (4.6.10) and Theorem 4.6.3,

(4.6.20) we have

$$\begin{aligned}
&\exp\left(-\frac{\sigma_1}{2}\right)\frac{x^{\frac{1}{2}mn_1-1}}{2^{\frac{1}{2}mn_1}\Gamma\left(\frac{mn_1}{2}\right)}\sum_{k=0}^{\infty}\frac{1}{k!\left(\frac{mn_1}{2}\right)_k}\left(-\frac{x}{2}\right)^k L_k^{\frac{1}{2}mn_1-1}\left(\frac{\sigma_1}{2}\right)\\
&=\exp\left\{-\frac{1}{2}(x+\sigma_1)\right\}\frac{x^{\frac{1}{2}mn_1-1}}{2^{\frac{1}{2}mn_1}\Gamma\left(\frac{mn_1}{2}\right)}\sum_{k=0}^{\infty}\frac{1}{k!\left(\frac{mn_1}{2}\right)_k}\left(\frac{x\sigma_1}{4}\right)^k\\
&=g_{mn_1}\left(x;\frac{1}{2}\sigma_1\right).
\end{aligned}$$

The term of order $1/n_2$ in (5.5.14) With the help of Corollary 4.6.2, (4.6.17) and Theorem 4.6.3, (4.6.22), (4.6.23) and (4.6.24), we have

$$\begin{aligned}
&\exp\left(-\frac{\sigma_1}{2}\right)\frac{x^{\frac{1}{2}mn_1-1}}{2^{\frac{1}{2}mn_1}\Gamma\left(\frac{mn_1}{2}\right)}\sum_{k=0}^{\infty}\frac{1}{k!\left(\frac{mn_1}{2}\right)_k}\left(-\frac{x}{2}\right)^k\\
&\qquad\times\sum_{\kappa}a_1(\kappa)L_k^{\frac{1}{2}(n-m-1)}\left(\frac{1}{2}\Sigma^{-1}MM'\right)
\end{aligned}$$

$$\begin{aligned}
&=\exp\left(-\frac{\sigma_1}{2}\right)\frac{x^{\frac{1}{2}mn_1-1}}{2^{\frac{1}{2}mn_1}\Gamma\left(\frac{mn_1}{2}\right)}\sum_{k=2}^{\infty}\frac{1}{(k-2)!\left(\frac{mn_1}{2}\right)_k}\left(-\frac{x}{2}\right)^k\\
&\qquad\times\left[\frac{1}{4}\left\{mn_1(m+n_1+1)L_{k-2}^{\frac{1}{2}mn_1+1}\left(\frac{\sigma_1}{2}\right)-2(n_1+m+1)\right.\right.\\
&\qquad\qquad\left.\left.\times\,\sigma_1 L_{k-2}^{\frac{1}{2}mn_1+2}\left(\frac{\sigma_1}{2}\right)+\sigma_2 L_{k-2}^{\frac{1}{2}mn_1+3}\left(\frac{\sigma_1}{2}\right)\right\}\right]
\end{aligned}$$

$$\begin{aligned}
&=\frac{1}{4}\left\{mn_1(n_1+m+1)-2(n_1+m+1)\sigma_1+\sigma_2\right\}g_{mn_1+4}\\
&\qquad+\frac{1}{2}\left\{(n_1+m+1)\sigma_1-\sigma_2\right\}g_{mn_1+6}+\frac{1}{4}\sigma_2 g_{mn_1+8};
\end{aligned}$$

$$\begin{aligned}
&\exp\left(-\frac{\sigma_1}{2}\right)\frac{x^{\frac{1}{2}mn_1-1}}{2^{\frac{1}{2}mn_1}\Gamma\left(\frac{mn_1}{2}\right)}\sum_{k=0}^{\infty}\frac{1}{k!\left(\frac{mn_1}{2}\right)_k}\left(-\frac{x}{2}\right)^k\\
&\qquad\times\sum_{\kappa}kL_\kappa^{\frac{1}{2}(n_1-m-1)}\left(\frac{1}{2}\Sigma^{-1}MM'\right)
\end{aligned}$$

$$=\frac{1}{2}(\sigma_1-mn_1)g_{mn_1+2}-\frac{1}{2}\sigma_1 g_{mn_1+4}.$$

Thus the first term is given by

$$\frac{1}{4}[mn_1(n_1-m-1)g_{mn_1}-2n_1(mn_1-\sigma_1)g_{mn_1+2}$$
$$+\{mn_1(n_1+m+1)-2(2n_1+m+1)\sigma_1+\sigma_2\}\,g_{mn_1+4}$$
$$+2\{(n_1+m+1)\sigma_1-\sigma_2\}\,g_{mn_1+6}+\sigma_2 g_{mn_1+8}]\,.$$

The derivation of the term of order $1/n_2^2$ of (5.5.14) is left to the reader as an exercise (Exercise 5.14).

Remark 5.5.3 It is of interest to know the development of the derivation of the asymptotic expansion of the probability density function (or cumulative distribution function) of T_0^2. An asymptotic expansion of the distribution of the generalized Hotelling's T_0^2 under the nonnull case was obtained up to order $1/n_2$ by Siotani (1957) and Ito (1960) using a perturbation technique; the expansion was extended up to order $1/n_2^2$ by Siotani (1971) by the same technique, and by Fujikoshi (1970) and Hayakawa (1972b) using weighted sums of generalized Laguerre polynomials. Muirhead (1972) obtained the expansion by applying the partial differential equation method. Lee (1971) also obtained this expansion with the help of a representation of the characteristic function due to Hsu (1940). The expansion for the central case was obtained by Ito (1956) and Muirhead (1970b). Davis (1968) obtained it by making use of the ordinal differential equation method.

Fujikoshi (1971) gave a similar result for the asymptotic expansion of the distribution of T_0^2 with complex argument.

5.6 Anderson's Linear Discriminant Function

Definition 5.6.1 *Anderson's linear discriminant function* Anderson's (1951) linear discriminant function is defined as

$$W=n\left\{\boldsymbol{x}-\frac{1}{2}(\bar{\boldsymbol{x}}_1+\bar{\boldsymbol{x}}_2)\right\}' S^{-1}(\bar{\boldsymbol{x}}_1-\bar{\boldsymbol{x}}_2)\,, \tag{5.6.1}$$

where $\bar{\boldsymbol{x}}_1$ and $\bar{\boldsymbol{x}}_2$ are the sample means of two independent training samples of sizes N_1 and N_2 drawn from m-variate normal populations Π_1 and Π_2, with identical covariance matrices, $N(\boldsymbol{\mu}_1,\Sigma)$ and $N(\boldsymbol{\mu}_2,\Sigma)$, respectively. Here S is the pooled sum of squares and products matrix on $n=N_1+N_2-2$ degrees of freedom, and $\boldsymbol{x}$ is a new independent observation to be classified into one of the populations.

Here we consider the asymptotic expansion of the distribution function of W along the lines of Davis (1987).

Lemma 5.6.1 If $\boldsymbol{x}$ is drawn from Π_1, Anderson's W is expressed as

$$W=n\,\mathrm{tr}\,(\Theta Y'S^{-1}Y)\,, \tag{5.6.2}$$

where $S \sim W_m(n, \Sigma)$, $n = N_1+N_2-2$, the $m\times 2$ matrix Y is distributed as $N_{m,2}(\Omega, \Sigma, A)$ and

$$\Theta = \begin{bmatrix} 2 & 1 \\ 1 & 0 \end{bmatrix}, \quad \Omega = \left[\frac{1}{2}(\mu_1 - \mu_2), 0\right],$$

$$A = \begin{bmatrix} \frac{1}{4}\left(\frac{1}{N_1} + \frac{1}{N_2}\right) & -\frac{1}{2N_1} \\ -\frac{1}{2N_1} & 1 + \frac{1}{N_1} \end{bmatrix}.$$

The proof is left to the reader (Exercise 5.15).

Theorem 5.6.1 The f-th moment μ_f of $\tilde{W} = \frac{\nu}{n}W$, $\nu = n - m - 1$ is

$$\mu_f = \nu^f \sum_\phi P_\phi(iZ, \tilde{A}) / \left(-\frac{\nu}{2}\right)_\phi, \quad f < \frac{\nu}{2} + 1 \tag{5.6.3}$$

where

$$\tilde{A} = A^{1/2}\Theta A^{1/2}, \qquad Z = (2\Sigma)^{-1/2}\Omega A^{-1/2}$$

and ϕ is a partition of f into not more than m parts. With the help of invariant polynomials in Theorem A.8 (Appendix),

$$P_\phi(iZ, \tilde{A}) = (-1)^f \left(\frac{m}{2}\right)_\phi \times \sum_{\kappa,\lambda} \sum_{\phi' \equiv \phi} \binom{f}{k} \theta_{\phi'}^{\kappa;\lambda} C_{\phi'}^{\kappa;\lambda}(B, D) / \left(\frac{m}{2}\right)_\lambda \tag{5.6.4}$$

where

$$B = A\Theta, \quad D = \frac{1}{2}\Omega'\Sigma^{-1}\Omega\Theta. \tag{5.6.5}$$

The sum over $\phi \equiv \phi'$ indicates that there may be more than one polynomial defined for given κ, λ and ϕ.

PROOF Substituting $U = (2\Sigma)^{-1/2} Y A^{-1/2}$, the f-th moment $\mu_f = E\left[\{\frac{\nu}{n}W\}^f\right]$ is given by

$$\mu_f = \frac{(2\nu)^f}{\pi^m} \int_U \operatorname{etr}\{-(U - Z)(U - Z)'\} E_V[\{\operatorname{tr}(Q)\}^f]\, dU$$

where

$$Q = U\tilde{A}U'V^{-1}, \quad \tilde{A} = A^{1/2}\Theta A^{1/2}, \quad Z = (2\Sigma)^{-1/2}\Omega A^{-1/2}$$

and $V \sim W_m(n, I_m)$. It should be noted that since the rank of Q is 2, only partitions into at most two nonzero parts need be considered. With the help of Exercise 4.8 and for $f < \frac{\nu}{2} + 1$,

$$E_V[\{\operatorname{tr}(Q)\}^f] = \left(-\frac{1}{2}\right)^f \sum_\phi C_\phi(U\tilde{A}U') / \left(-\frac{1}{2}\nu\right)_\phi.$$

The result may be obtained using Definition 4.6.4.

$$\mu_f = \nu^f \frac{(-1)^f}{\pi^m} \sum_{\phi} \int_U \operatorname{etr}\{-(U-Z)(U-Z)'\} \times C_\phi(U\tilde{A}U') / \left(-\frac{1}{2}\nu\right)_\phi dU$$

$$= \nu^f \sum_{\phi} P_\phi(iZ, \tilde{A}) / \left(-\frac{\nu}{2}\right)_\phi , \qquad f < \frac{\nu}{2} + 1 .$$

(5.6.4) is given by Theorem A.8 (Appendix).

Corollary 5.6.1 Let $b_1 = \operatorname{tr}(B)$, $d_1 = \operatorname{tr}(D)$, $b_2 = \operatorname{tr}(B^2)$, $d_2 = \operatorname{tr}(D^2)$ and $c = \operatorname{tr}(BD)$ with B and D as in (5.6.5), then the following reductions occur.

$$\operatorname{tr}(B^3) = -\frac{1}{2}b_1^3 + \frac{3}{2}b_1 b_2 , \quad \operatorname{tr}(B^4) = -\frac{1}{2}b_1^4 + b_1^2 b_2 + \frac{1}{2}b_2^2 ,$$

$$\operatorname{tr}(B^2 D^2) = \frac{1}{2}\left(-b_1^2 d_1^2 + 2b_1 d_1 c + b_2 d_2\right) ,$$

$$\operatorname{tr}(B^2 D) = -\frac{1}{2}b_1^2 d_1 + b_1 c + \frac{1}{2}b_2 d_1 ,$$

$$\operatorname{tr}(B^3 D) = \frac{1}{2}\left(-b_1^3 d_1 + b_1^2 c + b_1 b_2 d_1 + b_2 c\right) ,$$

$$\operatorname{tr}(BDBD) = \frac{1}{2}\left(-b_1^2 d_1^2 + b_1^2 d_2 + b_2 d_1^2 - b_2 d_2 + 2c^2\right) .$$

PROOF Note that B and D are 2×2 matrices. Exact expressions of P_ϕ up to order 4 are listed in Table 5.6.1 (see the end of this chapter).

Corollary 5.6.2 The mean and variance of νW are respectively

$$\tilde{\mu}_1 = -2P_1(iZ, \tilde{A}) = mb_1 + 2d_1$$

and

$$\frac{1}{2}(\nu-2)(\nu+1)\tilde{\mu}_2 = m(\nu+m)(b_1^2 + \nu b_2) + 4(\nu+m)(b_1 d_1 + \nu c) + 4(d_1^2 + \nu d_2) .$$

The third and fourth central moments of νW are given below.

$$\frac{1}{4}(\nu-4)(\nu-2)(\nu+1)\tilde{\mu}_3$$
$$= m(\nu+m)(\nu+2m)\left\{-(\nu-2)b_1^3 + 3\nu b_1 b_2\right\}$$
$$+ 6(\nu+m)(\nu+2m)\left\{-(\nu-2)b_1^2 d_1 + 2\nu b_1 c + \nu b_2 d_1\right\}$$
$$+ 24(\nu+m)\left\{-(\nu-2)b_1 d_1^2 + \nu b_1 d_2 + 2\nu d_1 c\right\}$$
$$+ 16\left\{-(\nu-2)d_1^3 + 3\nu d_1 d_2\right\} ;$$

$$\begin{aligned}
&\frac{1}{12}(\nu-6)(\nu-4)(\nu-2)(\nu-1)(\nu+1)\tilde{\mu}_4 \\
&\quad = (\nu+m)\left[-\left\{m(\nu+m)\xi_2 + 2\nu^2(\nu^2-2\nu+2)\right\}(mb_1^4 + 8b_1^3 d_1)\right. \\
&\qquad + 2\nu\left\{m(\nu+m)\xi_1 + 2\nu^3\right\}(mb_1^2 b_2 + 4b_1 b_2 d_1 + 4b_1^2 c) \\
&\qquad \left. +\nu^3(m+2)(\nu+m-2)(mb_2^2 + 8b_2 c)\right] \\
&\quad + 8(\nu+m)\left[-(3m+2)\xi_2 b_1^2 d_1^2 + 4\nu(m\xi_1 + 7\nu^2 - 6\nu)b_1 d_1 c\right. \\
&\qquad + \left\{m\nu\xi_1 + 4\nu^2(2\nu-3)\right\}(b_1^2 d_2 + b_2 d_1^2) \\
&\qquad \left. +\nu^3(m+4)b_2 d_2 + 2\nu^3(\nu+m-2)c^2\right] \\
&\quad + 32(\nu+m)\left\{-\xi_2 b_1 d_1^3 + \nu\xi_1(b_1 d_1 d_2 + d_1^2 c) + \nu^3 d_2 c\right\} \\
&\quad + 16\left(-\xi_2 d_1^4 + 2\nu\xi_1 d_1^2 d_2 + \nu^3 d_2^2\right),
\end{aligned}$$

where $\xi_1 = 11\nu - 12$, $\xi_2 = 10\nu^2 - 23\nu + 12$.

Theorem 5.6.2 The characteristic function of νW is given by

$$\varphi(t) = \sum_{f=0}^{\infty} \frac{(it)^f}{f!} \sum_{\phi} \frac{\nu^f}{\left(-\frac{\nu}{2}\right)_\phi} P_\phi(iZ, \tilde{A}). \tag{5.6.6}$$

When ν is large, $\varphi(t)$ is expanded in inverse powers of ν as

$$\varphi(t) = \psi(\sigma)\left[1 + \ell_1/\nu + \ell_2/\nu^2 + \ell_3/\nu^3 + O(1/\nu^4)\right] \tag{5.6.7}$$

where

$$\psi(\sigma) = |I - \sigma B|^{-\frac{1}{2}m} \operatorname{etr}\left[\sigma D(I-\sigma B)^{-1}\right],$$

$$\begin{aligned}
\ell_1 &= \sigma^2(\tilde{P}_{(2)} - \frac{1}{2}\tilde{P}_{(1^2)}), \\
\ell_2 &= \frac{\sigma^2}{2}\left(4\tilde{P}_{(2)} + \tilde{P}_{(1^2)}\right) + \frac{2}{3}\sigma^3\left(-4\tilde{P}_{(3)} + \tilde{P}_{(21)}\right) \\
&\quad + \frac{1}{24}\sigma^4\left(12\tilde{P}_{(4)} - 2\tilde{P}_{(31)} + 7\tilde{P}_{(2^2)}\right), \\
\ell_3 &= \sigma^2\left(d_1^2 + 3d_2\right) + 4\sigma^3\left(-d_1^3 + 5d_1 d_2\right) + \sigma^4\left(-5d_1^4 + 11d_1^2 d_2 + 6d_2^2\right) \\
&\quad + \frac{4}{3}\sigma^5 d_1 d_2\left(-d_1^2 + 3d_2\right) + \frac{1}{6}\sigma^6 d_2^3
\end{aligned}$$

and

$$\begin{aligned}
\sigma &= 2it, \\
\tilde{P}_\phi &= P_\phi\left(iZ(I-\sigma\tilde{A})^{-1/2}, (I-\sigma\tilde{A})^{-1/2}\tilde{A}(I-\sigma\tilde{A})^{-1/2}\right).
\end{aligned}$$

The cumulant generating function of νW is

$$\begin{aligned}
K(t) &= \log\varphi(t) \\
&= \log\psi(\sigma) + \frac{\tilde{\ell}_1}{\nu} + \frac{\tilde{\ell}_2}{\nu^2} + \frac{\tilde{\ell}_3}{\nu^3} + O\left(\frac{1}{\nu^4}\right)
\end{aligned} \tag{5.6.8}$$

where

$$\log \psi(\sigma) = \sum_{r=1}^{\infty} \sigma^r \left\{ \frac{m}{2r} \operatorname{tr}(B^r) + \operatorname{tr}(B^{r-1}D) \right\},$$

$$\tilde{\ell}_1 = \sigma^2 \left\{ \frac{1}{4} m \tilde{b}_1^2 + \frac{1}{4} m(m+1)\tilde{b}_2 + \tilde{b}_1 \tilde{d}_1 + \tilde{d}_2 + (m+1)\tilde{c} \right\},$$

$$\begin{aligned} \tilde{\ell}_2 = {} & \sigma^2 \left\{ (m+1)\tilde{b}_1 \tilde{d}_1 + \tilde{d}_1^2 + \tilde{d}_2 + (m+3)\tilde{c} \right\} \\ & + 2\sigma^3 \left\{ -(m+2)\tilde{b}_1 \tilde{d}_1^2 + (m+4)\tilde{b}_1 \tilde{d}_2 - \frac{2}{3}\tilde{d}_1^3 \right. \\ & \left. \qquad + 2\tilde{d}_1 \tilde{d}_2 + 2(m+3)\tilde{d}_1 \tilde{c} \right\} \\ & + 2\sigma^4 \left\{ -\tilde{b}_1 \tilde{d}_1^3 + \tilde{b}_1 \tilde{d}_1 \tilde{d}_2 + \tilde{d}_1^2 \tilde{c} + \tilde{d}_2 \tilde{c}_2 + \frac{1}{2}(m+3)\tilde{c}^2 \right\}, \end{aligned}$$

$$\tilde{\ell}_3 = \sigma^2(d_1^2 + 3d_2) + 4\sigma^3(-d_1^3 + 5d_1 d_2) + 5\sigma^4(-d_1^4 + 2d_1^2 d_2 + d_2^2)$$

and

$$\tilde{b}_1 = \operatorname{tr}(\tilde{B}), \quad \tilde{b}_2 = \operatorname{tr}(\tilde{B}^2), \quad \tilde{d}_1 = \operatorname{tr}(\tilde{D}), \quad \tilde{d}_2 = \operatorname{tr}(\tilde{D}^2),$$
$$\tilde{c} = \operatorname{tr}(\tilde{B}\tilde{D}),$$
$$\tilde{B} = B(I - \sigma B)^{-1}, \quad \tilde{D} = D(I - \sigma B)^{-2}.$$

PROOF Noting that

$$\begin{aligned} \nu^f / \left(-\frac{\nu}{2}\right)_\phi = (-2)^f \Big[1 + \frac{a_1(\phi)}{\nu} + \frac{1}{6\nu^2} \{3a_1^2(\phi) + a_2(\phi) - f\} \\ + \frac{1}{6\nu^3} \{a_1^3(\phi) + a_1(\phi)a_2(\phi) + 2a_3(\phi) - (f+2)a_1(\phi)\} \\ + O(1/\nu^4) \Big] \end{aligned}$$

and writing $\sigma = 2it$, $s_r = \operatorname{tr}(U\tilde{A}U')^r$, we have with the help of Definition 4.6.4 and Lemma 4.4.5,

$$\begin{aligned} \varphi(t) = {} & \sum_{f=0}^{\infty} \frac{\sigma^f}{f!} \sum_{\phi} \Big[1 + \frac{a_1(\phi)}{\nu} + \frac{1}{6\nu^2} \{3a_1^2(\phi) + a_2(\phi) - f\} \\ & + \frac{1}{6\nu^3} \{a_1^3(\phi) + a_1(\phi)a_2(\phi) + 2a_3(\phi) - (f+2)a_1(\phi)\} + O\left(\frac{1}{\nu^4}\right) \Big] \\ & \times (-1)^f P_\phi(iZ, \tilde{A}) \\ = {} & \frac{1}{\pi^m} \int_U \operatorname{etr}\{-(U-Z)(U-Z)'\} \exp(\sigma s_1) \\ & \times \left[1 + \frac{\sigma^2}{\nu} s_2 + \frac{1}{\nu^2} \left\{ \sigma^2(s_1^2 + s_2) + \frac{8}{3}\sigma^3 s_3 + \frac{1}{2}\sigma^4 s_2^2 \right\} \right. \end{aligned}$$

$$+\frac{1}{\nu^3}\left\{\sigma^2(s_1^2+3s_2)+8\sigma^3(s_1s_2+s_3)\right.$$

$$\left.+\sigma^4(s_1^2s_2+s_2^2+10s_4)+\frac{8}{3}\sigma^5 s_2s_3+\frac{1}{6}\sigma^6 s_2^3\right\}+O\left(\frac{1}{\nu^4}\right)\Bigg]\,dU\,.$$

Here the products of s_r are expressed in terms of $C_\phi(U\tilde{A}U')$. Thus we have

$$\frac{1}{\pi^m}\int_U \operatorname{etr}\{-(U-Z)(U-Z)'\}\exp(\sigma s_1)C_\phi(U\tilde{A}U')dU$$
$$=(-1)^f\psi(\sigma)P_\phi\left\{iZ(I-\sigma\tilde{A})^{-1/2},\right.$$
$$\left.(I-\sigma\tilde{A})^{-1/2}\tilde{A}(I-\sigma\tilde{A})^{-1/2}\right\}. \tag{5.6.9}$$

Note that P_ϕ in (5.6.9) will be denoted by $\tilde{P}_\phi$.

The asymptotic expansions of $\varphi(t)$ and $K(t)$ in inverse powers of ν are useless because b_1 and b_2 ($\tilde{b}_1$ and $\tilde{b}_2$) are functions of $1/N_1$ and $1/N_2$. It should be noted that

$$\lim_{N_1,N_2\to\infty}\psi(\sigma)=\exp\left(\frac{\Delta^2}{2}it-\frac{\Delta^2}{2}t^2\right), \tag{5.6.10}$$

where $\Delta^2=(\boldsymbol{\mu}_1-\boldsymbol{\mu}_2)'\Sigma^{-1}(\boldsymbol{\mu}_1-\boldsymbol{\mu}_2)$ is the squared Mahalanobis distance between the two populations. This implies that the limiting distribution of $\tilde{W}$ is $N\left(\frac{\Delta^2}{2},\,\Delta^2\right)$.

We consider the distribution of

$$\tilde{Z}=\frac{\left(\tilde{W}-\frac{1}{2}\Delta^2\right)}{\Delta} \tag{5.6.11}$$

and give the second order expansion of the distribution function of $\tilde{Z}$ with respect to $1/N_1$, $1/N_2$ and $1/\nu$.

Lemma 5.6.2 With the help of the second order expansion of the cumulant generating function of $\tilde{W}$, the cumulants $\tilde{\kappa}_r$ of $\tilde{W}$ are listed in Table 5.6.2 (see the end of this chapter).

Theorem 5.6.3 The second order expansion of the distribution function of $\tilde{Z}=\left(\tilde{W}-\frac{1}{2}\Delta^2\right)/\Delta$ when $\boldsymbol{x}$ comes from $\Pi_1\sim N(\boldsymbol{\mu}_1,\,\Sigma)$ is given below.

$$\Pr\left\{\tilde{Z}\le z\,|\,\boldsymbol{x}\in\Pi_1\right\}$$
$$=\Phi(z)+\left[\frac{a_1}{N_1}+\frac{a_2}{N_2}+\frac{a_3}{\nu}+\frac{b_{11}}{N_1^2}+\frac{b_{22}}{N_2^2}+\frac{b_{33}}{\nu^2}\right.$$
$$\left.+\frac{b_{12}}{N_1N_2}+\frac{b_{13}}{N_1\nu}+\frac{b_{23}}{N_2\nu}\right]\Phi(z)+O_3\,, \tag{5.6.12}$$

where δ stands for the differential operator $\frac{d}{dz}$, $\Phi(z)$ for the distribution function of

$N(0,1)$, and

$$a_1 = \frac{1}{2\Delta^2}\left\{\delta^4 + m\delta^2 + m\Delta\delta\right\},$$
$$a_2 = \frac{1}{2\Delta^2}\left\{\delta^4 - 2\Delta\delta^3 + (m+\Delta^2)\delta^2 - m\Delta\delta\right\},$$
$$a_3 = \frac{1}{4}\left\{4\delta^4 - 4\Delta\delta^3 + (2m+2+\Delta^2)\delta^2\right\},$$
$$b_{11} = \frac{1}{8\Delta^4}\{\delta^8 + 2(m+2)\delta^6 + 2\Delta(m+2)\delta^5 + m(m+2)(\delta^4 + 2\Delta\delta^3 + \Delta^2\delta^2)\},$$
$$b_{12} = \frac{1}{4\Delta^4}[\delta^8 - 2\Delta\delta^7 + (2m+4+\Delta^2)\delta^6 - 2\Delta(m+2)\delta^5 + (m^2+2m-m\Delta^2)\delta^4 + m\Delta^3\delta^3 - m^2\Delta^2\delta^2],$$
$$b_{13} = \frac{1}{8\Delta^2}[4\delta^8 - 4\Delta\delta^7 + (6m+34+\Delta^2)\delta^6 - 16\Delta\delta^5 + (2m^2+26m+40-3m\Delta^2)\delta^4 + \Delta\left\{2m^2+2m+m\Delta^2\right\}\delta^3 + 4(m^2+m-\Delta^2)\delta^2],$$
$$b_{22} = \frac{1}{8\Delta^4}\{\delta^8 - 4\Delta\delta^7 + 2(m+2+3\Delta^2)\delta^6 - 2\Delta(3m+6+2\Delta^2)\delta^5 + (m^2+2m+6(m+2)\Delta^2+\Delta^4)\delta^4 - 2\Delta(m^2+2m+(m+2)\Delta^2)\delta^3 + \Delta^2(m^2+2m)\delta^2\},$$
$$b_{23} = \frac{1}{8\Delta^2}\{4\delta^8 - 12\Delta\delta^7 + (6m+34+13\Delta^2)\delta^6 - 2\Delta(6m+34+3\Delta^2)\delta^5 + (2m^2+26m+40+(7m+42)\Delta^2+\Delta^4)\delta^4 - \Delta(2m^2+26m+40+8\Delta+\Delta^2)\delta^3 + 4(m^2+m+(m+2)\Delta^2)\delta^2\},$$
$$b_{33} = \frac{1}{96}[48\delta^8 - 96\Delta\delta^7 + 8(6m+62+9\Delta^2)\delta^6 - 24((2m+26)\Delta+\Delta^3)\delta^5 + 3(4m^2+80m+284+(4m+84)\Delta^2+\Delta^4)\delta^4 - 32(3(m+5)\Delta+\Delta^3)\delta^3 + 48(m+3+\Delta^2)\delta^2].$$

PROOF With the help of Table 5.6.2, the characteristic function $\varphi(t)$ of $\tilde{Z} = \left(\tilde{W} - \frac{1}{2}\Delta^2\right)/\Delta$ which has cumulants $\tilde{\lambda}_1 = (\tilde{\kappa}_1 - \Delta^2/2)/\Delta$, $\tilde{\lambda}_r = \tilde{\kappa}_r/\Delta^r$, $r = 2, 3, \ldots$, is expanded to the second-order terms with respect to $1/N_1$, $1/N_2$ and $1/\nu$. Noting that

$$\int_{-\infty}^{\infty} \exp(itx)\delta^{(\nu)}\Phi(x)\mathrm{d}x = (-it)^\nu \exp\left(-\frac{t^2}{2}\right),$$

where $\Phi(x)$ is the distribution function of $N(0,1)$, we have (5.6.12).

Remark 5.6.1 Okamoto (1963, 1968) obtained the second order expansion of $Z = (W - \frac{1}{2}\Delta^2)/\Delta$ and Siotani and Wang (1977) extended Okamoto's expansion to the third order. The result given above suggests that Okamoto's expansion may be derived as the Edgeworth expansion (Exercise 5.16).

5.7 Multivariate Calibration

In the multivariate regression model it is assumed that

$$Y = \boldsymbol{a}\mathbf{1}' + BX + E, \tag{5.7.1}$$

where $Y = [\boldsymbol{y}_1, \boldsymbol{y}_2, \ldots, \boldsymbol{y}_N]$ is a $p \times N$ observation matrix, $X = [\boldsymbol{x}_1, \boldsymbol{x}_2, \ldots, \boldsymbol{x}_N]$ is a known $q \times N$ design matrix with rank q, $\mathbf{1}$ is an $N \times 1$ vector of ones, $\boldsymbol{a}$ is an unknown $p \times 1$ location vector, and B is a $p \times q$ unknown regression matrix. $E = [\boldsymbol{e}_1, \boldsymbol{e}_2, \ldots, \boldsymbol{e}_N]$ is a $p \times N$ error matrix whose column vectors are independently distributed as $N(\mathbf{0}, \Omega)$. Without loss of generality we assume that $\bar{\boldsymbol{x}} = \sum_{\alpha=1}^{N} \boldsymbol{x}_\alpha / N = \mathbf{0}$.

Lemma 5.7.1 The usual estimators of $\boldsymbol{a}$, B and Ω are respectively

$$\hat{\boldsymbol{a}} = \bar{\boldsymbol{y}} = \frac{1}{N}\sum_{\alpha=1}^{N} \boldsymbol{y}_\alpha, \quad \hat{B} = YX'(XX')^{-1}$$

and

$$\hat{\Omega} = \frac{1}{n}S, \quad S = Y\left[I_N - \frac{1}{N}\mathbf{1}\mathbf{1}' - X'(XX')^{-1}X\right]Y',$$

where $n = N - q - 1$ and I_N is the $N \times N$ identity matrix. Vec$(\hat{B})$ and S are respectively distributed as follows:

$$\begin{aligned} &\mathrm{Vec}(\hat{B}) \sim N_{pq}(\mathrm{Vec}(B), \Omega \otimes (XX')^{-1}), \\ &S \sim W_p(n, \Omega), \end{aligned}$$

where Vec(B) is the $pq \times 1$ vector formed by stacking the rows of B under each other.

The proof is left to the reader (Exercise 5.19).

In the calibration problem, a new p-dimensional vector $\boldsymbol{y}$ having the same structure as (5.7.1) is observed, that is,

$$\boldsymbol{y} = \boldsymbol{a} + B\boldsymbol{x} + \boldsymbol{e}, \tag{5.7.2}$$

where $\boldsymbol{x}$ is an unknown $q \times 1$ vector, and $\boldsymbol{e}$ is distributed as $N(\mathbf{0}, \Omega)$, independently of E, and we estimate $\boldsymbol{x}$ or construct a confidence region for $\boldsymbol{x}$.

Hotelling's T^2 statistic is decomposed as

$$T^2 = (\boldsymbol{y} - \bar{\boldsymbol{y}} - \hat{B}\boldsymbol{x})'S^{-1}(\boldsymbol{y} - \bar{\boldsymbol{y}} - \hat{B}\boldsymbol{x})$$

$$= (\hat{x} - x)'\hat{B}'S^{-1}\hat{B}(\hat{x} - x) + (y - \bar{y} - \hat{B}\hat{x})'S^{-1}(y - \bar{y} - \hat{B}\hat{x})$$
$$= Q + R, \tag{5.7.3}$$

and $\hat{x} = (\hat{B}'S^{-1}\hat{B})^{-1}\hat{B}'S^{-1}(y - \bar{y})$ is the natural estimator of x. $Q = (\hat{x} - x)'\hat{B}'S^{-1}\hat{B}(\hat{x} - x)$ is the basis for a confidence region for x when $p > q$ and $R = (y - \bar{y} - \hat{B}\hat{x})'S^{-1}(y - \bar{y} - \hat{B}\hat{x})$ is a test statistic for the consistency of y with the model (5.7.1). (When $p = q$, $R \equiv 0$).

Let $P_X = X'(XX')^{-1}X$ with the complement $\bar{P}_X = I_N - P_X$, then

$$T^2 = (S^{-1/2}u)'(S^{-1/2}u),$$
$$Q = (S^{-1/2}u)'P_{\hat{B}'S^{-1/2}}(S^{-1/2}u), \tag{5.7.4}$$
$$R = (S^{-1/2}u)'\bar{P}_{\hat{B}'S^{-1/2}}(S^{-1/2}u), \tag{5.7.5}$$

where $u = y - \bar{y} - \hat{B}x$.

Theorem 5.7.1 Let $B_0 = \Omega^{-1/2}B$, $\bar{B} = \Omega^{-1/2}\hat{B}$, and

$$\tilde{Q} = Q/(1 + 1/N), \quad \tilde{R} = R/(1 + 1/N),$$
$$\tilde{x} = x/\sqrt{1 + 1/N}. \tag{5.7.6}$$

Then the conditional distribution of $\{n - (p - q) + 1\}\tilde{R}/(p - q)$ given $\bar{B}$ is a noncentral F distribution with $p - q$ and $n - (p - q) + 1$ degrees of freedom and noncentrality parameter

$$\mu = \tilde{x}'(\bar{B} - B_0)'\bar{P}_{\bar{B}'}(\bar{B} - B_0)\tilde{x}. \tag{5.7.7}$$

The conditional distribution of $\{(n - p + 1)/q\}\tilde{Q}/(1 + \tilde{R})$ given $\bar{B}$ and $\tilde{R}$ is a noncentral F with q and $n - p + 1$ degrees of freedom and noncentrality parameter

$$\lambda = (1 + \tilde{R})^{-1}\tilde{x}'(\bar{B} - B_0)'P_{\bar{B}'}(\bar{B} - B_0)\tilde{x}$$

$$= (1 + \tilde{R})^{-1}\lambda^*. \tag{5.7.8}$$

PROOF Multiply (5.7.1) by $\Omega^{-1/2}$ on the left-hand side and set $B_0 = \Omega^{-1/2}B$ and $\bar{B} = \Omega^{-1/2}\hat{B}$, then without loss of generality we may assume that $\Omega = I_p$. If the column vectors of the $p \times (p - q)$ matrix H_2 form an orthogonal basis of the orthogonal complement of $\hat{B}$, then

$$R = (S^{-1/2}u)'P_{H_2'S^{1/2}}(S^{-1/2}u) = u'H_2(H_2'SH_2)^{-1}H_2'u.$$

Let $H = [H_1, H_2]$ be a $p \times p$ orthogonal matrix, then $\tilde{u} = (1 + 1/N)^{-1/2}$.

$H'u$ given $\bar{B}$ is distributed as $N(H'(B_0 - \bar{B})\tilde{x}, I_p)$, where $\tilde{x} = (1 + 1/N)^{-1/2}x$, and

$\tilde{S} = H'SH$ is independently distributed as $W_p(n, I_p)$. Partitioning $\tilde{\boldsymbol{u}}$ and $\tilde{S}$ as

$$\tilde{\boldsymbol{u}} = \begin{bmatrix} \tilde{\boldsymbol{u}}_1 \\ \tilde{\boldsymbol{u}}_2 \end{bmatrix} \begin{matrix} q \\ p-q \end{matrix} = (1+1/N)^{-1/2} \begin{bmatrix} H_1'\boldsymbol{u} \\ H_2'\boldsymbol{u} \end{bmatrix},$$

$$\tilde{S} = \begin{bmatrix} \tilde{S}_{11} & \tilde{S}_{12} \\ \tilde{S}_{21} & \tilde{S}_{22} \end{bmatrix} \begin{matrix} q \\ p-q \end{matrix}, \quad \tilde{S}_{ij} = H_i'SH_j, \quad i,j = 1,2,$$

we have

$$\tilde{R} = R/(1+1/N) = \tilde{\boldsymbol{u}}_2' S_{22}^{-1} \tilde{\boldsymbol{u}}_2,$$

where $\tilde{S}_{22} \sim W_{p-q}(n, I_{p-q})$. With the help of Theorem 6.7a.1 in Mathai and Provost (1992), the conditional distribution of $\{(n-(p-q)+1)/(p-q)\}\tilde{R}$ given $\bar{B}$ is a noncentral F distribution with $p-q$ and $n-(p-q)+1$ degrees of freedom and noncentrality parameter $\mu = \tilde{\boldsymbol{x}}'\,(\bar{B}-B_0)'\bar{P}_{\bar{B}'}(\bar{B}-B_0)\tilde{\boldsymbol{x}}$, noting that the column vectors of H_2 span the orthogonal complement of $\bar{B}$. Furthermore

$$\begin{aligned}\tilde{Q} = Q/(1+1/N) &= \tilde{\boldsymbol{u}}'\tilde{S}^{-1}\tilde{\boldsymbol{u}} - \tilde{\boldsymbol{u}}_2' S_{22}^{-1}\tilde{\boldsymbol{u}}_2 \\ &= (\tilde{\boldsymbol{u}}_1 - \tilde{S}_{12}\tilde{S}_{22}^{-1}\tilde{\boldsymbol{u}}_2)'\tilde{S}_{11\cdot 2}(\tilde{\boldsymbol{u}}_1 - \tilde{S}_{12}\tilde{S}_{22}^{-1}\tilde{\boldsymbol{u}}_2).\end{aligned}$$

It is well known that

$$\tilde{S}_{11\cdot 2} \sim W_q(n-(p-q), I_q),$$
$$\tilde{S}_{22} \sim W_{p-q}(n, I_{p-q}),$$

and $G = \tilde{S}_{22}^{-1/2}\tilde{S}_{21}$ are independently distributed, the $q(p-q)$ elements of G being independently distributed as $N(0,1)$. Hence

$$\mathrm{E}[\tilde{\boldsymbol{u}}_1 - \tilde{S}_{12}\tilde{S}_{22}^{-1}\tilde{\boldsymbol{u}}_2 \,|\, \bar{B}, \tilde{\boldsymbol{u}}_2, \tilde{S}_{22}] = H_1'(B_0 - \bar{B})\tilde{\boldsymbol{x}}$$

and

$$\mathrm{Cov}[\tilde{\boldsymbol{u}}_1 - \tilde{S}_{12}\tilde{S}_{22}^{-1}\tilde{\boldsymbol{u}}_2 \,|\, \bar{B}, \tilde{\boldsymbol{u}}_2, \tilde{S}_{22}] = (1+\tilde{R})I_q.$$

This yields that the conditional distribution of $\{(n-p+1)/q\}(\tilde{Q}/(1+\tilde{R}))$, given $\bar{B}$ and $\tilde{R}$, is a noncentral F with q and $n-p+1$ degrees of freedom and noncentrality parameter (5.7.8).

Lemma 5.7.2 (Tiku (1965)) The cumulative distribution function of a noncentral F random variable with ν_1 and ν_2 degrees of freedom and noncentrality parameter τ is given by

$$\begin{aligned}&\Pr\{\, F_{\nu_1,\nu_2}(\tau) \le F \,\} \\ &= \Phi_{\nu_1,\nu_2}(F) + \phi_{\nu_1,\nu_2}(F) \sum_{r=1}^{\infty} \frac{(-\tau/2)^r}{r!} \sum_{j=0}^{r-1} (-1)^j \binom{r-1}{j} \\ &\qquad \times \frac{((\nu_1+\nu_2)/2)_j (\nu_1/\nu_2)^j F^{j+1}}{(\nu_1/2)_{j+1}\{1+(\nu_1/\nu_2)F\}^j},\end{aligned} \tag{5.7.9}$$

where $\Phi_{\nu_1,\nu_2}(F)$ and $\phi_{\nu_1,\nu_2}(F)$ are the distribution function and probability density function of a central F variable with ν_1 and ν_2 degrees of freedom respectively.

The proof is left to the reader (Exercise 5.20).

In order to obtain the unconditional distribution of $\tilde{R}$ from (5.7.9), it is required to evaluate moments $\overline{\mu^r} = E_{\bar{B}}[\mu^r]$. Similarly, the unconditional distribution of $\tilde{Q}$ is obtained as

$$\Pr\left\{\frac{n-p+1}{q}\tilde{Q} \le F\right\} = E_{\bar{B}}\left[\Pr\left\{F_{q,n-p+1}(\lambda) \le F + F\Delta \,|\, \bar{B}, \tilde{R}\right\}\right], \tag{5.7.10}$$

where $\Delta = (1+\tilde{R})^{-1} - 1$. Substituting $F + F\Delta$ in (5.7.9) and carrying out a Taylor expansion in terms of Δ,

$$E\left[\frac{\lambda^r}{(1+\tilde{R})^s}\right] = E_{\bar{B}}\left[(\lambda^*)^r E\left[\frac{1}{(1+\tilde{R})^{r+s}} \,|\, \bar{B}, \tilde{R}\right]\right].$$

It should be noted that the following holds.

$$E\left[\frac{1}{(1+\tilde{R})^s} \,|\, \bar{B}\right] = \frac{((n-(p-q)+1)/2)_s}{((n+1)/2)_s}\, {}_1F_1\left(s; \frac{n+1}{2}+s; -\frac{1}{2}\mu\right), \tag{5.7.11}$$

(Exercise 5.21). Hence, the product moments

$$\xi_{rs} = E_{\bar{B}}\left[(\lambda^*)^r \mu^s\right], \qquad r, s = 0, 1, 2, \ldots$$

are required in order to obtain the unconditional distribution.

Lemma 5.7.3 Let $U = \bar{B}(XX')^{1/2}$, $U_0 = B_0(XX')^{1/2}$ and $\Psi = (XX')^{1/2}\tilde{x}\tilde{x}'(XX')^{1/2}$, then $\xi_{rs} = E_{\bar{B}}[(\lambda^*)^r \mu^s]$ is a function of $U_0'U_0$.

PROOF By Lemma 5.7.1, ξ_{rs} can be expressed as

$$\begin{aligned}
&\frac{1}{(2\pi)^{pq/2}}|XX'|^{\frac{p}{2}} \int_{\bar{B}} \operatorname{etr}\left[-\frac{1}{2}(\bar{B}-B_0)(XX')(\bar{B}-B_0)'\right](\lambda^*)^r \mu^s d\bar{B} \\
&= \frac{1}{(2\pi)^{pq/2}} \int_U \operatorname{etr}\left[-\frac{1}{2}(U-U_0)(U-U_0)'\right] \\
&\qquad \times C_{(r)}(P_{U'}(U-U_0)\Psi(U-U_0)') \\
&\qquad \times C_{(s)}(\bar{P}_{U'}U_0\Psi U_0')dU,
\end{aligned}$$

where $C_{(r)}(\cdot)$ is the top ordered zonal polynomials corresponding to the partition (r) of the integer r, noting that the rank Ψ is 1. Let $U_0 = H_1(U_0'U_0)^{1/2}$, $H = [H_1, H_2] \in O(p)$

and $V = H'U = [V_1', V_2']'$, then

$$\xi_{rs} = \frac{1}{(2\pi)^{pq/2}} \operatorname{etr}\left(-\frac{1}{2}U_0'U_0\right) \int_V \operatorname{etr}\left\{(U_0'U_0)^{1/2}V_1 - \frac{1}{2}VV'\right\}$$
$$\times C_{(r)}\left[P_{V'}\left\{V - \begin{bmatrix}(U_0'U_0)^{1/2}\\ 0\end{bmatrix}\right\}\Psi\left\{V - \begin{bmatrix}(U_0'U_0)^{1/2}\\ 0\end{bmatrix}\right\}'\right]$$
$$\times C_{(s)}\left[\bar{P}_{V'}\begin{bmatrix}I_q\\ 0\end{bmatrix}(U_0'U_0)^{1/2}\,\Psi(U_0'U_0)^{1/2}[I_q : 0]\right] \mathrm{d}V, \qquad (5.7.12)$$

which is a function of $U_0'U_0$.

Theorem 5.7.2 The Laplace transform $L_{r,s}(W/2)$ of $|U_0'U_0|^{\frac{1}{2}(p-q-1)}\xi_{rs}$ with respect to $U_0'U_0$ is

$$L_{r,s}(W/2)$$
$$= \Gamma_q\left(\frac{p}{2}\right) 2^{r+s}\left(\frac{q}{2}\right)_r\left(\frac{p-q}{2}\right)_s\left|\frac{1}{2}W\right|^{-\frac{p}{2}} C_{(s)}(\Psi(I+W)^{-1})$$
$$\times \sum_{j=0}^{r}\binom{r}{j}\left(\frac{p}{2}\right)_j C_{(r)}^{(r-j),(j)}(\Psi(I+W)^{-1}, \Psi W(I+W)^{-1}) \Big/ \left(\frac{q}{2}\right)_j. \qquad (5.7.13)$$

PROOF Note that the Laplace transform is written as an integral with respect to U_0.

$$L_{r,s}(W/2) = \frac{\Gamma_q\left(\frac{p}{2}\right)}{\pi^{\frac{1}{2}pq}} \int_{U_0} \operatorname{etr}\left(-\frac{1}{2}WU_0'U_0\right)\xi_{rs}(U_0'U_0)\mathrm{d}U_0$$
$$= \frac{\Gamma_q\left(\frac{p}{2}\right)}{(2\pi)^{pq/2}\,\pi^{\frac{1}{2}pq}} \int_{U_0}\int_U \operatorname{etr}\left(-\frac{1}{2}WU_0'U_0\right.$$
$$\left. - \frac{1}{2}(U-U_0)(U-U_0)'\right) C_{(r)}(\cdot)C_{(s)}(\cdot)\mathrm{d}U_0\mathrm{d}U,$$

where the arguments of the zonal polynomials are as in (5.7.12). Let

$$Z = \{U_0 - U(I+W)^{-1}\}(I+W)^{1/2}, \quad V = UW^{1/2}(I+W)^{-1/2},$$

$$\Psi^* = (I+W)^{-1/2}\Psi(I+\Psi)^{-1/2},$$

then

$$L_{r,s}(W/2) = \frac{\Gamma_q\left(\frac{p}{2}\right)}{(2\pi)^{pq/2}\,\pi^{\frac{1}{2}pq}}|W|^{-\frac{1}{2}p} \int\int \operatorname{etr}\left(-\frac{1}{2}Z'Z - \frac{1}{2}V'V\right)$$
$$\times C_{(r)}(P_{V'}(Z - VW^{1/2})\Psi^*(Z - VW^{1/2})')$$
$$\times C_{(s)}(\bar{P}_{V'}Z\Psi^*Z')\mathrm{d}Z\mathrm{d}V.$$

Making the transformation $V = H_1(V'V)^{1/2}$, $H = [H_1, H_2] \in O(p)$, we have $P_{V'} =$

H_1H_1', $\bar{P}_{V'} = H_2H_2'$ and

$$L_{r,s}(W/2)$$
$$= \frac{1}{(2\pi)^{pq/2}}|W|^{-\frac{1}{2}p} \int\int\int \operatorname{etr}\left(-\frac{1}{2}Z'Z - \frac{1}{2}V'V\right)|V'V|^{\frac{1}{2}(p-q-1)}$$
$$\times C_{(r)}\left[\left\{H_1'Z - (V'V)^{1/2}W^{1/2}\right\}\Psi^*\left\{H_1'Z - (V'V)^{1/2}W^{1/2}\right\}'\right]$$
$$\times C_{(s)}(H_2'Z\Psi^*Z'H_2)\,dZ\,d(V'V)\,d(H).$$

Now letting

$$X = H'Z = \begin{bmatrix} H_1'Z \\ H_2'Z \end{bmatrix} = \begin{bmatrix} X_1 \\ X_2 \end{bmatrix} \begin{matrix} q \\ p-q \end{matrix},$$

the integral is decomposed into three factors.

$$L_{r,s}(W/2)$$
$$= \frac{1}{(2\pi)^{pq/2}|W|^{\frac{1}{2}p}} \int \operatorname{etr}\left(-\frac{1}{2}V'V\right)|V'V|^{\frac{1}{2}(p-q-1)}d(V'V)$$
$$\times \int \operatorname{etr}\left(-\frac{1}{2}X_2'X_2\right)C_{(s)}(X_2'X_2\Psi^*)dX_2 \int \operatorname{etr}\left(-\frac{1}{2}X_1'X_1\right)$$
$$\times C_{(r)}\left[\{X_1 - (V'V)^{1/2}W\}\Psi^*\{X_1 - (V'V)^{1/2}W^{1/2}\}'\right]dX_1.$$

With the help of (4.4.6) and (4.6.43), we have

$$L_{r,s}(W/2)$$
$$= 2^{r+s}\left(\frac{p-q}{2}\right)_s |W|^{-p/2}C_{(s)}(\Psi^*)\int \operatorname{etr}\left(-\frac{1}{2}V'V\right)$$
$$\times |V'V|^{\frac{1}{2}(p-q-1)}(-1)^r P_{(r)}\left(\frac{1}{\sqrt{2}i}(V'V)^{1/2}W^{1/2}, \Psi^*\right)d(V'V).$$

Since

$$(-1)^r P_{(r)}\left(\frac{1}{\sqrt{2}i}(V'V)^{1/2}W^{1/2}, \Psi^*\right)$$
$$= \left(\frac{q}{2}\right)_r \sum_{j=0}^{r}\binom{r}{j} C_{(r)}^{(r-j),(j)}\left(\Psi^*, \frac{1}{2}W^{1/2}(V'V)W^{1/2}\Psi^*\right)\Big/\left(\frac{q}{2}\right)_j,$$

the integration with respect to $V'V$ yields (5.7.13) by using (A.3.9) in the Appendix.

Theorem 5.7.3 Under the assumption

$$\frac{1}{m}XX' = \Theta = O(1), \qquad m = n - p - 1, \tag{5.7.14}$$

the distribution functions of $\tilde{R}$ and $\tilde{Q}$ are given up to the order $O(1/m)$ below.

$$\Pr\left\{\frac{n-(p-q)+1}{p-q}\tilde{R} \le F\right\} = \Phi_{\nu_1,\nu_2}(F) + \phi_{\nu_1,\nu_2}(F) \times \left[-\frac{\bar{\mu}}{\nu_1}F + \frac{\overline{\mu^2}}{4}\left\{\frac{F}{\nu_1} - \frac{(\nu_1+\nu_2)F^2}{\nu_2(\nu_1+2)\left(1+\frac{\nu_1}{\nu_2}F\right)}\right\}\right] + O(1/m^2), \quad (5.7.15)$$

where $\nu_1 = p-q$, $\nu_2 = n-(p-q)+1$, and the following are listed up to the order $1/m^2$.

$$\begin{aligned}\bar{\mu} = &\frac{1}{m}(p-q)\tilde{x}'\Theta^{-1}\tilde{x} - \frac{1}{m^2}(p-q)(p-q-1)\\ &\times \tilde{x}'(\Theta B'\Omega^{-1}B\Theta)^{-1}\tilde{x}\\ &+\frac{1}{m^3}(p-q)(p-q-1)\\ &\times\left[(p-q-2)\tilde{x}'(\Theta B'\Omega^{-1}B\Theta B'\Omega^{-1}B\Theta)^{-1}\tilde{x}\right.\\ &\left.-\tilde{x}'(\Theta B'\Omega^{-1}B\Theta)^{-1}\tilde{x}\,\mathrm{tr}\,(B'\Omega^{-1}B\Theta^{-1})\right],\end{aligned}$$

$$\begin{aligned}\overline{\mu^2} = &\frac{1}{m^2}(p-q)(p-q+2)(\tilde{x}'\Theta^{-1}\tilde{x})^2\\ &-\frac{1}{m^3}(p-q)(p-q+2)(p-q-1)\\ &\times\tilde{x}'(\Theta B'\Omega^{-1}B\Theta)^{-1}\tilde{x}\tilde{x}'\Theta^{-1}\tilde{x},\end{aligned}$$

$$\overline{\mu^3} = \frac{1}{m^3}(p-q)(p-q+2)(p-q+4)(\tilde{x}'\Theta^{-1}\tilde{x})^3.$$

By using a Cornish-Fisher type expansion, see Hill and Davis (1968), the upper α-point of $\tilde{R}$ is given by

$$\frac{\nu_2}{\nu_1}\tilde{R}_\alpha = F_{1\alpha}\left\{1+\frac{1}{m}\tilde{x}'\Theta^{-1}\tilde{x} - \frac{1}{m^2}(p-q-1)\tilde{x}'(\Theta B'\Omega^{-1}B\Theta)^{-1}\tilde{x}\right\} + O(1/m^3) \quad (5.7.16)$$

where $F_{1,\alpha} = F_{\nu_1,\nu_2}(\alpha)$.

$$\begin{aligned}\Pr\left\{\frac{\tilde{\nu}_2}{\tilde{\nu}_1}\tilde{Q} \le F\right\} = &\Phi_{\tilde{\nu}_1,\tilde{\nu}_2}(F)\\ &+\phi_{\tilde{\nu}_1,\tilde{\nu}_2}(F)\left[F\left(\bar{\Delta}-\frac{\bar{\lambda}}{\tilde{\nu}_1}\right) + F\left(\frac{\bar{\lambda}^2}{4} - \overline{\lambda\Delta}\right)\Big/\tilde{\nu}_1\right.\\ &\left.+F^2\left\{\frac{\phi'}{\phi}\left(\frac{\bar{\Delta}^2}{2} - \frac{\overline{\lambda\Delta}}{\tilde{\nu}_1}\right) - \frac{\bar{\lambda}^2(\tilde{\nu}_1+\tilde{\nu}_2)}{4\tilde{\nu}_2(\tilde{\nu}_1+2)\left(1+\frac{\tilde{\nu}_1}{\tilde{\nu}_2}F\right)}\right\}\right]\\ &+O\left(\frac{1}{m^3}\right)\end{aligned} \quad (5.7.17)$$

where $\tilde{\nu}_1 = q$, $\tilde{\nu}_2 = n - q + 1$, and $\phi = \phi_{\tilde{\nu}_1,\tilde{\nu}_2}(F)$ and $\phi' = \frac{d\phi}{dF}$. The following are listed up to the order $1/m^2$.

$$
\begin{aligned}
\lambda &= \frac{1}{m} q\tilde{x}'\Theta^{-1}\tilde{x} + \frac{1}{m^2}(p-q) \\
&\qquad \times \left\{-q\tilde{x}'\Theta^{-1}\tilde{x} + (p-q-1)\tilde{x}'(\Theta B'\Omega^{-1}B\Theta)^{-1}\tilde{x}\right\}, \\
\overline{\lambda^2} &= \frac{1}{m^2} q(q+2)(\tilde{x}'\Theta^{-1}\tilde{x})^2, \\
\bar{\Delta} &= -\frac{1}{m}(p-q) + \frac{1}{m^2}(p-q)(p+2-\tilde{x}'\Theta^{-1}\tilde{x}), \\
\overline{\Delta^2} &= \frac{1}{m^2}(p-q)(p-q+2), \\
\overline{\lambda\Delta} &= \bar{\lambda}\bar{\Delta} = -\frac{1}{m^2} q(p-q)\tilde{x}'\Theta^{-1}\tilde{x}.
\end{aligned}
$$

The upper α-point of $\tilde{Q}$ is given by

$$
\begin{aligned}
\frac{\tilde{\nu}_2}{\tilde{\nu}_1}\tilde{Q}_\alpha = F_{2,\alpha}\Bigg[1 &+ \frac{1}{m}\{p-q+\tilde{x}'\Theta^{-1}\tilde{x}\} \\
&+ \frac{1}{m^2}(p-q)\Bigg\{\tilde{x}'\Theta^{-1}\tilde{x} + \frac{q}{2}(F_{2,\alpha}-3) \\
&+ \frac{p-q-1}{q}x'(\Theta B'\Omega B\Theta)^{-1}\tilde{x}\Bigg\}\Bigg] \\
&+ O\left(\frac{1}{m^3}\right),
\end{aligned} \tag{5.7.18}
$$

where $F_{2,\alpha} = F_{q,n-p+1}(\alpha)$.

PROOF It should be noted that (5.7.14) implies that $\Psi = \Theta^{-1/2}\tilde{x}\tilde{x}'\Theta^{-1/2}/m$ and $U_0'U_0 = m\Theta^{1/2}B'\Omega^{-1}B\Theta^{1/2}$. The moment $\overline{\mu^s} = E_{\tilde{B}}[\mu^s]$ is obtained by taking the inverse Laplace transform of $L_{0,s}(W/2)$ which is expressed with the help of (A.3.26) in the Appendix as

$$
\begin{aligned}
L_{0,s}(W/2) &= \Gamma_q\left(\frac{p}{2}\right)2^s\left(\frac{p-q}{2}\right)_s |W/2|^{-p/2}C_{(s)}(\Psi(I+W)^{-1}) \\
&= \Gamma_q\left(\frac{p}{2}\right)2^s\left(\frac{p-q}{2}\right)_s |W/2|^{-p/2} \\
&\quad \times \sum_{k=0}^{\infty}\sum_{\kappa}\sum_{\phi\in\kappa\cdot s} \psi_s^{\phi}\theta_{\phi}^{\kappa,(s)}C_{\phi}^{\kappa,(s)}(-W,\Psi)/k!
\end{aligned} \tag{5.7.19}
$$

where for a partition $\phi = (f_1, f_2, \ldots, f_q)$, $f_1 \geq f_2 \geq \cdots \geq f_q \geq 0$

$$
\psi_s^{\phi} = \lim_{\beta\to 0}\frac{(\beta)_\phi}{(\beta)_s} = \frac{(f_1-1)!}{(s-1)!}\prod_{i=1}^{q-1}\left(-\frac{i}{2}\right)_{f_{i+1}}, \tag{5.7.20}
$$

which is zero if $f_3 \geq 2$. Inverting (5.7.19) with the help of Exercise 4.7 and Exercise 4.8, the asymptotic expansion of $\overline{\mu^s} = E_{\tilde{B}}[\mu^s]$ for large m is obtained as

$$\begin{aligned}\overline{\mu^s} &= 2^s \left(\frac{p-q}{2}\right)_s \sum_{k=0}^{\infty} \sum_{\kappa} \sum_{\phi \in \kappa \cdot (s)} 2^k \left(\frac{q+1-p}{2}\right)_{\kappa} \psi_s^{\phi} \theta_{\phi}^{\kappa,(s)} \\ &\quad \times C_{\phi}^{\kappa,(s)}((\Theta^{1/2}B'\Omega^{-1}B\Theta^{1/2})^{-1}, \Theta^{-1/2}\tilde{\boldsymbol{x}}\tilde{\boldsymbol{x}}'\Theta^{-1/2})/(m^{k+s}k!) \\ &= O(1/m^s).\end{aligned}$$

Tables A-1 in Davis (1979) give the terms of lower order of $\overline{\mu^s}$. The distribution function of $\frac{\nu_2}{\nu_1}\tilde{R}$, $\nu_1 = p-q$, $\nu_2 = n-(p-q)+1$, is given by

$$\begin{aligned}\Pr\left\{\frac{\nu_2}{\nu_1}\tilde{R} \leq F\right\} &= \Phi_{\nu_1,\nu_2}(F) \\ &\quad + \phi_{\nu_1,\nu_2}(F) \sum_{r=1}^{\infty} \frac{\left(-\frac{1}{2}\right)^r}{r!} E_{\tilde{B}}(\mu^r) \sum_{j=0}^{r-1} (-1)^j \binom{r-1}{j} \\ &\qquad \times \frac{((\nu_1+\nu_2)/2)_j \left(\frac{\nu_1}{\nu_2}\right)^j F^{j+1}}{\left(\frac{\nu_1}{2}\right)_{j+1} \left(1+\frac{\nu_1}{\nu_2}F\right)^j}.\end{aligned}$$

Noting that the order $\overline{\mu^r} = O(1/m^r)$, the distribution function is expressed as (5.7.15) up to the order $1/m^2$. The derivation of (5.7.17) is left to the reader (Exercise 5.22).

Remark 5.7.1 The content of this section is based on Davis and Hayakawa (1987). Fujikoshi and Nishii (1984) gave the asymptotic expansion of the distribution of $\tilde{Q}$ in terms of central chi-square distributions (Exercise 5.21). For the fundamental discussions on multivariate calibration, see Brown (1982), Brown and Sundberg (1987), Lieftinck-Koeijers (1988), Williams (1959) and Wood (1982).

5.8 Asymptotic Expansions of the Distribution of a Quadratic Form

As seen earlier the representation of the exact distribution of a quadratic form is complicated. It is therefore desirable to have asymptotic expressions. We give in this section two types of asymptotic expansions with the help of the following lemma.

Lemma 5.8.1

$$\begin{aligned}&\sum_{k=r}^{\infty} \sum_{\kappa} \frac{x^k}{(k-r)!} P_{\kappa}(T, A) \\ &\qquad = d(x, T, A) x^r \sum_{\tau} P_{\tau}(\tilde{T}, \tilde{A}), \quad r = 0, 1, 2, \ldots \qquad (5.8.1)\end{aligned}$$

and

$$\sum_{k=0}^{\infty}\sum_{\kappa}\frac{x^k}{k!}a_1(\kappa)P_\kappa(T,A)$$

$$= d(x,T,A)x^2\left[P_{(2)}(\tilde{T},\tilde{A})-\frac{1}{2}P_{(1^2)}(\tilde{T},\tilde{A})\right], \qquad (5.8.2)$$

where $\tilde{T}=T(I+xA)^{-1/2}$, $\tilde{A}=(I+xA)^{-1/2}A(I+xA)^{-1/2}$ for $\|xA\|<1$,

$$d(x,T,A)=|I+xA|^{-\frac{m}{2}}\operatorname{etr}\{T(I-(I+xA)^{-1})T'\},$$

τ is a partition of r and $\sum_\tau P_\tau(\tilde{T},\tilde{A})$ is given by (4.6.50).

PROOF We obtain the two formulae with the help of Lemma 4.4.5 and Definition 4.6.4.

Let X be an $m\times n$ matrix with $N_{m,n}(M,\Sigma,I_n)$, then from (5.3.9) and (5.3.10) the mixture type representation of the joint probability density function of the latent roots of $Z=n^{-1}\Sigma^{-1/2}(X-M)A(X-M)'\Sigma^{-1/2}$ is given by

$$\sum_{k=0}^{\infty}\sum_{\kappa}R_\kappa f_\kappa(\Lambda), \qquad (5.8.3)$$

where

$$f_\kappa(\Lambda)=\left(\frac{n}{2q}\right)^{\frac{1}{2}mn}\frac{\pi^{\frac{1}{2}m^2}}{\Gamma_m\left(\frac{m}{2}\right)\Gamma_m\left(\frac{n}{2};\kappa\right)C_\kappa(I_m)}$$
$$\times\operatorname{etr}\left(-\frac{n}{2q}\Lambda\right)|\Lambda|^{\frac{1}{2}(n-m-1)}C_\kappa\left(\frac{n}{2q}\Lambda\right)\prod_{i<j}(\lambda_i-\lambda_j),$$

$$k!R_\kappa=|B|^{-m/2}\operatorname{etr}(-\Omega)(-1)^k$$
$$\times P_\kappa\left(\frac{i}{\sqrt{2}}\Sigma^{-1/2}M(B-I)^{-1/2},I-B^{-1}\right), \qquad (5.8.4)$$

and

$$B=Aq,\quad A=\operatorname{diag}(a_1,a_2,\ldots,a_n),\quad a_1\ge a_2\ge\cdots\ge a_n>0,$$

$\Omega=\frac{1}{2}M'\Sigma^{-1}M$ and $q>0$.

Lemma 5.8.2 Let R_κ's be defined as in (5.8.4) and

$$t_j=\operatorname{tr}(B-I_n)^j,\quad s_j=\operatorname{tr}\left(B(B-I)^{j-1}\Omega\right),\quad j=1,2,3,$$

and $\omega=\operatorname{tr}(B\Omega)^2$, then the following equalities hold.

$$F_1 = 2\sum_{k=1}^{\infty}\sum_{\kappa} kR_\kappa = mt_1 + s_1 \ , \tag{5.8.5}$$

$$\begin{aligned} F_2 &= 2\sum_{k=2}^{\infty}\sum_{\kappa} k(k-1)R_\kappa \\ &= m\left(\frac{m}{2}t_1^2 + t_2\right) + mt_1 s_1 + \frac{1}{2}s_1^2 + s_2 \ , \end{aligned} \tag{5.8.6}$$

$$\begin{aligned} F_3 &= 4\sum_{k=3}^{\infty}\sum_{\kappa} k(k-1)(k-2)R_\kappa \\ &= \frac{m^3}{2}t_1^3 + 3m^2 t_1 t_2 + mt_3 + \frac{3}{2}\left(m^2 t_1^2 + 2mt_2\right)s_1 \\ &\quad + \frac{3}{2}mt_1(s_1^2 + 4s_2) + 12s_3 + 6s_1 s_2 + \frac{1}{2}s_1^3 \ , \end{aligned} \tag{5.8.7}$$

$$\begin{aligned} F_4 &= 4\sum_{k=0}^{\infty}\sum_{\kappa} a_1(\kappa)R_\kappa \\ &= mt_1^2 + m(m+1)t_2 + 2(m+1)s_2 + 2t_1 s_1 + \omega \,. \end{aligned} \tag{5.8.8}$$

PROOF These are obtained by using Lemma 5.8.1.

Theorem 5.8.1 Let X be an $m \times n$ matrix distributed as $N_{m,n}(M, \Sigma, I_n)$, M an $m \times n$ matrix with rank m, A an $n \times n$ diagonal matrix, $\mathrm{diag}(a_1, a_2, \ldots, a_n)$, $a_1 \geq a_2 \geq \cdots \geq a_n > 0$. Let $nZ = \Sigma^{-1/2}(X - M) \times A(X - M)'\Sigma^{-1/2}$. Then the asymptotic expansion of the distribution of

$$Q = \sqrt{\frac{n}{2m}}\,\log|Z/q| \tag{5.8.9}$$

is given below under the conditions $\mathrm{tr}\,(A - qI_n) = O(1)$ and $\mathrm{tr}\,(\Omega) = O(1)$, for large n.

$$\begin{aligned} \Pr\{Q \leq x\} &= \Phi(x) + \frac{1}{\sqrt{2mn}}\sum_{\alpha=1}^{2} \ell_{1\alpha}\Phi^{(2\alpha-1)}(x) \\ &\quad + \frac{1}{2mn}\sum_{\alpha=1}^{3} \ell_{2\alpha}\Phi^{(2\alpha)}(x) \\ &\quad + \frac{1}{(2mn)^{3/2}}\sum_{\alpha=1}^{5} \ell_{3\alpha}\Phi^{(2\alpha-1)}(x) + o\left(\frac{1}{n^{3/2}}\right), \end{aligned} \tag{5.8.10}$$

where

$$\ell_{11} = mp - F_1\,, \qquad \ell_{12} = \frac{1}{3}\,,$$
$$\ell_{21} = F_2 - mpF_1 + \frac{1}{2}mp(mp+2)\,,$$
$$\ell_{22} = \frac{1}{3}(mp+1-F_1)\,, \qquad \ell_{23} = \frac{1}{18}\,,$$
$$\ell_{31} = \frac{1}{6}m^2(2m^2+3m-1) + mF_4$$
$$\ell_{32} = \frac{1}{6}mp(mp+2)(mp+4) - \frac{1}{2}mp(mp+2)F_1 + mpF_2 - \frac{1}{3}F_3\,,$$
$$\ell_{33} = \frac{1}{30}(5m^2p^2+20mp+12) - \frac{1}{3}(mp+1)F_1 + \frac{1}{3}F_2\,,$$
$$\ell_{34} = \frac{1}{18}(mp+2-F_1)\,, \qquad \ell_{35} = \frac{1}{162}\,,$$

$p = \frac{1}{2}(m+1)$ and F_α, $\alpha = 1,2,3,4$ are given in Lemma 5.8.2.

PROOF The characteristic function of Q is given by

$$\varphi(t) = \varphi_0(t)\varphi_{Q,A}(t)\,,$$

where

$$\varphi_0(t) = \left(\frac{2}{n}\right)^{itm\sqrt{n/2m}} \frac{\Gamma_m\left(\frac{n}{2} + it\sqrt{\frac{n}{2m}}\right)}{\Gamma_m\left(\frac{n}{2}\right)}$$

and

$$\varphi_{Q,A}(t) = \sum_{k=0}^{\infty}\sum_{\kappa} R_\kappa \frac{\left(\frac{n}{2} + it\sqrt{\frac{n}{2m}}\right)_\kappa}{\left(\frac{n}{2}\right)_\kappa}\,.$$

With the help of (4.4.30), (4.4.31) and Lemma 5.8.2, we have

$$\varphi(t) = \exp\left(-\frac{t^2}{2}\right)\left[1 - \frac{1}{\sqrt{2mn}}\sum_{\alpha=1}^{2}\ell_{1\alpha}(it)^{2\alpha-1} + \frac{1}{2mn}\sum_{\alpha=1}^{3}\ell_{2\alpha}(it)^{2\alpha} - \frac{1}{(2mn)^{3/2}}\sum_{\alpha=1}^{5}\ell_{3\alpha}(it)^{2\alpha-1} + o\left(\frac{1}{n^{3/2}}\right)\right].$$

Inversion gives (5.8.10).

The mixture type representation of the probability density function of $mnT = \operatorname{tr}(\Sigma^{-1}(X-M)A(X-M)')$ is given by

$$\left(\frac{mn}{2q}\right)^{\frac{1}{2}mn} \frac{1}{\Gamma\left(\frac{mn}{2}\right)}\exp\left(-\frac{mn}{2q}T\right)T^{\frac{1}{2}mn-1}$$

$$\times \sum_{k=0}^{\infty} \frac{1}{\left(\frac{mn}{2}\right)_k k!} \left(\frac{mn}{2q}T\right)^k \sum_{\kappa} R_\kappa, \tag{5.8.11}$$

where R_κ's are given as (5.8.4).

Theorem 5.8.2 Let T be distributed with probability density function (5.8.11), then the asymptotic expansion of the probability density function of

$$x = \sqrt{\frac{mn}{2}} \log(T/q)$$

is given by the following under the assumption stated as in Theorem 5.8.1.

$$\phi(x)\left[1 + \frac{B_1}{\sqrt{2mn}} + \frac{B_2}{2mn} + \frac{B_3}{(2mn)^{3/2}}\right] + o\left(\frac{1}{n^{3/2}}\right), \tag{5.8.12}$$

where

$$\begin{aligned}
\phi(x) &= \frac{1}{\sqrt{2\pi}} \exp\left(-\frac{x^2}{2}\right), \\
B_1 &= -\left\{\frac{x^3}{3} - xF_1\right\}, \\
B_2 &= \frac{x^6}{18} - \frac{x^4}{6}(1 + 2F_1) + x^2(F_1 + F_2) - \frac{1}{3}(1 + 3F_2), \\
B_3 &= -\left[\frac{x^9}{162} - \frac{x^7}{18}(1 + F_1) + \frac{x^5}{30}(2 + 15F_1 + 10F_2) \right. \\
&\quad \left. - \frac{x^3}{9}(1 + 6F_1 + 21F_2 + 3F_3) + \frac{x}{3}(F_1 + 12F_2 + 3F_3)\right]
\end{aligned}$$

and the F_α's are defined in Lemma 5.8.2.

PROOF By expanding the probability density function of $T = q\exp\left(\sqrt{\frac{2}{mn}}x\right)$, we have (5.8.12).

Let X_i, $i = 1, 2, \ldots, n$ be independent $N(0,1)$, $A = \mathrm{diag}(a_1, \ldots, a_n)$, $a_i > 0$, $i = 1, 2, \ldots, n$ and $\boldsymbol{\mu} = (\mu_1, \ldots, \mu_n)$, where a_i and μ_i are constants. It is well known (Mathai and Provost (1992), Theorem 3.3.2) that the r-th cumulant of

$$Q = \sum_{i=1}^{n} a_i (X_i - \mu_i)^2$$

is $\kappa_r = 2^{r-1}(r-1)!\, m_r$, where $m_r = \sum_{i=1}^{n} a_i^r \left(1 + r\mu_i^2\right)$. It would be useful to assume that $E[Q] = m_1 = \sum_{i=1}^{n} a_i(1 + \mu_i^2)$ becomes large as n increases.

Theorem 5.8.3 Under the conditions

$$w_j = m_j/m_1 = O(1), \quad j = 2, 3, \ldots, \tag{5.8.13}$$

an asymptotic expansion of the distribution of Q for large m_1 is given by

$$\Pr\left[\sqrt{m_1}\left\{\left(\frac{Q}{m_1}\right)^h - 1 - \frac{1}{m_1}h(h-1)w_2\right\}\Big/\sqrt{2h^2w_2} \le x\right]$$
$$= \Phi(x) - \phi(x)\left[\frac{a_2}{m_1} + \frac{a_3}{m_1\sqrt{m_1}}\right] + o\left(\frac{1}{m_1\sqrt{m_1}}\right), \qquad (5.8.14)$$

where

$$a_2 = \frac{1}{w_2^3}\left\{\left(\frac{1}{2}w_4w_2 - \frac{20}{27}w_3^2 + \frac{2}{9}w_3w_2^2\right)H_3(x) + w_3\left(-\frac{2}{3}w_3 + \frac{2}{3}w_2^2\right)H_1(x)\right\},$$

$$a_3 = \frac{\sqrt{2}}{w_2^{9/2}}\left\{\left(\frac{2}{5}w_5w_2^2 - \frac{4}{3}w_4w_3w_2 + \frac{76}{81}w_3^3 + \frac{1}{9}w_3^2w_2^2 - \frac{1}{9}w_3w_2^4\right)H_4(x)\right.$$
$$+ w_3\left(-2w_4w_2 + \frac{184}{81}w_3^2 + \frac{4}{9}w_3w_2^2 - \frac{2}{3}w_2^4\right)H_2(x)$$
$$\left.+ w_3\left(\frac{2}{9}w_3^2 + \frac{1}{9}w_3w_2^2 - \frac{1}{3}w_2^4\right)\right\},$$

$$h = 1 - 2w_3/w_2^2,$$

and $H_k(x)$ is the Hermite polynomial of degree k.

PROOF Let $f(y)$ be a strictly monotone and twice differentiable function in the neighborhood of $y = \alpha$. Expanding $f\left(\frac{\alpha Q}{m_1}\right)$ at α and normalizing $f\left(\frac{\alpha Q}{m_1}\right)$ as

$$T = \sqrt{m_1}\left[f\left(\frac{\alpha Q}{m_1}\right) - f(\alpha) - \frac{c}{m_1}\right]\Big/\tau,$$

where $\tau = (2w_2\alpha^2 f'(\alpha)^2)^{1/2}$, an asymptotic expansion of the distribution of T when m_1 is large is given up to the order $1/\sqrt{m_1}$ as follows:

$$\Pr\{T \le x\} = \Phi(x) - \frac{1}{\sqrt{m_1}}\left[-\frac{\sqrt{2}}{3}w_3w_2^{-3/2} - \frac{c}{\tau}\right.$$
$$\left. + \frac{\sqrt{2}}{6}\left\{2w_3w_2^{-3/2} + 3w_2^{1/2}\alpha f''(\alpha)f'(\alpha)^{-1}\right\}x^2\right]\phi(x) + o(1/\sqrt{m_1}).$$

The function $f(\alpha)$ which makes the coefficient of x^2 in the terms of order $1/\sqrt{m_1}$ vanish for all values x is a solution of the differential equation

$$2w_3w_2^{-3/2} + 3w_2^{1/2}\alpha f''(\alpha)/f'(\alpha) = 0.$$

The solution of this equation is proportional to α^h, that is,

$$f(\alpha) \propto \alpha^{1-\frac{2}{3}\frac{w_3}{w_2^2}} = \alpha^h .$$

By setting $\alpha = 1$, we have the following transformation of Q

$$\left(\frac{Q}{m_1}\right)^{1-\frac{2}{3}\frac{w_3}{w_2^2}} = \left(\frac{Q}{m_1}\right)^h$$

and we choose c such that

$$c = -\frac{\sqrt{2}}{3}\frac{w_3}{w_2^{3/2}}\tau = h(h-1)w_2 .$$

Then the asymptotic expansion of the distribution of T is

$$\Pr\{T \le x\} = \Phi(x) + O(1/m_1) .$$

Extending this procedure up to the order $1/(m_1\sqrt{m_1})$, (5.8.14) is obtained.

Remark 5.8.1 Konishi, Niki and Gupta (1988) gave the expansion of the distribution up to the order $1/m_1^3$. For the normalization method referred to in this chapter, see Konishi (1978, 1981).

EXERCISES

5.1 Evaluate the h-th moment of $|S|$ for S given in (5.1.2).

5.2 Consider the q-parameter linear regression model

$$y = X\beta + e$$

where X is an $n \times q$ matrix of rank q $(< n)$. By using the estimated residual $\hat{e} = Me$, $M = I_n - X(XX')^{-1}X'$, the Durbin-Watson statistic for testing serial correlation is expressed as

$$d = \hat{e}'A_1\hat{e}/\hat{e}'\hat{e} ,$$

$$A_1 = \begin{bmatrix} 1 & -1 & 0 & & & & \\ -1 & 2 & -1 & & & 0 & \\ & -1 & 2 & & & & \\ \cdot & & & \cdot & & & \\ \cdot & & & & \cdot & & \\ \cdot & & & & & \cdot & \\ & 0 & & & -1 & 2 & -1 \\ & & & & & -1 & 1 \end{bmatrix} .$$

Let Q be an $n \times m$ matrix of rank $m = n - q$ such that $M = QQ'$, $Q'Q = I_m$ and let e be distributed as $N(0, \sigma^2 I_n)$.

(i) Put $Q'e = \omega = rh$, $r = (\omega'\omega)^{1/2}$ and $h'h = 1$. Show that

$$d = h'Ah\,, \qquad A = Q'A_1Q\,,$$

and that h is uniformly distributed over the surface of the unit sphere with center $\mathbf{0}$.

(ii) Show that

$$E[d^k] = \frac{\left(\frac{1}{2}\right)_k C_{(k)}(A)}{\left(\frac{m}{2}\right)_k}.$$

5.3 Let x be an $n \times 1$ normal random vector with mean $\mathbf{0}$ and covariance matrix Σ. Let A and B be $n \times n$ symmetric matrices. Show that

$$\begin{aligned} &E\left[(x'Ax)^k(x'Bx)^\ell\right] \\ &\quad = 2^{k+\ell}\left(\frac{n}{2}\right)_{k+\ell}\sum_{\kappa}\sum_{\lambda}\sum_{\phi\in\kappa\cdot\lambda}\theta_\phi^{\kappa,\lambda}C_\phi^{\kappa,\lambda}(A\Sigma, B\Sigma)/C_\phi(I) \end{aligned}$$

where ϕ is a partition of $k + \ell$ and $C_\phi^{\kappa,\lambda}(\cdot,\cdot)$ is an invariant polynomial defined in the Appendix.

5.4 Let x be an $n \times 1$ vector distributed as $N(\mathbf{0}, \Omega)$ and let r denote a ratio of quadratic forms

$$r = (x'Ax)^p(x'Bx)^{-q}$$

where p and q are nonnegative real numbers and $p \neq q$. A and B are $n \times n$ symmetric matrices. Without loss of generality we assume that $\Omega = I_n$. Let $x = s^{1/2}h$, where $s = (x'x)$ and $h'h = 1$. By making use of (A.3.1) show that $E(r)$ can be expressed as follows:

$$\begin{aligned} E(r) &= 2^{p-q}\frac{\Gamma\left(\frac{n}{2}+p-q\right)}{\Gamma\left(\frac{n}{2}\right)}\alpha^{-q}\beta^{q} \\ &\times\sum_{k=0}^{\infty}\sum_{\ell=0}^{\infty}\frac{1}{k!\ell!}\frac{(-p)_k(q)_\ell\left(\frac{n}{2}\right)_{k+\ell}}{\left(\frac{1}{2}\right)_{k+\ell}}C_{(k+\ell)}^{(\kappa),(\ell)}(I-\alpha A, I-\beta B) \end{aligned}$$

where $0 < \alpha < 2/\|A\|$ and $0 < \beta < 2/\|B\|$ (Smith (1989)).

5.5 *Fractional matrix operator* Let X be any $m \times m$ matrix and let ∂X be defined as $\partial X = \left(\frac{\partial}{\partial x_{ij}}\right)$. If f is an analytic function of X and α is any number for which $\alpha > \frac{1}{2}(m-1)$, the fractional matrix operator $D_X \equiv |\partial X|$ is defined by

$$D_X^{-\alpha}f(X) = \frac{1}{\Gamma_m(\alpha)}\int_{S>0} f(X-S)\,|S|^{\alpha-\frac{1}{2}(m+1)}dS\,.$$

(i) Show that for $\alpha, \beta > \frac{1}{2}(m-1)$,

$$D_X^{-\alpha}D_X^{-\beta} = D_X^{-\alpha-\beta}.$$

Let μ be an any number for which $\mu \geq -\frac{1}{2}(m-1)$. Define

$$n = \left[\mu + \frac{1}{2}(m-1)\right] + 1, \qquad \alpha = n - \mu,$$

where $[\ \cdot\ ]$ denotes the integer part of its argument. The operator D_X^μ is defined by

$$D_X^\mu f(X) = D^{-\alpha}[D^n f(X)].$$

(ii) Show that for $\mu = n_1 - \alpha$ and $\nu = n_2 - \beta$ with n_1 and n_2 integers

$$D_X^\mu D_X^\nu = D_X^{\mu+\nu}.$$

(iii) Show that

$$\begin{aligned}
D_X^\mu \operatorname{etr}(AX) &= \operatorname{etr}(AX)|A|^\mu, \\
D_X^\nu[\,|I-X|^{-a}] &= \frac{\Gamma_m(a+\mu)}{\Gamma_m(a)}|I-X|^{-a-\mu}, \\
& \quad a > \frac{1}{2}(m-1), \quad a+\mu > \frac{1}{2}(m-1), \\
D_X^\mu\, {}_1F_1(\alpha;\beta;X) &= \frac{\Gamma_m(\alpha+\mu)\Gamma_m(\beta)}{\Gamma_m(\alpha)\Gamma_m(\beta+\mu)}\, {}_1F_1(\alpha+\mu;\beta+\mu;X), \\
& \quad \alpha > \frac{1}{2}(m-1), \quad \beta > \frac{1}{2}(m-1), \\
& \quad \alpha+\mu > \frac{1}{2}(m-1), \quad \beta+\mu > \frac{1}{2}(m-1).
\end{aligned}$$

Note : The fractional matrix operator $D_X^{-\alpha}$ is extended to the complex case by replacing α with a complex number α for which $\mathcal{R}e(\alpha) > \frac{1}{2}(m-1)$. D_X^μ is also defined for a complex number μ such that

$$\begin{aligned}
&\mathcal{R}e(\mu) \geq -\frac{1}{2}(m-1), \\
&n = [\mathcal{R}e(\mu) + \frac{1}{2}(m-1)] + 1, \\
&\mathcal{I}m(\alpha) = -\mathcal{I}m(\mu), \quad \mathcal{R}e(\alpha) = n - \mathcal{R}e(\mu)
\end{aligned}$$

(Phillips (1985)).

5.6 Consider the multivariate linear model

$$Y = BX + E$$

where $Y = [\boldsymbol{y}_1, \boldsymbol{y}_2, \ldots, \boldsymbol{y}_T]$ is an $n \times T$ dependent matrix, B is an $n \times p$ parameter matrix, $X = [\boldsymbol{x}_1, \boldsymbol{x}_2, \ldots, \boldsymbol{x}_T]$ is a $p \times T$ non-random independent matrix and $E = [\boldsymbol{e}_1, \boldsymbol{e}_2, \ldots, \boldsymbol{e}_T]$ is an $n \times T$ random matrix with $N(\boldsymbol{0}, \Sigma, I_T)$. Consider the hypothesis

$$H_0 : D\text{vec}(B) = \boldsymbol{d} \quad \text{vs} \quad H_1 : D\text{vec}(B) - \boldsymbol{d} = \boldsymbol{b} \neq \boldsymbol{0}$$

where D is a $q \times np$ known constant matrix of rank q, $\boldsymbol{d}$ is a known vector and $\text{vec}(B)$ stacks rows of B. The Wald statistic W for testing the hypothesis is

$$W = (D\text{vec}(\hat{B}) - \boldsymbol{d})'(D(\hat{\Sigma} \otimes M)D')^{-1}(D\text{vec}(\hat{B}) - \boldsymbol{d})$$

where

$$\begin{aligned} \hat{B} &= YX'(XX')^{-1}, \quad \hat{\Sigma} = S/N, \quad S = Y(I - P_X)Y', \\ P_X &= X'(XX')^{-1}X, \quad M = (XX')^{-1}. \end{aligned}$$

Let

$$\begin{aligned} \boldsymbol{z} &= \Omega^{-1/2}\{D\text{vec}(\hat{B}) - \boldsymbol{d}\}, \quad \Omega = D(\Sigma \otimes M)D' \\ G^{-1} &= \Omega^{-1/2}\{D(S \otimes M)D'\}\Omega^{-1/2} \quad \boldsymbol{\mu} = \Omega^{-1/2}\boldsymbol{b}. \end{aligned}$$

Show the following:

(i) $\boldsymbol{z} \sim N(\boldsymbol{\mu}, I_q)$ under H_1;

(ii) The conditional probability density function of $u = \boldsymbol{z}'G\boldsymbol{z}$ given S is expressed as

$$\begin{aligned} &\frac{1}{2^{\frac{q}{2}}\Gamma\left(\frac{q}{2}\right)}|G|^{-\frac{1}{2}}\exp\left(-\frac{1}{2}\boldsymbol{\mu}'\boldsymbol{\mu}\right)u^{\frac{q}{2}-1} \\ &\qquad \times \sum_{f=0}^{\infty}\frac{\left(\frac{1}{2}\right)^f u^f}{f!\left(\frac{q}{2}\right)_f}P_f\left(\frac{1}{\sqrt{2}}\boldsymbol{\mu}', G^{-1}\right). \end{aligned}$$

By noting that $\left(\frac{m}{2}\right)_\phi C_\phi(I_n) = \left(\frac{n}{2}\right)_\phi C_\phi(I_m)$ the density function can be expressed as

$$\begin{aligned} &\frac{1}{2^{\frac{q}{2}}\Gamma\left(\frac{q}{2}\right)}|G|^{-\frac{1}{2}}\exp\left(-\frac{1}{2}\boldsymbol{\mu}'\boldsymbol{\mu}\right)u^{\frac{q}{2}-1} \\ &\qquad \times \sum_{k,\ell=0}^{\infty}\frac{u^f}{k!\ell!\left(\frac{1}{2}\right)_\ell}C_{(f)}^{(k),(\ell)}\left(-\frac{1}{2}G^{-1}, \frac{1}{4}G^{-1}\boldsymbol{\mu}\boldsymbol{\mu}'\right)\Big/ C_{(f)}(I_q) \end{aligned}$$

where $f = k + \ell$;

(iii) With the help of Exercise 5.5, the probability density function of $w = N\boldsymbol{z}'G\boldsymbol{z}$ is given by

$$\begin{aligned} &\frac{1}{(2N)^{\frac{q}{2}}\Gamma\left(\frac{q}{2}\right)}w^{\frac{q}{2}-1}\sum_{f=0}^{\infty}\sum_{k+\ell=f}\frac{\left(-\frac{1}{2}\right)^k\left(\frac{1}{4}\right)^\ell w^f}{k!\ell!N^f\left(\frac{1}{2}\right)_\ell C_{(f)}(I_q)} \\ &\quad \times \left[|\Omega^{-1/2}D(\partial Z \otimes M)D'\Omega^{-1/2}|^{1/2}\right] \\ &\quad \times C_{(f)}^{(k),(\ell)}\left(\Omega^{-1/2}D(\partial Z \otimes M)D'\Omega^{-1/2},\right. \\ &\qquad\qquad \left.\Omega^{-1/2}D(\partial Z \otimes M)D'\Omega^{-1/2}\boldsymbol{\mu}\boldsymbol{\mu}'\right) \\ &\quad \times \left.|I - 2\Sigma Z|^{-\frac{1}{2}N}\right|_{Z=0}. \end{aligned}$$

This series converges for $0 \le w < N$ (Phillips (1986)).

5.7 Show that the integrations of (5.3.7) and (5.3.8) over $\lambda_1 > \lambda_2 > \ldots > \lambda_m > 0$ are unity, respectively.

5.8 Prove Corollary 5.3.1.

5.9 Let Z be an $m \times p$ matrix and V an $m \times m$ symmetric matrix. Show that if $m \le p \le n$,

$$\sum_{k=0}^{\infty}\sum_{\kappa} \frac{\left(\frac{m+n}{2}\right)_\kappa}{\left(\frac{p}{2}\right)_\kappa k!} P_\kappa(Z, I_p, V)$$
$$= |I+V|^{-\frac{1}{2}(m+n)} {}_1F_1\left(\frac{1}{2}(m+n); \frac{p}{2}; ZZ'V(I+V)^{-1}\right).$$

5.10 Let X be distributed as $N_{p,m}(O, \Sigma_1, \Sigma_2)$ and let A be a $p \times p$ matrix independently distributed as $W_p(n, \Sigma)$, $m \le p$. Put

$$T = \Sigma_1^{1/2} A^{-1/2} \Sigma_1^{-1/2} X + M.$$

Show that the probability density function of T is given by

$$\frac{\Gamma_p\left(\frac{m+n}{2}\right)}{\Gamma_p\left(\frac{n}{2}\right)\pi^{\frac{1}{2}pm}} |\Sigma_1\Sigma^{-1}|^{-\frac{m}{2}} |\Sigma_2|^{-\frac{1}{2}p}$$
$$\times |I + \Sigma_1^{-1/2}\Sigma\Sigma_1^{-1/2}(T-M)\Sigma_2^{-1}(T-M)'|^{-\frac{1}{2}(m+n)},$$

and that the probability density function of $R = T'BT$ is given by

$$\frac{\Gamma_p\left(\frac{m+n}{2}\right)}{\Gamma_p\left(\frac{m}{2}\right)\Gamma_p\left(\frac{n}{2}\right)} |B\Sigma_1|^{-\frac{1}{2}m} |\Sigma_2|^{-\frac{1}{2}p} |\Sigma|^{-\frac{1}{2}n} |\Sigma_0|^{\frac{1}{2}(m+n)}$$
$$\times \operatorname{etr}\left(-\frac{q}{2}\Sigma_2^{-1}R\right) |R|^{\frac{1}{2}(p-m-1)}$$
$$\times \sum_{k=0}^{\infty}\sum_{\kappa} \frac{1}{k!\left(\frac{p}{2}\right)_\kappa} E_W\left[P_\kappa\left(\frac{1}{\sqrt{2}}M^*, A^*, \frac{1}{2}\Sigma_2^{-1/2}R\Sigma_2^{-1/2}\right)\right],$$

where

$$\Sigma_0^{-1} = \Sigma^{-1} + \Sigma_1^{-1/2} M \Sigma_2^{-1} M' \Sigma_1^{-1/2},$$
$$M^* = \Sigma_2^{-1/2} M' \Sigma_1^{-1/2} W \Sigma_1^{-1/2}\left(\Sigma_1^{-1/2} W \Sigma_1^{-1/2} - qB\right)^{-1/2},$$
$$A^* = \left(\Sigma_1^{-1/2} W \Sigma_1^{-1/2} - qB\right)^{-1/2} B^{-1} \left(\Sigma_1^{-1/2} W \Sigma_1^{-1/2} - qB\right)^{-1/2}$$

and the expectation is taken over $W_p(m+n, \Sigma_0)$.

5.11 (i) Show that

$$|d_\phi^{\kappa,\lambda}| \le n_1^k \theta_\phi^{\kappa,\lambda} C_\phi^{\kappa,\lambda}(I_m),$$

where $d_\phi^{\kappa,\lambda}$ is given by (5.5.9).

(ii) With the help of the equality

$$\frac{1}{\Gamma_m(a)}\int_{U>0} \operatorname{etr}(-U)\,|U|^{a-\frac{1}{2}(m+1)}C_\kappa(U)C_\lambda(U)dU$$
$$= \sum_{\phi\in\kappa\cdot\lambda} (\theta_\phi^{\kappa,\lambda})^2(a)_\phi C_\phi(I_m),$$

show that the representation (5.5.8) of the probability density function of $\operatorname{tr}(XX'S_2^{-1})$ converges everywhere for $T>0$ (Phillips (1987)).

5.12 With the help of (4.6.41) show that

$$|I-Z|^{-a}\,{}_1F_1^{(m)}\left(a;\gamma+\frac{1}{2}(m+1);X,-Z(I-Z)^{-1}\right)$$
$$=\sum_{k=0}^{\infty}\sum_{\kappa}\frac{(a)_\kappa L_\kappa^\gamma(X)C_\kappa(Z)}{\left(\gamma+\frac{1}{2}(m+1)\right)_\kappa k!C_\kappa(I_m)}$$

for $X>0$, $\|Z\|<1$, $\gamma>-1$. By comparing this formula and (5.5.10), show that (5.5.3) is obtained by using Lemma 5.5.1.

5.13 Show that

$$|I+A|^{-a}\,{}_1F_1\left(a;u;B(I+A)^{-1}\right)$$
$$=\sum_{\kappa,\lambda:\phi}\frac{(a)_\phi\theta_\phi^{\kappa,\lambda}}{k!\ell!(u)_\lambda}C_\phi^{\kappa,\lambda}(-A,B)$$
$$=\sum_{f=0}^{\infty}\sum_{\phi}\frac{(-1)^f(a)_\phi}{f!(u)_\phi}L_\phi^{u-\frac{1}{2}(m+1)}(BA^{-1},A)$$

where $\sum_{\kappa,\lambda:\phi}$ stands for $\sum_{k=0}^{\infty}\sum_{\ell=0}^{\infty}\sum_\kappa\sum_\lambda\sum_{\phi\in\kappa\cdot\lambda}$.

5.14 Find the term $1/n_2^2$ in (5.5.14).

5.15 Show that Lemma 5.6.1 holds.

5.16 (i) Find the cumulants of W of (5.6.1) with the help of Table 5.6.2 when x comes from $\Pi_1 : N(\mu_1,\Sigma)$.

(ii) Show that the asymptotic expansion of the distribution of W is given as follows.

$$\Pr\{W\le x\,|\,\Pi_1\}$$
$$=\Phi(x)+\left[\frac{\ell_1}{N_1}+\frac{\ell_2}{N_2}+\frac{\ell_3}{n}+\frac{\ell_{11}}{N_1^2}+\frac{\ell_{22}}{N_2^2}+\frac{\ell_{33}}{n^2}\right.$$
$$\left.+\frac{\ell_{12}}{N_1N_2}+\frac{\ell_{13}}{N_1n}+\frac{\ell_{23}}{N_2n}\right]\Phi(x)+O_3\,,$$

where

$$\ell_1 = \frac{1}{2\Delta^2}\left\{\delta^4 + m\delta^2 + m\Delta\delta\right\} ,$$
$$\ell_2 = \frac{1}{2\Delta^2}\left\{\delta^4 - 2\Delta\delta^3 + (m+\Delta^2)\delta^2 - m\Delta\delta\right\} ,$$
$$\ell_3 = \frac{1}{4}\left\{4\delta^4 - 4\Delta\delta^3 + \left\{6(p+1) + \Delta^2\right\}\delta^2 - 2(p+1)\Delta\delta\right\} ,$$

$$\begin{aligned}\ell_{11} = \frac{1}{8\Delta^4}&\{\delta^8 + 2(m+2)\delta^6 + 2(m+2)\Delta\delta^5\\ &+m(m+2)\left\{\delta^4 + 2\Delta\delta^3 + \Delta^2\delta^2\right\}\} ,\end{aligned}$$
$$\begin{aligned}\ell_{22} = \frac{1}{8\Delta^4}&\{\delta^8 - 4\Delta\delta^7 + 2(m+2+3\Delta^2)\delta^6\\ &- 2\Delta(3m+6+2\Delta^2)\delta^5\\ &+ (m^2+2m+6(m+2)\Delta^2+\Delta^4)\delta^4\\ &- 2\Delta(m^2+2m+(m+2)\Delta^2)\delta^3\\ &+(m^2+2m)\Delta^2\delta^2\} ,\end{aligned}$$

$$\begin{aligned}\ell_{33} = \frac{1}{96}&\{48\delta^8 - 96\Delta\delta^7 + 8\left\{18p+74+9\Delta^2\right\}\delta^6\\ &- 24\left\{8(m+4)\Delta + \Delta^3\right\}\delta^5\\ &+ 3\left\{4(9m^2+76m+119) + 4(7m+27)\Delta^2 + \Delta^4\right\}\delta^4\\ &- 4\left\{2(9m^2+78m+117)\Delta + (3m+11)\Delta^3\right\}\delta^3\\ &+ 12\left\{4(6m^2+13m+9) + (m^2+8m+11)\Delta^2\right\}\delta^2\\ &-48(m+1)^2\Delta\delta\} ,\end{aligned}$$

$$\begin{aligned}\ell_{12} = \frac{1}{4\Delta^4}&\{\delta^8 - 2\Delta\delta^7 + (2m+4+\Delta^2)\delta^6 - 2(m+2)\Delta\delta^5\\ &+(m^2+2m-m\Delta^2)\delta^4 + m\Delta^3\delta^3 - m^2\Delta^2\delta^2\} ,\end{aligned}$$
$$\begin{aligned}\ell_{13} = \frac{1}{8\Delta^2}&\{4\delta^8 - 4\Delta\delta^7 + (10m+38+\Delta^2)\delta^6 - 2(m+9)\Delta\delta^5\\ &+ (6m^2+48m+56-3m\Delta^2)\delta^4\\ &+ \left\{4m(m+1)\Delta + m\Delta^3\right\}\delta^3\\ &+ 2\left\{6m(m+1) - (m^2+m+2)\Delta^2\right\}\delta^2\\ &+ 4m(m+1)\Delta\delta\} ,\end{aligned}$$

$$\begin{aligned}
\ell_{23} = \frac{1}{8\Delta^2} & \{4\delta^8 - 12\Delta\delta^7 + (10m + 38 + 13\Delta^2)\delta^6 \\
& - 2\{(11m + 39)\Delta + 3\Delta^3\}\delta^5 \\
& + \{2(3m^2 + 23m + 28) + 5(3m + 10)\Delta^2 + \Delta^4\}\delta^4 \\
& - \{8(m^2 + 7m + 8)\Delta + (3m + 10)\Delta^3\}\delta^3 \\
& + 2\{6m(m + 1) + (m^2 + 7m + 8)\Delta^2\}\delta^2 \\
& - 4m(m + 1)\Delta\delta\}
\end{aligned}$$

(Okamoto (1963, 1968)).

5.17 (i) For the statistics $\tilde{Q}$ and $\tilde{R}$ in Theorem 5.7.1, show that

$$\begin{aligned}
& \Pr\left\{n\tilde{Q} \le u\right\} \\
& = E_{\tilde{R},\hat{B}}\left[\Pr\left\{\left(1 + \frac{2}{m}\right)^{-1} q\, F_{q,m+2}(\lambda) \le (1 + \tilde{D})\tilde{u} \,\middle|\, \tilde{R}, \hat{B}\right\}\right],
\end{aligned}$$

where $m = n - p - 1$, $\tilde{u} = \left(1 + \frac{2}{m}\right)^{-1} u$,

$$\tilde{D} = \left(1 + \tilde{R}\right)^{-1} \frac{m + 2}{m + p + 1} - 1,$$

and $F_{r,s}(\lambda)$ is a noncentral F random variable with r and s degrees of freedom and noncentrality parameter λ (5.7.8).

(ii) Show that the distribution function of $\left(1 + \frac{2}{m}\right)^{-1} F_{q,m+2}(\lambda)$ under the condition $\tilde{R}$ and $\hat{B}$ fixed can be expanded as follows.

$$\begin{aligned}
G_q(v, \lambda) & + \frac{1}{4m} q(q + 2)\{G_q(v, \lambda) - 2G_{q+2}(v, \lambda) + G_{q+4}(v, \lambda)\} \\
& + \frac{q + 2}{2m} \lambda \{G_{q+2}(v, \lambda) - 2G_{q+4}(v, \lambda) + G_{q+6}(v, \lambda)\} \\
& + \frac{1}{96m^2} q(q + 2) \sum_{j=0}^{4} h_j G_{q+j}(v, \lambda) + O\left(\frac{1}{m^3}\right),
\end{aligned}$$

where $G_q(v, \lambda)$ is the distribution function of a noncentral chi-square random variable with q degrees of freedom and noncentrality parameter λ, and

$$\begin{aligned}
& h_0 = (q - 2)(3q + 4), \quad h_1 = -12q(q + 2), \\
& h_2 = 6(q + 2)(3q + 8), \quad h_3 = -4(q + 2)(3q + 10), \\
& h_4 = 3(q + 4)(q + 6).
\end{aligned}$$

(iii) Under the condition $\frac{1}{m}XX' = \Theta = O(1)$ and (5.7.14), show the following:

$$E\left[\tilde{R}\right] = \frac{1}{m}(p-q) + \frac{1}{m^2}\left\{-q(p-q) + (p-q)\boldsymbol{x}'\Theta^{-1}\boldsymbol{x}\right\} + O\left(\frac{1}{m^3}\right),$$

$$E\left[\tilde{R}^2\right] = \frac{1}{m^2}\left\{(p-q)^2 + 2(p-q)\right\} + O\left(\frac{1}{m^2}\right),$$

$$E\left[\tilde{D}\right] = -\frac{1}{m}(2p-q-1) + \frac{1}{m^2}\left\{3p^2 - (2q-1)p - q - 1 - (p-q)\boldsymbol{x}'\Theta^{-1}\boldsymbol{x}\right\} + O\left(\frac{1}{m^3}\right),$$

$$E\left[\tilde{D}^2\right] = \frac{1}{m^2}\left\{4p^2 - 2(2q+1)p + q^2 + 1\right\} + O\left(\frac{1}{m^3}\right),$$

$$E\left[\tilde{D}\tilde{\lambda}\right] = -\frac{1}{m^2}q(2p-q-1)\boldsymbol{x}'\Theta\boldsymbol{x} + O\left(\frac{1}{m^3}\right).$$

(iv) Show that

$$\begin{aligned}
&E\left[G_k(u+\tilde{D}u,\lambda)\right] \\
&= G_k(u) + g_k(u)\Bigg[\frac{1}{m}\left\{-(2p-q-1)u - \frac{q}{k}\left(\boldsymbol{x}'\Theta^{-1}\boldsymbol{x}\right)u\right\} \\
&\quad + \frac{1}{m^2}\Bigg\{\frac{1}{4}\left[k(4p^2 - 2(2q+1)p + q^2 + 1) + 2(2p^2 + 4p - q^2 - 2q - 3)\right]u \\
&\quad - \frac{1}{4}\left\{4p^2 - 2(2q+1)p + q^2 + 1\right\}u^2 \\
&\quad + \frac{1}{2k}\boldsymbol{x}'\Theta^{-1}\boldsymbol{x}[\{2(q-k)(p-q) + qk(2p-q-1)\}u - q(2p-q-1)u^2] \\
&\quad + \frac{q(q+2)}{4k(k+2)}\left(\boldsymbol{x}'\Theta^{-1}\boldsymbol{x}\right)^2\left\{(k+2)u - u^2\right\} \\
&\quad - \frac{1}{k}\boldsymbol{x}'\left(\Theta B'\Omega^{-1}B\Theta\right)\boldsymbol{x}(p-q)(p-q-1)u\Bigg\}\Bigg] \\
&\quad + O\left(\frac{1}{m^3}\right).
\end{aligned}$$

Hint : Note $E[\mu^\ell] = O(1/m^\ell)$, $E[\lambda^\ell] = O(1/m^\ell)$ and $E[\tilde{D}^\ell] = O(1/m^\ell)$, $\ell \geq 1$, and

$$E\left[G_k(u+\tilde{D}u,\lambda)\right]$$

$$= E\left[\sum_{j=0}^{2} \frac{\left(\tilde{D}u\right)^j}{j!} \frac{\partial^j}{\partial u^j} G_k(u,\lambda)\right] + O\left(\frac{1}{m^3}\right)$$

$$= E\left[\sum_{\ell=0}^{2} e^{-\frac{1}{2}\lambda} \frac{\left(\frac{\lambda}{2}\right)^\ell}{\ell!} G_{k+2\ell}(u) + \tilde{D}u \sum_{\ell=0}^{1} e^{-\frac{1}{2}\lambda} \frac{\left(\frac{\lambda}{2}\right)^\ell}{\ell!} g_{k+2\ell}(u) + \frac{1}{2}(Du)^2 g'_{k+2\ell}(u)\right] + O\left(\frac{1}{m^3}\right),$$

where $g_k(u)$ is the probability density function of the central chi-square random variable with k degrees of freedom and

$$u g_k(u) = \frac{k}{2}\left[G_k(u) - G_{k+2}(u)\right].$$

(v) Show that the distribution function of $n\tilde{Q}$ can be expanded as follows.

$$\Pr\left\{n\tilde{Q} \le u\right\} = G_q(u) + g_q(u)\left[\frac{1}{m}a_1(u) + \frac{1}{m^2}a_2(u)\right] + O\left(\frac{1}{m^2}\right),$$

where

$$a_1(u) = -\frac{1}{2}\left\{(4p - 3q)u + u^2\right\} - (\boldsymbol{x}'\Theta^{-1}\boldsymbol{x})\, u,$$

$$\begin{aligned} a_2(u) = &\left\{(q+1)p^2 - \frac{1}{2}(3q^2 - q - 4)p + \frac{1}{48}(27q^3 - 56q^2 - 72q + 8)\right\} u \\ &- \left\{p^2 - \frac{1}{2}(4q+1)p + \frac{1}{48}(45q^2 + 2q - 28)\right\} u^2 \\ &+ \frac{1}{48}(-24p + 21q + 10)u^3 - \frac{1}{16}u^4 \\ &+ \frac{1}{4}\left(\boldsymbol{x}'\Theta^{-1}\boldsymbol{x}\right)\left\{q(4p - 3q)u - 2(2p - 2q - 1)u^2 - u^3\right\} \\ &+ \frac{1}{4}\left(\boldsymbol{x}'\Theta^{-1}\boldsymbol{x}\right)^2\left\{(q+2)u - u^2\right\} \\ &- \boldsymbol{x}'\left(\Theta B'\Omega^{-1}B\Theta\right)\boldsymbol{x}\, \frac{1}{q}(p-q)(p-q-1)u \end{aligned}$$

(Fujikoshi and Nishii (1984)).

5.18 When $\boldsymbol{x}$ is observed, the atypicality of $\boldsymbol{x}$ with respect to the population Π_i distributed as $N(\boldsymbol{\mu}_i, \Sigma)$ is measured by

$$e_i(\boldsymbol{x}) = (\boldsymbol{x} - \boldsymbol{\mu}_i)'\Sigma^{-1}(\boldsymbol{x} - \boldsymbol{\mu}_i), \quad i = 1, 2.$$

The log odd ratio $\theta(x)$ of x coming from Π_1 is defined as

$$\theta(\boldsymbol{x}) = \frac{1}{2}[e_2(\boldsymbol{x}) - e_1(\boldsymbol{x})] .$$

(i) Show that the minimum variance unbiased estimator $\hat{\theta}(\boldsymbol{x})$ of $\theta(\boldsymbol{x})$ is given by

$$\hat{\theta}(\boldsymbol{x}) = \frac{1}{2}[\hat{e}_2(\boldsymbol{x}) - \hat{e}_1(\boldsymbol{x})] ,$$

where

$$\hat{e}_i(\boldsymbol{x}) = \nu(\boldsymbol{x} - \bar{\boldsymbol{x}}_i)' S^{-1}(\boldsymbol{x} - \bar{\boldsymbol{x}}_i) - m/Ni , \quad i = 1, 2 ,$$

and $\bar{\boldsymbol{x}}_1$ and $\bar{\boldsymbol{x}}_2$ are sample means of two independent samples of sizes N_1 and N_2 drawn from Π_1 and Π_2, respectively. S is the pooled sum of squares and products matrix on $n = N_1 + N_2 - 2$ degrees of freedom, and $\nu = n - m - 1$.

(ii) Show that $\hat{\theta}(\boldsymbol{x})$ with the fixed $\boldsymbol{x}$ is expressed as

$$\hat{\theta}(\boldsymbol{x}) = \nu \operatorname{tr}(\Theta Y' S^{-1} Y) + \frac{m}{2}\left(\frac{1}{N_2} - \frac{1}{N_1}\right) ,$$

with

$$\Theta = \frac{1}{2}\begin{bmatrix} 0 & 1 \\ 1 & 0 \end{bmatrix} , \quad \Omega = \left[\boldsymbol{x} - \frac{1}{2}(\boldsymbol{\mu}_1 + \boldsymbol{\mu}_2), \boldsymbol{\mu}_1 - \boldsymbol{\mu}_2\right] ,$$

$$A = \begin{bmatrix} \frac{1}{4}\left(\frac{1}{N_1} + \frac{1}{N_2}\right) , & \frac{1}{2}\left(\frac{1}{N_2} - \frac{1}{N_1}\right) \\ \frac{1}{2}\left(\frac{1}{N_2} - \frac{1}{N_1}\right) , & \frac{1}{N_1} + \frac{1}{N_2} \end{bmatrix} .$$

(iii) Find B and D corresponding to (5.6.5) and give b_1, b_2, d_1, d_2 and c.

(iv) Find $E[\hat{\theta}(\boldsymbol{x})]$, $\mathrm{Var}[\hat{\theta}(\boldsymbol{x})]$ and further cumulants of it and give the Edgeworth expansion of the distribution of $\left(\hat{\theta}(\boldsymbol{x}) - \theta(\boldsymbol{x})\right) / \sqrt{\mathrm{Var}[\hat{\theta}(\boldsymbol{x})]}$ (Davis (1987)).

5.19 Show that the least squares estimators of $\boldsymbol{a}$, B and Ω are given as in Lemma 5.7.1.

5.20 Show Lemma 5.7.2.

5.21 Show (5.7.11).

5.22 Show that the distribution function of $\tilde{Q}$ is expressed in terms of central F distributions as (5.7.17).

5.23 Consider the seemingly unrelated regression models

$$Y_i = X_i B_i + E_i , \quad i = 1, 2$$

where the regression matrices B_i are $r_i \times p_i$ $(i = 1, 2)$ and X_i are $n \times r_i$ design matrices with rank r_i $(i = 1, 2)$. Assume that the rank of $X = [X_1, X_2]$ is r_0. Y_i and E_i are $n \times p_i$ matrices and the rows of $E = [E_1, E_2]$, $p = p_1 + p_2$, are p-variate normal random vectors with mean zero and covariance matrix

$$\Sigma = \begin{bmatrix} \Sigma_{11} & \Sigma_{12} \\ \Sigma_{21} & \Sigma_{22} \end{bmatrix} .$$

(i) Define $Q_i = I_n - X_i(X_i'X_i)^{-1}X_i'$, $i = 1, 2$ and Q_0 to be the projection matrix onto the orthogonal complement of $X = [X_1, X_2]$. Let $L(A)$ denote the linear subspace generated by the columns of matrix A. Let $\bar{Q}_i$ be the orthogonal projection onto $L(X) \cap L(Q_i)$, $i = 1, 2$. Write

$$n_i = n - r_i\,, \quad p_i = r_0 - r_i\,, \quad i = 1, 2.$$

Show that there exist matrices $Z_0, \bar{Z}_1, \bar{Z}_2, Z_1$ and Z_2 such that

$$\begin{aligned} Q_0 &= Z_0 Z_0'\,, \quad Z_0' Z_0 = I_{n_0}\,, \\ \bar{Q}_i &= \bar{Z}_i \bar{Z}_i'\,, \quad \bar{Z}_i' \bar{Z}_i = I_{p_i}\,, \quad \bar{Z}_i Z_0 = 0\,, \quad i = 1, 2 \end{aligned}$$

and if $Z_i = [\bar{Z}_i, Z_0]_{n \times n_i}$, $i = 1, 2$ show that

$$Q_i = Z_i Z_i'\,, \quad Z_i' Z_i = I_{n_i}\,, \quad n = 1, 2.$$

(ii) Let $W_i = Z_i' Y_i$, $n_i \times p_i$, $i = 1, 2$ and define the $p \times p$ nonnegative definite matrix

$$S = \begin{bmatrix} S_{11} & S_{12} \\ S_{21} & S_{22} \end{bmatrix} = \begin{bmatrix} W_1' W_1 & W_1' Z_1' Z_2 W_2 \\ W_2' Z_2' Z_1 W_1 & W_2' W_2 \end{bmatrix}.$$

Show that

$$\begin{aligned} &S_{ii} \sim W_{p_i}(n - r_i, \Sigma_{ii})\,, \quad i = 1, 2\,, \\ &W_1 \text{ and } W_2 \text{ are independent if } \Sigma_{12} = 0\,. \end{aligned}$$

(iii) For testing the hypothesis $H_0 : \Sigma_{12} = O$ the following three statistics are proposed :

$$\begin{aligned} T_1 &= -n_0 \log |I - R|\,, \\ T_2 &= n_0 \operatorname{tr}(R)\,, \\ T_3 &= \frac{1}{n_0}\{n_1 n_2 \operatorname{tr}(R) - n_1 p_2 \operatorname{tr}(S_{11}^{-1} W_1' Z_1' Z_2 Z_2' Z_1 W_1) \\ &\qquad - n_2 p_1 \operatorname{tr}(S_{22}^{-1} W_2' Z_2' Z_1 Z_1' Z_2 W_2)\}\,, \end{aligned}$$

where $R = S_{11}^{-1} S_{12} S_{22}^{-1} S_{21}$.

(iv) Show that there is no loss of generality in taking

$$\Sigma = \begin{bmatrix} I_{p_1} & \Lambda \\ \Lambda' & I_{p_2} \end{bmatrix},$$

and that W_i is decomposed under $H_0 : \Lambda = O$ as follows:

$$W_i = H_i T_i\,, \quad H_i' H_i = I_{p_i}\,, \quad T_i' T_i = S_{ii}$$

where T_i is a $p_i \times p_i$ upper triangular matrix and H_i is an $n_i \times p_i$ matrix with the uniform distribution over $H_i' H_i = I_{p_i}$.

(v) Show that the characteristic functions of T_1, T_2 and T_3 are given as

$$G_1(t) = {}_3F_2\left(itn_0, \frac{1}{2}p_1, \frac{1}{2}p_2; \frac{1}{2}n_1, \frac{1}{2}n_2; MM'\right),$$

$$G_2(t) = {}_2F_2\left(\frac{1}{2}p_1, \frac{1}{2}p_2; \frac{1}{2}n_1, \frac{1}{2}n_2; itn_o MM'\right),$$

$$G_3(t) = \operatorname{etr}\left(-\frac{itp_1p_2}{n_0}MM'\right)\sum_{k=0}^{\infty}\sum_{\kappa}\frac{1}{k!}\left(\frac{itp_1p_2}{n_0}\right)^k \times \frac{C_\kappa(MM')C_\kappa(I_{n_1}+U_1)C_\kappa(I_{n_2}+U_2)}{C_\kappa(I_{n_1})C_\kappa(I_{n_2})},$$

where

$$M = Z_1'Z_2 = \begin{bmatrix} K & 0 \\ 0 & I_{n_0} \end{bmatrix}, \quad K = \bar{Z}_1'Z_2,$$

$$\Gamma = \begin{bmatrix} KK' - I_{\rho_1} & 0 \\ 0 & 0 \end{bmatrix},$$

$$U_i = -\frac{n_i}{p_i}\begin{bmatrix} I_{p_i} & 0 \\ 0 & 0 \end{bmatrix}, \quad i = 1, 2.$$

(vi) By noting

$$\left(\frac{p_1}{2}\right)_\kappa \Big/ \left(\frac{n_1}{2}\right)_\kappa = C_\kappa(I_{p_1})/C_\kappa(I_{n_1}),$$

show that $G_1(t)$ can be expressed as

$$G_1(t) = \sum_{s=0}^{\infty}\sum_{\sigma}\frac{C_\sigma(\Gamma)}{C_\sigma(I_{n_1})}Q_\sigma,$$

$$Q_\sigma = \sum_{k=s}^{\infty}\sum_{\kappa}\frac{1}{k!}\binom{\kappa}{\sigma}\frac{(itn_0)_\kappa}{\left(\frac{n_2}{2}\right)_\kappa}\left(\frac{1}{2}p_2\right)_\kappa C_\kappa(I_{p_1}).$$

Show that $G_1(t)$ can be asymptotically expanded up to the order n_0^{-1} for large n_0, n_1 and n_2 as follows.

$$G_1(t) = \frac{1}{(1-2it)^{\frac{1}{2}p_1p_2}}\left[1 + \frac{1}{n_0}\frac{p_1p_2}{2}\frac{2it}{1-2it}\left\{\frac{1}{2}(p+1) + \operatorname{tr}(KK') - (\rho_1+\rho_2)\right\} + o\left(\frac{1}{n_0}\right)\right].$$

Hint : Use Lemma 4.4.6 and Lemma 4.5.3, and note that

$$C_\sigma(\Gamma)/C_\sigma(I_{n_1}) = O(n_0^{-s}).$$

(vii) Show that $G_2(t)$ in (v) can be asymptotically expanded as follows.

$$G_2(t) = \frac{1}{(1-2it)^{\frac{1}{2}p_1p_2}}\left[1 + \frac{1}{n_0}\left\{-\frac{1}{4}(p+1)\left(\frac{2it}{1-2it}\right)^2\right.\right.$$

$$\left.\left.+\frac{1}{2}\left(\frac{2it}{1-2it}\right)(\operatorname{tr}(KK') - (\rho_1+\rho_2))\right\} + o\left(\frac{1}{n_0}\right)\right]$$

(Davis (1989)).

5.24 Let X be an $m \times k$ ($m \geq k$) matrix satisfying $X'X = I_k$ and F an $m \times k$ matrix. The probability density function of the matrix Langevin distribution denoted by $L(m, k; F)$ is defined as

$$\operatorname{etr}(F'X)/ {}_0F_1\left(\frac{m}{2}; \frac{1}{4}F'F\right).$$

(i) Let $X_1, X_2, \ldots, X_n$ be a random sample from $L(m, k; F)$ and let

$$Z = \sqrt{\frac{m}{n}}\sum_{j=1}^{n} X_j.$$

Show that the characteristic function of Z is given by

$$\psi(T) = \left\{ {}_0F_1\left(\frac{m}{2}; \frac{1}{4}\left(F + i\sqrt{\frac{m}{n}}T\right)'\left(F + i\sqrt{\frac{m}{n}}T\right)\right)\right.$$

$$\left. \Big/ {}_0F_1\left(\frac{m}{2}; \frac{1}{4}F'F\right)\right\}^n.$$

(ii) Show that under the assumption

$$F = F_0/\sqrt{n},$$

the inversion formula of the characteristic function $\psi(T)$

$$(2\pi)^{-km}\left\{ {}_0F_1\left(\frac{m}{2}; \frac{1}{4}F'F\right)\right\}^{-n} \operatorname{etr}\left(\frac{1}{\sqrt{m}}F'Z\right)$$

$$\times \int \operatorname{etr}(-iZ'T)\left\{ {}_0F_1\left(\frac{m}{2}; -\frac{1}{4}\frac{m}{n}T'T\right)\right\}^n dT$$

yields the asymptotic expansion of the probability density function $f(Z)$ of Z

with the help of Table 4.3.2(a) and (4.6.34) as follows.

$$f(Z) = \varphi^{(mk)}\left(Z - \frac{1}{\sqrt{m}}F_0\right)$$
$$\times\left[1 + \frac{1}{n}\sum_{\sigma\vdash 2} a_\sigma \left\{4H_\sigma\left(\frac{1}{\sqrt{2}}Z\right) - C_\sigma\left(\frac{1}{m}F_0'F_0\right)\right\}\right.$$
$$+\frac{1}{n^2}\left\{-4\left(\sum_{\sigma\vdash 2} a_\sigma C_\sigma\left(\frac{1}{m}F_0'F_0\right)\right)\sum_{\sigma\vdash 2} a_\sigma H_\sigma\left(\frac{1}{\sqrt{2}}Z\right)\right.$$
$$\left.\left.+\sum_{\sigma\vdash 3,4} a_\sigma \left\{2^s H_\sigma\left(\frac{1}{\sqrt{2}}Z\right) + (-1)^s C_\sigma\left(\frac{1}{m}F_0'F_0\right)\right\}\right\}\right] + O\left(\frac{1}{n^3}\right),$$

where $\varphi^{(mk)}(U)$ is the probability density function of an $m \times k$ matrix U distributed as $N(O, I_m, I_k)$, and $\sum_{\sigma\vdash s}$ denotes the summation with respect to the partitions σ into not more than k parts of s. a_σ, $\sigma \vdash 2, 3, 4$ are given as follows.

$$a_{(2)} = -\frac{1}{4(m+2)}, \quad a_{(1^2)} = \frac{1}{8(m-1)},$$
$$a_{(3)} = \frac{1}{3(m+2)(m+4)}, \quad a_{(21)} = -\frac{1}{12(m-1)(m+2)},$$
$$a_{(1^3)} = \frac{1}{12(m-1)(m-2)},$$
$$a_{(4)} = \frac{1}{16(m+2)^2}, \quad a_{(31)} = -\frac{m+6}{96(m+2)^2(m-1)},$$

$$a_{(2^2)} = \frac{7m^2 - 4m + 12}{192(m+2)^2(m-1)^2},$$
$$a_{(21^2)} = -\frac{m-3}{96(m+2)(m-1)^2},$$
$$a_{(1^4)} = \frac{1}{64(m-1)^2}.$$

(iii) By noting that $H_\sigma(Z)$ corresponding to a partition σ of s is a function of $Z'Z$, show that

$$\int_{Z'Z=W} \varphi^{(mk)}(Z+U)H_\sigma(Z)dZ$$
$$= \frac{1}{2^{\frac{1}{2}mk}\Gamma_m\left(\frac{k}{2}\right)} \operatorname{etr}\left(-\frac{1}{2}U'U\right) \operatorname{etr}\left(-\frac{1}{2}W\right) |W|^{\frac{1}{2}(m-k-1)}$$
$$\times {}_0F_1\left(\frac{m}{2}; \frac{1}{4}U'UW\right)(-2)^s L_\sigma^{\frac{1}{2}(m-k-1)}\left(\frac{1}{2}W\right).$$

(iv) Show that under the condition $F = \frac{1}{\sqrt{n}}F_0$, the probability density function of

$S = Z'Z$ can be expanded as

$$w\left(S; m, I_k, \frac{1}{2}\frac{F_0'F_0}{m}\right)$$
$$\times \left[1 + \frac{1}{n}\sum_{\sigma \vdash 2} a_\sigma \left\{4L_\sigma^{\frac{1}{2}(m-k-1)}\left(\frac{1}{2}S\right) - C_\sigma\left(\frac{1}{m}F_0'F_0\right)\right\}\right.$$
$$+ \frac{1}{n^2}\left\{-4\left(\sum_{\sigma \vdash 2} a_\sigma C_\sigma\left(\frac{1}{m}F_0'F_0\right)\right)\sum_{\sigma \vdash 2} a_\sigma L_\sigma^{\frac{1}{2}(m-k-1)}\left(\frac{1}{2}S\right)\right.$$
$$+ \sum_{\sigma \vdash 3,4} a_\sigma \left\{(-2)^s L_\sigma^{\frac{1}{2}(m-k-1)}\left(\frac{1}{2}S\right) + C_\sigma\left(\frac{1}{m}F_0'F_0\right)\right\}$$
$$\left. + O\left(\frac{1}{n^3}\right)\right],$$

where $w(S; m, I_k, \Omega)$ is the probability density function of the noncentral Wishart distribution $W(m, I_k, \Omega)$ (Chikuse (1991)).

5.25 Let X be an $m \times k$ $(m \geq k)$ matrix distributed as $L(m, k; F)$ with the singular value decomposition $F = \Gamma\Delta\Theta'$ where Γ and Θ are an $m \times p$ matrix and a $k \times p$ matrix satisfying $\Gamma'\Gamma = I_p$ and $\Theta'\Theta = I_p$, and $\Delta = \mathrm{diag}(\lambda_1, \lambda_2, \ldots, \lambda_p)$.

(i) Show that the probability density function of $Y = \sqrt{m}\,\Gamma' X$ for large m can be expressed as

$$\varphi^{(pk)}(Y)\left[1 + \sum_{\alpha=1}^{4}\frac{M_\alpha}{(m)^{\alpha/2}} + o\left(\frac{1}{m^2}\right)\right],$$

where

$$M_1 = \mathrm{tr}\,(\Omega Y'),$$
$$M_2 = 4\sum_{\sigma \vdash 2} a_\sigma H_\sigma\left(\frac{1}{\sqrt{2}}Y\right) + \frac{1}{2}(\mathrm{tr}\,(\Omega Y'))^2 - \frac{1}{2}\mathrm{tr}\,(\Delta^2),$$
$$M_3 = \mathrm{tr}\,(\Omega Y')(\left[4\sum_{\sigma \vdash 2} a_\sigma H_\sigma\left(\frac{1}{\sqrt{2}}Y\right) + \frac{1}{6}(\mathrm{tr}\,(\Omega Y'))^2 - \frac{1}{2}\mathrm{tr}\,(\Delta^2)\right],$$
$$M_4 = \sum_{\sigma \vdash 3,4} 2^s H_\sigma\left(\frac{1}{\sqrt{2}}Y\right)$$
$$+ 2\sum_{\sigma \vdash 2}\left\{a_\sigma(\mathrm{tr}\,(\Omega Y'))^2 - a_\sigma\,\mathrm{tr}\,(\Delta^2) + 2b_\sigma\right\} H_\sigma\left(\frac{1}{\sqrt{2}}Y\right)$$
$$+ \frac{1}{24}(\mathrm{tr}\,(\Omega Y'))^4 - \frac{1}{4}\mathrm{tr}\,(\Delta^2)(\mathrm{tr}\,(\Omega Y'))^2 + \frac{1}{8}(\mathrm{tr}\,(\Delta^2))^2,$$

$$a_{(2)} = -\frac{1}{4}, \quad a_{(1^2)} = \frac{1}{8},$$
$$b_{(2)} = \frac{1}{2}, \quad b_{(1^2)} = \frac{1}{8},$$
$$b_{(3)} = \frac{1}{3}, \quad b_{(21)} = -\frac{1}{12}, \quad b_{(1^3)} = \frac{1}{12},$$
$$b_{(4)} = \frac{1}{32}, \quad b_{(32)} = b_{(21^2)} = -\frac{1}{192}, \quad b_{(2^2)} = \frac{7}{384},$$
$$b_{(1^4)} = \frac{1}{128},$$

and $\Omega = \Delta\Theta'$.

(ii) Show that the probability density function of $W = YY'$ for large m can be expressed as

$$w(W; n, I_k)\left[1 + \frac{1}{m}N_1 + \frac{1}{m^2}N_2 + O\left(\frac{1}{m^3}\right)\right],$$

where

$$N_1 = 4\sum_{\sigma\vdash 2} a_\sigma L_\sigma^{\frac{1}{2}(k-p-1)}\left(\frac{1}{2}W\right) + \frac{1}{2p}\operatorname{tr}(\Delta^2 W) - \frac{1}{2}\operatorname{tr}(\Delta^2),$$

$$\begin{aligned} N_2 = {} & \sum_{\sigma\vdash 3,4} (-2)^s b_\sigma L_\sigma^{\frac{1}{2}(k-p-1)}\left(\frac{1}{2}W\right) \\ & + \sum_{\sigma\vdash 2}\left\{\left(\frac{a_\sigma}{2p}\operatorname{tr}(\Delta^2 W) - 2a_\sigma \operatorname{tr}(\Delta^2) + 4b_\sigma\right)\right. \\ & \qquad \left. \times L_\sigma^{\frac{1}{2}(k-p-1)}\left(\frac{1}{2}W\right) + \frac{1}{32\left(\frac{1}{2}p\right)_\sigma} C_\sigma(\Delta^2 W)\right\} \\ & - \left(\operatorname{tr}(\Delta^2)\right)\left\{\frac{1}{4p}\operatorname{tr}(\Delta^2 W) - \frac{1}{8}\operatorname{tr}(\Delta^2)\right\} \end{aligned}$$

(Chikuse (1993)).

Table 5.6.1 Exact expressions for P_ϕ up to order 4.

ϕ	$(-1)^f P_\phi$
1	$\frac{1}{2}mb_1 + d_1$
2	$\frac{1}{3}\left\{\frac{1}{4}m(m+2)(b_1^2+2b_2)+(m+2)(b_1d_1+2c)+d_1^2+2d_2\right\}$
1^2	$\frac{1}{3}\left\{\frac{1}{2}m(m-1)(b_1^2-b_2)+2(m-1)(b_1d_1-c)+2(d_1^2-d_2)\right\}$
3	$\frac{1}{5}\{\frac{1}{8}m(m+2)(m+4)(-b_1^3+6b_1b_2)+\frac{3}{4}(m+2)(m+4)(-b_1^2d_1+4b_1c+2b_2d_1)$ $+\frac{3}{2}(m+4)(-b_1d_1^2+2b_1d_2+4d_1c)-d_1^3+6d_1d_2\}$
21	$\frac{3}{5}\{\frac{1}{4}m(m-1)(m+2)(b_1^3-b_1b_2)+\frac{1}{2}(m-1)(m+2)(3b_1^2d_1-2b_1c-b_2d_1)$ $+3(m+2)b_1d_1^2-(m+4)b_1d_2-2(m-1)d_1c+2(d_1^3-d_1d_2)\}$
4	$\frac{1}{35}\{\frac{1}{16}m(m+2)(m+4)(m+6)(-13b_1^4+36b_1^2b_2+12b_2^2)+\frac{1}{2}(m+2)(m+4)$ $\times(m+6)(-13b_1^3d_1+18b_1^2c+18b_1b_2d_1+12b_2c)+\frac{3}{2}(m+4)(m+6)$ $\times(-13b_1^2d_1^2+6b_1^2d_2+24b_1d_1c+6b_2d_1^2+4b_2d_2+8c^2)$ $+2(m+6)(-13b_1d_1^3+18b_1d_1d_2+18d_1^2c+12d_2c)$ $-13d_1^4+36d_1^2d_2+12d_2^2\}$
31	$\frac{1}{7}[\frac{1}{4}m(m+2)(m+4)(m-1)(b_1^4+b_1^2b_2-2b_2^2)+(m+2)(m+4)(m-1)$ $\times(2b_1^3d_1+b_1^2c+b_1b_2d_1-4b_2c)+(m+4)\{(6m+1)b_1^2d_1^2+(m-8)b_1^2d_2$ $+4(m-1)b_1d_1c+(m+13)b_2d_1^2-8(m-1)c^2-2(2m+5)b_2d_2\}$ $+4\{(2m+5)b_1d_1^3+(m-8)b_1d_1d_2+(m+13)d_1^2c-2(2m+5)d_2c\}$ $+4(d_1^4+d_1^2d_2-2d_2^2)]$
2^2	$\frac{2}{5}[\frac{1}{8}m(m+2)(m-1)(m+1)(b_1^4-2b_1^2b_2+b_2^2)+(m+2)(m-1)(m+1)$ $\times(b_1^3d_1-b_1^2c-b_1b_2d_1+b_2c)+(m+1)\{(3m+2)b_1^2d_1^2-(m+4)b_1^2d_2$ $-4(m-1)b_1d_1c-(m+4)b_2d_1^2+(m+4)b_2d_2+2(m-1)c^2\}$ $+4(m+1)(b_1d_1^3-b_1d_1d_2-d_1^2c+d_2c)+2(d_1^4-2d_1^2d_2+d_2^2)]$

Table 5.6.2 Cumulants of $\tilde{W}$ to second order terms

	$\tilde{\kappa}_1$	$\tilde{\kappa}_2$	$\tilde{\kappa}_3/6$	$\tilde{\kappa}_4/24$	$\tilde{\kappa}_5/120$	$\tilde{\kappa}_6/720$
1	$\frac{1}{2}\Delta^2$	Δ^2				
N_1^{-1}	$-\frac{1}{2}m$	m		$\frac{1}{2}\Delta^2$		
N_2^{-1}	$\frac{1}{2}m$	$m+\Delta^2$	Δ^2	$\frac{1}{2}\Delta^2$		
ν^{-1}		$(m+1)\Delta^2+\frac{1}{2}\Delta^4$	Δ^4	Δ^4		
N_1^{-2}		$\frac{1}{2}m$	$-\frac{1}{2}m$	$\frac{1}{4}m$	$-\frac{1}{2}\Delta^2$	$\frac{1}{2}\Delta^2$
N_2^{-2}		$\frac{1}{2}m$	$\frac{1}{2}(m+\Delta^2)$	$\frac{1}{4}m+\frac{3}{2}\Delta^2$	$\frac{3}{2}\Delta^2$	$\frac{1}{2}\Delta^2$
$(N_1N_2)^{-1}$				$\frac{1}{2}m$	Δ^2	Δ^2
$(N_1\nu)^{-1}$		$m(m+1)-\Delta^2$		$(3m+5)\Delta^2$	$2\Delta^4$	$4\Delta^4$
$(N_2\nu)^{-1}$		$m(m+1)+(m+2)\Delta^2$	$(3m+5)\Delta^2+\Delta^4$	$(3m+5)\Delta^2+5\Delta^4$	$8\Delta^4$	$4\Delta^4$
ν^{-2}		$(m+3)\Delta^2+\Delta^4$	$(m+5)\Delta^4+\frac{1}{3}\Delta^6$	$\frac{1}{4}(9m+35)\Delta^4+\frac{5}{2}\Delta^6$	$6\Delta^6$	$\frac{14}{3}\Delta^6$

Appendix

Invariant Polynomials

In Chapter 4 zonal polynomials are defined in terms of elementary symmetric functions. Zonal polynomials of a matrix argument were originally introduced by using the theory of group representation, see James (1961a). Here we give a brief introduction to invariant polynomials of two matrix arguments. These results can be extended to the case of r matrix arguments, see Davis (1979, 1980a, 1981).

A.1 Representation of a Group

The right regular realization of the general linear group $G\ell(m,R)$

$$U \longrightarrow UL, \quad U, L \in G\ell(m,R)$$

induces the right regular representation

$$u(U) \longrightarrow T(L)u(U) = u(UL)$$

in the vector space U_{2k} of homogeneous polynomials $u(U)$ of degree $2k$ in the elements of the matrix U. Since $O(m)$ is compact, U_{2k} can be projected upon a subspace of polynomials constant on the coset $O(m)U$ by averaging over the orthogonal group

$$\begin{aligned} u(U) \stackrel{e}{\longrightarrow} \int_{O(m)} u(HU) \mathrm{d}(H) &= \varphi(U'U) \\ &= \varphi(X), \quad \text{if} \quad U'U = X. \end{aligned}$$

Thus, let X be an $m \times m$ real positive definite symmetric matrix and $P_k[X]$ the space of homogeneous polynomials $\varphi(X)$ of degree k in the elements of X with complex coefficients. The transitive group of congruence transformations

$$X \longrightarrow L'XL$$

induces transformations

$$\varphi(X) \longrightarrow T_{2k}(L)\varphi(X) = \varphi(L'XL)$$

of $P_k[X]$ which form a right regular representation of $G\ell(m, R)$ [Vilenkin (1965), Chapter I]. Here we consider the polynomial representation whose elements of a representation matrix $\tilde{T}_{2k}(L)$ are homogeneous polynomials of degree $2k$ in the elements of L. Similarly for the congruence transform

$$Y \longrightarrow L'YL$$

the right regular representation is defined as

$$\psi(Y) \longrightarrow T_{2\ell}(L)\psi(Y) = \psi(L'YL)$$

for the space $P_\ell[Y]$ of homogeneous polynomials $\psi(Y)$ of degree ℓ in the elements of Y with complex coefficients.

Let $\boldsymbol{\xi}_k$ be the vector of monomials $\prod_{i\leq j} x_{ij}^{k_{ij}}$ $(\sum_{i\leq j} k_{ij} = k,\ k_{ij} \geq 0)$, then $\boldsymbol{\xi}_k$ is a basis for $P_k[X]$. Define a similar basis $\boldsymbol{\eta}_\ell$ for $P_\ell[Y]$.

$P_{k,\ell}[X, Y]$ is a vector space defined as

$$P_{k,\ell}[X, Y] = P_k[X] \otimes P_\ell[Y]$$

in the sense that $\boldsymbol{\xi}_k \otimes \boldsymbol{\eta}_\ell$ is a basis for $P_{k,\ell}[X, Y]$.

The simultaneous congruence transforms

$$X \longrightarrow L'XL, \quad Y \longrightarrow L'YL$$

produce linear transforms with matrices $\tilde{T}_{2k}(L)$, $\tilde{T}_{2\ell}(L)$ and $\tilde{T}_{2k}(L) \otimes \tilde{T}_{2\ell}(L)$. For the right regular representation the basis transform is

$$\boldsymbol{\xi}_k \longrightarrow (\tilde{T}_{2k}(L))'\boldsymbol{\xi}_k, \quad \boldsymbol{\eta}_\ell \longrightarrow (\tilde{T}_{2\ell}(L))'\boldsymbol{\eta}_\ell,$$

$$\boldsymbol{\xi}_k \otimes \boldsymbol{\eta}_\ell \longrightarrow (\tilde{T}_{2k}(L)' \otimes \tilde{T}_{2\ell}(L)')(\boldsymbol{\xi}_k \otimes \boldsymbol{\eta}_\ell).$$

James (1961a) uses the following results of the representation theory of group to introduce zonal polynomials.

(i) Any vector space, in which a polynomial representation $\tilde{T}_{2k}(L)$ of $G\ell(m, R)$ is defined, may be decomposed into a direct sum of subspaces which are invariant under $T_{2k}(L)$ and irreducible [Boerner (1963), Chapter V, S1]. Choosing the basis vectors of the vector space to lie in the invariant subspaces, $\tilde{T}_{2k}(L)$ decomposes into a direct sum of irreducible representations of $G\ell(m, R)$, each one defined in a corresponding irreducible invariant subspace.

The inequivalent irreducible representations of $G\ell(m, R)$ as a polynomial of degree n may be indexed by the ordered partitions ν of n into not more than m parts. This is denoted by $\tilde{T}_\nu(L)$. Here we are concerned with polynomial representations $\tilde{T}_{2k}(L)$, $\tilde{T}_{2\ell}(L)$ and $\tilde{T}_{2k,2\ell}(L)$ of even degrees $n = 2k, 2\ell, 2f$ $(f = k + \ell)$.

(ii) $\tilde{T}_\nu(L)$, where ν is an ordered partition of $2k$, occurs in the decomposition of $\tilde{T}_{2k}(L)$

if and only if $\tilde{T}_\nu(L)$ contains the identity representation of $O(m)$ when it is restricted to $O(m)$. The representation of $O(m)$ obtained by restricting $\tilde{T}_\nu(L)$ to $L \in O(m)$ decomposes into a direct sum of irreducible representations of $O(m)$ which includes the identity representation: the 1×1 matrix (1).

(iii) The only such representations are $\tilde{T}_{2\kappa}(L)$ where $2\kappa = 2(k_1, k_2, \ldots, k_m) = (2k_1, 2k_2, \ldots, 2k_m)$, $k_1 + k_2 + \cdots + k_m = k$, $k_1 \geq k_2 \geq \cdots \geq k_m \geq 0$ and $\tilde{T}_{2\kappa}(L)$ occurs with multiplicity one in the decomposition of $\tilde{T}_{2k}(L)$ so that

$$\tilde{T}_{2k}(L) \equiv \bigoplus_{\kappa} \tilde{T}_{2\kappa}(L), \qquad \tilde{T}_{2\ell}(L) \equiv \bigoplus_{\lambda} \tilde{T}_{2\lambda}(L),$$

where κ and λ run through the ordered partitions of k and ℓ, respectively, and "$\equiv$" denotes equivalence of representations [Thrall (1942), Theorem III].

(iv) $\tilde{T}_{2\kappa}(L)$ contains the identity representation exactly once when it is restricted to $O(m)$ [James and Constantine (1974), (11, 10); Weyl (1946), p. 158].

$$\tilde{T}_{2\kappa}(H) = \begin{bmatrix} 1 & 0 & \cdots & 0 \\ 0 & & & \\ \vdots & & H_2 & \\ 0 & & & \end{bmatrix}, \qquad H \in O(m).$$

(v) The corresponding decompositions into the irreducible subspaces are given as

$$P_k[X] = \bigoplus_{\kappa} V_\kappa[X], \qquad P_\ell[Y] = \bigoplus_{\lambda} V_\lambda[Y],$$

where $V_\kappa[X]$ contains exactly one-dimensional subspace generated by a polynomial in X invariant under $X \to H'XH$, $H \in O(m)$, which when suitably normalized is the zonal polynomial $C_\kappa(X)$.

The representation is decomposed as

$$\tilde{T}_{2k,2\ell}(L) = \bigoplus_{\kappa} \bigoplus_{\lambda} \left\{ \tilde{T}_{2\kappa}(L) \otimes \tilde{T}_{2\ell}(L) \right\},$$

and

$$P_{k,\ell}[X, Y] = \bigoplus_{\kappa} \bigoplus_{\lambda} \{ V_\kappa[X] \otimes V_\lambda[Y] \}.$$

$\tilde{T}_{2\kappa}(L) \otimes \tilde{T}_{2\lambda}(L)$ are representations of $G\ell(m, R)$ having polynomials of degree $2f$, but are not in general irreducible. They may be decomposed into direct sums of the representation $\tilde{T}_\Phi(L)$, Φ being an ordered partition of $2f$ into not more than m parts.

$$P_{k,\ell}[X, Y] = \bigoplus_{\kappa} \bigoplus_{\lambda} \bigoplus_{\Phi} V_\Phi^{\kappa,\lambda}[X, Y]. \tag{A.1.1}$$

Φ runs through the irreducible representations in the decomposition of $\tilde{T}_{2\kappa} \otimes \tilde{T}_{2\lambda}$. Table A.1 gives the decomposition for some low order κ and λ.

Table A.1 Decomposition of Kronecker products of irreducible representations

$2f$	2κ	2λ	Φ					
4	2	2	4	31	2^2			
6	4	2	6	51	42			
	2^2	2	42	321	2^3			
8	6	2	8	71	62			
	42	2	62	53	521	4^2	431	42^2
	2^3	2	42^2	32^21	2^4			

Davis (1980a) proved the following fundamental results.

(a) By (iv) above, if $\Phi = 2\phi$, $\phi = (f_1, \ldots, f_m)$, $f_1 + \cdots + f_m = f = k + \ell$, then $V_{2\phi}^{\kappa,\lambda}[X, Y]$ contains a single one-dimensional subspace, on which the identity representation of $O(m)$ is defined when T_Φ is restricted to $O(m)$. This subspace consists of multiples of a single polynomial in X and Y which is invariant under the simultaneous transformations

$$X \longrightarrow H'XH, \quad Y \longrightarrow H'YH, \quad H \in O(m),$$

and is a 2-matrix analogue of $C_\kappa(X)$. We denote the invariant polynomial by $\Gamma_\phi^{\kappa,\lambda}(X, Y)$.

(b) Subspaces $V_\Phi^{\kappa,\lambda}$ occur with Φ not of the form 2ϕ; these contain no invariant polynomials, since T_Φ contains no identity representation when L is restricted on $O(m)$.

(c) A representation 2ϕ may occur with multiplicity greater than 1 for given κ, λ, so that an additional subscript should be used: for notational simplicity we shall omit this, and write for example $\phi \equiv \phi'$ to indicate equivalent representations. It should be noted that the invariant subspaces $V_{2\phi}^{\kappa,\lambda}[X, Y]$, and hence their invariant polynomials $\Gamma_\phi^{\kappa,\lambda}(X, Y)$, cannot be uniquely defined when 2ϕ occurs with multiplicity greater than one for given κ, λ. This non-uniqueness first occurs when $k = \ell = 3$, $f = 6$. In the decomposition of $\tilde{T}_{2\kappa} \otimes \tilde{T}_{2\lambda}$ when $\kappa = \lambda = [21]$, the irreducible representation $\tilde{T}_{2\phi}$, $\phi = [321]$, occurs three times. However, the direct sum of equivalent subspaces

$$U_{2\phi}^{\kappa,\lambda} = \bigoplus_{\phi \equiv \phi'} V_{2\phi}^{\kappa,\lambda} \tag{A.1.2}$$

is uniquely defined.

A.2 Integration of the Representation Matrix over the Orthogonal Group

Lemma A.1 If $\tilde{T}(H)$ is the matrix of an irreducible non-identity representation of $O(m)$ in a vector space V, then

$$\int_{O(m)} \tilde{T}(H)\mathrm{d}(H) = O\,. \tag{A.2.1}$$

PROOF For $H_1 \in O(m)$ and the invariance of $\mathrm{d}(H)$,

$$\int_{O(m)} \tilde{T}(H_1 H)\mathrm{d}(H) = \int_{O(m)} \tilde{T}(H)\mathrm{d}(H) = M \text{ (a constant matrix)}.$$

The left-hand side is expressed as

$$M = \tilde{T}(H_1)\int_{O(m)} \tilde{T}(H)\mathrm{d}(H) = \tilde{T}(H_1)M\,.$$

Similarly

$$M = \int_{O(m)} \tilde{T}(HH_1)\mathrm{d}(H) = M\tilde{T}(H_1)\,.$$

From Schur's lemma, this implies that

$$M = cI_d \text{ for some } c \in C\,.$$

Thus

$$O = (\tilde{T}(H_1) - I)M = c(\tilde{T}(H_1) - I)\,.$$

As $\tilde{T}(H_1)$ is not an identity representation, we have $c = 0$, which implies $M = O$.

Lemma A.2 Choose $\Gamma_{\phi}^{\kappa,\lambda}(X,Y)$ as the first basis vector in $V_{2\phi}^{\kappa,\lambda}[X,Y]$, and the remainder in subspaces invariant under the non-identity representations of $O(m)$. Then the (1,1)-element of $\tilde{T}_{2\phi}(L)$ is $C_{\phi}(L'L)/\{C_{\phi}(I)\}$ and

$$\int_{O(m)} \tilde{T}_{2\phi}(H)\mathrm{d}(H) = \begin{bmatrix} 1 & 0 & \cdots & 0 \\ 0 & & & \\ \vdots & & O & \\ 0 & & & \end{bmatrix}. \tag{A.2.2}$$

PROOF Let the representation matrix be denoted as

$$\tilde{T}_{2\phi}(L) = \begin{bmatrix} t_{11}(L), \ldots, t_{1d}(L) \\ \vdots \qquad \vdots \\ t_{d1}(L), \ldots, t_{dd}(L) \end{bmatrix},$$

where d is the dimension of the representation matrix and the $t_{ij}(L)$'s are homogeneous polynomials of degree $2f$ in the elements of L. With the help of (iv) the representation matrix $\tilde{T}_{2\phi}(H)$, $H \in O(m)$, can be expressed as

$$\tilde{T}_{2\phi}(H) = \begin{bmatrix} 1 & 0 & \cdots & 0 \\ 0 & & & \\ \vdots & & H_2 & \\ 0 & & & \end{bmatrix},$$

where H_2 is not an identity matrix. Then

$$\begin{aligned}(t_{ij}(HL)) &= \begin{bmatrix} 1 & 0 & \cdots & 0 \\ 0 & & & \\ \vdots & & H_2 & \\ 0 & & & \end{bmatrix} \begin{bmatrix} t_{11}(L), \ldots, t_{1d}(L) \\ \vdots \qquad\qquad \vdots \\ t_{d1}(L), \ldots, t_{dd}(L) \end{bmatrix} \\ &= \begin{bmatrix} t_{11}(L), \ldots, t_{1d}(L) \\ \\ \vdots \quad * \end{bmatrix}.\end{aligned}$$

Thus we have

$$t_{1j}(HL) = t_{1j}(L) \quad \text{for all} \quad H \in O(m),$$

which implies that the elements $t_{1j}(L)$ of the first row of $\tilde{T}_{2\phi}(L)$ are polynomials of degree f in $L'L$, say

$$t_{1j}(L) = g_j(L'L), \quad j = 1, 2, \ldots, d.$$

Similarly, considering $(t_{ij}(LH))$ we may show that the elements of the first column of $\tilde{T}_{2\phi}(L)$ are invariant under the transformation $L \to LH$. Thus they are functions of LL'. In particular,

$$g_1(L'L) = g_1(H'L'LH) \quad \text{for all} \quad H \in O(m),$$

which implies that $t_{11}(L)$ is a homogeneous symmetric function of the latent roots of $L'L$. Now fix $L = L_1 \in G\ell(m, R)$ and let $Z = L_1'L_1$. Then $t_{1j}(L_1) = g_j(Z)$ is in $P_f(Z)$, $j = 1, 2, \ldots, d$, and for any $L \in G\ell(m, R)$

$$\begin{aligned} g_j(L'ZL) &= t_{1j}(L_1L) \\ &= \sum_{k=1}^{d} t_{1k}(L_1)t_{kj}(L) \\ &= \sum_{k=1}^{d} t_{kj}(L)g_k(Z). \end{aligned}$$

Hence, if V denotes the linear subspace of $P_f(Z)$ generated by the $g_j(Z)$'s , $j = 1, 2, \ldots, d$, and $p(Z) = \sum_{j=1}^{d} a_j g_j(Z)$ is any polynomial in V, then

$$p(L'ZL) = \sum_{j=1}^{d} \tilde{a}_j(L) g_j(Z)$$

where

$$\begin{bmatrix} \tilde{a}_1(L) \\ \vdots \\ \tilde{a}_d(L) \end{bmatrix} = \tilde{T}_{2\phi}(L) \begin{bmatrix} a_1 \\ \vdots \\ a_d \end{bmatrix}.$$

Thus V is an invariant subspace of $P_f(Z)$ and the irreducible representation $\tilde{T}_{2\phi}(L)$ is defined on it. But $V_\phi(Z)$ is the unique invariant subspace of $P_f(Z)$ with this property, so that $V = V_\phi(Z)$. This implies that the elements of the first row of $\tilde{T}_{2\phi}(L), t_{1j}(L_1) = g_j(Z)$, $j = 1, 2, \ldots, d$, constitute a basis for $V_\phi(Z)$. In particular $g_1(Z) = t_{11}(L'L)$ is invariant under $Z \to H'ZH$ and therefore it must be proportional to $C_\phi(Z)$. Since $\tilde{T}_{2\phi}(I_m) = I_d$, we have $t_{11}(I_m) = 1$, and hence

$$t_{11}(L) = C_\phi(L'L)/\{C_\phi(I_m)\}$$

where $\phi = (f_1, \ldots, f_m)$ is a partition of f. With the help of Lemma A.1,

$$\int_{O(m)} H_2 \mathrm{d}(H) = O,$$

which gives (A.2.2).

A.3 Fundamental Properties of $C_\phi^{\kappa,\lambda}(X,Y)$

Let $\Gamma_\phi^{\kappa,\lambda}(X,Y)$ be an invariant polynomial in $V_{2\phi}^{\kappa,\lambda}[X,Y]$ and let $C_\phi^{\kappa,\lambda}(X,Y)$ be a normalized polynomial defined as

$$C_\phi^{\kappa,\lambda}(X,Y) = \sqrt{z_\phi}\,\Gamma_\phi^{\kappa,\lambda}(X,Y),$$

where $z_\phi = C_\phi(I_m)/\{2^f(\frac{m}{2})_\phi\}$. The following properties hold.

(i) $C_\phi^{\kappa,\lambda}(X,X) = \theta_\phi^{\kappa,\lambda} C_\phi(X)$, $\theta_\phi^{\kappa,\lambda} = C_\phi^{\kappa,\lambda}(I,I)/\{C_\phi(I)\}$.

(ii) $C_\phi^{\kappa,\lambda}(X,I) = \left\{\theta_\phi^{\kappa,\lambda} C_\phi(I)/\{C_\kappa(I)\}\right\} C_\kappa(X)$ with a corresponding result for $C_\phi^{\kappa,\lambda}(I,Y)$.

(iii) $C_\kappa^{\kappa,0}(X,Y) \stackrel{\text{def}}{=} C_\kappa(X)$, $C_\lambda^{0,\lambda}(X,Y) = C_\lambda(Y)$.

Theorem A.1

$$\int_{O(m)} C_\phi^{\kappa,\lambda}(AH'XH,\ AH'YH)\mathrm{d}(H) = C_\phi^{\kappa,\lambda}(X,Y) C_\phi(A)/\{C_\phi(I)\}.$$

PROOF Choose $\Gamma_\phi^{\kappa,\lambda}(X,Y)$ as the first basis vector in $V_{2\phi}^{\kappa,\lambda}[X,Y]$ and the remainder in a subspace invariant under the non-identity representation of $O(m)$. The representation for the simultaneous congruence transformation $X \to L'XL$, and $Y \to L'YL$, $L \in G\ell(m,R)$ produces $(T_{2\phi}(L))\Gamma_\phi^{\kappa,\lambda}(X,Y) = \Gamma_\phi^{\kappa,\lambda}(L'XL, L'YL)$. Let $L = HA^{1/2}$ where

$H \in O(m)$ and $L'L = A$. Then

$$\left(T_{2\phi}(HA^{1/2})\right)\Gamma_\phi^{\kappa,\lambda}(X,Y) = \Gamma_\phi^{\kappa,\lambda}(A^{1/2}H'XHA^{1/2},\ A^{1/2}H'YHA^{1/2}).$$

Let $\tilde{T}_{2\phi}(L)$ denote the representation matrix of the representation $T_{2\phi}(L)$, then

$$\tilde{T}_{2\phi}(HA^{1/2}) = \tilde{T}_{2\phi}(H)\,\tilde{T}_{2\phi}(A^{1/2}).$$

By noting that the vector of basis polynomial transforms according to $(\tilde{T}_{2\phi}(L))'$, and by using Lemma A.2, one has that the first component of the basis vector corresponds to the one obtained by integrating over $O(m)$ as follows:

$$\int_{O(m)} \Gamma_\phi^{\kappa,\lambda}(A^{1/2}H'XHA^{1/2},\ A^{1/2}H'YHA^{1/2})\mathrm{d}(H)$$

$$\longrightarrow \left\{\begin{bmatrix} 1 & 0 & \cdots & 0 \\ 0 & & & \\ \vdots & & 0 & \\ 0 & & & \end{bmatrix}\begin{bmatrix} C_\phi(A^{1/2}A^{1/2})/\{C_\phi(I)\}, \cdots \\ \vdots \end{bmatrix}\right\}'\begin{bmatrix} \Gamma_\phi^{\kappa,\lambda}(X,Y) \\ \vdots \end{bmatrix}$$

$$= \begin{bmatrix} \frac{C_\phi(A)}{C_\phi(I)}\Gamma_\phi^{\kappa,\lambda}(X,Y) \\ \vdots \end{bmatrix}.$$

Thus this correspondence gives

$$\begin{aligned}\int_{O(m)} &\Gamma_\phi^{\kappa,\lambda}(A^{1/2}H'XHA^{1/2},\ A^{1/2}H'YHA^{1/2})\mathrm{d}(H) \\ &= \int_{O(m)} \Gamma_\phi^{\kappa,\lambda}(AH'XH,\ AH'YH)\mathrm{d}(H) \\ &= \frac{C_\phi(A)}{C_\phi(I)}\Gamma_\phi^{\kappa,\lambda}(X,Y).\end{aligned}$$

It should be noted that no result corresponding to the above holds if A is replaced by B in one argument of $\Gamma_\phi^{\kappa,\lambda}$.

Theorem A.2

$$E_W[C_\phi^{\kappa,\lambda}(XW,YW)] = 2^f\left(\frac{n}{2}\right)_\phi C_\phi^{\kappa,\lambda}(X\Sigma,Y\Sigma),$$

where $W \sim W_m(n,\Sigma,O)$.

PROOF Let $W = \Sigma^{1/2}U\Sigma^{1/2}$, then the distribution of U is invariant under the transformation $U \to H'UH$, $H \in O(m)$. Thus we have with the help of Theorem A.1

$$E_U\,[C_\phi^{\kappa,\lambda}(X\Sigma^{1/2}U\Sigma^{1/2},\ Y\Sigma^{1/2}U\Sigma^{1/2})]$$

$$= E_U\left[\int_{O(m)} C_\phi^{\kappa,\lambda}(UH\Sigma^{1/2}X\Sigma^{1/2}H',\, UH\Sigma^{1/2}Y\Sigma^{1/2}H')\mathrm{d}(H)\right]$$

$$= \frac{C_\phi^{\kappa,\lambda}(\Sigma^{1/2}X\Sigma^{1/2},\, \Sigma^{1/2}Y\Sigma^{1/2})}{C_\phi(I)} E_U[C_\phi(U)]$$

$$= 2^f\left(\frac{n}{2}\right)_\phi C_\phi^{\kappa,\lambda}(X\Sigma, Y\Sigma).$$

Theorem A.3

$$\int_{O(m)} C_\kappa(AH'XH)C_\lambda(BH'YH)\mathrm{d}(H) \tag{A.3.1}$$

$$= \sum_{\phi\in\kappa\cdot\lambda} C_\phi^{\kappa,\lambda}(A,B)C_\phi^{\kappa,\lambda}(X,Y)/\{C_\phi(I)\}.$$

PROOF Since $C_\kappa(AX)C_\lambda(BY)$ is invariant under the congruence transformation $X \to L'XL$, $Y \to L'YL$, together with the contragradient transformation $A \to L^{-1}AL^{-1'}$, $B \to L^{-1}BL^{-1'}$, $L \in G\ell(m,R)$, with the help of an argument developed by James (1960, section 4), see also Hannan (1965, section 10.2), the following form holds:

$$\int_{O(m)} C_\kappa(AH'XH)C_\lambda(BH'YH)\mathrm{d}(H)$$

$$= \sum_{\phi\in\kappa\cdot\lambda}\sum_{\phi\equiv\phi'} q_{\phi,\phi'}^{\kappa,\lambda}\Gamma_\phi^{\kappa,\lambda}(A,B)\Gamma_{\phi'}^{\kappa,\lambda}(X,Y),$$

where $\phi \in \kappa\cdot\lambda$ is an abbreviation for $2\phi \in 2\kappa \otimes 2\lambda$. Since $C_\kappa(AX)C_\lambda(BY)$ may span $V_\kappa[X]\otimes V_\lambda[Y]$ as A and B vary over the symmetric matrices, the integral of the left-hand side must span the polynomials in the space invariant under

$$X \longrightarrow H'XH, \quad Y \longrightarrow H'YH, \quad H \in O(m),$$

and hence the bilinear form in the right-hand side must have a non-singular matrix

$$Q_{\kappa,\lambda} = \left(q_{\phi,\phi'}^{\kappa,\lambda}\right) \qquad \left(q_{\phi,\phi'}^{\kappa,\lambda} = 0 \text{ unless } \phi \equiv \phi'\right).$$

Let $U = (u_{ij})$ be an $m \times m$ matrix of independent standard normal variables and let $U = W^{1/2}H$, $W \sim W(m, I_m, O)$, $H \in O(m)$, then

$$g = E_U[\operatorname{etr}(\alpha AU'XU + \beta BU'YU)]$$

$$= \sum_{k=0}^{\infty}\sum_{\ell=0}^{\infty} \frac{\alpha^\kappa\beta^\ell}{k!\,\ell!}E_{k,\ell},$$

where

$$E_{k,\ell} = \sum_{\kappa,\lambda} E_W \left[\int_{O(m)} C_\kappa \left(AH'W^{1/2}XW^{1/2}H \right) \right.$$
$$\left. \times C_\lambda \left(BH'W^{1/2}YW^{1/2}H \right) \mathrm{d}(H) \right]$$
$$= \boldsymbol{\Gamma}_{k,\ell}(X,Y)' Q_{k,\ell} \boldsymbol{\Gamma}_{k,\ell}(A,B).$$

Here $\boldsymbol{\Gamma}_{k,\ell}(X,Y)$ denotes the vector of all components $\Gamma_\phi^{\kappa,\lambda}(X,Y)$ for fixed k and ℓ and

$$Q_{k,\ell} = \mathrm{diag}\,(Q_{\kappa,\lambda}), \quad Q_{\kappa,\lambda} = \left(2^f \left(\frac{m}{2} \right)_\phi q_{\phi,\phi'}^{\kappa,\lambda} \right).$$

On the other hand, writing $G = \alpha A \otimes X + \beta B \otimes Y$,

$$g = |\, I_{m^2} - G\,|^{-1/2} = \sum_{f=0}^{\infty} \frac{1 \cdot 3 \cdots (2f-1)}{2^f f!} C_f(G),$$

whence

$$E_{k,\ell} = \boldsymbol{\Pi}_{k,\ell}(A,B)' \Delta_{k,\ell} \boldsymbol{\Pi}_{k,\ell}(X,Y),$$

where $\Delta_{k,\ell}$ is a diagonal matrix with positive diagonal elements and $\boldsymbol{\Pi}_{k,\ell}(X,Y)$ denotes the vector of all distinct products of traces

$$(\,\mathrm{tr}\,(X^{a_1}Y^{b_1}X^{c_1}\cdots))^{r_1}(\,\mathrm{tr}\,(X^{a_2}Y^{b_2}X^{c_2}\cdots))^{r_2}\cdots \tag{A.3.2}$$

of total degree k and ℓ in the elements of symmetric matrices X and Y, respectively. Thus it follows from the above that

(a) (A.3.2) constitutes an elementary basis for the invariant polynomials. The number of such items should equal the sum of the multiplicities of the irreducible representation $\tilde{T}_{2\phi}$ occurring in $T_{2k} \otimes T_{2\ell}$;

(b) $E_{k,\ell} = \boldsymbol{\sigma}'_{k,\ell}(A,B)\boldsymbol{\sigma}_{k,\ell}(X,Y)$ if and only if the components of $\boldsymbol{\sigma}_{k,\ell}$ (X,Y) are $\Delta_{k,\ell}$-orthonormal linear combinations of (A.3.2) in the sense that

$$\boldsymbol{\sigma}_{k,\ell}(X,Y) = \Delta_{k,\ell}^{1/2} \boldsymbol{\Pi}_{k,\ell}(X,Y);$$

(c) $\Gamma_\phi^{\kappa,\lambda}$ are $\Delta_{k,\ell}$-orthogonal for inequivalent ϕ;

(d) For ϕ' equivalent to ϕ, choose any set of $\Delta_{k,\ell}$-orthonormal polynomial $\Gamma_{\phi'}^{\kappa,\lambda}(X,Y)$ in the unique invariant subspace $U_{2\phi}^{\kappa,\lambda}$.

We have

$$\sum_{\phi \in \kappa \cdot \lambda} \Gamma_{\phi'}^{\kappa,\lambda}(A,B) \Gamma_\phi^{\kappa,\lambda}(X,Y) / \left\{ 2^f \left(\frac{m}{2} \right)_\phi \right\}.$$

It should be noted that the denominators are constant for equivalent ϕ. Renormalizing to

$$C_\phi^{\kappa,\lambda}(X,Y) = \sqrt{z_\phi}\,\Gamma_\phi^{\kappa,\lambda}(X,Y)$$

where $z_\phi = C_\phi(I_m)/\{2^f\left(\frac{m}{2}\right)_\phi\}$ and the sign of the square root is chosen to make the coefficient of $(\operatorname{tr}(X))^k(\operatorname{tr}(Y))^\ell$ in $C_\phi^{\kappa,\lambda}(X,Y)$ positive if it is non-zero, we have

$$\int_{O(m)} C_\kappa(AH'XH)C_\lambda(BH'YH)\mathrm{d}(H)$$
$$= \sum_{\phi\in\kappa\cdot\lambda} C_\phi^{\kappa,\lambda}(A,B)C_\phi^{\kappa,\lambda}(X,Y)/\{C_\phi(I_m)\}\,.$$

Corollary A.1

$$C_\kappa(X)C_\lambda(Y) = \sum_{\phi\in\kappa\cdot\lambda} \theta_\phi^{\kappa,\lambda}C_\phi^{\kappa,\lambda}(X,Y), \tag{A.3.3}$$

$$E_W[C_\kappa(XW)C_\lambda(YW)] = \sum_{\phi\in\kappa\cdot\lambda} \theta_\phi^{\kappa,\lambda}2^f\left(\frac{n}{2}\right)_\phi C_\phi^{\kappa,\lambda}(X\Sigma, Y\Sigma), \tag{A.3.4}$$

where $W \sim W_m(n,\Sigma,O)$.

$$(\operatorname{tr}(X))^k(\operatorname{tr}(Y))^\ell = \sum_{\kappa,\lambda;\phi\in\kappa\cdot\lambda} \theta_\phi^{\kappa,\lambda}C_\phi^{\kappa,\lambda}(X,Y). \tag{A.3.5}$$

PROOF (A.3.3) is obtained by setting $A = B = I_m$ in (A.3.1). (A.3.4) is obtained with the help of Theorem A.2 and (A.3.3).

Remark A.1 (A.3.3) gives, for $X = Y$,

$$C_\kappa(X)C_\lambda(X) = \sum_{\phi\in\kappa\cdot\lambda}\sum_{\phi\equiv\phi'} (\theta_\phi^{\kappa,\lambda})^2 C_\phi(X).$$

This corresponds to (4.3.65) and

$$a_{\kappa,\lambda}^\phi = \sum_{\phi\equiv\phi'} (\theta_\phi^{\kappa,\lambda})^2\,. \tag{A.3.6}$$

This implies that $a_{\kappa,\lambda}^\phi \geq 0$. If $a_{\kappa,\lambda}^\phi > 0$ so that not all $\theta_{\phi'}^{\kappa;\lambda} = 0$ $(\phi' \equiv \phi)$, choose the first $C_\phi^{\kappa,\lambda}$ in $U_{2\phi}^{\kappa,\lambda}$ to be proportional to the component of $(\operatorname{tr}(X))^k(\operatorname{tr}(Y))^\ell$ in this space with $\theta_\phi^{\kappa,\lambda} = +\sqrt{a_{\kappa,\lambda}^\phi}$, the remaining $C_{\phi'}^{\kappa;\lambda}$ having $\theta_{\phi'}^{\kappa;\lambda} = 0$ $(\phi' \equiv \phi)$.

Remark A.2

$$\int_{O(m)} \operatorname{etr}(AH'XH + BH'YH)\mathrm{d}(H) \tag{A.3.7}$$

$$= \sum_{\kappa,\lambda;\phi} C_\phi^{\kappa,\lambda}(A,B) C_\phi^{\kappa,\lambda}(X,Y) \Big/ \{k!\,\ell!\, C_\phi(I)\},$$

where $\sum_{\kappa,\lambda;\phi}$ stands for $\sum_{k=0}^{\infty}\sum_{\ell=0}^{\infty}\sum_{\kappa}\sum_{\lambda}\sum_{\phi\in\kappa\cdot\lambda}$. The right-hand side will be denoted as ${}_0F_0^{(m)}(A,B;X,Y)$.

$$\int_{O(m)} C_\phi^{\kappa,\lambda}(A'H'XHA,B)\mathrm{d}(H) \tag{A.3.8}$$
$$= C_\phi^{\kappa,\lambda}(A'A,B) C_\kappa(X)/\{C_\kappa(I)\}.$$

PROOF Replacing A with AA' and $Y = I$ in (A.3.7), the left-hand side can be expressed as

$$\int_{O(m)} \mathrm{etr}(AA'H'XH + B)\mathrm{d}(H)$$
$$= \sum_{k=0}^{\infty}\sum_{\ell=0}^{\infty} \frac{1}{k!\,\ell!} \sum_{\kappa,\lambda} C_\lambda(B) \int_{O(m)} C_\kappa(AA'H'XH)\mathrm{d}(H)$$
$$= \sum_{k,\ell=0}^{\infty}\sum_{\kappa,\lambda} \frac{1}{k!\,\ell!} C_\lambda(B) \frac{C_\kappa(AA')C_\kappa(X)}{C_\kappa(I)}$$
$$= \sum_{\kappa,\lambda;\phi} \frac{1}{k!\,\ell!} \theta_\phi^{\kappa,\lambda} C_\phi^{\kappa,\lambda}(A'A,B) C_\kappa(X)/\{C_\kappa(I)\}.$$

The left-hand side is also expressed as

$$\sum_{\kappa,\lambda;\phi} \frac{1}{k!\,\ell!} \theta_\phi^{\kappa,\lambda} \int_{O(m)} C_\phi^{\kappa,\lambda}(A'H'XHA,B)\mathrm{d}(H).$$

Then (A.3.8) is obtained by comparing the coefficient of $\theta_\phi^{\kappa,\lambda}$.

Theorem A.4 *Laplace transform*

Let W be a complex symmetric matrix with $\mathcal{R}e(W) > 0$. Then

$$\int_{R>0} \mathrm{etr}\,(-RW)|\,R\,|^{t-\frac{1}{2}(m+1)} C_\phi^{\kappa,\lambda}(ARA',B)\mathrm{d}R \tag{A.3.9}$$
$$= \Gamma_m(t;\phi)|\,W\,|^{-t} C_\phi^{\kappa,\lambda}(AW^{-1}A',B).$$

$$\int_{R>0} \mathrm{etr}\,(-RW)|\,R\,|^{t-\frac{1}{2}(m+1)} C_\phi^{\kappa,\lambda}(AR,BR)\mathrm{d}R \tag{A.3.10}$$
$$= \Gamma_m(t;\phi)|\,W\,|^{-t} C_\phi^{\kappa,\lambda}(AW^{-1},BW^{-1}).$$

$$\int_{R>0} \operatorname{etr}(-RW)|R|^{t-\frac{1}{2}(m+1)} C_\phi^{\kappa,\lambda}(AR^{-1}, BR^{-1}) \mathrm{d}R \tag{A.3.11}$$
$$= (-1)^f \frac{\Gamma_m(t)}{\left(-t+\frac{1}{2}(m+1)\right)_\phi} |W|^{-t} C_\phi^{\kappa,\lambda}(AW, BW).$$

PROOF Let $R = W^{-1/2}UW^{-1/2}$, then the left-hand side of (A.3.9) is given by

$$|W|^{-t} \int_{U>0} \operatorname{etr}(-U)|U|^{t-\frac{1}{2}(m+1)} C_\phi^{\kappa,\lambda}(AW^{-1/2}UW^{-1/2}A', B)\mathrm{d}U$$
$$= |W|^{-t} \int_{U>0} \operatorname{etr}(-U)|U|^{t-\frac{1}{2}(m+1)}$$
$$\times \int_{O(m)} C_\phi^{\kappa,\lambda}(AW^{-1/2}H'UHW^{-1/2}A', B)\mathrm{d}(H)\mathrm{d}U$$
$$= |W|^{-t} \int_{U>0} \operatorname{etr}(-U)|U|^{t-\frac{1}{2}(m+1)}$$
$$\times C_\phi^{\kappa,\lambda}(AW^{-1}A', B)C_\kappa(U)\mathrm{d}U/\{C_\kappa(I)\}$$
$$= \Gamma_m(t;\phi)|W|^{-t} C_\phi^{\kappa,\lambda}(AW^{-1}A', B).$$

(A.3.10) is obtained with the help of Theorem A.2. (A.3.11) is similarly obtained as (A.3.10).

Theorem A.5 *Binomial expansion*

$$C_\phi(X+Y) = \sum_{\kappa,\lambda(\phi\in\kappa\cdot\lambda)} \sum_{\phi'\equiv\phi} \binom{f}{k} \theta_{\phi'}^{\kappa,\lambda} C_{\phi'}^{\kappa,\lambda}(X,Y). \tag{A.3.12}$$

Let R and S be $r\times r$ and $s\times s$ symmetric matrices, $r+s=m$, then

$$C_\phi^{\kappa,\lambda}\left(\begin{bmatrix} R & O \\ O & O \end{bmatrix}, \begin{bmatrix} O & O \\ O & S \end{bmatrix}\right) = \theta_\phi^{\kappa,\lambda} \frac{z_\phi}{z_\kappa z_\lambda} C_\kappa(R)C_\lambda(S), \tag{A.3.13}$$

$$C_\phi\left(\begin{bmatrix} R & O \\ O & S \end{bmatrix}\right) = z_\phi \sum_{\kappa,\lambda(\phi\in\kappa\cdot\lambda)} \binom{f}{k} a_{\kappa,\lambda}^\phi \frac{C_\kappa(R)C_\lambda(S)}{z_\kappa z_\lambda}, \tag{A.3.14}$$

where $z_\phi = C_\phi(I_m)/\{2^f \left(\frac{m}{2}\right)_\phi\}$. $a_{\kappa,\lambda}^\phi$ and $b_{\kappa,\lambda}^\phi$ are defined as in (4.3.65) and (4.3.66), respectively.

$$b_{\kappa,\lambda}^\phi = \binom{f}{k} a_{\kappa,\lambda}^\phi z_\phi/\{z_\kappa z_\lambda\}. \tag{A.3.15}$$

PROOF Comparing the term for the partition ϕ of f on both sides of the expansion of $\operatorname{etr}(X+Y)$ gives (A.3.12).

Let $J = \operatorname{diag}(J_1, J_2)$, where J_1 and J_2 are orthogonal matrices of order r and s, respectively, and $m = r + s$. By the invariance of $C_\phi^{\kappa,\lambda}$,

$$C_\phi^{\kappa,\lambda}\left(\begin{bmatrix} R & O \\ O & O \end{bmatrix}, \begin{bmatrix} O & O \\ O & S \end{bmatrix}\right) = \int_{O(r)}\int_{O(s)} C_\phi^{\kappa,\lambda}\left(\begin{bmatrix} J_1'RJ_1 & O \\ O & O \end{bmatrix}, \begin{bmatrix} O & O \\ O & J_2'SJ_2 \end{bmatrix}\right) d(J_1)d(J_2).$$

Let the orthogonal matrix H be partitioned as $H = [H_1, H_2]$, where H_1 is an $m \times r$ matrix and H_2 is an $m \times s$ matrix, respectively. Then,

$$\begin{aligned}
&\int_{O(r)}\int_{O(s)}\int_{O(m)} \operatorname{etr}\left[J'\begin{bmatrix} R & O \\ O & S \end{bmatrix} J\left\{\begin{bmatrix} I_r & O \\ O & O \end{bmatrix} H'XH + \begin{bmatrix} O & O \\ O & I_s \end{bmatrix} H'YH\right\}\right] d(J_1)d(J_2)d(H) \\
&= \int_{O(r)}\int_{O(s)}\int_{O(m)} \operatorname{etr}(J_1'RJ_1H_1'XH_1 + J_2'SJ_2H_2'YH_2)d(J_1)d(J_2)d(H) \\
&= \sum_{k=0}^{\infty}\sum_{\ell=0}^{\infty}\sum_{\kappa}\sum_{\lambda} \frac{C_\kappa(R)C_\lambda(S)}{k!\ell!C_\kappa(I_r)C_\lambda(I_s)} \times \int_{O(m)} C_\kappa(H_1'XH_1)C_\lambda(H_2'YH)d(H) \\
&= \sum_{k=0}^{\infty}\sum_{\ell=0}^{\infty}\sum_{\kappa}\sum_{\lambda} \frac{C_\kappa(R)C_\lambda(S)}{k!\ell!C_\kappa(I)C_\lambda(I)} \times \int_{O(m)} C_\kappa\left(\begin{bmatrix} I_r & O \\ O & O \end{bmatrix} H'XH\right) C_\lambda\left(\begin{bmatrix} O & O \\ O & I_s \end{bmatrix} H'YH\right) d(H) \\
&= \sum_{\kappa,\lambda;\phi} \frac{C_\kappa(R)C_\lambda(S)}{k!\ell!C_\kappa(I)C_\lambda(I)} C_\phi^{\kappa,\lambda}\left(\begin{bmatrix} I_r & O \\ O & O \end{bmatrix}, \begin{bmatrix} O & O \\ O & I_s \end{bmatrix}\right) \frac{C_\phi^{\kappa,\lambda}(X,Y)}{C_\phi(I)}.
\end{aligned}$$

On the other hand this integral can also be expressed as

$$\begin{aligned}
&\sum_{\kappa,\lambda;\phi} \frac{C_\phi^{\kappa,\lambda}(X,Y)}{k!\ell!C_\phi(I)} \int_{O(r)}\int_{O(s)} C_\phi^{\kappa,\lambda}\left(J'\begin{bmatrix} R & O \\ O & S \end{bmatrix} J \begin{bmatrix} I_r & O \\ O & O \end{bmatrix}, J'\begin{bmatrix} R & O \\ O & S \end{bmatrix} J \begin{bmatrix} O & O \\ O & I_s \end{bmatrix}\right) d(J_1)d(J_2) \\
&= \sum_{\kappa,\lambda;\phi} \frac{C_\phi^{\kappa,\lambda}(X,Y)}{k!\ell!C_\phi(I)} C_\phi^{\kappa,\lambda}\left(\begin{bmatrix} R & O \\ O & O \end{bmatrix}, \begin{bmatrix} O & O \\ O & S \end{bmatrix}\right).
\end{aligned}$$

Comparing the coefficient of $C_\phi^{\kappa,\lambda}(X,Y)/\{k!\ell!C_\phi(I_m)\}$, we have

$$C_\phi^{\kappa,\lambda}\left(\begin{bmatrix} R & O \\ O & O \end{bmatrix}, \begin{bmatrix} O & O \\ O & S \end{bmatrix}\right)$$
$$= C_\phi^{\kappa,\lambda}\left(\begin{bmatrix} I_r & O \\ O & O \end{bmatrix}, \begin{bmatrix} O & O \\ O & I_s \end{bmatrix}\right) \frac{C_\kappa(R)C_\lambda(S)}{C_\kappa(I_r)C_\lambda(I_s)}.$$

This shows that

$$C_\phi^{\kappa,\lambda}\left(\begin{bmatrix} R & O \\ O & O \end{bmatrix}, \begin{bmatrix} O & O \\ O & S \end{bmatrix}\right) = \psi_\phi^{\kappa,\lambda} C_\kappa(R)C_\lambda(S),$$

where $\psi_\phi^{\kappa,\lambda}$ is a constant. With the help of (4.3.40) $C_\kappa(X)$ is expressed as $C_\kappa(X) = z_\kappa Z_\kappa(X)$, where the coefficient of $(\operatorname{tr}(X))^k$ in $Z_\kappa(X)$ is one. Thus the coefficient of $(\operatorname{tr}(R))^k(\operatorname{tr}(S))^\ell$ in $C_\kappa(R)C_\lambda(S)$ is $z_\kappa z_\lambda$. In $C_\phi^{\kappa,\lambda}(X,Y)$ the coefficient of $(\operatorname{tr}(X))^k(\operatorname{tr}(Y))^\ell$ equals to the coefficient of $(\operatorname{tr}(X))^f$, $f = k+\ell$, in $C_\phi^{\kappa,\lambda}(X,X) = \theta_\phi^{\kappa,\lambda}C_\phi(X)$, that is, $\theta_\phi^{\kappa,\lambda}z_\phi$. Hence

$$\psi_\phi^{\kappa,\lambda} = \theta_\phi^{\kappa,\lambda}\frac{z_\phi}{z_\kappa z_\lambda},$$

which establishes (A.3.13). (A.3.14) is obtained from (A.3.12) and (A.3.13).

Remark A.3 Table 4.3.2(a) and Table 4.3.2(b) are obtained by making use of Table A.1, (A.3.6) and (A.3.15), respectively. Table A.1 is given at the end of this Appendix.

Definition A.1 *Binomial Coefficient* Let X and Y be $m\times m$ symmetric matrices. The binomial coefficients $\begin{pmatrix} \rho,\sigma:\tau \\ \kappa,\lambda:\phi \end{pmatrix}$ and $\left(\begin{matrix} \rho:\tau \\ \kappa:\phi \end{matrix}\middle|\lambda\right)$ are defined as follows:

$$C_\tau^{\rho,\sigma}(I+X, I+Y)/\{C_\tau(I)\} \tag{A.3.16}$$
$$= \sum_{k=0}^{r}\sum_{\ell=0}^{s}\sum_{\kappa}\sum_{\lambda}\sum_{\phi\in\kappa\cdot\lambda}\begin{pmatrix} \rho,\sigma:\tau \\ \kappa,\lambda:\phi \end{pmatrix} C_\phi^{\kappa,\lambda}(X,Y)/\{C_\phi(I)\},$$

$$C_\tau^{\rho,\lambda}(I+X, Y)/\{C_\tau(I)\} \tag{A.3.17}$$
$$= \sum_{k=0}^{r}\sum_{\kappa,\phi\in\kappa\cdot\lambda}\left(\begin{matrix} \rho:\tau \\ \kappa:\phi \end{matrix}\middle|\lambda\right) C_\phi^{\kappa,\lambda}(X,Y)/\{C_\phi(I)\},$$

where ρ,σ and τ are partitions of r,s and $r+s$, and κ,λ and ϕ are partitions of k,ℓ and $k+\ell$, respectively.

The following equalities hold.

$$2^{r+s}\,\theta_\tau^{\rho,\sigma} = \sum_{k=0}^{r}\sum_{\ell=0}^{s}\sum_{\kappa}\sum_{\lambda}\sum_{\phi\in\kappa\cdot\lambda}\begin{pmatrix}\rho,\sigma:\tau\\ \kappa,\lambda:\phi\end{pmatrix}\theta_\phi^{\kappa,\lambda}\,.$$

$$2^{r}\theta_\tau^{\rho,\lambda} = \sum_{k=0}^{r}\sum_{\kappa,\phi\in\kappa\cdot\lambda}\left(\begin{matrix}\rho:\tau\\ \kappa:\phi\end{matrix}\middle|\lambda\right)\theta_\phi^{\kappa,\lambda}\,.$$

Theorem A.6 Let ρ,σ and τ be partitions of r,s and $r+s$ and let κ,λ and ϕ be the ones of k,ℓ and $k+\ell$, respectively. Then

$$\sum_{r=k}^{\infty}\sum_{s=\ell}^{\infty}\sum_{\rho}\sum_{\sigma}\sum_{\tau\in\rho\cdot\sigma}\begin{pmatrix}\rho,\sigma:\tau\\ \kappa,\lambda:\phi\end{pmatrix}\frac{C_\tau^{\rho,\sigma}(X,Y)}{r!s!} = C_\phi^{\kappa,\lambda}(X,Y)\,\mathrm{etr}\,(X+Y)/\{k!\,\ell!\}\,, \tag{A.3.18}$$

$$\sum_{r=k}^{\infty}\sum_{s=\ell}^{\infty}\sum_{\rho}\sum_{\sigma}\sum_{\tau\in\rho\cdot\sigma}\begin{pmatrix}\rho,\sigma:\tau\\ \kappa,\lambda:\phi\end{pmatrix}(a)_\tau C_\tau^{\rho,\sigma}(X,Y)/\{r!s!\} = |\,I-Z\,|^{-a}(a)_\phi C_\phi^{\kappa,\lambda}(X(I-Z)^{-1},Y(I-Z)^{-1})/\{k!\,\ell!\}\,, \tag{A.3.19}$$

where $Z = X+Y$.

Proof Noting that

$$\begin{aligned}&\mathrm{tr}\,(AHXH') + \mathrm{tr}\,(BHYH')\\ &\quad = \mathrm{tr}\,((aI+A)H(xI+X)H') + \mathrm{tr}\,((bI+B)H(yI+Y)H')\\ &\quad + \mathrm{tr}\,(AX) + \mathrm{tr}\,(BY) - \mathrm{tr}\,((aI+A)(xI+X)) - \mathrm{tr}\,((bI+B)(yI+Y))\end{aligned}$$

we have

$$\begin{aligned}&\mathrm{etr}\,(-AX-BY)\,{}_0F_0^{(m)}(A,B;X,Y)\\ &\quad = \mathrm{etr}\,\{-(aI+A)(xI+X)-(bI+B)(yI+Y)\}\\ &\qquad \times {}_0F_0^{(m)}(aI+A,bI+B;\,xI+X,yI+Y)\,.\end{aligned}$$

Setting $x=y=0$ and $a=b=1$, we have

$$\begin{aligned}&\mathrm{etr}\,(X+Y)\sum_{\kappa,\lambda;\phi} C_\phi^{\kappa,\lambda}(A,B)C_\phi^{\kappa,\lambda}(X,Y)/\{k!\,\ell!\}\,C_\phi(I)\\ &\quad = \sum_{\rho,\sigma;\tau} C_\tau^{\rho,\sigma}(X,Y)C_\tau^{\rho,\sigma}(I+A,I+B)/\{r!\,\ell!\,C_\tau(I)\}.\end{aligned}$$

Thus

$$\sum_{\kappa,\lambda;\phi}\left\{\mathrm{etr}\,(X+Y)C_\phi^{\kappa,\lambda}(X,Y)/\{k!\,\ell!\}\right\}C_\phi^{\kappa,\lambda}(A,B)/\{C_\phi(I)\}$$

$$= \sum_{\rho,\sigma;\tau} C_\tau^{\rho,\sigma}(X,Y)/\{r!s!\}$$
$$\times \left\{ \sum_{k=0}^{r} \sum_{\ell=0}^{s} \sum_{\kappa} \sum_{\lambda} \sum_{\phi\in\kappa\cdot\lambda} \binom{\rho,\sigma:\tau}{\kappa,\lambda:\phi} C_\phi^{\kappa,\lambda}(A,B)/\{C_\phi(I)\} \right\}$$
$$= \sum_{\kappa,\lambda;\phi} \left\{ \sum_{r=k}^{\infty} \sum_{s=\ell}^{\infty} \sum_{\rho} \sum_{\sigma} \sum_{\tau\in\rho\cdot\sigma} \binom{\rho,\sigma:\tau}{\kappa,\lambda:\phi} C_\tau^{\rho,\sigma}(X,Y)/\{r!s!\} \right\}$$
$$\times C_\phi^{\kappa,\lambda}(A,B)/\{C_\phi(I)\}.$$

On comparing the coefficient of $C_\phi^{\kappa,\lambda}(A,B)/\{C_\phi(I)\}$, we obtain (A.3.18). Replace X and Y by XS and YS in (A.3.18), multiply by $\operatorname{etr}(-S) \times |S|^{a-\frac{1}{2}(m+1)}$ on both sides, and then integrate over $S > 0$ to obtain (A.3.19).

Theorem A.7 *Incomplete gamma function*

$$\int_{0<S<X} \operatorname{etr}(-AS)|S|^{t-\frac{1}{2}(m+1)} C_\lambda(BS)\mathrm{d}S \tag{A.3.20}$$
$$= \frac{\Gamma_m(t)\Gamma_m\left(\frac{m+1}{2}\right)}{\Gamma_m\left(t+\frac{1}{2}(m+1)\right)} |X|^t \sum_{k=0}^{\infty} \sum_{\kappa;\phi\in\kappa\cdot\lambda} (t)_\phi \theta_\phi^{\kappa,\lambda}$$
$$\times \frac{C_\phi^{\kappa,\lambda}(-AX, BX)}{k!\left(t+\frac{1}{2}(m+1)\right)_\phi}.$$

Incomplete beta function

$$\int_{0<S<X} |S|^{t-\frac{1}{2}(m+1)} |I-S|^{u-\frac{1}{2}(m+1)} C_\lambda(AS)\mathrm{d}S \tag{A.3.21}$$
$$= \frac{\Gamma_m(t)\Gamma_m\left(\frac{m+1}{2}\right)}{\Gamma_m\left(t+\frac{1}{2}(m+1)\right)} |X|^t$$
$$\times \sum_{k=0}^{\infty} \sum_{\kappa;\phi\in\kappa\cdot\lambda} \theta_\phi^{\kappa,\lambda} \frac{(t)_\phi \left(-u+\frac{1}{2}(m+1)\right)_\kappa}{\left(t+\frac{1}{2}(m+1)\right)_\phi} \frac{C_\phi^{\kappa,\lambda}(X, AX)}{k!}.$$

PROOF Let $S = X^{1/2}UX^{1/2}$, then the left-hand side of (A.3.14) is obtained by using Corollary 4.4.1.

$$|X|^t \sum_{k=0}^{\infty} \sum_{\kappa} \frac{1}{k!} \int_{0<U<I} |U|^{t-\frac{1}{2}(m+1)}$$
$$\times C_\kappa(-AX^{1/2}UX^{1/2}) C_\lambda(BX^{1/2}UX^{1/2})\mathrm{d}U$$
$$= |X|^t \sum_{k=0}^{\infty} \sum_{\kappa} \frac{1}{k!} \int_{0<U<I} |U|^{t-\frac{1}{2}(m+1)}$$

$$\times \int_{O(m)} C_\kappa(-X^{1/2}AX^{1/2}H'UH)$$
$$\times C_\lambda(X^{1/2}BX^{1/2}H'UH)\mathrm{d}(H)\mathrm{d}U$$
$$= |X|^t \sum_{k=0}^{\infty} \sum_{\kappa;\phi\in\kappa\cdot\lambda} \frac{1}{k!}\theta_\phi^{\kappa,\lambda} C_\phi^{\kappa,\lambda}(-X^{1/2}AX^{1/2}, X^{1/2}BX^{1/2})/\{C_\phi(I)\}$$
$$\times \int_{0<U<I} |U|^{t-\frac{1}{2}(m+1)} C_\phi(U)\mathrm{d}U$$
$$= |X|^t \sum_{k=0}^{\infty} \sum_{\kappa;\phi\in\kappa\cdot\lambda} \frac{1}{k!} \frac{\Gamma_m(t;\phi)\Gamma_m\left(\frac{m+1}{2}\right)}{\Gamma_m\left(t+\frac{1}{2}(m+1);\phi\right)} \theta_\phi^{\kappa,\lambda} C_\phi^{\kappa,\lambda}(-AX, BX).$$

(A.3.21) is obtained in a similar way.

Corollary A.2

$$\int_{0<S<I} |S|^{t-\frac{1}{2}(m+1)} |I-S|^{u-\frac{1}{2}(m+1)} C_\phi^{\kappa,\lambda}(S, I-S)\mathrm{d}S$$
$$= \frac{\Gamma_m(t;\kappa)\Gamma_m(u;\lambda)}{\Gamma_m(t+u;\phi)} \theta_\phi^{\kappa,\lambda} C_\phi(I). \tag{A.3.22}$$

PROOF Applying (A.3.9) to each argument, we have

$$\Gamma_m(t;\kappa)\Gamma_m(u;\lambda) C_\phi^{\kappa,\lambda}(I,I)$$
$$= \int_{X>0}\int_{Y>0} \operatorname{etr}(-X-Y) |X|^{t-\frac{1}{2}(m+1)} |Y|^{u-\frac{1}{2}(m+1)}$$
$$\times C_\phi^{\kappa,\lambda}(X,Y)\mathrm{d}X\mathrm{d}Y.$$

Letting $R = X + Y$, the right-hand side is

$$\int_{R>0}\int_{0<X<R} \operatorname{etr}(-R) |X|^{t-\frac{1}{2}(m+1)} |R-X|^{u-\frac{1}{2}(m+1)}$$
$$\times C_\phi^{\kappa,\lambda}(X, R-X)\mathrm{d}R\mathrm{d}X.$$

Taking $X = R^{1/2}SR^{1/2}$ we have

$$\int_{0<S<I} |S|^{t-\frac{1}{2}(m+1)} |I-S|^{u-\frac{1}{2}(m+1)}\mathrm{d}S$$
$$\times \int_{R>0} \operatorname{etr}(-R) |R|^{t+u-\frac{1}{2}(m+1)} C_\phi^{\kappa,\lambda}(RS, R(I-S))\,\mathrm{d}R$$
$$= \Gamma_m(t+u;\phi) \int_{0<S<I} |S|^{t-\frac{1}{2}(m+1)}$$
$$\times |I-S|^{u-\frac{1}{2}(m+1)} C_\phi^{\kappa,\lambda}(S, I-S)\mathrm{d}S,$$

which completes the proof.

Theorem A.8

$$P_\phi(T,A) = \left(\frac{m}{2}\right)_\phi \sum_{\kappa,\lambda(\phi\in\kappa\cdot\lambda)} \sum_{\phi\equiv\phi'} \binom{f}{k} \theta_\phi^{\kappa,\lambda} \frac{C_\phi^{\kappa,\lambda}(-A, T'TA)}{\left(\frac{m}{2}\right)_\lambda}, \tag{A.3.23}$$

$$L_\phi^t(X,A) = \left(t+\frac{1}{2}(m+1)\right)_\phi \sum_{\kappa,\lambda} \sum_{\phi\in\kappa\cdot\lambda} \sum_{\phi\equiv\phi'} \binom{f}{k} \frac{\theta_\phi^{\kappa,\lambda} C_{\phi'}^{\kappa,\lambda}(A,-XA)}{\left(t+\frac{1}{2}(m+1)\right)_\lambda}. \tag{A.3.24}$$

PROOF The generating function of $P_\phi(T,A)$ is given as

$$\int_{O(m)}\int_{O(n)} \operatorname{etr}\,(-UH_2AH_2'U' + 2H_1UH_2A^{1/2}T')\mathrm{d}(H_1)\mathrm{d}(H_2)$$
$$= \sum_{f=0}^{\infty}\sum_{\phi} \frac{P_\phi(T,A)C_\phi(U'U)}{f!\left(\frac{n}{2}\right)_\phi C_\phi(I_m)}.$$

The left-hand side can also be expressed as

$$\int_{O(n)} \operatorname{etr}(-UH_2AH_2'U')\,{}_0F_1\left(\frac{m}{2}; U'UH_2A^{1/2}T'TA^{1/2}H_2'\right)\mathrm{d}(H_2)$$
$$= \sum_{k,\ell}\sum_{\kappa,\lambda} \frac{1}{k!\,\ell!\left(\frac{m}{2}\right)_\lambda} \int_{O(n)} C_\kappa(-U'UH_2AH_2')$$
$$\times C_\lambda(U'UH_2A^{1/2}T'TA^{1/2}H_2')\mathrm{d}(H_2)$$
$$= \sum_{\kappa,\lambda;\phi}\sum_{\phi\equiv\phi'} \frac{1}{k!\,\ell!\left(\frac{m}{2}\right)_\lambda} \theta_\phi^{\kappa,\lambda} \frac{C_\phi^{\kappa,\lambda}(-A,T'TA)}{C_\phi(I_n)} C_\phi(U'U).$$

Noting that $\left(\frac{m}{2}\right)_\phi C_\phi(I_n) = \left(\frac{n}{2}\right)_\phi C_\phi(I_m)$ and comparing the coefficient of $C_\phi(U'U)$, we have (A.3.23). The Laplace transform of $L_\phi^t(X,A)|\,X\,|^t$ is given by

$$\Gamma_m\left(t+\frac{1}{2}(m+1);\phi\right)|\,Z\,|^{-t-\frac{1}{2}(m+1)} C_\phi\left((I-Z^{-1})A\right)$$
$$= \Gamma_m\left(t+\frac{1}{2}(m+1);\phi\right)|\,Z\,|^{-t-\frac{1}{2}(m+1)}$$
$$\times \sum_{\kappa,\lambda(\phi\in\kappa\cdot\lambda)}\sum_{\phi'\equiv\phi} \binom{f}{k} \theta_\phi^{\kappa,\lambda} C_\phi^{\kappa,\lambda}(A,-AZ^{-1}).$$

Thus the inverse Laplace transform of $\Gamma_m(t+\frac{1}{2}(m+1);\lambda)|\,Z\,|^{-t-\frac{1}{2}(m+1)}C_\phi^{\kappa,\lambda}(A,-AZ^{-1})$ is $|\,X\,|^t C_\phi^{\kappa,\lambda}(A,-AX)$.

Corollary A.3

$$\begin{aligned} &| I + Z |^{-a} \; {}_1F_1(a; u; Y(I + Z)^{-1}) \qquad (A.3.25) \\ &\quad = \sum_{\kappa,\lambda;\phi} \frac{(a)_\phi \theta_\phi^{\kappa,\lambda}}{k!\,\ell!\,(u)_\lambda} C_\phi^{\kappa,\lambda}(-Z, Y) \\ &\quad = \sum_{f=0}^{\infty} \sum_{\phi} \frac{(-1)^f (a)_\phi}{f!(u)_\phi} L_\phi^{u-\frac{1}{2}(m+1)}(YZ^{-1}, Z). \end{aligned}$$

Proof Expanding both sides of

$$| I - (X + Y) |^{-a} = | I + Z |^{-a} | I - (X + Y + Z)(I + Z)^{-1} |^{-a}$$

one gets

$$\begin{aligned} &\sum_{\kappa,\lambda;\phi} \frac{(a)_\phi \theta_\phi^{\kappa,\lambda}}{k!\,\ell!} C_\phi^{\kappa,\lambda}(X, Y) \\ &\quad = | I + Z |^{-a} \sum_{\kappa,\lambda;\phi} \frac{(a)_\phi \theta_\phi^{\kappa,\lambda}}{k!\,\ell!} C_\phi^{\kappa,\lambda}\left((X + Z)(I + Z)^{-1}, Y(I + Z)^{-1}\right). \end{aligned}$$

Replacing Y by Y^{-1}, multiplying by $| Y |^{-u}$ and inverting the Laplace transform yield

$$\begin{aligned} &\sum_{\kappa,\lambda;\phi} \frac{(a)_\phi \theta_\phi^{\kappa,\lambda}}{k!\,\ell!\,(u)_\lambda} C_\phi^{\kappa,\lambda}(X, Y) \\ &\quad = | I + Z |^{-a} \sum_{\kappa,\lambda;\phi} \frac{(a)_\phi \theta_\phi^{\kappa,\lambda}}{k!\,\ell!\,(u)_\lambda} C_\phi^{\kappa,\lambda}\left((X + Z)(I + Z)^{-1}, Y(I + Z)^{-1}\right). \end{aligned}$$

Taking $X = -Z$ and using (A.3.24), we have

$$\begin{aligned} &| I + Z |^{-a} \, {}_1F_1\left(a; u; Y(I + Z)^{-1}\right) \\ &\quad = \sum_{\kappa,\lambda;\phi} \frac{(a)_\phi \theta_\phi^{\kappa,\lambda}}{k!\,\ell!\,(u)_\lambda} C_\phi^{\kappa,\lambda}(-Z, Y) \\ &\quad = \sum_{f=0}^{\infty} \frac{(-1)^f}{f!} \sum_{\phi} (a)_\phi \left\{ (u)_\phi \sum_{\kappa,\lambda(\phi \in \kappa\cdot\lambda)} \binom{f}{k} \frac{\theta_\phi^{\kappa,\lambda} C_\phi^{\kappa,\lambda}(-Z, Y)}{(u)_\lambda} \right\} \frac{1}{(u)_\phi} \\ &\quad = \sum_{f=0}^{\infty} \frac{(-1)^f}{f!} \sum_{\phi} \frac{(a)_\phi}{(u)_\phi} L_\phi^{u-\frac{1}{2}(m+1)}(YZ^{-1}, Z). \end{aligned}$$

Corollary A.4 Let $\kappa = (k_1, \ldots, k_m)$ and $\lambda = (\ell_1, \ldots, \ell_m)$ be partitions of k and ℓ

and let $\phi = (f_1, \dots, f_m)$ be a partition of $f = k + \ell$, respectively then

$$C_\lambda((I-Z)^{-1}) = \sum_{k=0}^{\infty} \sum_{\kappa,\phi\in\kappa\cdot\lambda} \psi_\lambda^\phi \left(\theta_\phi^{\kappa,\lambda}\right)^2 \frac{C_\phi(I)}{C_\kappa(I)} \frac{C_\kappa(Z)}{k!}, \tag{A.3.26}$$

where

$$\psi_\lambda^\phi = \lim_{a\to 0} \frac{(a)_\phi}{(a)_\lambda} = \frac{(f_1-1)!}{(\ell_1-1)!} \prod_{i=1}^{m-1} \frac{\left(-\frac{i}{2}\right)_{f_{i+1}}}{\left(-\frac{i}{2}\right)_{\ell_{i+1}}}, \tag{A.3.27}$$

which is zero for $f_3 \geq 2$ and, undefined when $\ell_3 \geq 2$.

PROOF Let $a = u$, and replace Z by $-Z^{-1}$ in (A.3.25), then

$$|I-Z|^{-a} \operatorname{etr}(Y(I-Z)^{-1}) = \sum_{\kappa,\lambda;\phi} \frac{(a)_\phi \theta_\phi^{\kappa,\lambda}}{k!\,\ell!\,(a)_\lambda} C_\phi^{\kappa,\lambda}(Z,Y).$$

By setting $Y = A^{1/2}HXH'A^{1/2}$, $H \in O(m)$ and integrating over $O(m)$, we have

$$|I-Z|^{-a} \sum_{\ell=0}^{\infty}\sum_{\lambda} \frac{C_\lambda(A(I-Z)^{-1})}{\ell!\,C_\lambda(I_m)} C_\lambda(X) = \sum_{\kappa,\lambda;\phi} \frac{(a)_\phi \theta_\phi^{\kappa,\lambda}}{k!\,\ell!\,(a)_\lambda} \frac{C_\phi^{\kappa,\lambda}(Z,A)}{C_\lambda(I_m)} C_\lambda(X).$$

Comparing the coefficient of $C_\lambda(X)/\{C_\lambda(I)\}$, we have

$$|I-Z|^{-a} C_\lambda(A(I-Z)^{-1}) = \sum_{k=0}^{\infty} \sum_{\kappa,\phi\in\kappa\cdot\lambda} \frac{\theta_\phi^{\kappa,\lambda}}{k!} \frac{(a)_\phi}{(a)_\lambda} C_\phi^{\kappa,\lambda}(Z,A).$$

Setting $A = I_m$ and letting a equal zero, we have (A.3.26).

As mentioned earlier, the results in this Appendix can be extended to the case of r matrix arguments. Additional material on invariant polynomials may be found in Chikuse (1980, 1981, 1986, 1987, 1992a, 1992b), Chikuse and Davis (1986a, 1986b), Davis (1976, 1979, 1980a, 1980b, 1981, 1982a, 1982b), Hayakawa (1981, 1982), Hillier (1985) and Hillier, Kinal and Srivastava (1984).

Table A.1 *Invariant polynomials*

The invariant polynomials $C_{\phi}^{\kappa,\lambda}(X,Y)$ defined in Davis (1979, 1980a) are given below with some modifications. Table A.1 reads as

$$C_{31}^{21,1}(X,Y) = \frac{2\sqrt{10}}{15\sqrt{7}}\{(X)^3(Y) + 4(XY)(X)^2 + (X^2)(X)(Y) - 2(X^2)(XY) + 6(X^2Y)(X) - 2(X^3)(Y) - 8(X^3Y)\}.$$

$k = 1, \ell = 1$

$\kappa \ \lambda$ / ϕ	1 1 / 2	1 1 / 1^2
$c_{\phi}^{\kappa,\lambda}$	1/3	2/3
$\theta_{\phi}^{\kappa,\lambda}$	1	1
$(X)(Y)$	1	1
(XY)	2	−1

$k = 2, \ell = 1$

$\kappa \ \lambda$ / ϕ	2 1 / 3	2 1 / 21	1^2 1 / 21	1^2 1 / 1^3
$c_{\phi}^{\kappa\lambda}$	1/15	2/5	$1/\sqrt{5}$	1/3
$\theta_{\rho}^{\kappa\lambda}$	1	2/3	$\sqrt{5}/3$	1
$(X)^2(Y)$	1	1	1	1
$(XY)(X)$	4	−1	2	−2
$(X^2)(Y)$	2	2	−1	−1
(X^2Y)	8	−2	−2	2

$k=3,\ \ell=1$

$\kappa\ \ \lambda$ ϕ	$3\ \ 1$ 4	$3\ \ 1$ 31	$21\ \ 1$ 31	$21\ \ 1$ 2^2	$21\ \ 1$ 21^2	$1^3\ \ 1$ 21^2	$1^3\ \ 1$ 1^4
$c_\phi^{\kappa,\lambda}$	$\frac{1}{105}$	$\frac{2\sqrt{10}}{35\sqrt{3}}$	$\frac{2\sqrt{10}}{15\sqrt{7}}$	$\frac{2}{15}$	$\frac{2\sqrt{2}}{3\sqrt{5}}$	$\frac{2\sqrt{2}}{5\sqrt{3}}$	$\frac{2}{15}$
$\theta_\phi^{\kappa,\lambda}$	1	$\sqrt{\frac{3}{10}}$	$\sqrt{\frac{7}{10}}$	1	$\frac{\sqrt{10}}{4}$	$\frac{\sqrt{6}}{4}$	1
$(X)^3(Y)$	1	1	1	1	1	1	1
$(XY)(X)^2$	6	−1	4	1	−2	2	−3
$(X^2)(X)(Y)$	6	6	1	1	1	−3	−3
$(X^2)(XY)$	12	−2	−2	7	−2	−2	3
$(X^2Y)(X)$	24	−4	6	−6	0	−4	6
$(X^3)(Y)$	8	8	−2	−2	−2	2	2
(X^3Y)	48	−8	−8	−2	4	4	−6

$k = 2, \ell = 2$

$\kappa \;\; \lambda$ ϕ	$2 \;\; 2$ 4	$2 \;\; 2$ 31	$2 \;\; 2$ 2^2	$2 \;\; 1^2$ 31	$2 \;\; 1^2$ 21^2	$1^2 \;\; 2$ 31	$1^2 \;\; 2$ 21^2
$c_\phi^{\kappa,\lambda}$	$\frac{1}{105}$	$\frac{4\sqrt{2}}{63}$	$\frac{4}{45}$	$\frac{2\sqrt{2}}{9\sqrt{7}}$	$\frac{4\sqrt{2}}{9\sqrt{5}}$	$\frac{2\sqrt{2}}{9\sqrt{7}}$	$\frac{4\sqrt{2}}{9\sqrt{5}}$
$\theta_\phi^{\kappa,\lambda}$	1	$\frac{\sqrt{2}}{3}$	$\frac{2}{3}$	$\frac{\sqrt{7}}{3\sqrt{2}}$	$\frac{\sqrt{5}}{3\sqrt{2}}$	$\frac{\sqrt{7}}{3\sqrt{2}}$	$\frac{\sqrt{5}}{3\sqrt{2}}$
$(X)^2(Y)^2$	1	1	1	1	1	1	1
$(X)^2(Y^2)$	2	2	2	−1	−1	2	2
$(XY)(X)(Y)$	8	1	−2	4	−2	4	−2
$(X^2)(Y)^2$	2	2	2	2	2	−1	−1
$(XY)^2$	8	−6	3	0	0	0	0
$(X^2)(Y^2)$	4	4	4	−2	−2	−2	−2
$(XY^2)(X)$	16	2	−4	−4	2	8	−4
$(X^2Y)(Y)$	16	2	−4	8	−4	−4	2
$(XYXY)$	16	−12	6	0	0	0	0
(X^2Y^2)	32	4	−8	−8	4	−8	4

$1^2 \quad 1^2$ 2^2	$1^2 \quad 1^2$ 21^2	$1^2 \quad 1^2$ 1^4
$\frac{2}{9\sqrt{5}}$	$\frac{16}{45}$	$\frac{2}{15}$
$\frac{\sqrt{5}}{3}$	$\frac{2}{3}$	1
1	1	1
−1	−1	−1
4	1	−4
−1	−1	−1
6	−3	2
1	1	1
−4	−1	4
−4	−1	4
−6	3	−2
4	1	−4

Table A.2 Let $[\kappa, \lambda : \phi]$ stand for $\theta_\phi^{\kappa,\lambda} C_\phi^{\kappa,\lambda}(X, Y)$. Table A.2 reads as

$$(XY) = \frac{1}{2}\{2[1,1:2] - [1,1:1^2]\}$$
$$(X^2Y) = \frac{1}{4}\{4[2,1:3] - [2,1:21] - [1^2,1:21] + [1^2,1:1^3]\}\ .$$

Multiplier = 1/2	$(X)(Y)$	(XY)
$[1,1:2]$	2	2
$[1,1:1^2]$	2	−1

Multiplier = 1/4	$(X)^2(Y)$	$(XY)(X)$	$(X^2)(Y)$	(X^2Y)
$[2,1:3]$	4	4	4	4
$[2,1:21]$	4	−1	4	−1
$[1^2,1:21]$	4	2	−2	−1
$[1^2,1:1^3]$	4	−2	−2	1

Multiplier = 1/24	$(X)^3(Y)$	$(XY)(X)^2$	$(X^2)(X)(Y)$	$(X^2)(XY)$	$(X^2Y)(X)$	$(X^3)(Y)$	(X^3Y)
$[3,1:4]$	24	24	24	24	24	24	24
$[3,1:31]$	24	−4	24	−4	−4	24	−4
$[21,1:31]$	24	16	4	−4	6	−6	−4
$[21,1:2^2]$	24	4	4	14	−6	−6	−1
$[21,1:21^2]$	24	−8	4	−4	0	−6	2
$[1^3,1:21^2]$	24	8	−12	−4	−4	6	2
$[1^3,1:1^4]$	24	−12	−12	6	6	6	−3

Multiplier $= 1/32$	$(X)^2(Y)^2$	$(X)^2(Y^2)$	$(XY)(X)(Y)$	$(X^2)(Y)^2$	$(XY)^2$	$(X^2)(Y^2)$	$(XY^2)(X)$	$(X^2Y)(Y)$	$(XYXY)$	(X^2Y^2)
$[2,2:4]$	32	32	32	32	32	32	32	32	32	32
$[2,2:31]$	32	32	4	32	−24	32	4	4	−24	4
$[2,2:2^2]$	32	32	−8	32	12	32	−8	−8	−12	−8
$[2,1^2:31]$	32	−16	16	32	0	−16	−8	16	0	−8
$[2,1^2:21^2]$	32	−16	−8	32	0	−16	4	−8	0	4
$[1^2,2:31]$	32	32	16	−16	0	−16	16	−8	0	−8
$[1^2,2:21^2]$	32	32	−8	−16	0	−16	−8	4	0	4
$[1^2,1^2:2^2]$	32	−16	16	−16	24	8	−8	−8	−12	4
$[1^2,1^2:21^2]$	32	−16	4	−16	−12	8	−2	−2	6	1
$[1^2,1^2:1^4]$	32	−16	−16	−16	8	8	8	8	−4	−4

Table A.3 Binomial coefficients for

$$C_\tau^{\rho,\sigma}(I+X,I+Y)/C_\tau(I) = \sum_{k=0}^{r}\sum_{\ell=0}^{s}\sum_{\kappa}\sum_{\lambda}\sum_{\phi\in\kappa\cdot\lambda}\binom{\rho,\sigma:\tau}{\kappa,\lambda:\phi}C_\phi^{\kappa,\lambda}(X,Y)/C_\phi(I),$$

where ρ, σ and τ are partitions of r, s and $r+s$, and κ, λ and ϕ are partitions of k, ℓ and $k+\ell$, respectively.

$\kappa,\lambda:\phi$ \ $\rho,\sigma;\tau$	$1,1:2$	$1,1:1^2$	$2,1:3$	$2,1:21$	$1^2,1:21$	$1^2,1:1^3$
$0,0:0$	1	1	1	2/3	$\sqrt{5}/3$	1
$1,0:1$	1	1	2	4/3	$2\sqrt{5}/3$	2
$0,1:1$	1	1	1	2/3	$\sqrt{5}/3$	1
$2,0:2$			1	2/3		
$1^2,0:1^2$					$\sqrt{5}/3$	1
$1,1:2$	1		2	2/9	$4\sqrt{5}/9$	
$1,1:1^2$		1		10/9	$2\sqrt{5}/9$	2
$2,1:3$			1			
$2,1:21$				1		
$1^2,1:21$					1	
$1^2,1:1^3$						1

$\diagdown \; \rho,\sigma;\tau$ $\kappa,\lambda:\phi$	$3,1:4$	$3,1:31$	$21,1:31$	$21,1:2^2$	$21,1:21^2$
$0,0:0$	1	$\sqrt{30}/10$	$\sqrt{70}/10$	1	$\sqrt{10}/4$
$1,0:1$	3	$3\sqrt{30}/10$	$3\sqrt{70}/10$	3	$3\sqrt{10}/4$
$0,1:1$	1	$\sqrt{30}/10$	$\sqrt{70}/10$	1	$\sqrt{10}/4$
$2,0:2$	3	$3\sqrt{30}/10$	$2\sqrt{70}/15$	$4/3$	$\sqrt{10}/3$
$1^2,0:1^2$			$\sqrt{70}/6$	$5/3$	$5\sqrt{10}/12$
$1,1:2$	3	$2\sqrt{30}/30$	$7\sqrt{70}/30$	$4/3$	$\sqrt{10}/12$
$1,1:1^2$		$7\sqrt{30}/30$	$\sqrt{70}/15$	$5/3$	$2\sqrt{10}/3$
$3,0:3$	1	$\sqrt{30}/10$	$\sqrt{70}/10$		
$21,0:21$				1	$\sqrt{10}/4$
$1^3,0:1^3$					
$2,1:3$	3	$\sqrt{30}/50$	$3\sqrt{70}/25$		
$2,1:21$		$21\sqrt{30}/50$	$\sqrt{70}/50$	2	$\sqrt{10}/2$
$1^2,1:21$			$\sqrt{14}/2$	$\sqrt{5}$	$\sqrt{2}/8$
$1^2,1:1^3$					$3\sqrt{10}/8$
$3,1:4$	1				
$3,1:31$		1			
$21,1:31$			1		
$21,1:2^2$				1	
$21,1:21^2$					1
$1^3,1:21^2$					
$1^3,1:1^4$					

$1^3,1:21^2$	$1^3,1:1^4$
$\sqrt{6}/4$	1
$3\sqrt{6}/4$	3
$\sqrt{6}/4$	1
$3\sqrt{6}/4$	3
$5\sqrt{6}/12$	
$\sqrt{6}/3$	3
$\sqrt{6}/4$	1
$3\sqrt{30}/8$	
$\sqrt{6}/8$	3
1	
	1

$\backslash \ \rho,\sigma;\tau$ $\kappa,\lambda:\phi$	$2,2:4$	$2,2:31$	$2,2:2^2$	$2,1^2:31$	$2,1^2:21^2$	$1^2,2:31$
$0,0:0$	1	$\sqrt{2}/3$	$2/3$	$\sqrt{14}/6$	$\sqrt{10}/6$	$\sqrt{14}/6$
$1,0:1$	2	$2\sqrt{2}/3$	$4/3$	$\sqrt{14}/3$	$\sqrt{10}/3$	$\sqrt{14}/3$
$0,1:1$	2	$2\sqrt{2}/3$	$4/3$	$\sqrt{14}/3$	$\sqrt{10}/3$	$\sqrt{14}/3$
$2,0:2$	1	$\sqrt{2}/3$	$2/3$	$\sqrt{14}/6$	$\sqrt{10}/6$	
$0,2:2$	1	$\sqrt{2}/3$	$2/3$			$\sqrt{14}/6$
$1^2,0:1^2$						$\sqrt{14}/6$
$0,1^2:1^2$				$\sqrt{14}/6$	$\sqrt{10}/6$	
$1,1:2$	4	$5\sqrt{2}/9$	$4/9$	$4\sqrt{14}/9$	$\sqrt{10}/9$	$4\sqrt{14}/9$
$1,1:1^2$		$7\sqrt{2}/9$	$20/9$	$2\sqrt{14}/9$	$5\sqrt{10}/9$	$2\sqrt{14}/9$
$2,1:3$	2	$\sqrt{2}/5$		$\sqrt{14}/5$		
$2,1:21$		$7\sqrt{2}/10$	2	$\sqrt{14}/5$	$\sqrt{10}/2$	
$1,2:3$	2	$\sqrt{2}/5$				$\sqrt{14}/5$
$1,2:21$		$7\sqrt{2}/10$	2			$\sqrt{14}/5$
$1^2,1:21$						$\sqrt{70}/5$
$1,1^2:21$				$\sqrt{70}/5$	$\sqrt{2}/4$	
$1^2,1:1^3$					$\sqrt{10}/4$	
$1,1^2:1^3$						
$2,2:4$	1					
$2,2:31$		1				
$2,2:2^2$			1			
$2,1^2:31$				1		
$2,1^2:21^2$					1	
$1^2,2:31$						1
$1^2,2:21^2$						
$1^2,1^2:2^2$						
$1^2,1^2:21^2$						
$1^2,1^2:1^4$						

$1^2, 2 : 21^2$	$1^2, 1^2 : 2^2$	$1^2, 1^2 : 21^2$	$1^2, 1^2 : 1^4$
$\sqrt{10}/6$	$\sqrt{5}/3$	$2/3$	1
$\sqrt{10}/3$	$2\sqrt{5}/3$	$4/3$	2
$\sqrt{10}/3$	$2\sqrt{5}/3$	$4/3$	2
$\sqrt{10}/6$			
$\sqrt{10}/6$	$\sqrt{5}/3$	$2/3$	1
	$\sqrt{5}/3$	$2/3$	1
$\sqrt{10}/9$	$8\sqrt{5}/9$	$10/9$	
$5\sqrt{10}/9$	$4\sqrt{5}/9$	$14/9$	4
$\sqrt{10}/2$			
$\sqrt{2}/4$	2	$\sqrt{5}/2$	
	2	$\sqrt{5}/2$	
$\sqrt{10}/4$		$1/2$	2
		$1/2$	2
1			
	1		
		1	
			1

It is worth noting that the following identity holds for the partitions in Table A.3

$$\begin{pmatrix} \rho, \sigma : \tau \\ \kappa, \sigma : \phi \end{pmatrix} = \left(\begin{matrix} \rho : \tau \\ \kappa : \phi \end{matrix} \middle| \, \sigma \right).$$

Bibliography

Anderson, T. W. (1951). Classification by multivariate analysis. *Psychometrika*, **16**, 31-50.

Anderson, T. W. (1984). *An Introduction to Multivariate Statistical Analysis.* John Wiley and Sons, New York.

Anderson, T.W., and Fang, K.-T. (1987). Cochran's theorem for elliptically contoured distributions. *Sankhyā, Series A*, **49**, 305–315.

Anderson, T.W., Fang, K.-T., and Hsu, H. (1986). Maximum likelihood estimates and likelihood ratio criteria for multivariate elliptically contoured distributions. *The Canadian Journal of Statistics*, **14**, 55-59.

Barnes, E. W. (1899). The theory of gamma function. *Messeng. Math.*, **29**, 64-129.

Beckett, J. III., Schucany, W. R., and Broffit, J. D. (1980). A special distributional result for bilinear forms. *Journal of the American Statistical Association (Theory and Methods)*, **75**, 466-468.

Bingham, C. (1974). An identity involving partial generalized binomial coefficients. *Journal of Multivariate Analysis*, **4**, 210-223.

Bochner, S. (1944). Group invariance of Cauchy's formula in several variables. *Annals of Mathematics*, **45**, 686-707.

Bochner, S. (1952). Bessel functions and modular relations of higher type and hyperbolic differential equations. *Communications du séminaire mathématique de l'université de Lund, tome supplémentaire, dédié à Marcel Riesz*, 12-20.

Bochner, S. and Martin, W. T. (1948). *Several Complex Variables.* Princeton University Press.

Boerner, H. (1963). *Representations of Groups.* North Holland Publishing Company, Amsterdam.

Brown, P. J. (1982). Multivariate calibration. *Journal of the Royal Statistical Society*, B, **44**, 287-321.

Brown, P. J. and Sundberg, R. (1987). Confidence and conflict in multivariate calibration. *Journal of the Royal Statistical Society*, B, **49**, 46-57.

Cacoullos, T., and Koutraz, M. (1984). Quadratic forms in spherical random variables, generalized noncentral χ^2 - distribution. *Nan. Res. Logst. Quart.*, **31**, 447-461.

Cambanis, S., Huang, S., and Simons, G. (1981). On the theory of elliptically contoured distributions. *Journal of Multivariate Analysis*, **11**, 368-385.

Chemielewst, M.A., (1981). Elliptically symmetric distributions: A review and a bibliography. *International Statistical Review*, **49**, 67-74.

Chikuse, Y. (1976). Partial differential equations for hypergeometric functions of complex argument matrices and their applications. *Annals of the Institute of Statistical Mathematics*, **28**, 187-199.

Chikuse, Y. (1980). Invariant polynomials with matrix arguments. *Multivariate Analysis*, Gupta, R. P. Ed., North Holland, 53-68.

Chikuse, Y. (1981). Distributions of some matrix variates and latent roots in multivariate Behrens-Fisher discriminant analysis. *Annals of Statistics*, **9**, 401-407.

Chikuse, Y. (1986). Multivariate Meixrer classes of invariant distributions. *Linear Algebra and its Applications*, **52**, 177-200.

Chikuse, Y. (1987). Methods for constructing top order invariant polynomials. *Econometric Theory*, **3**, 195-207.

Chikuse, Y. (1991). Asymptotic expansions for distributions of the large sample matrix resultant and related statistics on the Stiefel manifold. *Journal of Multivariate Analysis*, **39**, 270-283.

Chikuse, Y. (1992a). Properties of Hermite and Laguerre polynomials in matrix argument and their applications. *Linear Algebra and its Applications*, **176**, 237-260.

Chikuse, Y. (1992b). Generalized Hermite and Laguerre polynomials in multiple symmetric matrix argument and their applications. *Linear Algebra and its Applications*, **176**, 261-287.

Chikuse, Y. (1993). High dimensional asymptotic expansions for the matrix Langevin distributions on the Stiefel manifold. Journal of Multivariate Analysis, **44**, 82-101.

Chikuse, Y. and Davis, A. W. (1986a). A survey on the invariant polynomials with matrix arguments in relation to econometric distribution theory. *Econometric Theory*, **2**, 232-248.

Chikuse, Y. and Davis, A. W. (1986b). Some properties of invariant polynomials with matrix arguments and their applications in econometrics. *Annals of the Institute of Statistical Mathematics*, **39**, 109-122.

Conradie, W. J. and Troskie, C. G. (1982). Calculation of coefficients appearing in expansions of zonal polynomials of Hermitian matrices. *South African Statistical Journal*, **16**, 113-131.

Constantine A. G. (1963). Some noncentral distribution problems in multivariate analysis. *The Annals of Mathematical Statistics*, **34**, 1270-1285.

Constantine A. G. (1966). The distribution of Hotelling's generalized T_0^2. *The Annals of the Mathematical Statistics*, **37**, 215-225.

Constantine, A. G. and Muirhead, R. J. (1972). Partial differential equations for hypergeometric functions of two argument matrices. *Journal of Multivariate Analysis*, **2**, 332-338.

Crowther, N. A. S. (1975). The exact non-central distribution of a quadratic form in normal vectors. *South African Statistical Journal*, **9**, 27-36.

Das Gupta, S., Eaton, M.L., Olkin, I., Perlman, M., Savage, L.J. and Sobel, M. (1972). Inequalities on the probability content of convex regions for elliptically contoured distributions. *Proceedings of the Sixth Berkeley Symposium on Mathematical Statistics and Probability*, **2**, 241-264.

David, F. N. , Kendall, M. G. and Barton, D. E. (1966). *Symmetric Function and Allied Tables*. Biometrika Trustees.

Davis, A. W. (1968). A system of linear differential equations for the distribution of Hotelling's generalized T_0^2. *The Annals of Mathematical Statistis*, **39**, 815-832.

Davis, A. W. (1976). Statistical distributions in univariate and multivariate Edgeworth populations. *Biometrika*, **63**, 661-670.

Davis, A. W. (1979). Invariant polynomials with two matrix arguments extending the zonal polynomials: Applications to multivariate distribution theory. *Annals of the Institute of Statistical Mathemacis*, **31**, 465-485.

Davis, A. W. (1980a). Invariant polynomials with two matrix arguments, extending the zonal polynomials. *Multivariate Analysis-V*, Krishnaiah, P. R. Ed., North Holland, 287-299.

Davis, A. W. (1980b). On the effects of moderate multivariate nonnormality on Wilks's likelihood ratio criterion. *Biometrika*, **62**, 419-427.

Davis, A. W. (1981). On the construction of a class of invariant polynomials in several matrices, extending the zonal polynomials. *Annals of the Institute of Statistical Mathematics*, **33**, 297-313.

Davis, A. W. (1982a). On a result of Roy and Gnanadesikan concerning multivariate variance components. *Annals of the Institute of Statistical Mathematics*, **34**, 517-521.

Davis, A. W. (1982b). On the distribution of Hotelling's one sample T^2 under moderate non-normality. *Journal of Applied Probability*, **19**, 207-216.

Davis, A. W. (1987). Moment of linear discriminant functions and an asymptotic confidence interval for the log odds ratio. *Biometrika*, **74**, 829-840.

Davis, A. W. (1989). Distribution theory for some tests of independence of seemingly unrelated regressions. *Journal of Multivariate Analysis*, **31**, 69-82.

Davis, A. W. and Hayakawa, T. (1987). Some distribution theory relating to confidence regions in multivariate calibration. *Annals of the Institute of Statistical Mathematics*, **39**, 141-152.

Dawid, A. P. (1977). Spherical matrix distributions and a multivariate model. *Journal of the Royal Statistical Society, Series B*, **39**, 254-261.

Deemer, W. L. and Olkin, I. (1951). The Jacobians of certain matrix transformations useful in multivariate analysis. *Biometrika*, **38**, 345-367.

Driscoll, M. F. and Gundberg, W. R. (1986). A history of the development of Craig's theorem. *The American Statistician*, **40**, 65-70.

Erdélyi, A., Magnus, W., Obergettinger, T. and Tricomi, F. G. (1953a). *Higher Transcendental Functions*, Vol. I. McGraw-Hill, New York.

Erdélyi, A., Magnus, W., Obergettinger, T. and Tricomi, F. G. (1953b). *Higher Transcendental Functions*, Vol. II. McGraw-Hill, New York.

Fan, J. (1986). Distributions of quadratic forms and non-central Cochran's theorem. *Acta Mathematicae Sinica* (New Series), **2**, 185-198.

Fang, K.-T. and Anderson, T.W. (1990). *Statistical Inference in Elliptically and Related Distributions.* Allerton Press Inc., New York.

Fang, K.-T., Fan, J.-Q., and Xu, J.-L. (1987). The distributions of quadratic forms of random idempotent matrices with their applications. *Chinese Journal of Applied Probability and Statistics*, **3**, 289-297.

Fang, K.-T., Kotz, S., and Ng, K.-W. (1990). *Symmetric Multivariate and Related Distributions.* Chapman and Hall, New York.

Fang, K.-T. and Wu, Y. (1984). Distribution of quadratic forms and Cochran's theorem. *Mathematics in Economics*, **1**, 29-48.

Farrel, R. H. (1980). Calculation of complex zonal polynomials. *Multivariate Analysis-V*, P. R. Krishnaiah, Ed., North Holland, New York, 301-320.

Farrel R. H. (1985). *Multivariate Calculation, Use of the Continuous Groups*. Springer-Verlag, Berlin.

Fraser, D.A.S. and Ng, K.W. (1980). Multivariate regression analysis with spherical error. *Multivariate Analysis-V*, P.R. Krishnaiah, Ed., North Holland, New York, 369-386.

Fujikoshi, Y. (1970). Asymptotic expansions of the distributions of test statistics in multivariate analysis. *Journal of Science of the Hiroshima University*. Series A-1, **34**, 73-144.

Fujikoshi, Y. (1971). Asymptotic expansions of the non-null distributions of two criteria for the linear hypothesis concerning complex multivariate normal populations. *Annals of the Institute of Statistical Mathematics*, **23**, 477-490.

Fujikoshi, Y. (1973). Asymptotic formulas for the distributions of three statistics for multivariate linear hypothesis. *Annals of the Institute of Statistical Mathematics*, **25**, 423-437.

Fujikoshi, Y. (1975). Partial differential equations for hypergeometric functions ${}_3F_2$ of matrix argument. *The Canadian Journal of Statistics, Theory and Methods*, **3**, 153-163.

Fujikoshi, Y. and Nishii, R. (1984). On the distribution of a statistic in multivariate inverse regression analysis. *Hiroshima Mathematical Journal*, **14**, 215-225.

Gang, L. (1987). Moments of a random vector and its quadratic forms. *Journal of Statistics and Applied Probability*, **2**, 219-229.

Gradshteyn, I. S. and Ryzhik, I. M. (1980). *Tables of Integrals, Series and Products*. Academic Press, New York.

Graybill, F.A. and Milliken, G. (1969). Quadratic forms and idempotent matrices with random elements. *Annals of Mathematical Statistics*, **40**, 1430-1438.

Hannan, E. J. (1965). Group representations and applied probability. *Journal of Applied Probability*, **2**, 1-68.

Hayakawa, T. (1967). On the distribution of the maximum latent root of a positive definite symmetric random matrix. *Annals of the Institute of Statistical Mathematics*, **19**, 1-17.

Hayakawa, T. (1969). On the distribution of the latent roots of a positive definite random symmetric matrix I. *Annals of the Institute of Statistical Mathematics*, **21**, 1-21.

Hayakawa, T. (1972a). On the distribution of the multivariate quadratic form in multivariate normal samples. *Annals of the Institute of Statistical Mathematics*, **24**, 205-230.

Hayakawa, T. (1972b). On the derivation of the asymptotic distribution of the generalized Hotelling's T_0^2. *Annals of the Institute of Statistical Mathematics*, **24**, 19-32.

Hayakawa, T. (1972c). On the distributions of the latent roots of a complex Wishart matrix (Non-central case). *Annals of the Institute of Statistical Mathematics*, **24**, 1-17.

Hayakawa, T. (1972d). The asymptotic distributions of the statistics based on the complex Gaussian distribution. *Annals of the Institute of Statistical Mathematics*, **24**, 231-244.

Hayakawa, T. (1981). Asymptotic distribution of a generalized Hotelling's T_0^2 in the doubly noncentral case. *Annals of the Institute of Statistical Mathematics*, **33**, 15-25.

Hayakawa, T. (1982). On some formulas for weighted sum of invariant polynomials of two matrix arguments. *Journal of Statistical Planning and Inference*, **6**, 105-114.

Hayakawa, T. (1986). On testing hypotheses of covariance matrix under an elliptical population. *Journal of Statistical Planning and Inference*, **13**, 193-202.

Hayakawa. T. (1987). Normalizing and variance stabilizing transformations of multivariate statistics under an elliptical population. *Annals of the Institute of Statistical Mathematics*, **39**, 299-306.

Hayakawa, T. (1989). On the distributions of the functions of the F-matrix under an elliptical population. *Journal of Statistical Planning and Inference*, **21**, 41-52.

Hayakawa, T. and Kikuchi, Y. (1978). On the distribution of a Wald type statistic. *Research Memorandum of The Institute of Statistical Mathematics*, No. **128**.

Hayakawa, T. and Kikuchi, Y. (1979). The moments of a function of traces of a matrix with a multivariate symmetric normal distribution. *South African Statistical Journal*, **13**, 71-82.

Hayakawa, T. and Puri, M. L. (1985). Asymptotic distributions of likelihood ratio criteria for testing latent roots and latent vectors of a covariance matrix under an elliptical population. *Biometrika*, **72**, 331-338.

Herz, C. S. (1955). Bessel functions of matrix argument. *Annals of Mathematics*. **61**, 474-523.

Hill, G. W. and Davis, A. W. (1968). Generalized asymptotic expansions of Cornish-Fisher type. *The Annals of Mathematical Statistics*, **39**, 1264-1273.

Hillier, G. H. (1985). On the joint and the marginal densities of instrumental variable estimators in a general structural equation. *Econometric Theory*, **1**, 53-72.

Hillier, G. H., Kinal, T. W. and Srivastava, V. K. (1984). On the moments of ordinary least squares and instrumental variables estimators in general structural equation. *Econometrica*, **52**, 185-202.

Hsu, H. (1990). Noncentral distributions of quadratic forms for elliptically contoured distributions. *Statistical Inference in Elliptically Contoured and Related Distributions*, Fang, K.-T. and Anderson, T.W., Eds., 97-102.

Hsu, P. L. (1940). On generalized analysis of variance. *Biometrika*, **31**, 221-237.

Hua, L. K. (1959). *Harmonic analysis of functions of several complex variables in the classical domain*. Moscow. Translated into English, *American Mathematical Society*, 1963.

Ito, K. (1956). Asymptotic formulae for the distribution of Hotelling's generalized T_0^2 statistic. *The Annals of Mathematical Statistics*, **27**, 1091-1195.

Ito, K. (1960). Asymptotic formulae for the distribution of Hotelling's generalized T_0^2 statistic II. *The Annals of Mathematical Statistics*, **31**, 1148-1153.

Jack, H. (1966). Jacobians of transformations involving orthogonal matrices. *The Proceeding of the Royal Society of Edinburgh, Series A*, **47**, 81-103

James, A. T. (1954). Normal multivariate analysis and the orthogonal group. *The Annals of Mathematical Statistics*, **25**, 40-75.

James, A. T. (1960). The distribution of the latent roots of the covariance matrix. *The Annals of Mathematical Statistics*, **31**, 151-158.

James, A. T. (1961a). Zonal polynomials of the real positive definite symmetric matrices. *Annal of Mathematics*, **74**, 456-469.

James, A. T. (1961b). The distribution of noncentral means with known covariances. *Annals of Mathematical Statistics*, **32**, 874-882.

James, A. T. (1964). Distributions of matrix variates and latent roots derived from normal samples. *The Annals of Mathematical Statistics*, **35**, 475-501.

James, A. T. (1968). Calculation of zonal polynomial coefficients by use of the Laplace-Beltrami operator. *The Annals of Mathematical Statistics* **39**, 1711-1718.

James, A. T. (1976). Special functions of matrix and single argument in statistics, *Theory and Applications of Special Functions*. Askey, R. A. Ed., 497-520, Academic Press, New York.

James, A. T. and Constantine, A. G. (1974). Generalized Jacobi polynomials as spherical functions of the Grassmann manifold. *Proceedings of The London Mathematical Society*, **29**, 174-192.

Johnson, N. L. and Kotz, S. (1970). *Continuous univariate distributions-2*. Houghton Mifflin Company, Boston.

Kariya, T. and Eaton, M. L. (1977). Robust tests for spherical symmetry. *Annals of Statistics*, **5**, 700-709.

Kawada, Y. (1950). Independence of quadratic forms in normally correlated variables. *Annals of Mathematical Statistics*, **21**, 614-615.

Kelker, D. (1970). Distribution theory of spherical distributions and a location-scale parameter generalization. *Sankhyā*, Series A, **32**, 419-430.

Khatri, C. G. (1966). On certain distribution problems based on positive definite quadratic functions in normal vectors. *The Annals of Mathematical Statistics*, **37**, 468-479.

Khatri, C. G. (1970). On the moments of traces of two matrices in three situations for complex multivariate normal populations. *Sankhyā, Series A*, **32**, 65-80.

Khatri, C. G. (1971). Series representations of distributions of a quadratic forms in normal vectors and generalized variance. *Journal of Multivariate Analysis*, **2**, 199-214.

Khatri, C. G. (1975). Distribution of a quadratic form in normal vectors (Multivariate non-central case). *A Modern Course on Statistical Distributions in Scientific Work, Vol. I, Models and Structures*. Edited by G. P. Patel, S. Kotz and J. K. Ord. D. Reidel Publishing Co., Boston, 345-354.

Khatri, C. G. (1977). Distribution of a quadratic form in noncentral normal vectors using generalized Laguerre polynomials. *South African Statistical Journal*, **11**, 167-179.

Khatri, C. G. and Pillai, K. C. S. (1968). On the non-central distributions of two test criteria in multivariate analysis of variance. *The Annals of Mathematical Statistics*, **39**, 215-226.

Konishi, S. (1978). An approximation to the distribution of the sample correlation coefficient. *Biometrika*, **65**, 654-656.

Konishi, S. (1981). Normalizing transformations of some statistics in multivariate analysis. *Biometrika*, **68**, 647-651.

Konishi, S., Niki, N. and Gupta, A.K. (1988). Asyptotic expansions for the distribution of quadratic forms in normal variables. *Annals of the Institute of Statistical Mathematics*, **40**, 279-296.

Kushner, H.B. (1985). On the expansion of $C_\rho^*(V+I)$ as a sum of zonal polynomials. *Journal of Multivariate Analysis*, **17**, 84-98.

Kushner, H.B., Lebow, A., and Meisner, M. (1981). Eigenfunctions of expected value operations in the Wishart distribution II. *Journal of Multivariate Analysis*, **11**, 418-433.

Kushner, H.B. and Meisner, M. (1980). Eigenfunctions of expected value operators in the Wishart distribution. *The Annals of Statistics*, **8**, 977-988.

Kushner, H.B. and Meisner, M. (1984). Formulas for zonal polynomials. *Journal of Multivariate Analysis*, **14**, 336-347.

Laha, R.G. (1956). On the stochastic independence of two second degree polynomial statistics in normally distributed variates. *Annals of Mathematical Statistics*, **27**, 790-796.

Lee, T.S. (1971). Distribution of the canonical correlations and asymptotic expansions for distributions of certain independence test statistics. *The Annals of Mathematical Statistics*, **42**, 526-537.

Lieftinck-Koeijers, C.A.J. (1988). Multivariate calibration: A generalization of the classical estimator. *Journal of Multivariate Analysis*, **25**, 31-44.

Lukacs, E. (1956). Characterization of populations of suitable statistics. *Proceedings of the Third Berkeley Symposium on Mathematical Statistics and Probability* **2**, University of California Press, Los Angeles and Berkeley, 195-214.

Macdonald, I.G. (1979). *Symmetric Functions and Hall Polynomials*. Oxford Mathematical Monographs, Clarendon Press, Oxford.

Mathai, A.M. (1992). On bilinear forms in random variables. *Annals of the Institute of Statistical Mathematics*, **44**, 769-779.

Mathai, A.M. (1993a). On noncentral generalized Laplacianness of quadratic forms in normal variables. *Journal of Multivariate Analysis*, **45**, 239-246.

Mathai, A.M. (1993b). Generalized Laplace distribution with applications. *Journal of Applied Statistical Science*, **1**(**2**), 169-178.

Mathai, A.M. (1993c). The residual effect of a growth-decay mechanism and the distributions of covariance structures. *Canadian Journal of Statistics*, **21**(**3**), 277-283.

Mathai, A.M. (1993d). On noncentral generalized Laplacian distribution. *Journal of Statistical Research*, **27**(**1&2**), 57-80.

Mathai, A.M. (1993e). *A Handbook of Generalized Special Functions for Statistical and Physical Sciences*. Clarendon Press, Oxford.

Mathai, A.M. and Haubold, H.J. (1988). *Modern Problems in Nuclear and Neutrino Astrophysics*. Akademie-Verlag, Berlin.

Mathai, A.M. and Provost, S.B. (1992). *Quadratic Forms in Random Variables : Theory and Applications*. Marcel Dekker Inc., New York.

Mathai, A.M. and Saxena, R.K. (1973). *Generalized Hypergeometric Functions with Applications in Statistics and Physical Sciences*. Springer-Verlag, Lecture Notes No. 348, Heidelberg.

Mathai, A.M. and Saxena, R.K. (1978). *The H-function with Applications in Statistics and Other Disciplines*. Wiley Halsted, New York.

Menzefricke, U. (1981). On positive definite quadratic forms in correlated t variables. *Annals of the Institute of Statistical Mathematics*, **33**, 385-390.

Muirhead, R. J. (1970a). Systems of partial differential equations for hypergeometric functions of matrix arguments. *The Annals of Mathematical Statistics*, **41**, 991-1001.

Muirhead, R. J. (1970b). Asymptotic distributions of some multivariate tests. *The Annals of Mathematical Statistics*, **41**, 1002-1010.

Muirhead, R. J. (1972). The asymptotic noncentral distribution of Hotelling's generalized T_0^2. *The Annals of Mathematical Statistics*, **43**, 1671-1677.

Muirhead, R. J. (1974). On the calculation of generalized binomial coefficients. *Journal of Multivariate Analyis*, **4**, 341-346.

Muirhead, R. J. (1982). *Aspects of Multivariate Statistical Theory*. John Wiley and Sons, New York.

Muirhead, R. J. and Chikuse, Y. (1975). Asymptotic expansions for the joint and marginal distributions of the latent roots of the covariance matrix. *The Annals of Statistics*, **3**, 1011-1017.

Nagao, H. (1973). On some test criteria for covariance matrix. *Annals of Statistics*, **4**, 700-709.

Okamoto, M. (1963). An asymptotic expansion for the distribution of the linear discriminant function. *The Annals of Mathematical Statistics*, **34**, 1286-1301.

Okamoto, M. (1968). Correction to "An asymptotic expansion for the distribution of the linear discriminant function". *The Annals of Mathematical Statistics*, **39**, 1358-1359.

Olkin, I. and Sampson, A. R. (1972). Jacobians of matrix transformations and induced functional equations. *Linear Algebra and its Applications*, **5**, 257-276.

Parkhurst, A. M. and James, A. T. (1974). Zonal polynomials of order 1 through 12. *Selected Tables in Mathematical Statistics* (Harter, H. L. and Owen, D. B. Eds), 199-388, American Mathematical Society, Providence, R.I.

Phillips, P. C. B. (1984). An everywhere convergent series representation of the distribution of Hotelling's generalized T_0^2. *Cowles Foundation Discussion Paper at Yale Univesity*, No. 723.

Phillips, P. C. B. (1985). The exact distribution of the SUR estimator. *Econometrica*, **53**, 745-756.

Phillips, P. C. B. (1986). The exact distribution of the Wald statistics. *Econometrica*, **54**, 881-895.

Phillips, P. C. B. (1987). An everywhere convergent series representation of the distribution of Hotelling's generalized T_0^2. *Journal of Multivariate Analysis*, **21**, 238-249.

Pillai, K. C. S. and Jouris, G. M. (1969). On the moments of elementary symmetric functions of the roots of two matrices. *Annals of the Institute of Statistical Mathematics*, **21**, 309-320.

Provost, S. (1988). The exact density of a general linear combination of gamma variates. *Metron*, **46**, 61-69.

Provost, S. (1989). The distribution function of a statistics for testing the equality of scale parameters in two gamma populations. *Metrika*, **36**, 337-345.

Provost, S. (1994). On the Craig-Sakamoto lemma. *Pakistan Journal of Statistics*, **10**, 205-212.

Provost, S. (1995). The distribution of positive definite quadratic forms in elliptically contoured random vectors. Technical Report 95-03, Department of Statistical and Actuarial Sciences, The University of Western Ontario.

Provost, S. and Kawczak, J. (1995). A distributional result for quadratic forms in spherically symmetric random vectors. Technical Report 95-01, Department of Statistical and Actuarial Sciences, The University of Western Ontario.

Provost, S. and Rudiuk, E. (1992). The exact distribution function of the ratio of two quadratic forms in noncentral normal variables. *Metron* **50**, 33-58.

Provost, S. and Rudiuk, E. (1993). The exact distribution function of indefinite quadratic forms in noncentral normal vectors. Technical Report 93-04, Department of Statistical and Actuarial Sciences, The University of Western Ontario.

Provost, S. and Rudiuk, E. (1994). The exact density of the ratio of two dependent linear combinations of of chi-square variables. *The Annals of the Institute of Statistical Mathematics*, **46**, 557-571.

Richards, D. St. P. (1979). Calculation of zonal polynomials of 3×3 positive definite symmetric matrices. *Annals of the Institute of Statistical Mathematics*, **31**, 207-213.

Richards, D. St. P. (1982a). Differential operators associated with zonal polynomials I. *Annals of the Institute of Statistical Mathematics*, **34**, 111-117.

Richards, D. St. P. (1982b). Differential operators associated with zonal polynomials II. *Annals of the Institute of Statistical Mathematics*, **34**, 119-121.

Roy, S. N. (1957). *Some Aspects of Multivariate Analysis*. John Wiley and Sons, New York.

Saw, J. G. (1977). Zonal polynomials: An alternative approach. *Journal of Multivariate Analysis*, **7**, 461-467.

Shah, B. K. (1970). Distribution theory of a positive definite quadratic form with matrix argument. *The Annals of Mathematical Statistics*, **41**, 692-697.

Siotani, M. (1957). Note on the utilization of the generalized student ratio in the analysis of variance or dispersion. *Annals of the Institute of Mathematical Statistics*, **9**, 157-171.

Siotani, M. (1971). An asymptotic expansion of the non-central distribution of Hotelling's generalized T_0^2-statistic. *The Annals of Mathematical Statistics*, **42**, 560-571.

Siotani, M., Hayakawa, T. and Fujikoshi, Y. (1985). *Modern Multivariate Statistical Analysis: A Graduate Course and Handbook*. American Sciences Press, Ohio.

Siotani, M. and Wang, R. H. (1977). Asymptotic expansions for error rates and comparison of W-procedure and the Z-procedure in discriminant analysis. *Multivariate Analysis-IV*, P. R. Krishnaiah Ed., North Holland Publishing Company, 523-545.

Smith, M. D. (1989). On the expectation of a ratio of quadratic form in normal variables. *Journal of Multivariate Analysis*, **31**, 244-257.

Srivastava, M. S. and Khatri, C. G. (1979). *An Introduction to Multivariate Statistics*. North Holland, New York.

Sugiura, N. (1971). Note on some formulas for weighted sums of zonal polynomials. *Annals of Mathematical Statistics*, **42**, 768-772.

Sugiura, N. (1972). Asymptotic solutions of hypergeometric function ${}_1F_1$ of matrix argument, useful in multivariate analysis. *Annals of the Institute of Statistical Mathematics*, **24**, 517-524.

Sugiura, N. (1973). Derivatives of the characteristic root of a symmetric or Hermitian matrix with two applications in multivariate analysis. *Communications in Statistics - Theory and Methods*, **1**, 393-417.

Sugiura, N. (1974). Asymptotic formulas for the hypergeometric function ${}_2F_1$ of matrix argument, useful in multivariate analysis. *Annals of the Institute of Statistical Mathematics*, **26**, 117-125.

Sugiura, N. and Fujikoshi, Y. (1969). Asymptotic expansions of the non-null distributions of the likelihood ratio criteria for multivariate linear hypothesis and independence. *The Annals of Mathematical Statistics*, **40**, 942-952.

Sugiyama, T. (1967a). On the distribution of the largest latent root of the covariance matrix. *The Annals of Mathematical Statistics*, **38**, 1148-1151.

Sugiyama, T. (1967b). Distribution of the largest latent root and the smallest latent root of the generalized B statistic and F statistic in multivariate analysis. *The Annals of Mathematical Statistics*, **38**, 1152-1159.

Takemura, A. (1982). A Statistical Approach to Zonal Polynomials. Ph.D. Dissertation of Stanford University.

Takemura, A. (1984). *Zonal Polynomials*. Institute of Mathematical Statistics Lecture Note - Monograph Series. Vol. **4**.

Takeuchi, K. and Takemura, A. (1985). Eigenfunction of association algebra of pairings and zonal polynomials. *Discussion Paper 85-F-5*, Faculty of Economics, University of Tokyo.

Thrall, R. M. (1942). On symmetrized Kronecker powers and the structure of the free Lie ring. *American Journal of Mathematics*, **64**, 371-388.

Tiku, M. L. (1965). Laguerre series forms of non-central χ^2 and F distributions. *Biometrika*, **52**, 415-427.

Tukey, J.W. (1949). One degree of freedom for nonadditivity. *Biometrics*, **5**, 232-242.

Vilenkin, N. Ja. (1965). Special functions and theory of group representations. Moscow. Translated into English, *American Mathematical Society*, 1968.

Wakaki, H. (1994). Discriminant analysis under elliptical populations. *Hiroshima Mathematical Journal*, **24**, 257-298.

Weyl, H. (1946). *The Classical Groups, their Invariants and Representations*. Princeton University Press.

Williams, E. J. (1959). *Regression Analysis*. John Wiley and Sons, New York.

Wood, J. T. (1982). Estimating the age of an animal; an application of multivariate calibration. *The Proceeding of 11th International Biometrics Conference*.

Glossary of Symbols

All the vectors appearing in this book are denoted by bold characters, the random vectors being capitalized.

A'	Transpose of the matrix A	Notation 1.1.1
$A > 0$	A is positive definite	Notation 1.1.1
$A \geq 0$	A is positive semidefinite	Notation 1.1.1
$A < 0$	A is negative definite	Notation 1.1.1
$A \leq 0$	A is negative semidefinite	Notation 1.1.1
$\|A\|$	Determinant of the square matrix A	Notation 1.1.1
$\mathrm{tr}(A)$	Trace of the square matrix A	Notation 1.1.1
$\int_X$	Integral over the elements of the matrix X	Notation 1.1.1
$\int_{Y>0}$	Integral over the positive definite matrix Y	Notation 1.1.1
$\int_{A<Y<B} = \int_A^B$	Integral over Y such that $A > 0,\ Y - A > 0,\ B - Y > 0$	Notation 1.1.1
$\mathrm{Re}(\cdot)$	Real part of $(\cdot)$	Notation 1.1.1
$\mathrm{diag}(\lambda_1, \ldots, \lambda_p)$	Diagonal matrix with diagonal elements $\lambda_1, \ldots, \lambda_p$	Notation 1.1.1
$A^{\frac{1}{2}}$	Symmetric square root of the matrix A	Notation 1.1.1
$(\mathrm{d}\mathbf{X}) = (\mathrm{d}x_{ij})$	Matrix of differentials	Notation 1.1.1
$\mathrm{d}X$	Wedge product	Notation 1.1.1

$N_p(\boldsymbol{\mu}, \Sigma)$	Multivariate normal distribution with mean $\boldsymbol{\mu}$ and covariance matrix Σ	Section 1.2a
$\sim$	is distributed as	Section 1.2a
$M_{\boldsymbol{X}}(\boldsymbol{t})$	Moment generating function of the random vector $\boldsymbol{X}$	Def. 1.2.3
$\Gamma_p(\alpha)$	Matrix-variate gamma function (generalized gamma function)	Notation 1.4.1
$W_p(n, \Sigma)$	Wishart distribution	Def. 1.4b.2
$M_X(T)$	Moment generating function of the random matrix X	Section 1.4c
$B_p(\alpha, \beta)$	Matrix-variate beta function	Def. 1.4d.1
${}_rF_s((a); (b); x)$	Hypergeometric series in the real scalar case	Def. 1.5.1
$M_{Q_1,Q_2}(t_1, t_2)$	The joint m.g.f. of Q_1 and Q_2	Section 2.4a
K_{r_1,r_2}	r_1-th and r_2-th joint cumulant	Notation 2.4.1
$\sum_{(r_1,r_2,r_3)}(E_1E_2E_3)$	The sum of product of permutations of E_1, E_2, E_3	Notation 2.4.2
$K_r^{Q_2}$	The r-th cumulant of Q_2	Corollary 2.4.1
NS	Necessary and sufficient	Section 2.5
NGL	Noncentral generalized Laplace (Laplacianness)	Section 2.5
$\mathcal{C}_p(\xi; \boldsymbol{\mu}, \Sigma)$	Elliptically contoured distribution	Def. 3.1.1
$\mathcal{S}_p(\xi)$	Spherical distribution	Def. 3.1.2
$\boldsymbol{U}^{(p)}$	Uniform distribution on the p-dimensional sphere	Eq. (3.1.8)
$\boldsymbol{X} \simeq \boldsymbol{Y}$	$\boldsymbol{X}$ and $\boldsymbol{Y}$ are identically distributed	Notation 3.1.1

$\xi(\boldsymbol{t}'\boldsymbol{t})$	Characteristic function of a spherical distribution	Theorem 3.1.1
$V_{p,q}$	Stieffel manifold	Def. 3.1.3
$O(p)$	Full orthogonal group	Def. 3.1.3
$\mathcal{D}_\ell(a_1, \ldots, a_\ell)$	Dirichlet distribution (type 1)	Def. 3.1.4
$\text{Beta}(a_1, a_2)$	Beta distribution (type 1)	Def. 3.1.4
$\text{Gamma}(\alpha, \beta)$	Gamma distribution	Def. 3.1.5
$\chi^2(\xi; p_1, \ldots, p_\ell)$	Generalized chi-square distribution	Def. 3.1.6
$\chi^2(\xi; p; \delta^2)$	Generalized noncentral chi-square distribution	Def. 3.1.7
$F(\xi; m, n; \delta^2)$	Generalized F-distribution	Def. 3.1.8
$\otimes$	Kronecker product	Def. 3.2.1
K_{pp}	Permutation matrix	Def. 3.2.1 (vii)
$M_k(\boldsymbol{X})$	k-th moment of the random vector $\boldsymbol{X}$	Def. 3.2.3
$\text{B}(a, b)$	$(\Gamma(a)\Gamma(b))/\Gamma(a+b)$	Theorem 3.3.1
$I_{\{A\}}$	Indicator function of the set A	Section 3.4
$LS_{n\times p}(\xi)$	Left spherical distribution	Section 3.5
$\mathcal{W}_\ell(\xi; n_1, \ldots, n_\ell)$	Generalized Wishart distribution	Def. 3.5.1
Φ_k	A class of characteristic functions	Notation 3.6.1
$\chi^2(\xi; p_1, \ldots, p_\ell; \delta_1^2, \ldots, \delta_\ell^2)$	Multivariate generalized noncentral chi-square distribution	Def. 3.6.1
$\text{etr}\,(A)$	Exponential of trace of A	Eq. (4.1.1)
$\Gamma_m(a)$	Generalized gamma function	Eq. (4.1.3)
$O(m)$	Orthogonal group of order m	Eq. (4.1.5)

$\mathrm{d}(H)$	Normalized invariant measure of orthogonal group	Eq. (4.1.6)
$\mathrm{d}(L)$	Normalized invariant measure of a Stieffel manifold	Eq. (4.1.7)
$\mathrm{d}(\tilde{L})$	Invariant measure of a Stieffel manifold $\leftrightarrow \left\{ \prod_{j=1}^{q} \lvert U_{(j)} \rvert \right\} \mathrm{d}U$ [cf 3.1.22], see also Theorem 1.1a.11	Eq. (4.1.7)
$N_{m,n}(M, \Sigma, B)$	Multinormal distribution of a certain structure	Eq. (4.1.11)
χ_p^2	Chi-square variable with p degrees of freedom	Lemma 4.1.2
κ	Partition of k into not more than m parts	Section 4.2
$\kappa > \tau$	Lexicographic ordering for partitions κ and τ	Def. 4.2.1
$m_\kappa(\alpha)$	Monomial symmetric function associated with the partition κ	Def. 4.2.2
a_r	Elementary symmetric function of order r	Def. 4.2.3
$a_\kappa(\alpha)$	Symmetric function due to the product of elementary functions	Def. 4.2.3
s_r	Weighted sum of r-th power of m arguments	Eq. (4.2.8)
$S_\kappa(\alpha)$	Symmetric function resulting from the product of weighted sums	Eq. (4.2.7)
$A_{i_1,\ldots,i_k}$	$k \times k$ matrix formed from A by deleting all but the $i_1, \ldots, i_k$-th rows and columns	Eq. (4.3.1)
$a_\kappa(A)$	Symmetric function resulting from the product of elementary symmetric functions of latent roots of A	Eq. (4.3.2)

$\Gamma_m(a;\kappa)$	Generalized gamma function associated with a partition κ	Eq. (4.3.5)
$(a)_\kappa$	Generalized hypergeometric coefficient associated with a partition κ	Eq. (4.3.6)
$(a)_k$	Hypergeometric coefficient	Lemma 4.3.1
$Y_\kappa(A)$	Zonal polynomial for a partition κ	Def. 4.3.1
$Z_\kappa(A)$	Zonal polynomial whose coefficient of $(\operatorname{tr} A)^k$ is unity	Def. 4.3.2
$C_\kappa(A)$	Zonal polynomial satisfying $(\operatorname{tr} A)^k = \sum_\kappa C_\kappa(A)$	Def. 4.3.3
$\mathcal{R}e(Z)$	Real part of a complex symmetric matrix	Remark 4.3.2
∂	Symmetric differential operator matrix	Lemma 4.3.6
D_Λ	Laplace-Beltrami operator	Eq. (4.3.61)
$\kappa \succ \lambda$	Partition λ is majorized by partition κ	Remark 4.3.6
∂_κ	Differential operator associated with κ	Def. 4.3.4
$a^\phi_{\kappa,\lambda}$	Coefficient associated with a product of zonal polynomials	Eq. (4.3.65)
$b^\phi_{\kappa,\lambda}$	Coefficient for zonal polynomials whose argument is a direct sum of two matrices	Eq. (4.3.66)
${}_pF_q(\cdot,\cdot;Z)$	Hypergeometric function of matrix argument Z	Eq. (4.4.12)
$\lVert Z \rVert$	The maximum of the absolute values of the latent roots of Z	Example 4.4.1
$B_r(h)$	Bernoulli polynomial of degree r	Eq. (4.4.30)
${}_pF_q^{(m)}$	Hypergeometric function of two matrix arguments	Eq. (4.5.3)
$\binom{\kappa}{\sigma}$	Binomial coefficient	Def. 4.5.1
$\binom{\sigma}{P}$	Generalized binomial coefficient	Def. 4.5.2

$A_\gamma(S)$	Bessel function of a matrix argument	Def. 4.6.1
$L_\kappa^\gamma(S)$	Laguerre polynomial of a matrix argument	Def. 4.6.2
$L_k^a(x)$	Univariate Laguerre polynomial	Eq. (4.6.10)
$g_\alpha(x,z)$	Probability density function of a non-central chi-square random variable with α degrees of freedom and noncentrality parameter z	Theorem 4.6.3
$\Vert L_\kappa^\gamma \Vert^2$	L^2-norm of L_κ^γ	Eq. (4.6.32)
$H_\kappa(T)$	Hermite polynomial of a rectangular matrix	Def. 4.6.3
$P_\kappa(T,A,B)$	Polynomial of three matrix arguments	Def. 4.6.4
$P_\kappa(T,A)$	Polynomial of two matrix arguments	Def. 4.6.4
$L_\kappa^\gamma(S,B)$	Laguerre polynomial of two matrix arguments	Def. 4.6.5
$L_\kappa^\gamma(T,S,B)$	Laguerre polynomial of three matrix arguments	Def. 4.6.5
$\Gamma_m(a;-\kappa)$	Generalized gamma function associated with $-\kappa$	Exercise 4.7
$H_\kappa^{(m)}(X)$	Hermite polynomial of a symmetric matrix of order m	Exercise 4.32
XAX'	Quadratic form of a matrix argument	Eq. (5.1.2)
$Q_m(A,\Sigma_1,\Sigma_2,M,n)$	Distribution of a quadratic form with certain parameters	Eq. (5.1.2)
$\Pr\{S<\Omega\}$	Probability that $\Omega - S$ is positive definite	Eq. (5.1.10)
$\varphi(t)$	Characteristic function	Eq. (5.6.6)

Δ^2	Mahalanobis distance	Eq. (5.6.10)
$\Phi(x)$	Standard normal distribution function	Theorem 5.6.3
$\mathrm{Vec}(B)$	Vector formed by stacking the rows of B	Lemma 5.7.1
$F_{m,n}(\tau)$	Non-central F random variable with m and n degrees of freedom and noncentrality parameter τ	Eq. (5.7.9)
D_X	Fractional matrix operator	Exercise 5.5
$L(m,k;F)$	Matrix Langevin distribution	Exercise 5.24
$G\ell(m,R)$	Generalized linear group of order m	Appendix A.1
$T(L)$	Right regular representation	Appendix A.1
$P_k[X]$	Vector space of homogeneous polynomials of order k in the elements of X	Appendix A.1
$\tilde{T}_{2k}(L)$	Representation matrix whose elements are homogeneous polynomials of degree $2k$ in the elements of L	Appendix A.1
$P_{k,\ell}[X,Y]$	Vector space defined by the tensor product of $P_k[X]$ and $P_\ell[Y]$	Appendix A.1
$\equiv$	Equivalence of representations	Appendix A.1
$C_\phi^{\kappa,\lambda}(X,Y)$	Invariant polynomial associated with partitions κ, λ and ϕ	Appendix A.3
$\theta_\phi^{\kappa,\lambda}$	Constant associated with partitions κ, λ and ϕ	Appendix A.3
$\begin{pmatrix} \rho,\sigma : \tau \\ \kappa,\lambda : \phi \end{pmatrix}$	Binomial coefficient	Eq. (A.3.16)
$\left(\begin{matrix} \rho : \tau \\ \kappa : \phi \end{matrix}\middle\vert \lambda\right)$	Binomial coefficient	Eq. (A.3.17)

Author Index

Subject Index

Lecture Notes in Statistics

For information about Volumes 1 to 15
please contact Springer-Verlag

Vol. 16: J.P. Florens, M. Mouchart, J.P. Raoult, L. Simar, and A.F.M. Smith (Editors), Specifying Statistical Models: From Parametric to Non-Parametric, Using Bayesian or Non-Bayesian Approaches. xi, 204 pages, 1983.

Vol. 17: I.V. Basawa and D.J. Scott, Asymptotic Optimal Inference for Non-Ergodic Models. ix, 170 pages, 1983.

Vol. 18: W. Britton, Conjugate Duality and the Exponential Fourier Spectrum. v, 226 pages, 1983.

Vol. 19: L. Fernholz, von Mises Calculus For Statistical Functionals. viii, 124 pages, 1983.

Vol. 20: U. Herkenrath, D. Kalin, W. Vogel (Editors), Mathematical Learning Models — Theory and Algorithms: Proceedings of a Conference. xiv, 226 pages, 1983.

Vol. 21: H. Tong, Threshold Models in Non-linear Time Series Analysis. x, 323 pages, 1983.

Vol. 22: S. Johansen, Functional Relations, Random Coefficients and Nonlinear Regression with Application to Kinetic Data, viii, 126 pages, 1984.

Vol. 23: D.G. Saphire, Estimation of Victimization Prevalence Using Data from the National Crime Survey. v, 165 pages, 1984.

Vol. 24: T.S. Rao, M.M. Gabr, An Introduction to Bispectral Analysis and Bilinear Time Series Models. viii, 280 pages, 1984.

Vol. 25: E. Parzen (Editor), Time Series Analysis of Irregularly Observed Data. Proceedings, 1983. vii, 363 pages, 1984.

Vol. 26: J. Franke, W. Härdle and D. Martin (Editors), Robust and Nonlinear Time Series Analysis. Proceedings, 1983. ix, 286 pages, 1984.

Vol. 27: A. Janssen, H. Milbrodt, H. Strasser, Infinitely Divisible Statistical Experiments. vi, 163 pages, 1985.

Vol. 28: S. Amari, Differential-Geometrical Methods in Statistics. v, 290 pages, 1985.

Vol. 29: B.J.T. Morgan and P.M. North (Editors), Statistics in Ornithology. xxv, 418 pages, 1985.

Vol 30: J. Grandell, Stochastic Models of Air Pollutant Concentration. v, 110 pages, 1985.

Vol. 31: J. Pfanzagl, Asymptotic Expansions for General Statistical Models. vii, 505 pages, 1985.

Vol. 32: R. Gilchrist, B. Francis and J. Whittaker (Editors), Generalized Linear Models. Proceedings, 1985. vi, 178 pages, 1985.

Vol. 33: M. Csörgo, S. Csörgo, L. Horváth, An Asymptotic Theory for Empirical Reliability and Concentration Processes. v, 171 pages, 1986.

Vol. 34: D.E. Critchlow, Metric Methods for Analyzing Partially Ranked Data. x, 216 pages, 1985.

Vol. 36: B. Matérn, Spatial Variation. Second Edition. 151 pages, 1986.

Vol. 37: R. Dykstra, T. Robertson and F.T. Wright (Editors), Advances in Order Restricted Statistical Inference. Proceedings, 1985. viii, 295 pages, 1986.

Vol. 38: R.W. Pearson and R.F. Boruch (Editors), Survey Research Designs: Towards a Better Understanding of Their Costs and Benefits. v, 129 pages, 1986.

Vol. 39: J.D. Malley, Optimal Unbiased Estimation of Variance Components. ix, 146 pages, 1986.

Vol. 40: H.R. Lerche, Boundary Crossing of Brownian Motion. v, 142 pages, 1986.

Vol. 41: F. Baccelli, P. Brémaud, Palm Probabilities and Stationary Queues. vii, 106 pages, 1987.

Vol. 42: S. Kullback, J.C. Keegel, J.H. Kullback, Topics in Statistical Information Theory. ix, 158 pages, 1987.

Vol. 43: B.C. Arnold, Majorization and the Lorenz Order: A Brief Introduction. vi, 122 pages, 1987.

Vol. 44: D.L. McLeish, Christopher G. Small, The Theory and Applications of Statistical Inference Functions. vi, 124 pages, 1987.

Vol. 45: J.K. Ghosh (Editor), Statistical Information and Likelihood. 384 pages, 1988.

Vol. 46: H.-G. Müller, Nonparametric Regression Analysis of Longitudinal Data. vi, 199 pages, 1988.

Vol. 47: A.J. Getson, F.C. Hsuan, {2}-Inverses and Their Statistical Application. viii, 110 pages, 1988.

Vol. 48: G.L. Bretthorst, Bayesian Spectrum Analysis and Parameter Estimation. xii, 209 pages, 1988.

Vol. 49: S.L. Lauritzen, Extremal Families and Systems of Sufficient Statistics. xv, 268 pages, 1988.

Vol. 50: O.E. Barndorff-Nielsen, Parametric Statistical Models and Likelihood. vii, 276 pages, 1988.

Vol. 51: J. Hüsler, R.-D. Reiss (Editors). Extreme Value Theory, Proceedings, 1987. x, 279 pages, 1989.

Vol. 52: P.K. Goel, T. Ramalingam, The Matching Methodology: Some Statistical Properties. viii, 152 pages, 1989.

Vol. 53: B.C. Arnold, N. Balakrishnan, Relations, Bounds and Approximations for Order Statistics. ix, 173 pages, 1989.

Vol. 54: K.R. Shah, B.K. Sinha, Theory of Optimal Designs. viii, 171 pages, 1989.

Vol. 55: L. McDonald, B. Manly, J. Lockwood, J. Logan (Editors), Estimation and Analysis of Insect Populations. Proceedings, 1988. xiv, 492 pages, 1989.

Vol. 56: J.K. Lindsey, The Analysis of Categorical Data Using GLIM. v, 168 pages, 1989.

Vol. 57: A. Decarli, B.J. Francis, R. Gilchrist, G.U.H. Seeber (Editors), Statistical Modelling. Proceedings, 1989. ix, 343 pages, 1989.

Vol. 58: O.E. Barndorff-Nielsen, P. Blæsild, P.S. Eriksen, Decomposition and Invariance of Measures, and Statistical Transformation Models. v, 147 pages, 1989.

Vol. 59: S. Gupta, R. Mukerjee, A Calculus for Factorial Arrangements. vi, 126 pages, 1989.

Vol. 60: L. Györfi, W. Härdle, P. Sarda, Ph. Vieu, Nonparametric Curve Estimation from Time Series. viii, 153 pages, 1989.

Vol. 61: J. Breckling, The Analysis of Directional Time Series: Applications to Wind Speed and Direction. viii, 238 pages, 1989.

Vol. 62: J.C. Akkerboom, Testing Problems with Linear or Angular Inequality Constraints. xii, 291 pages, 1990.

Vol. 63: J. Pfanzagl, Estimation in Semiparametric Models: Some Recent Developments. iii, 112 pages, 1990.

Vol. 64: S. Gabler, Minimax Solutions in Sampling from Finite Populations. v, 132 pages, 1990.

Vol. 65: A. Janssen, D.M. Mason, Non-Standard Rank Tests. vi, 252 pages, 1990.

Vol 66: T. Wright, Exact Confidence Bounds when Sampling from Small Finite Universes. xvi, 431 pages, 1991.

Vol. 67: M.A. Tanner, Tools for Statistical Inference: Observed Data and Data Augmentation Methods. vi, 110 pages, 1991.

Vol. 68: M. Taniguchi, Higher Order Asymptotic Theory for Time Series Analysis. viii, 160 pages, 1991.

Vol. 69: N.J.D. Nagelkerke, Maximum Likelihood Estimation of Functional Relationships. V, 110 pages, 1992.

Vol. 70: K. Iida, Studies on the Optimal Search Plan. viii, 130 pages, 1992.

Vol. 71: E.M.R.A. Engel, A Road to Randomness in Physical Systems. ix, 155 pages, 1992.

Vol. 72: J.K. Lindsey, The Analysis of Stochastic Processes using GLIM. vi, 294 pages, 1992.

Vol. 73: B.C. Arnold, E. Castillo, J.-M. Sarabia, Conditionally Specified Distributions. xiii, 151 pages, 1992.

Vol. 74: P. Barone, A. Frigessi, M. Piccioni, Stochastic Models, Statistical Methods, and Algorithms in Image Analysis. vi, 258 pages, 1992.

Vol. 75: P.K. Goel, N.S. Iyengar (Eds.), Bayesian Analysis in Statistics and Econometrics. xi, 410 pages, 1992.

Vol. 76: L. Bondesson, Generalized Gamma Convolutions and Related Classes of Distributions and Densities. viii, 173 pages, 1992.

Vol. 77: E. Mammen, When Does Bootstrap Work? Asymptotic Results and Simulations. vi, 196 pages, 1992.

Vol. 78: L. Fahrmeir, B. Francis, R. Gilchrist, G. Tutz (Eds.), Advances in GLIM and Statistical Modelling: Proceedings of the GLIM92 Conference and the 7th International Workshop on Statistical Modelling, Munich, 13-17 July 1992. ix, 225 pages, 1992.

Vol. 79: N. Schmitz, Optimal Sequentially Planned Decision

Vol. 80: M. Fligner, J. Verducci (Eds.), Probability Models and Statistical Analyses for Ranking Data. xxii, 306 pages, 1992.

Vol. 81: P. Spirtes, C. Glymour, R. Scheines, Causation, Prediction, and Search. xxiii, 526 pages, 1993.

Vol. 82: A. Korostelev and A. Tsybakov, Minimax Theory of Image Reconstruction. xii, 268 pages, 1993.

Vol. 83: C. Gatsonis, J. Hodges, R. Kass, N. Singpurwalla (Editors), Case Studies in Bayesian Statistics. xii, 437 pages, 1993.

Vol. 84: S. Yamada, Pivotal Measures in Statistical Experiments and Sufficiency. vii, 129 pages, 1994.

Vol. 85: P. Doukhan, Mixing: Properties and Examples. xi, 142 pages, 1994.

Vol. 86: W. Vach, Logistic Regression with Missing Values in the Covariates. xi, 139 pages, 1994.

Vol. 87: J. Møller, Lectures on Random Voronoi Tessellations. vii, 134 pages, 1994.

Vol. 88: J. E. Kolassa, Series Approximation Methods in Statistics. viii, 150 pages, 1994.

Vol. 89: P. Cheeseman, R.W. Oldford (Editors), Selecting Models From Data: AI and Statistics IV. xii, 487 pages, 1994.

Vol. 90: A. Csenki, Dependability for Systems with a Partitioned State Space: Markov and Semi-Markov Theory and Computational Implementation. x, 241 pages, 1994.

Vol. 91: J.D. Malley, Statistical Applications of Jordan Algebras. viii, 101 pages, 1994.

Vol. 92: M. Eerola, Probabilistic Causality in Longitudinal Studies. vii, 133 pages, 1994.

Vol. 93: Bernard Van Cutsem (Editor), Classification and Dissimilarity Analysis. xiv, 238 pages, 1994.

Vol. 94: Jane F. Gentleman and G.A. Whitmore (Editors), Case Studies in Data Analysis. viii, 262 pages, 1994.

Vol. 95: Shelemyahu Zacks, Stochastic Visibility in Random Fields. x, 175 pages, 1994.

Vol. 96: Ibrahim Rahimov, Random Sums and Branching Stochastic Processes. viii, 195 pages, 1995.

Vol. 97: R. Szekli, Stochastic Ordering and Dependence in Applied Probablility, viii, 194 pages, 1995.

Vol. 98: Philippe Barbe and Patrice Bertail, The Weighted Bootstrap, viii, 230 pages, 1995.

Vol. 99: C.C. Heyde (Editor), Branching Processes: Proceedings of the First World Congress, viii, 185 pages, 1995.

Vol. 100: Wlodzimierz Bryc, The Normal Distribution: Characterizations with Applications, viii, 139 pages, 1995.

Vol. 101: H.H. Andersen, M. Højbjerre, D. Sørensen, P.S. Eriksen, Linear and Graphical Models: for the Multivariate complex Normal Distribution, x, 184 pages, 1995.

Vol. 102: A.M. Mathai, Serge B. Provost, Takesi Hayakawa, Bilinear Forms and Zonal Polynomials, x, 378 pages, 1995.